THE EIGHTFOLD WAY

Advanced Book Classics

DAVID PINES, Series Editor

Anderson, P. W.	**BASIC NOTIONS OF CONDENSED MATTER PHYSICS**
Bethe, H. and Jackiw, R.	**INTERMEDIATE QUANTUM MECHANICS**, *Third Edition*
Cowan, G. and Pines, D.	**COMPLEXITY:** *Metaphors, Models, and Reality*
de Gennes, P. G.	**SUPERCONDUCTIVITY OF METALS AND ALLOYS**
d'Espagnat, B.	**CONCEPTUAL FOUNDATIONS OF QUANTUM MECHANICS**, *Second Edition*
Feynman, R.	**PHOTON-HADRON INTERACTIONS**
Feynman, R.	**QUANTUM ELECTRODYNAMICS**
Feynman, R.	**STATISTICAL MECHANICS**
Feynman, R.	**THE THEORY OF FUNDAMENTAL PROCESSES**
Gell-Mann, M. and Ne'eman, Y.	**THE EIGHTFOLD WAY**
Khalatnikov, I. M.	**AN INTRODUCTION TO THE THEORY OF SUPERFLUIDITY**
Ma, S-K.	**MODERN THEORY OF CRITICAL PHENOMENA**
Migdal, A. B.	**QUALITATIVE METHODS IN QUANTUM THEORY**
Negele, J. W. and Orland, H.	**QUANTUM MANY-PARTICLE SYSTEMS**
Nozières, P.	**THEORY OF INTERACTING FERMI SYSTEMS**
Nozières, P. and Pines, D.	**THE THEORY OF QUANTUM LIQUIDS**
Parisi, G.	**STATISTICAL FIELD THEORY**
Pines, D.	**ELEMENTARY EXCITATIONS IN SOLIDS**
Pines, D.	**THE MANY-BODY PROBLEM**

Quigg, C.	**GAUGE THEORIES OF THE STRONG, WEAK, AND ELECTROMAGNETIC INTERACTIONS**
Schrieffer, J. R.	**THEORY OF SUPERCONDUCTIVITY**, *Revised*
Schwinger, J.	**PARTICLES, SOURCES, AND FIELDS**, *Volume I*
Schwinger, J.	**PARTICLES, SOURCES, AND FIELDS**, *Volume II*
Schwinger, J.	**PARTICLES, SOURCES, AND FIELDS**, *Volume III*
Schwinger, J.	**QUANTUM KINEMATICS AND DYNAMICS**
Wyld, H. W.	**MATHEMATICAL METHODS FOR PHYSICS**

THE EIGHTFOLD WAY

With new contributions from the authors

MURRAY GELL-MANN
Santa Fe Institute

YUVAL NE'EMAN
Tel Aviv University

CRC Press
Taylor & Francis Group
Boca Raton London New York

CRC Press is an imprint of the
Taylor & Francis Group, an **informa** business

Advanced Book Program

First published 1964 by Westview Press

Published 2018 by CRC Press
Taylor & Francis Group
6000 Broken Sound Parkway NW, Suite 300
Boca Raton, FL 33487-2742

CRC Press is an imprint of the Taylor & Francis Group, an informa business

Visit the Taylor & Francis Web site at
http://www.taylorandfrancis.com

and the CRC Press Web site at
http://www.crcpress.com

Catalog number available from the Library of Congress.

ISBN 13: 978-0-7382-0299-0 (pbk)

CONTENTS

REMARKS ON THE HISTORY OF STRONG-INTERACTION SYMMETRIES

YUVAL NE'EMAN

Quantum mechanics had just about succeeded in bringing order and understanding to the atomic world—in principle at least—when chaos was again unleashed† by the discovery of the neutron. Clearly, there in the nucleus lay a new force, an interaction whose coupling could be neither electric charge e nor the gravitational constant G.

Some foundations for a new theory aiming at these strong interactions were laid by Yukawa. Since the advent of relativity and quantum mechanics, action at a distance had been forsaken, and interactions were supposed to be mediated by particles with Bose statistics. Yukawa conceived an exchange mechanism with a meson mediating the new interaction. Yet it was at this early stage that the complementary approach—symmetries—produced its first dividends in this new field of physics by requiring the mesons to form a triplet, including a neutral component. Heisenberg had suggested using a representation of the unitary unimodular group in two dimensions, SU(2), to describe the nucleon

$$N = \begin{pmatrix} p \\ n \end{pmatrix}$$

To make a scalar interaction Lagrangian of the trilinear type suggested by electromagnetism—that is, with single meson emission or

†The author's view is probably based upon Babylonian lore, where primordial chaos is represented by Ti-amath, the Beast.

absorption at each vertex—meant using a pion triplet. Isospin was thus introduced from the very start to explain the variegation in the species and predict multiplicities—a role which was later extended to strangeness and is now fulfilled by SU(3). At the same time, Kemmer's symmetric Lagrangian, later written as

$$\mathcal{L} = ig\bar{N}\gamma_5 \boldsymbol{\tau}\cdot\boldsymbol{\pi} N \tag{1}$$

had produced a mathematical rendering of the charge independence of nuclear forces observed by Breit. The group—that is, the three reduction matrices $\boldsymbol{\tau}$ used to pick a vector triplet out of the four components of the outer product $\bar{N}N$ required by Dirac theory—supplied a set of coefficients

$$g_{\bar{p}p\pi^0} = -g_{\bar{n}n\pi^0} = 2^{-1/2} g_{\bar{p}n\pi^+} = 2^{-1/2} g_{\bar{n}p\pi^-} \tag{2}$$

Such coefficients form in fact the SU(2) generalization of electric charge in the Coulomb law, the latter corresponding to a purely additive system based on a U(1) or phase group. The set (Eq. 2), when summed over the two vertices corresponding in each case to all possible exchanges, yields a force proportional to g^2, similar to the Coulomb e^2.

Hadrons (the strongly interacting particles) participate in the nonstrong interactions too. Their electric charge coincides—except for the strength scale and an additive constant—with a particular component of the strong-interaction symmetry SU(2). Electric couplings mark a preferred direction τ_3 in the space of the generator algebra of SU(2); the symmetry is thus broken, though in a rather weak way, because the strength ratio between the two interactions is of the order of 10^{-3}.

Similarly, beta decay, in the more recent conserved-vector-current theory of the weak interactions, again displayed a pair of components of the SU(2) algebra, τ^+ and τ^-, which fixed the composition of the weak hadron current interacting with the leptons or with the intermediate bosons. This time the coupling is much weaker; nevertheless, had isospin not been conceived to account for strong interaction results, it would have emerged as the hadron symmetry whose conserved currents carry the weak and electromagnetic "charges" of baryons and mesons. We return to this aspect in Part VIII.

The 1947–1949 discovery of the pions confirmed the use of isospin to predict charge multiplets. It was immediately followed by the discovery of series of "heavy events," better known by their latter-day name of "strange" particles: This was again a case of "variety of the species." The puzzle of the long lifetimes of the new particles was solved in 1953 by Gell-Mann and Nishijima. The answer lay in the conservation of an additional additive quantum number, a U(1) symmetry S (for strangeness), defined and conserved through its dependence upon other previously recognized quantum numbers,

$$Q = I_3 + \frac{B + S}{2} \tag{3}$$

B is the baryon number; B + S later came to be known as hypercharge Y. Particles were assigned values of strangeness and isospin, providing certain predictions about undiscovered hadrons. This was symmetry in yet another aspect of its uses: the determination of selection rules for reactions.

It is at this stage that the search for a "global" symmetry came in. The motivation stemmed from these same notions that had brought about the conception of isospin or strangeness. In the nucleon case, observation of rough charge independence had led to Kemmer's symmetric Lagrangian as an abstracted theory; nucleons inhabit nuclei and the fact that n-n, n-p, and p-p forces are roughly equal could be readily noticed. With the new crop of particles, one had to postulate a symmetry first and then check it, because there was no easy way to observe the over-all forces ratio. Most textbooks written during 1956–61 carry the sentence

$$\begin{aligned}\mathcal{L}_{\text{strong}} &= g_1 \overline{N} \boldsymbol{\tau} N \cdot \boldsymbol{\pi} + g_2 \overline{\boldsymbol{\Sigma}} \Lambda \cdot \boldsymbol{\pi} + \text{hc} + g_3 \overline{\boldsymbol{\Sigma}} \wedge \overline{\boldsymbol{\Sigma}} \cdot \boldsymbol{\pi} \\ &\quad + g_4 \overline{\Xi} \boldsymbol{\tau} \Xi \cdot \boldsymbol{\pi} + g_5 \overline{N} K \Lambda + \text{hc} + g_6 \overline{N} \boldsymbol{\tau} K \cdot \boldsymbol{\Sigma} \\ &\quad + \text{hc} + g_7 \overline{\Xi} \overline{K} \Lambda + \text{hc} + g_8 \overline{\Xi} \boldsymbol{\tau} K \cdot \boldsymbol{\Sigma} + \text{hc}\end{aligned} \tag{4}$$

with eight independent couplings and nothing fixed with respect to the Dirac bilinears, because relative parities were as yet unknown except for the N-π. This was the most general description of a Yukawa-like strong interaction in terms of the eight baryons and seven mesons that had been discovered. Here was one task for the "global" symmetry: to relate these eight $g_1 \ldots g_8$, to fix parities, to point to any further missing terms.

The new symmetry should account for the appearance of just these particles; it should also account for the mass spectrum. When Gell-Mann and Schwinger proposed a scheme of that type, this last point was one of their main aims: could the hadron mass spectrum be visualized as another case of orderly broken symmetry? Would the breaking emerge as some preferred direction in the algebra space? Since there was no observed interaction half-way between the strong and the electromagnetic, the best physical model seemed to be a distinction between highly symmetrical π mediated and unsymmetrical K mediated interactions. This was the suggestion, and it yielded a first-order mass formula

$$2(m_N + m_\Xi) = 3m_\Sigma + m_\Lambda \tag{5}$$

that was only some 4% off the experimental values.

The fatal blow to global symmetry was dealt by Salam at the Kiev Conference in 1959, when he pointed out that the S-wave phase shifts deduced from experiments were considerably different in π-N and π-hyperon scattering, contrary to the theory.

At about the same time, two other lines of work, originating from different approaches, happened to converge with the search for a global symmetry; both had, in fact, been in the background from the very start.

One was the "model" approach—the most recent incarnation of Democritos' atomistic concept. The question raised here was: Is there a "fundamental" set of particles or fields from which all particles can be constructed? Looking again at the pion-nucleon Lagrangian (Eq. 1), we notice that making $\mathcal{L}$ scalar implied picking in the reduction of the $N\bar{N}$ tensor a vector that would have the exact transformation properties of the pion triplet, including spin, parity, and charge parity,

$$\bar{N}\gamma_5 \boldsymbol{\tau} N \sim \pi$$

Fermi and Yang tried to replace this homomorphy relation by a true equation—that is, introducing a physical model of the pion as an extremely bound state of nucleons and antinucleons; the binding would then have to account for 92% of the original energy involved. When strangeness came in, Goldhaber and Christy added the Kaon to the list of building blocks; Sakata then propounded the (N-Λ) triplet model. Since the dynamical theory of the binding was obscure, the whole approach made very little progress until, in 1959–60, Ikeda Ogawa and Ohnuki in Japan and Thirring, Wess, and Yamaguchi in Europe found a proper mathematical phrasing of the symmetry aspect of the construction process in this model. This was the group SU(3), with an eight-parameter algebra, whose representations would correspond to the structures allowed within the Sakata model. It assumed invariance of the particle-building process with respect to unitary unimodular transformations of the three "fundamental" complex fields, identified with Sakata's (p,n,Λ) set. The mass spectrum was assumed to derive from the propagation of the Λ-N mass difference.

Another line of development had been started by Yang and Mills in 1954; it was a guess at a closer analogy with electric charge conservation than was available in the isospin formalism of the day. The U(1) symmetry of electric charge holds even for a locally dependent phase or parameter $e\alpha(x)$; because the matter spinor obeys the Dirac equation, this involves a derivative $e\partial_\mu \alpha$, and invariance is maintained only through the existence of a vector-potential A_μ and a vector type of interaction where $e\partial_\mu \alpha$ is canceled out by $eA_\mu \to e(A_\mu - \partial_\mu \alpha)$. This is related to the "universality" of electric charge, because gauge invariance ensures, in field theory for example, that not only will the bare charge be universal, but also the renormalized coupling constant, because the renormalization will come about from photon self-energy contributions only.

The isospin homologue would be given by an isotriplet of vector fields. Because this had not been observed and the local gauge idea in pure form led to massless vector fields, the whole vector field approach lay dormant for a few years except for ideas by Teller, Lee and Yang, and Fujii, with respect to a similar treatment of the baryon charge. Interest in the Yang-Mills approach was renewed in 1959, mainly because of two unrelated facts: on the one hand, the weak interactions had turned out to be of the vector-pseudovector type, with perhaps even a spin-one intermediate field resembling A_μ; on the other hand, nucleon structure experiments had been shown by Nambu to indicate the existence of a vector meson.

Utiyama had studied the generalized group formalism in connection with the Yang-Mills type of assumption; it was further analyzed by Gell-Mann and Glashow and by Ionides, who showed that the algebraic condition involved was the emergence of a universal coupling f_{abc}, real and triply antisymmetric. This coefficient would then correspond to the structure constants of a compact Lie group. Currents and vector mesons have to lie in the "adjoint" representation of the group—that representation whose dimensionality and transformation properties are the same as those of the generator algebra—that is, one meson and one current per group parameter.

The way now lay open to a renewed search for a "global" symmetry, this time systematized from the algebraic viewpoint and not tied to a Fermi-Yang type of model. Speiser and Tarski, Lee and Yang, and Behrends and Sirlin used group systematics, trying to identify the global symmetry, but without using the gauge idea; Salam and Ward tried to adapt the Yang-Mills mechanism to existing models—first, to an R(8) symmetry and then to the Sakata model with U(3). The present authors set out looking for a formalism that would provide the greatest resemblance to phenomena as they stood; the result is given in Papers 1 and 2 in Part I. It consisted in the use of SU(3), with a rank-two algebra, so that the only additive quantum numbers for the hadron symmetry would be I_3 and Y as observed; it also picked out the eight-dimensional adjoint representation—not only for the vector particles and currents, but also for the baryons and pseudoscalar mesons as well. It is this representation, of course, that gives its name to the present book.

[illegible] of intro[illegible] current [illegible] and [illegible] to [illegible] universality [illegible] charge because gauge invariance equations in electrodynamics, for example, that not only will the bare charge be universal, but also the renormalized coupling constant, because the renormalization will come then from [illegible] self-energy [illegible].

The description above would be given by an isomultiplet of vector fields. Because this had not been observed and the local gauge idea in pure form led to massless vector fields, the whole vector field approach lay dormant for several years except for ideas by Teller, Lee and Yang, and [illegible] with respect to a similar treatment of the baryon number. Interest in the Yang-Mills approach was renewed in 1958, mainly because of two unrelated facts: on the one hand, the weak interactions had turned out to be of the vector [illegible] type, with perhaps even a similar intermediate field [illegible] coupling [illegible]; on the other hand, nucleon structure experiments had been shown by [illegible] to indicate the existence of a vector meson.

Utiyama has studied the generalized [illegible] formalism in connection with the Yang-Mills type of description; it was further analyzed by Gell-Mann and Glashow and by Ionides, who showed that the algebraic condition involved in the emergence of a universal coupling [illegible] real and totally antisymmetric. This coefficient would then correspond to the structure constants of a compact Lie group. [illegible] vector [illegible] representation whose dimensionality and [illegible] properties are the same as those of the generator algebra—that is, [illegible] mesons and the currents and group parameters.

The way now lay open to a renewed search for a global symmetry, this time systematically [illegible] the [illegible] viewpoint and not tied to a Fermi-Yang type of model. Speiser and Tarski, Lee and Yang, and [illegible] and [illegible] used group systematics, trying to identify the global symmetry but without using the gauge idea. Salam and Ward tried to adapt the Yang-Mills mechanism to existing [illegible] first to [illegible] symmetry and then to the Sakata model with [illegible]. The present authors set out looking for a [illegible] that would provide the greatest resemblance to phenomena as they [illegible]; the result is given in Papers 1 and 2 in Part [illegible] consisted in the use of SU(3) with a rank two algebra, so that the only additive quantum numbers for the hadron symmetry would be I_3 and Y as observed; it also picked out the eight-dimensional adjoint representation not only for the vector mesons and currents but [illegible] baryons and pseudoscalar mesons as well. It is this representation, of course, that gives its name to the present book.

I

THE BROKEN SYMMETRY AND THE MASS FORMULA

MURRAY GELL-MANN

A. ESSENTIAL FEATURES OF THE EIGHTFOLD-WAY SCHEME

1. It is suggested that the strong interactions, besides conserving the three components of isotopic spin (I_1, I_2, I_3) and the hypercharge Y, also approximately conserve four more operators F_4, F_5, F_6, F_7. Defining $F_1 = I_1$, $F_2 = I_2$, $F_3 = I_3$, and $F_8 = (3^{1/2}/2)Y$, we have an algebraic system of eight operators $F_1 \ldots F_8$, and it is proposed that these have the right commutation rules to form the eight generators of the algebra of SU(3), just as F_1, F_2, F_3 form the three generators of the algebra of SU(2). (The higher symmetry is thus chosen to be the simplest generalization of isotopic spin.) The new operators $F_4 \ldots F_7$ then obey the rules $|\Delta I| = \frac{1}{2}$, $|\Delta Y| = 1$. They tie together strongly interacting particle multiplets with different values of I and Y (but the same spin and parity) in approximately degenerate supermultiplets. The supermultiplets correspond, of course, to irreducible representations of SU(3), just as the isotopic spin multiplets correspond to irreducible representations of SU(2).

2. It is *not* proposed to identify any known strongly interacting particles with the fundamental three-dimensional representation of SU(3), which is analogous to the $I = \frac{1}{2}$ representation of SU(2), or with any other "spinor representations" of SU(3). Thus, all known

strongly interacting particles are to be assigned to "tensor representations," the smallest of which are: (a) the trivial one-dimensional representation, denoted by **1**, consisting of just a neutral singlet with I = 0, Y = 0 [**1** is analogous to the I = 0 representation of SU(2)]; (b) the "adjoint" eight-dimensional representation, denoted by **8**, consisting of a doublet with $I = \frac{1}{2}$, Y = 1, a doublet with $I = \frac{1}{2}$, Y = −1, a singlet with I = 0, Y = 0, and a triplet with I = 1, Y = 0 [**8** is analogous to the I = 1 representation of SU(2)].

It is suggested that the baryon multiplets N, Ξ, Λ, and Σ form an "octet" (representation **8**) with $J = \frac{1}{2}^+$. Likewise, the pseudoscalar mesons K, $\overline{K}$, η, and π are to form an octet; the existence of the I = 0 meson η (called χ by Gell-Mann and $\pi^{0\prime}$ by Ne'eman) is thus predicted in advance of its discovery in 1962. The existence of a similar octet of vector mesons is also suggested (called M, $\overline{M}$, ω, and ρ by Gell-Mann and Z, $\overline{Z}$, X, and V by Ne'eman), as well as a possible singlet vector meson (called B by Gell-Mann); these nine vector mesons are now firmly established experimentally and may be referred to according to their energies in MeV as $\kappa(885)$, $\overline{\kappa}(885)$, $\eta(1020)$, $\pi(750)$, and $\eta(790)$. It is notorious that the I = 0 member of the octet and the I = 0 singlet mix with each other so much that between $\eta(1020)$ and $\eta(790)$ it is not easy to say which is which. The most familiar names of these vector mesons are $K^* = \kappa(885)$, $\phi = \eta(1020)$, $\rho = \pi(750)$, and $\omega = \eta(790)$.

3. It is proposed that the part of the strong interaction that violates the higher symmetry transforms like a component of an octet; this must be the eighth component in order that isotopic spin and strangeness be conserved. To first order in the violation, the splitting of a supermultiplet is then subject to constraints. For an octet, we obtain a sum rule for the masses of the four multiplets involved; it is given by Gell-Mann for the baryon octet

$$\frac{m_N + m_\Xi}{2} = \frac{3m_\Lambda + m_\Sigma}{4}$$

and is obeyed to within a few MeV.

For the meson octets, it is more natural to apply this rule to the squares of the masses. A mass of 563 MeV would then be predicted for the pseudoscalar η on the basis of the K and π masses; the experimental value has turned out to be about 550 MeV. A corresponding prediction of the mass squared of the eighth component of the vector meson octet gives a value about midway between $(790\ \text{MeV})^2$ and $(1020\ \text{MeV})^2$. When the meson at 790 MeV was discovered in 1961, it was suggested by Gell-Mann that the other one might be found a couple of hundred MeV higher up and that each would be a roughly equal mixture of octet and singlet. This "ϕ-ω mixing," with the associated violation of the sum rule for the masses, is presumably to be attributed to an accidental near-degeneracy of the singlet and the octet.

Okubo in his paper generalizes the first-order mass formula from the octet representation to the general case, obtaining for any supermultiplet the following formula for the masses of the isotopic multiplets contained in it:

$$M = M_0 + AY + B\left[I(I+1) - \frac{Y^2}{4}\right]$$

(or the equivalent formula with mass squared for mesons). It is easy to see that for the octet case this result reduces to the sum rule quoted above.

A more compact derivation of the generalized mass formula has been given by Goldberg and Lehrer-Ilamed [J. Math. Phys., **4**, 501 (1963)].

4. Another proposal, which concerns the strong and weak interactions jointly, is a generalization of the conserved vector current hypothesis (CVC). The CVC idea is that the $\Delta Y = 0$ vector weak current of the strongly interacting particles is the current of a component of the isotopic spin. The generalization states that the whole vector weak current of the strongly interacting particles is the current of a component of the F-spin. There are then two possible terms: the strangeness-preserving isotopic spin current with $|\Delta \mathbf{I}| = 1$ and the strangeness changing current with $\Delta Y/\Delta Q = +1$, $|\Delta \mathbf{I}| = \frac{1}{2}$. The terms correspond to $F_1 + iF_2$ and $F_4 + iF_5$, respectively.

B. FEATURES THAT REQUIRE UPDATING

1. The emphasis in both papers on a generalized Yang-Mills field theory for the vector mesons is connected with a point of view that may soon become obsolete, according to which the eight $J = \frac{1}{2}^+$ baryons and some set of mesons are associated with "fundamental" fields in a Lagrangian theory of strong interactions.

If there are eight "fundamental" baryons and just one vector meson coupled to the baryon current, the system has the symmetry of SU(8); the symmetry is reduced to that of SU(3) if eight more vector mesons, obeying the generalized Yang-Mills equations, are introduced and coupled to the eight currents of the F-spin components.

Such a picture looks too complicated to many physicists today. Two possibilities are considered as replacements, a "bootstrap" theory or a "fundamental triplet theory."

The bootstrap theory, discussed especially by Chew and Frautschi, is usually formulated in terms of dispersion and unitarity relations on the mass shell for strongly interacting particles (sometimes called S-matrix theory). However, we can see no reason why it cannot be embedded in an abstract local field theory, especially since local field operators are required for the electromagnetic and weak current operators and for the stress-energy-momentum tensor (including the energy density as one component) that couples to gravitation. The whole system in any case is employed practically only with electromagnetism, the weak interaction, and gravitation treated as perturbations.

The idea of the bootstrap theory is that none of the "hadrons" or strongly interacting particles is fundamental; each is merely a dynamical bound state of various combinations of hadrons including itself. This notion, while we do not think it conflicts with abstract local field theory, may disagree with simple Lagrangian field theory with "fields" for sets of observed particles.

There is a part below in which the bootstrap idea is discussed, especially in connection with attempts to derive symmetry. We may note particularly the case in which a set of vector mesons is treated approximately, in isolation from the other hadrons, by the bootstrap method. Each vector meson is described roughly as a bound state of two vector mesons. The resulting mathematics, studied particularly by Cutkosky, resembles quite closely the mathematics of the Yang-Mills formalism and they both lead naturally to a simple Lie algebra.

2. The "fundamental triplet" theory involves two ideas, that there are triplets [or, in general, supermultiplets corresponding to spinor representations of SU(3)] among the hadrons, and that these triplets are basic objects represented by fields in a Lagrangian field theory. We note that even if triplets exist, they might be bootstrapped.

We discuss triplets more fully in a later part, but here it is worthwhile to point out that in "The Eightfold Way" paper, one possible triplet scheme is discussed at length for pedagogical purposes. A triplet ℓ of fermions is introduced (consisting of an isotopic doublet with one negative and one neutral member and a negative isotopic singlet) and a similar triplet L of bosons. Using ℓ and L, the baryon octet can be described as transforming like eight combinations $L\ell$ and meson octets like eight combinations $\bar{L}L$ or $\bar{\ell}\ell$.

The ℓ particles are called "leptons," although it is stated that they have nothing to do with real leptons and that the simple analogy between ℓ and real leptons would be spoiled if there were separate neutrinos for electron and muon. As is well known, this last turned out to be the case experimentally. Slightly different analogies between leptons and hadrons, as well as other possible patterns of triplets, are discussed in the later part.

THE EIGHTFOLD WAY: A THEORY OF STRONG INTERACTION SYMMETRY*

Murray Gell-Mann

March 15, 1961

(Second printing: April, 1962)
(Third printing: October, 1963)

(Preliminary version circulated Jan. 20, 1961)

*Research supported in part by the U. S. Atomic Energy Commission Contract No. AT(11-1)-68, and the Alfred P. Sloan Foundation.

We attempt once more, as in the global symmetry scheme, to treat the eight known baryons as a supermultiplet, degenerate in the limit of a certain symmetry but split into isotopic spin multiplets by a symmetry-breaking term. Here we do not try to describe the symmetry violation in detail, but we ascribe it phenomenologically to the mass differences themselves, supposing that there is some analogy to the μ-e mass difference.

The symmetry is called unitary symmetry and corresponds to the "unitary group" in three dimensions in the same way that charge independence corresponds to the "unitary group" in two dimensions. The eight infinitesimal generators of the group form a simple Lie algebra, just like the three components of isotopic spin. In this important sense, unitary symmetry is the simplest generalization of charge independence.

The baryons then correspond naturally to an eight-dimensional irreducible representation of the group; when the mass differences are turned on, the familiar multiplets appear. The pion and K meson fit into a similar set of eight particles, along with a predicted pseudoscalar meson χ^0 having $I = 0$. The pattern of Yukawa couplings of π, K, and χ is then nearly determined, in the limit of unitary symmetry.

The most attractive feature of the scheme is that it permits the description of eight vector mesons by a unified theory of the Yang-Mills type (with a mass term). Like Sakurai, we have a triplet ρ of vector mesons coupled to the isotopic spin current and a singlet vector meson ω^0 coupled to the hypercharge current. We also have a pair of doublets M and $\overline{M}$, strange vector mesons coupled to strangeness-changing currents that are conserved when the mass differences are turned off. There is only one coupling constant, in the symmetric limit, for the system of eight vector mesons. There is some experimental evidence for the existence of ω^0 and M, while ρ is presumably the famous $I = 1$, $J = 1$, π-π resonance.

A ninth vector meson coupled to the baryon current can be accommodated naturally in the scheme.

The most important prediction is the qualitative one that the eight baryons should all have the same spin and parity and that the pseudoscalar and vector mesons should form "octets", with possible additional "singlets".

If the symmetry is not too badly broken in the case of the renormalized coupling constants of the eight vector mesons, then numerous detailed predictions can be made of experimental results.

The mathematics of the unitary group is described by considering three fictitious "leptons", ν, e^-, and μ^-, which may or may not have something to do with real leptons. If there is a connection, then it may throw light on the structure of the weak interactions.

It has seemed likely for many years that the strongly interacting particles, grouped as they are into isotopic multiplets, would show traces of a higher symmetry that is somehow broken. Under the higher symmetry, the eight familiar baryons would be degenerate and form a supermultiplet. As the higher symmetry is broken, the Ξ, Λ, Σ, and N would split apart, leaving inviolate only the conservation of isotopic spin, of strangeness, and of baryons. Of these three, the first is partially broken by electromagnetism and the second is broken by the weak interactions. Only the conservation of baryons and of electric charge are absolute.

An attempt[1,2] to incorporate these ideas in a concrete model was the scheme of "global symmetry", in which the higher symmetry was valid for the interactions of the π meson, but broken by those of the K. The mass differences of the baryons were thus attributed to the K couplings, the symmetry of which was unspecified, and the strength of which was supposed to be significantly less than that of the π couplings.

The theory of global symmetry has not had great success in predicting experimental results. Also, it has a number of defects. The peculiar distribution of isotopic multiplets among the observed mesons and baryons is left unexplained. The arbitrary K couplings (which are not really particularly weak) bring in several adjustable constants. Furthermore, as admitted in Reference 1 and reemphasized recently by Sakurai[3,4] in his remarkable articles predicting vector mesons, the global model makes no direct connection between physical couplings and the currents of the conserved symmetry operators.

In place of global symmetry, we introduce here a new model of the higher symmetry of elementary particles which has none of these faults and a number of virtues.

We note that the isotopic spin group is the same as the group of all unitary 2x2 matrices with unit determinant. Each of these matrices can be written as exp(iA), where A is a hermitian 2x2 matrix. Since there are three independent hermitian 2x2 matrices (say, those of Pauli), there are three components of the isotopic spin.

Our higher symmetry group is the simplest generalization of isotopic spin, namely the group of all unitary 3x3 matrices with unit determinant. There are eight independent traceless 3x3 matrices and consequently the new "unitary spin" has eight components. The first three are just the components of the isotopic spin, the eighth is proportional to the hypercharge Y (which is +1 for N and K, -1 for Ξ and $\bar{K}$, 0 for Λ, Σ, π, etc.), and the remaining four are strangeness-changing operators.

Just as isotopic spin possesses a three-dimensional representation (spin 1), so the "unitary spin" group has an eight-dimensional irreducible representation, which we shall call simply $\underset{\sim}{8}$. In our theory, the baryon supermultiplet corresponds to this representation. When the symmetry is reduced, then $\underset{\sim}{I}$ and Y are still conserved but the four other components of unitary spin are not; the supermultiplet then breaks up into Ξ, Σ, Λ, and N. Thus the distribution of multiplets and the nature of strangeness or hypercharge are to some extent explained.

The pseudoscalar mesons are also assigned to the representation $\underset{\sim}{8}$. When the symmetry is reduced, they become the multiplets K, $\bar{K}$, π, and χ, where χ is a neutral isotopic singlet meson the existence of which we predict. Whether the PS mesons are regarded as fundamental or as bound states, their Yukawa couplings in the limit of "unitary" symmetry are describable in terms of only two coupling parameters.

The vector mesons are introduced in a very natural way, by an extension of the gauge principle of Yang and Mills.[5] Here too we have a supermultiplet of eight mesons, corresponding to the representation $\underset{\sim}{8}$. In the limit of unitary symmetry and with the mass of these vector mesons "turned off", we have a completely gauge-invariant and minimal theory, just like electromagnetism. When the mass is turned on, the gauge invariance is reduced (the gauge function may no longer be space-time dependent) but the conservation of unitary spin remains exact. The sources of the vector mesons are the conserved currents of the eight components of the unitary spin.[6]

When the symmetry is reduced, the eight vector mesons break up into a triplet ρ (coupled to the still-conserved isotopic spin current), a singlet ω (coupled to the still-conserved hypercharge current), and a pair of doublets M and $\bar{M}$ (coupled to a strangeness-changing current that is no longer conserved). The particles ρ and ω were both discussed by Sakurai. The ρ meson is presumably identical to the $I = 1$, $J = 1$, π-π resonance postulated by Frazer and Fulco[7] in order to explain the isovector electromagnetic form factors of the nucleon. The ω meson is no doubt the same as the $I = 1$, $J = 0$ particle or 3π resonance predicted by Nambu[8] and later by Chew[9] and others in order to explain the isoscalar form factors of the nucleon. The strange meson M may be the same as the K^* particle observed by Alston *et al.*[10]

Thus we predict that the eight baryons have the same spin and parity, that K is pseudoscalar and that χ exists, that ρ and ω exist with the properties assigned to them by Sakurai, and that M exists. But besides these qualitative predictions, there are also the many symmetry rules associated with the unitary spin. All of these are broken, though, by whatever destroys the unitary symmetry, and it is a delicate matter to find ways in which these effects of a broken symmetry can be explored experimentally.

Besides the eight vector mesons coupled to the unitary spin, there can be a ninth, which is invariant under unitary spin and is thus not degenerate with the other eight, even in the limit of unitary symmetry. We call this meson B^o. Presumably it exists too and is coupled to the baryon current. It is the meson predicted by Teller[11] and later by Sakurai[3] and explains most of the hard-core repulsion between nucleons and the attraction between nucleons and anti-nucleons at short distances.

We begin our exposition of the "eightfold way" in the next Section by discussing unitary symmetry using fictitious "leptons" which may have nothing to do with real leptons but help to fix the physical ideas in a rather graphic way. If there is a parallel between these "leptons" and the real ones, that would throw some light on the weak interactions, as discussed briefly in Section VI.

Section III is devoted to the $\underset{\sim}{8}$ representation and the baryons and Section IV to the pseudoscalar mesons. In Section V we present the theory of the vector mesons.

The physical properties to be expected of the predicted mesons are discussed in Section VII, along with a number of experiments that bear on those properties.

In Section VIII we take up the vexed question of the broken symmetry, how badly it is broken, and how we might succeed in testing it.

II The "Leptons" as a Model for Unitary Symmetry

For the sake of a simple exposition, we begin our discussion of unitary symmetry with "leptons", although our theory really concerns the baryons and mesons and the strong interactions. The particles we consider here for mathematical purposes do not necessarily have anything to do with real leptons, but there are some suggestive parallels. We consider three leptons, ν, e^-, and μ^-, and their antiparticles. The neutrino is treated on the same footing as the other two, although experience suggests that if it is treated as a four-component Dirac field, only two of the components have physical interaction. (Furthermore, there may exist two neutrinos, one coupled to the electron and the other to the muon.)

As far as we know, the electrical and weak interactions are absolutely symmetrical between e^- and μ^-, which are distinguished, however, from ν. The charged particles e^- and μ^- are separated by the mysterious difference in their masses. We shall not necessarily attribute this difference to any interaction, nor shall we explain it in any way. (If one insists on connecting it to an interaction, one might have to consider a coupling that becomes important only at exceedingly high energies and is, for the time being, only of academic interest.) We do, however, guess that the μ-e mass splitting is related to the equally mysterious mechanism that breaks the unitary symmetry of the baryons and mesons and splits the super-multiplets into isotopic multiplets. For practical purposes, we shall put all of these splittings into the mechanical masses of the particles involved.

It is well known that in present quantum electrodynamics, no one has succeeded in explaining the e-ν mass difference as an electromagnetic effect. Without prejudice to the question of its physical origin, we shall proceed with our discussion as if that mass difference were "turned on" along with the charge of the electron.

If we now "turn off" the μ-e mass difference, electromagnetism, and the weak interactions we are left with a physically vacuous theory of three exactly similar Dirac particles with no rest mass and no known couplings. This empty model is ideal for our mathematical purposes, however, and is physically motivated by the analogy with the strongly interacting particles, because it is at the corresponding stage of total unitary symmetry that we shall introduce the basic baryon mass and the strong interactions of baryons and mesons.

The symmetric model is, of course, invariant under all unitary transformations on the three states, ν, e^-, and μ^-.

Let us first suppose for simplicity that we had only two particles ν and e^-. We can factor each unitary transformation uniquely into one which multiplies both particles by the same phase factor and one (with determinant unity) which leaves invariant the product of the phase factors of ν and e^-. Invariance under the first kind of transformation corresponds to conservation of leptons ν and e^-. It may be considered separately from invariance under the class of transformations of the second kind (called by mathematicians the unitary unimodular group in two dimensions).

Each transformation of the first kind can be written as a matrix $e^{i\phi}1$, where 1 is the unit 2x2 matrix. The infinitesimal transformation is $1 + i(\delta\phi)1$ and so the unit matrix is the infinitesimal generator of these transformations. The transformations of the second kind are generated in the same way by the three independent traceless 2x2 matrices, which may be taken to be the three Pauli isotopic spin matrices τ_1, τ_2, τ_3. We thus have

$$1 + i \sum_{k=1}^{3} \delta\theta_k \frac{\tau_k}{2} \tag{2.1}$$

as the general infinitesimal transformation of the second kind. Symmetry under all the transformations of the second kind is the same as symmetry under τ_1, τ_2, τ_3, in other words charge independence or isotopic spin symmetry. The whole formalism of isotopic spin theory can then be constructed by considering the transformation properties of this doublet or spinor (ν, e^-) and of more complicated objects that transform like combinations of two or more such leptons.

The Pauli matrices τ_k are hermitian and obey the rules

$$\begin{aligned} \mathrm{Tr}\, \tau_i \tau_j &= 2\delta_{ij} \\ \left[\tau_i, \tau_j\right] &= 2ie_{ijk}\, \tau_k \\ \left\{\tau_i, \tau_j\right\} &= 2\delta_{ij}\, 1 \end{aligned} \tag{2.2}$$

We now generalize the idea of isotopic spin by including the third object μ^-. Again we factor the unitary transformations on the leptons into those which are generated by the 3x3 unit matrix 1 (and which correspond to lepton conservation) and those that are generated by the eight independent traceless 3x3 matrices (and which form the "unitary unimodular group" in three dimensions). We may construct a typical set of eight such matrices by analogy with the 2x2 matrices of Pauli. We call them $\lambda_1 \ldots\ldots \lambda_8$ and list them in Table I. They are hermitian and have the properties

$$\mathrm{Tr}\ \lambda_i \lambda_j = 2\delta_{ij}$$

$$\left[\lambda_i, \lambda_j\right] = 2if_{ijk}\lambda_k$$

$$\left\{\lambda_i, \lambda_j\right\} = \frac{4}{3}\delta_{ij}1 + 2d_{ijk}\lambda_k \quad , \tag{2.3}$$

where the f_{ijk} are real and totally antisymmetric like the Kronecker symbols e_{ijk} of Eq. (2.2), while the d_{ijk} are real and totally symmetric. These properties follow from the equations

$$\mathrm{Tr}\ \lambda_k \left[\lambda_i, \lambda_j\right] = 4if_{ijk}$$

$$\mathrm{Tr}\ \lambda_k \left\{\lambda_i, \lambda_j\right\} = 4d_{ijk} \tag{2.4}$$

derived from (2.3).

The non-zero elements of f_{ijk} and d_{ijk} are given in Table II for our choice of λ_i. Even and odd permutations of the listed indices correspond to multiplication of f_{ijk} by ± 1 respectively and of d_{ijk} by +1.

The general infinitesimal transformation of the second kind is, of course,

$$1 + i \sum_i \delta \theta_i \frac{\lambda_i}{2} \qquad (2.5)$$

by analogy with (2.1). Together with conservation of leptons, invariance under the eight λ_i corresponds to complete "unitary symmetry" of the three leptons.

It will be noticed that λ_1, λ_2, and λ_3 correspond to τ_1, τ_2, and τ_3 for ν and e^- and nothing for the muon. Thus, if we ignore symmetry between (ν, e^-) and the muon, we still have conservation of isotopic spin. We also have conservation of λ_8, which commutes with λ_1, λ_2, and λ_3 and is diagonal in our representation. We can diagonalize at most two λ's at the same time and we have chosen them to be λ_3 (the third component of the ordinary isotopic spin) and λ_8, which is like strangeness or hypercharge, since it distinguishes the isotopic singlet μ^- from the isotopic doublet (ν, e^-) and commutes with the isotopic spin.

Now the turning-on of the muon mass destroys the symmetry under λ_4, λ_5, λ_6, and λ_7 (i.e., under the "strangeness-changing" components of the "unitary spin") and leaves lepton number, "isotopic spin", and "strangeness" conserved. The electromagnetic interactions (along with the electron mass) then break the conservation of λ_1 and λ_2, leaving lepton number λ_3, and strangeness conserved. Finally, the weak interactions allow the strangeness to be changed (in muon decay) but continue to conserve the lepton number n_ℓ and the electric charge

$$Q = \frac{e}{2}\left(\lambda_3 + \frac{\lambda_8}{\sqrt{3}} - \frac{4}{3}\, n_\ell\right) \qquad (2.6)$$

where n_ℓ is the number of leptons minus the number of antileptons and equals 1 for ν, e^-, and μ^- (i.e., the matrix 1).

We see that the situation is just what is needed for the baryons and mesons. We transfer the symmetry under unitary spin to them and assign them strong couplings and basic symmetrical masses. Then we turn on the mass splittings, and the symmetry under the 4th, 5th, 6th, and 7th components of the unitary spin is lifted, leaving baryon number, strangeness, and isotopic spin conserved. Electromagnetism destroys the symmetry under the 1st and 2nd components of the spin, and the weak interactions destroy strangeness conservation. Finally, only charge and baryon number are conserved.

III Mathematical Description of the Baryons

In the case of isotopic spin $\underset{\sim}{I}$, we know that the various possible charge multiplets correspond to "irreducible representations" of the simple 2x2 matrix algebra described above for (ν, e^-). Each multiplet has $2I + 1$ components, where the quantum number I distinguishes one representation from another and tells us the eigenvalue $I(I + 1)$ of the operator $\sum_{i=1}^{3} I_i^2$, which commutes with all the elements of the isotopic spin group and in particular with all the infinitesimal group elements $1 + i \sum_{i=1}^{3} \delta\theta_i I_i$. The operators I_i are represented, within the multiplet, by hermitian $(2I + 1) \times (2I + 1)$ matrices having the same commutation rules

$$\left[I_i, I_j\right] = ie_{ijk} I_k \qquad (3.1)$$

as the 2x2 matrices $\tau_i/2$. For the case of $I = 1/2$, we have just $I_i = \tau_i/2$ within the doublet.

If we start with the doublet representation, we can build up all the others by considering superpositions of particles that transform like the original doublet. Thus, the antiparticles e^+, $-\bar{\nu}$ also form a doublet. (Notice the minus sign on the anti-neutrino state or field.) Taking $\frac{e^+e^- + \bar{\nu}\nu}{\sqrt{2}}$, we obtain a singlet, that is, a one-dimensional representation for which all the I_i are zero. Calling the neutrino and electron e_α with $\alpha = 1, 2$, we can describe the singlet by $\frac{1}{\sqrt{2}}\ \bar{e}_\alpha\, e_\alpha$ or, more concisely, $\frac{1}{\sqrt{2}}\ \bar{e}e$. The three components of a triplet can be formed by taking $e^+\nu = \frac{1}{2}\ \bar{e}(\tau_1 - i\tau_2)e$, $\frac{e^+e^- - \bar{\nu}\nu}{\sqrt{2}} = \frac{1}{\sqrt{2}}\ \bar{e}\,\tau_3\, e$, and $\nu e^- = \frac{1}{2}\ \bar{e}(\tau_1 + i\tau_2)e$. Rearranging these, we have just $\frac{1}{\sqrt{2}}\ \bar{e}\,\tau_j\, e$ with $j = 1, 2, 3$. Among these three states, the 3x3 matrices I_i^{jk} of the three components of I are given by

$$I_i^{jk} = -ie_{ijk} \quad . \qquad (3.2)$$

Now let us generalize these familiar results to the set of three states ν, e^-, and μ^-. Call them ℓ_α with $\alpha = 1, 2, 3$ and use $\bar{\ell}\ell$ to mean $\bar{\ell}_\alpha\, \ell_\alpha$, etc. For this system we define $F_i = \lambda i/2$ with $i = 1, 2, \ldots, 8$, just as $I_i = \tau i/2$ for isotopic spin. The F_i are the 8 components of the unitary spin operator $\underset{\sim}{F}$ in this case and we shall use the same notation in all representations. The first three components of $\underset{\sim}{F}$ are identical with the three components of the isotopic spin $\underset{\sim}{I}$ in all cases, while F_8 will always be $\frac{\sqrt{3}}{2}$ times the hypercharge Y (linearly related to the strangeness). In all representations, then, the components of $\underset{\sim}{F}$ will have the same commutation rules

$$\left[F_i, F_j\right] = if_{ijk} F_k \tag{3.3}$$

that they do in the simple lepton representation for which $F_i = \lambda_i/2$. (Compare the commutation rules in Eq. (2.3).) The trace properties and anticommutation properties will not be the same in all representations any more than they are for $\underset{\sim}{I}$. We see that the rules (3.1) are just a special case of (3.3) with indices 1, 2, 3, since the f's equal the e's for these values of the indices.

We must call attention at this point to an important difference between unitary or $\underset{\sim}{F}$ spin and isotopic or $\underset{\sim}{I}$ spin. Whereas, with a simple change of sign on $\bar{\nu}$, we were able to construct from $\bar{e}_\alpha$ a doublet transforming under $\underset{\sim}{I}$ just like e_α, we are not able to do the same thing for the $\underset{\sim}{F}$ spin when we consider the three anti-leptons $\bar{\ell}_\alpha$ compared to the three leptons ℓ_α. True, the anti-leptons do give a representation for $\underset{\sim}{F}$, but it is, in mathematical language, <u>inequivalent</u> to the lepton representation, even though it also has three dimensions. The reason is easy to see: when we go from leptons to anti-leptons the eigenvalues of the electric charge, the third component of $\underset{\sim}{I}$, and the lepton number all change sign, and thus the eigenvalues of F_8 change sign. But they were $\frac{1}{2\sqrt{3}}$, $\frac{1}{2\sqrt{3}}$, and $\frac{-1}{\sqrt{3}}$ for leptons and so they are a different set for anti-leptons and no similarity transformation can change one representation into the other. We shall refer to the lepton representation as 3 and the anti-lepton representation as $\underset{\sim}{\bar{3}}$.

Now let us consider another set of "particles" L_α transforming exactly like the leptons ℓ_α under unitary spin and take their anti-particles $\bar{L}_\alpha$. We follow the same procedure used above for the isotopic spin and the doublet e. We first construct the state $\frac{1}{\sqrt{3}}\, \bar{L}_\alpha\, \ell_\alpha$ or $\frac{1}{\sqrt{3}}\, \bar{L}\, \ell$. Just as $\frac{\bar{e}e}{\sqrt{2}}$ gave a one-dimensional representation of $\underset{\sim}{I}$ for which all the I_i were zero, so $\frac{\bar{L}\ell}{\sqrt{3}}$ gives a one-dimensional representation of $\underset{\sim}{F}$ for which all the F_i are zero. Call this one-dimensional representation $\underset{\sim}{1}$.

Now, by analogy with $\frac{\bar{e}\,\tau_i\, e}{\sqrt{2}}$ with $i = 1, 2, 3$, we form $\frac{\bar{L}\,\lambda_i\,\ell}{\sqrt{2}}$ with $i = 1, 2, \ldots, 8$. These states transform under unitary spin $\underset{\sim}{F}$ like an irreducible representation of dimension 8, which we shall call $\underset{\sim}{8}$. In this representation, the 8x8 matrices F_i^{jk} of the eight components F_i of the unitary spin are given by the relation

$$F_i^{jk} = -\, i f_{ijk} \quad , \tag{3.4}$$

analogous to Eq. (3.2).

When we formed an isotopic triplet from two isotopic doublets, in the discussion preceding Eq. (3.2), we had to consider linear combinations of the $\frac{\bar{e}\,\tau_i\, e}{\sqrt{2}}$ in order to get simple states with definite electric charges, etc. We must do the same here. Using the symbol ~ for "transforms like", we define

$$\begin{aligned}
\Sigma^+ &\sim \tfrac{1}{2}\, \bar{L}(\lambda_1 - i\lambda_2)\ell && \sim D^+\nu \\
\Sigma^- &\sim \tfrac{1}{2}\, \bar{L}(\lambda_1 + i\lambda_2)\ell && \sim D^0 e^- \\
\Sigma^0 &\sim \tfrac{1}{\sqrt{2}}\, \bar{L}\,\lambda_3\,\ell && \sim \frac{D^0\nu - D^+e^-}{\sqrt{2}} \\
p &\sim \tfrac{1}{2}\, \bar{L}(\lambda_4 - i\lambda_5)\ell && \sim S^+\nu \\
n &\sim \tfrac{1}{2}\, \bar{L}(\lambda_6 - i\lambda_7)\ell && \sim S^+e^- \\
\Xi^0 &\sim \tfrac{1}{2}\, \bar{L}(\lambda_6 + i\lambda_7)\ell && \sim D^+\mu^- \\
\Xi^- &\sim \tfrac{1}{2}\, \bar{L}(\lambda_4 + i\lambda_5)\ell && \sim D^0\mu^- \\
\Lambda &\sim \tfrac{1}{\sqrt{2}}\, \bar{L}\,\lambda_8\,\ell && \sim (D^0\nu + D^+e^- - 2S^+\mu^-)/\sqrt{6} \quad . \qquad (3.5)
\end{aligned}$$

The most graphic description of what we are doing is given in the last column, where we have introduced the notation D^0, D^+, and S^+ for the $\bar{L}$ particles analogous to the $\bar{\ell}$ particles $\bar{\nu}$, e^+, and μ^+ respectively. D stands for doublet and S for singlet with respect to isotopic spin. Using the last column, it is easy to see that the isotopic spins, electric charges, and hypercharges of the multiplets are exactly as we are accustomed to think of them for the baryons listed.

We say, therefore, that the eight known baryons form one degenerate supermultiplet with respect to unitary spin. When we introduce a perturbation that transforms like the μ-e mass difference, the supermultiplet will break up into exactly the known multiplets. (Of course, D will split from S at the same time as e^-, ν from μ^-.)

Of course, another type of baryon is possible, namely a singlet neutral one that transforms like $\frac{1}{\sqrt{3}}\ \bar{L}\ \ell$. If such a particle exists, it may be very heavy and highly unstable. At the moment, there is no evidence for it.

We shall attach no physical significance to the ℓ and $\bar{L}$ "particles" out of which we have constructed the baryons. The discussion up to this point is really just a mathematical introduction to the properties of unitary spin.

IV Pseudoscalar Mesons

We have supposed that the baryon fields N_j transform like an octet $\underset{\sim}{8}$ under $\underset{\sim}{F}$, so that the matrices of $\underset{\sim}{F}$ for the baryon fields are given by Eq. (3.4). We now demand that all mesons transform under $\underset{\sim}{F}$ in such a way as to have $\underset{\sim}{F}$-invariant strong couplings. If the 8 mesons π_i are to have Yukawa couplings, they must be coupled to $\bar{N}\ \Theta_i\ N$ for some matrices Θ_i, and we must investigate how such bilinear forms transform under $\underset{\sim}{F}$.

In mathematical language, what we have done in Section III is to look at the direct product $\underset{\sim}{\bar{3}} \times \underset{\sim}{3}$ of the representations $\underset{\sim}{\bar{3}}$ and $\underset{\sim}{3}$ and to find that it reduces to the direct sum of $\underset{\sim}{8}$ and $\underset{\sim}{1}$. We identified $\underset{\sim}{8}$ with the baryons and, for the time being, dismissed $\underset{\sim}{1}$. What we must do now is to look at $\underset{\sim}{\bar{8}} \times \underset{\sim}{8}$. Now it is easy to show that actually $\underset{\sim}{\bar{8}}$ is equivalent to $\underset{\sim}{8}$; this is unlike the situation for $\underset{\sim}{\bar{3}}$ and $\underset{\sim}{3}$. (We note that the values of Y, I_3, Q, etc., are symmetrically disposed about zero in the $\underset{\sim}{8}$ representation.) So the anti-baryons transform essentially like the baryons and we must reduce out the direct product $\underset{\sim}{8} \times \underset{\sim}{8}$. Standard group theory gives the result

$$\underset{\sim}{8} \times \underset{\sim}{8} = \underset{\sim}{1} + \underset{\sim}{8} + \underset{\sim}{8} + \underset{\sim}{10} + \underset{\sim}{\overline{10}} + \underset{\sim}{27} \quad , \tag{4.1}$$

where $\underset{\sim}{\overline{27}} = \underset{\sim}{27}$ (this can happen only when the dimension is the cube of an integer). The representation $\underset{\sim}{27}$ breaks up, when mass differences are turned on, into an isotopic singlet, triplet, and quintet with $Y = 0$, a doublet and a quartet with $Y = 1$, a doublet and a quartet with $Y = -1$, a triplet with $Y = 2$, and a triplet with $Y = -2$. The representation $\underset{\sim}{10}$ breaks up, under the same conditions, into a triplet with $Y = 0$, a doublet with $Y = -1$, a quartet with $Y = +1$, and a singlet with $Y = +2$. The conjugate representation $\underset{\sim}{\overline{10}}$ looks the same, of course, but with equal and opposite values of Y. None of these much resembles the pattern of the known mesons.

The $\underset{\sim}{8}$ representation, occurring twice, looks just the same for mesons as for baryons and is very suggestive of the known π, K, and $\overline{K}$ mesons plus one more neutral pseudoscalar meson with $I = 0$, $Y = 0$, which corresponds to Λ in the baryon case. Let us call this meson χ^0 and suppose it exists, with a fairly low mass. Then we have identified the known pseudoscalar mesons with an octet under unitary symmetry, just like the baryons. The representations $\underset{\sim}{1}$, $\underset{\sim}{10}$, $\underset{\sim}{\overline{10}}$, and $\underset{\sim}{27}$ may also correspond to mesons, even pseudoscalar ones, but presumably they lie higher in mass, some or all of them perhaps so high as to be physically meaningless.

To describe the eight pseudoscalar mesons as belonging to $\underset{\sim}{8}$, we put (very much as in (3.5))

$$\chi^{o} = \pi_8$$

$$\pi^{+} = (\pi_1 - i\pi_2)/\sqrt{2}$$

$$\pi^{-} = (\pi_1 + i\pi_2)/\sqrt{2}$$

$$\pi^{o} = \pi_3$$

$$K^{+} = (\pi_4 - i\pi_5)/\sqrt{2}$$

$$K^{o} = (\pi_6 - i\pi_7)/\sqrt{2}$$

$$\overline{K^{o}} = (\pi_6 + i\pi_7)/\sqrt{2}$$

$$K^{-} = (\pi_4 + i\pi_5)/\sqrt{2} \qquad (4.2)$$

and we know then that the matrices of $\underset{\sim}{F}$ connecting the π_j are just the same as those connecting the N_j, namely $F_i^{jk} = - if_{ijk}$.

To couple the 8 mesons invariantly to 8 baryons (say by γ_5), we must have a coupling

$$2i\, g_o\, \bar{N}\, \gamma_5\, \theta_i\, N\, \pi_i \qquad (4.3)$$

for which the relation

$$\left[F_i, \theta_j\right] = if_{ijk}\, \theta_k \qquad (4.4)$$

holds. Now the double occurrence of $\underset{\sim}{8}$ in Eq. (4.1) assures us that there are two independent sets of eight 8x8 matrices θ_i obeying (4.4). One of these sets evidently consists of the F_i themselves. It is not hard to find the other set if we go back to the commutators and anti-commutators of the λ matrices in the $\underset{\sim}{3}$ representation (Eq. (2.3)). Just as we formed

$F_i^{\ jk} = -if_{ijk}$, we define

$$D_i^{\ jk} = d_{ijk} \qquad (4.5)$$

and it is easy to show that the D's also satisfy Eq. (4.4). We recall that where the F matrices are imaginary and antisymmetric with respect to the basis we have chosen, the D's are real and symmetric.

Now what is the physical difference between coupling the pseudoscalar mesons π_i by means of D_i and by means of F_i? It lies in the symmetry under the operation

$$R:\ p \leftrightarrow \Xi^-,\ n \leftrightarrow \Xi^0,\ \Sigma^+ \leftrightarrow \Sigma^-,\ \Sigma^0 \leftrightarrow \Sigma^0,\ \Lambda \leftrightarrow \Lambda$$

$$K^+ \leftrightarrow \pm K^-,\ K^0 \leftrightarrow \pm \overline{K^0},\ \pi^+ \leftrightarrow \pm\pi^-,\ \pi^0 \leftrightarrow \pm\pi^0,\ \chi^0 \leftrightarrow \pm\chi^0 \quad , \qquad (4.6)$$

which is not a member of the unitary group, but a kind of reflection. In the language of N_i, we may say that R changes the sign of the second, fifth, and seventh particles; we note that λ_2, λ_5, and λ_7 are imaginary while the others are real. From Table II we can see that under these sign changes f_{ijk} is odd and d_{ijk} even.

It may be that in the limit of unitary symmetry the coupling of the pseudoscalar mesons is invariant under R as well as the unitary group. In that case, we choose either the plus sign in (4.6) and the D coupling or else the minus sign and the F coupling. The two possible coupling patterns are listed in Table III.

If only one of the patterns is picked out (case of R-invariance), it is presumably the D coupling, since that gives a large $\Lambda\pi\Sigma$ interaction (while the F coupling gives none) and the $\Lambda\pi\Sigma$ interaction is the best way of explaining the binding of Λ particles in hypernuclei.

In general, we may write the Yukawa coupling (whether fundamental or phenomenological, depending on whether the π_i are elementary or not) in the form

$$L_{int} = 2i\, g_o\, \bar{N}\, \gamma_5 \left[\alpha\, D_i + (1 - \alpha)\, F_i\right] N\, \pi_i \quad . \tag{4.7}$$

We note that in no case is it possible to make the couplings ΛKN and ΣKN both much smaller than the $N\pi N$ coupling. Since the evidence from photo-K production seems to indicate smaller effective coupling constants for ΛKN and ΣKN than for $N\pi N$ (indeed, that was the basis of the global symmetry scheme), we must conclude that our symmetry is fairly badly broken. We shall return to that question in Section VII.

A simple way to read off the numerical factors in Table III, as well as those in Table IV for the vector mesons, is to refer to the chart in Table V, which gives the transformation properties of mesons and baryons in terms of the conceptual "leptons" and "L particles" of Section III.

An interesting remark about the baryon mass differences may be added at this point. If we assume that they transform like the μ-e mass difference, that is, like the 8th component of the unitary spin, then there are only two possible mass-difference matrices, F_8 and D_8. That gives rise to a sum rule for baryon masses:

$$1/2\,(m_N + m_\Xi) = 3/4\, m_\Lambda + 1/4\, m_\Sigma \quad , \tag{4.8}$$

which is very well satisfied by the observed masses, much better than the corresponding sum rule for global symmetry.

There is no particular reason to believe, however, that the analogous sum rules for mesons are obeyed.

V Vector Mesons

The possible transformation properties of the vector mesons under $\underset{\sim}{F}$ are the same as those we have already examined in the pseudoscalar case. Again it seems that for low mass states we can safely ignore the representations $\underset{\sim}{27}$, $\underset{\sim}{10}$, and $\underset{\sim}{\overline{10}}$. We are left with 1 and the two cases of $\underset{\sim}{8}$.

A vector meson transforming according to $\underset{\sim}{1}$ would have $Q = 0$, $I = 0$, $Y = 0$ and would be coupled to the total baryon current $i \bar{N} \gamma_\mu N$, which is exactly conserved. Such a meson may well exist and be of great importance. The possibility of its existence has been envisaged for a long time.

We recall that the conservation of baryons is associated with the invariance of the theory under infinitesimal transformations

$$N \to (1 + i\epsilon)N \qquad , \tag{5.1}$$

where ϵ is a constant. This is gauge-invariance of the first kind. We may, however, consider the possibility that there is also guage invariance of the second kind, as discussed by Yang and Lee.[12] Then we could make ϵ a function of space-time. In the free baryon Lagrangian

$$L_N = - \bar{N}(\gamma_\alpha \partial_\alpha + m_o)N \tag{5.2}$$

this would produce a new term

$$L_N \to L_N - i \bar{N} \gamma_\alpha N \partial_\alpha \epsilon \tag{5.3}$$

which can be cancelled only if there exists a neutral vector meson field B_α coupled to the current $\bar{N} \gamma_\alpha N$:

$$L_B = - 1/4 (\partial_\alpha B_\beta - \partial_\beta B_\alpha)^2$$

$$L_{int} = if_o \bar{N} \gamma_\alpha N B_\alpha \qquad (5.4)$$

and which undergoes the gauge transformation

$$B_\alpha \rightarrow B_\alpha + 1/f_o \partial_\alpha \epsilon \quad . \qquad (5.5)$$

As Yang and Lee pointed out, such a vector meson is massless and if it existed with any appreciable coupling constant, it would simulate a kind of anti-gravity, for baryons but not leptons, that is contradicted by experiment.

We may, however, take the point of view that there are vector mesons associated with a gauge-invariant Lagrangian plus a mass term, which breaks the gauge invariance of the second kind while leaving inviolate the gauge invariance of the first kind and the conservation law. Such situations have been treated by Glashow,[13] Salam and Ward,[14] and others, but particularly in this connection by Sakurai.[3]

The vector meson transforming according to $\underset{\sim}{1}$ would then be of such a kind. Teller,[11] Sakurai,[3] and others have discussed the notion that such a meson may be quite heavy and very strongly coupled, binding baryons and anti-baryons together to make the pseudoscalar mesons according to the compound model of Fermi and Yang.[15] We shall leave this possibility open, but not consider it further here. If it is right, then the Yukawa couplings (4.7) must be treated as phenomenological rather than fundamental; from an immediate practical point of view, it may not make much difference.

We go on to consider the $\underset{\sim}{8}$ representation. An octet of vector mesons would break up into an isotopic doublet with $Y = 1$, which we shall call M (by analogy with K -- the symbol L is already used to mean π or μ); the corresponding doublet $\bar{M}$ analogous to $\bar{K}$; a triplet ρ with $Y = 0$ analogous to π; and a singlet ω^o with $Y = 0$ analogous to χ^o.

We may tentatively identify M with the K^* reported by Alston et al.[10] at 884 MeV with a width $\Gamma \approx 15$ MeV for break-up into $\pi + K$. Such a narrow width certainly points to a vector rather than a scalar state. The vector meson ρ may be identified, as Sakurai has proposed, with the $I = 1$, $J = 1$, π-π resonance discussed by Frazer and Fulco[7] in connection with the electromagnetic structure of the nucleon. The existence of ω^o has been postulated for similar reasons by Nambu,[8] Chew,[9] and others.

In principle, we have a choice again between couplings of the $\underset{\sim}{D}$ and the $\underset{\sim}{F}$ type for the vector meson octet. But there is no question which is the more reasonable theory. The current $i\,\bar{N} F_j \gamma_\alpha N$ is the current of the F-spin for baryons and in the limit of unitary symmetry the total F-spin current is exactly conserved. (The conservation of the strangeness-changing currents, those of F_4, F_5, F_6, and F_7, is broken by the mass differences, the conservation of F_2 and F_3 by electromagnetism, and that of F_3 and F_8 separately by the weak interactions. Of course, the current of the electric charge

$$Q = e\left(F_3 + \frac{F_8}{\sqrt{3}}\right) \tag{5.6}$$

is exactly conserved.)

Sakurai has already suggested that ρ is coupled to the isotopic spin current and ω to the hypercharge current. We propose in addition that the strange vector mesons M are coupled to the strangeness-changing components of the F-spin current and that the whole system is completely invariant under $\underset{\sim}{F}$ before the mass-differences have been turned on, so that the three coupling constants (suitably defined) are approximately equal even in the presence of the mass differences.

Now the vector mesons themselves carry F spin and therefore contribute to the current which is their source. The problem of constructing a nonlinear theory of this kind has been completely solved in the case of isotopic spin by Yang and Mills[5)] and by Shaw.[5)] We have only to generalize their result (for three vector mesons) to the case of F spin and eight vector mesons.

We may remark parenthetically that the Yang-Mills theory is irreducible, in the sense that all the 3 vector mesons are coupled to one another inextricably. We may always make a "reducible" theory by adjoining other, independent vector mesons like the field B_α discussed earlier in connection with the baryon current. It is an interesting mathematical problem to find the set of all irreducible Yang-Mills tricks. Glashow and the author[16)] have shown that the problem is the same as that of finding all the simple Lie algebras, one that was solved long ago by the mathematicians. The possible dimensions are 3, 8, 10, 14, 15, 21, and so forth. Our generalization of the Yang-Mills trick is the simplest one possible.

But let us "return to our sheep", in this case the 8 vector mesons. We first construct a completely gauge-invariant theory and then add a mass term for the mesons. Let us call the eight fields $\rho_{i\alpha}$, just as we denoted the eight pseudoscalar fields by π_i. We may think of the N_i, the π_i, and the $\rho_{i\alpha}$ as vectors in an 8-dimensional space. (The index α here refers to the four space-time components of a vector field.) We use our totally antisymmetric tensor f_{ijk} to define a cross product

$$(\underset{\sim}{A} \times \underset{\sim}{B})_i = f_{ijk} A_j B_k \tag{5.7}$$

The gauge transformation of the second kind analogous to Eqs. (5.1) and (5.5) is performed with an eight-component gauge function $\underset{\sim}{\phi}$:

$$\begin{aligned} &\underset{\sim}{N} \to \underset{\sim}{N} + \underset{\sim}{\phi} \times \underset{\sim}{N} \\ &\underset{\sim}{\rho}_\alpha \to \underset{\sim}{\rho}_\alpha + \underset{\sim}{\phi} \times \underset{\sim}{\rho}_\alpha - (2\gamma_0)^{-1} \ \partial_\alpha \underset{\sim}{\phi} \\ &\underset{\sim}{\pi} \to \underset{\sim}{\pi} + \underset{\sim}{\phi} \times \underset{\sim}{\pi} \end{aligned} \tag{5.8}$$

We have included the pseudoscalar meson field for completeness, treating it as elementary. We shall not write the π-N and possible π-π couplings in what follows, since they are not relevant and may simply be added in at the end. The bare coupling parameter is γ_0.

We define gauge-covariant field strengths by the relation

$$\underset{\sim}{G}_{\alpha\beta} = \partial_\alpha \underset{\sim}{\rho}_\beta - \partial_\beta \underset{\sim}{\rho}_\alpha + 2\gamma_0 \underset{\sim}{\rho}_\alpha \times \underset{\sim}{\rho}_\beta \tag{5.9}$$

and the gauge-invariant Lagrangian (to which a common vector meson mass term is presumably added) is simply

$$\begin{aligned} L = &- \tfrac{1}{4} \underset{\sim}{G}_{\alpha\beta} \cdot \underset{\sim}{G}_{\alpha\beta} - m_0 \overline{\underset{\sim}{N}} \cdot \underset{\sim}{N} - \overline{\underset{\sim}{N}} \gamma_\alpha \cdot (\partial_\alpha \underset{\sim}{N} + 2\gamma_0 \underset{\sim}{\rho}_\alpha \times \underset{\sim}{N}) \\ &- \tfrac{1}{2} \mu_0^{\ 2} \underset{\sim}{\pi} \cdot \underset{\sim}{\pi} - \tfrac{1}{2} (\partial_\alpha \underset{\sim}{\pi} + 2\gamma_0 \underset{\sim}{\rho}_\alpha \times \underset{\sim}{\pi}) \cdot (\partial_\alpha \underset{\sim}{\pi} + 2\gamma_0 \underset{\sim}{\rho}_\alpha \times \underset{\sim}{\pi}) \ . \end{aligned} \tag{5.10}$$

There are trilinear and quadrilinear interactions amongst the vector mesons, as usual, and also trilinear and quadrilinear couplings with the pseudoscalar mesons. All these, along with the basic coupling of vector mesons to the baryons, are characterized in the limit of no mass differences by the single coupling parameter γ_o. The symmetrical couplings of $\underset{\sim}{\rho}_\alpha$ to the bilinear currents of baryons and pseudoscalar mesons are listed in Table IV. In Section VII, we shall use them to predict a number of approximate relations among experimental quantities relevant to the vector mesons.

As in the case of the pseudoscalar couplings, the various vector couplings will have somewhat different strengths when the mass differences are included, and some couplings which vanish in (5.10) will appear with small coefficients. Thus, in referring to experimental renormalized coupling constants (evaluated at the physical masses of the vector mesons) we shall use the notation $\gamma_{N\Lambda M}$, $\gamma_{NN\rho}$, etc. In the limit of unitary symmetry, all of these that do not vanish are equal.

VI Weak Interactions

So far, the role of the leptons in unitary symmetry has been purely symbolic. Although we introduced a mathematical F spin for ν, e^-, and μ^-, that spin is not coupled to the eight vector mesons that take up the F spin gauge for baryons and mesons. If we take it seriously at all, we should probably regard it as a different spin, but one with the same mathematical properties.

Let us make another point, which may seem irrelevant but possibly is not. The photon and the charge operator to which it is coupled have not so far been explicitly included in our scheme. They must be put in as an afterthought, along with the corresponding gauge transformation, which was the model for the more peculiar gauge transformations we have treated. If the weak interactions are carried[17)] by vector bosons X_α and generated by a gauge transformation[18,19)] of their own, then these bosons and gauges have been ignored as well. Such considerations might cause us, if we are in a highly speculative frame of mind, to wonder about the possibility that each kind of interaction has its own type of gauge and its own set of vector particles and that the algebraic properties of these gauge transformations conflict with one another.

When we draw a parallel between the "F spin" of leptons and the F spin of baryons and mesons, and when we discuss the weak interactions at all, we are exploring phenomena that transcend the scheme we are using. Everything we say in this Section must be regarded as highly tentative and useful only in laying the groundwork for a possible future theory. The same is true of any physical interpretation of the mathematics in Sections II and III.

We shall restrict our discussion to charge - exchange weak currents and then only to the vector part. A complete discussion of the axial vector weak currents may involve more complicated concepts and even new mesons[20)] (scalar and/or axial vector) lying very high in energy.

The vector weak current of the leptons is just $\bar{\nu} \gamma_\alpha e + \bar{\nu} \gamma_\alpha \mu$. If we look at the abstract scheme for the baryons in Eq. (3.5), we see that a baryon current with the same transformation properties under F would consist of two parts: one, analogous to $\bar{\nu} \gamma_\alpha e$, would have $|\Delta \underset{\sim}{I}| = 1$ and $\Delta S = 0$, while the other, analogous to $\bar{\nu} \gamma_\alpha \mu$, would have $|\Delta \underset{\sim}{I}| = 1/2$ and

$\Delta S/\Delta Q = +1$. These properties are exactly the ones we are accustomed to associate with the weak interactions of baryons and mesons.

Now the same kind of current we have taken for the leptons can be assigned to the conceptual bosons L of Section III. Suppose it to be of the same strength. Then, depending on the relative sign of the lepton and L weak currents, the matrices in the baryon system may be F's or D's.

Suppose, in the $\Delta S = 0$ case, the relative sign is such as to give F. Then the resulting current is just one component of the isotopic spin current; and the same result will hold for mesons. Thus we will have the conserved vector current that has been proposed[17] to explain the lack of renormalization of the Fermi constant.

In the $\Delta S = 1$ case, by taking the same sign, we could get the almost-conserved strangeness-changing vector current, the current of $F_4 + iF_5$.

Further speculations along these lines might lead to a theory of the weak interactions.[21]

VII Properties of the New Mesons

The theory we have sketched is fairly solid only in the realm of the strong interactions, and we shall restrict our discussion of predictions to the interactions among baryons and mesons.

We predict the existence of 8 baryons with equal spin and parity following the pattern of N, Λ, Σ, and Ξ. Likewise, given the π and its coupling constant, we predict a pseudoscalar K and a new particle, the χ^o, both coupled (in the absence of mass differences) as in Eq. (4.7), and we predict pion couplings to hyperons as in the same equation.

Now in the limit of unitary symmetry an enormous number of selection and intensity rules apply. For example, for the reactions PS meson + baryon → PS meson + baryon, there are only 7 independent amplitudes. Likewise, baryon-baryon forces are highly symmetric. However, the apparent smallness of $g_1{}^2/4\pi$ for $NK\Lambda$ and $NK\Sigma$ compared to $N\pi N$ indicates that unitary symmetry is badly broken, assuming that it is valid at all. We must thus rely principally on qualitative predictions for tests of the theory; in Section VIII we take up the question of how quantitative testing may be possible.

The most clear-cut new prediction for the pseudoscalar mesons is the existence of χ^o, which should decay into 2γ like the π^o, unless it is heavy enough to yield $\pi^+ + \pi^- + \gamma$ with appreciable probability. (In the latter case, we must have $(\pi^+\pi^-)$ in an odd state.) $\chi^o \rightarrow 3\pi$ is forbidden by conservation of I and C. For a sufficiently heavy χ^o, the decay $\chi^o \rightarrow 4\pi$ is possible, but hampered by centrifugal barriers.

Now we turn to the vector mesons, with coupling pattern as given in Table IV. We predict, like Sakurai, the ρ meson, presumably identical with the resonance of Frazer and Fulco, and the ω meson, coupled to the hypercharge. In addition, we predict the strange vector meson M, which may be the same as the K^* of Alston et al.

Some of these are unstable with respect to the strong interactions and their physical coupling constants to the decay products are given by the decay widths. Thus, for $M \rightarrow K + \pi$, we have

$$\Gamma_M = 2 \frac{\gamma^2_{MK\pi}}{4\pi} \frac{k^3}{m_M^2} , \tag{7.1}$$

where k is the momentum of one of the decay mesons. We expect, of course, a $\cos^2\theta$ angular distribution relative to the polarization of M and a charge ratio of 2:1 in favor of $K^o + \pi^+$ or $K^+ + \pi^-$.

For the $I = 1$, $J = 1$, π-π resonance we have the decay $\rho \to 2\pi$ with width

$$\Gamma_\rho = \frac{8}{3} \frac{\gamma^2_{\rho\pi\pi}}{4\pi} \frac{k^3}{m_\rho^2} \tag{7.2}$$

Using a value $m_\rho = 4.5\ m_\pi$, we would have $\Gamma \approx m_\pi \frac{\gamma^2}{4\pi}$ and agreement with the theory of Bowcock et al.[7] would require a value of $\frac{\gamma^2}{4\pi}$ of the order of 2/3. If, now, we assume that the mass of M is really around 880 MeV, then Eq. (7.1) yields $\Gamma_M \approx \frac{\gamma^2}{4\pi} \cdot 50$ MeV. If the width is around 15 MeV, then the two values of $\gamma^2/4\pi$ are certainly of the same order.

We can obtain information about vector coupling constants in several other ways. If we assume, with Sakurai and Dalitz, that the Y^* of Alston et al.[22] (at 1380 MeV with decay $Y^* \to \pi + \Lambda$) is a bound state of $\bar{K}$ and N in a potential associated with the exchange of ω and ρ, then with simple Schrödinger theory we can roughly estimate the relevant coupling strengths. In the Schrödinger approximation (which is fairly bad, of course) we have the potential

$$V(\text{triplet}) \approx -3 \frac{\gamma_{NN\omega}\ \gamma_{KK\omega}}{4\pi} \frac{e^{-m_\omega r}}{r} + \frac{\gamma_{NN\rho}\ \gamma_{KK\rho}}{4\pi} \frac{e^{-m_\rho r}}{r} . \tag{7.3}$$

If ω has a mass of around 400 MeV (as suggested by the isoscalar form factor of the nucleon), then the right binding results with both $\gamma^2/4\pi$ of the order of 2/3.

A most important result follows if this analysis has any element of truth, since the singlet potential is

$$V(\text{singlet}) \approx -3 \frac{\gamma_{NN\omega}\gamma_{KK\omega}}{4\pi} \frac{e^{-m_\omega r}}{r} - 3 \frac{\gamma_{NN\rho}\gamma_{KK\rho}}{4\pi} \frac{e^{-m_\rho r}}{r} \quad . \qquad (7.4)$$

A singlet version of Y^* should exist considerably below the energy of Y^* itself. Call it Y_s^*. If it is bound by more than 100 MeV or so, it is metastable and decays primarily into $\Lambda + \gamma$, since $\Lambda + \pi$ is forbidden by charge independence. Thus, Y_s^* is a fake Σ^0, with $I = 0$ and different mass, and may have caused some difficulty in experiments involving the production of Σ^0 at high energy. If, because of level shifts due to absorption, Y_s^* is not very far below Y^*, then it should be detectable in the same way as Y^*; one should observe its decay into $\pi + \Sigma$.

Bound systems like Y^* and Y_s^* should occur not only for $\bar{K}N$ but also for $K\Xi$. (In the limit of unitary symmetry, these come to the same thing.)

The vector coupling constants occur also in several important poles. (For the unstable mesons, these are of course not true poles, unless we perform an analytic continuation of the scattering amplitude onto a second sheet, in which case they become poles at complex energies; they behave almost like true poles, however, when the widths of the vector meson states are small.) There is the pole at $q^2 = -m_M^2$ in the reactions $\pi^- + p \to \Lambda + K^0$ and $\pi^- + p \to \Sigma + K$; a peaking of K in the forward direction has already been observed in some of these reactions and should show up at high energies in all of them. Likewise, the pole at $q^2 = -m_\pi^2$ in the reaction $K + N \to M + N$ should be observable at high energies and its strength can be predicted directly from the width of M. In the reactions $\pi + N \to \Lambda + M$ and $\pi + N \to \Sigma + M$, there is a pole at $q^2 = -m_K^2$ and measurement of its strength can determine the coupling constants $g_{NK\Lambda}^2/4\pi$ and $g_{NK\Sigma}^2/4\pi$ for the K meson.

In πN scattering, we can measure the pole due to exchange of the ρ meson. In KN and $\bar{K}$N scattering, there are poles from the exchange of ρ and of ω; these can be separated since only the former occurs in the charge-exchange reaction. In NN scattering with charge-exchange, there is a ρ meson pole in addition to the familiar pion pole. Without charge exchange, the situation is terribly complicated, since there are poles from π, ρ, ω, χ, and B.

When the pole term includes a baryon vertex for the emission or absorption of a vector meson, we must remember that there is a "strong magnetic" term analogous to a Pauli moment as well as the renormalized vector meson coupling constant.

In a relatively short time, we should have a considerable body of information about the vector mesons.

VIII Violations of Unitary Symmetry

We have mentioned that within the unitary scheme there is no way that the coupling constants of K to both $N\Lambda$ and $N\Sigma$ can both be much smaller than 15, except through large violations of the symmetry. Yet experiments on photoproduction of K particles seem to point to such a situation. Even if unitary symmetry exists as an underlying pattern, whatever mechanism is responsible for the mass differences apparently produces a wide spread among the renormalized coupling constants as well. It is true that the binding of Λ particles in hypernuclei indicates a $\pi\Lambda\Sigma$ coupling of the same order of magnitude as the πNN coupling, but the anomalously small renormalized constants of the K meson indicate that a quantitative check of unitary symmetry will be very difficult.

What about the vector mesons? Let us discuss first the ρ and ω fields, which are coupled to conserved currents. For typical couplings of these fields, we have the relations

$$\gamma^2_{\rho\pi\pi} = \gamma_0{}^2 \, Z_3(\rho) \left[V_\pi{}^\rho(0)\right]^{-2} , \tag{8.1}$$

$$\gamma^2_{\rho NN} = \gamma_0{}^2 \, Z_3(\rho) \left[V_1{}^\rho(0)\right]^{-2} , \tag{8.2}$$

$$\gamma^2_{\omega NN} = \gamma_0{}^2 \, Z_3(\omega) \left[V_1{}^\omega(0)\right]^{-2} , \tag{8.3}$$

etc. Here, each renormalized coupling constant is written as a product of the bare constant, a vacuum polarization renormalization factor, and a squared form factor evaluated at zero momentum transfer. The point is that at zero momentum transfer there is no vertex renormalization because the source currents are conserved. To check, for example, the hypothesis that ρ is really coupled to the isotopic spin current, we must check that γ_0^2 in (8.1) is the same as γ_0^2 in (8.2). We can measure (say, by "pole experiments" and by the width of the π-π resonance) the renormalized constants on the left. The quantities V^2 are of the order unity in any case, and their ratios can be measured by studying electromagnetic form factors.[23]

The experimental check of "universality" between (8.1) and (8.2) is thus possible, but that tests only the part of the theory already proposed by Sakurai, the coupling of ρ to the isotopic spin current. To test unitary symmetry, we must compare (8.2) and (8.3); but then the ratio $Z_3(\rho)/Z_3(\omega)$ comes in to plague us. We may hope, of course, that this ratio is sufficiently close to unity to make the agreement striking, but we would like a better way of testing unitary symmetry quantitatively.

When we consider the M meson, the situation is worse, since the source current of M is not conserved in the presence of the mass differences. For each coupling of M, there is a vertex renormalization factor that complicates the comparison of coupling strengths.

An interesting possibility arises if the vector charge-exchange weak current is really given in the $|\Delta S| = 1$ case by the current of $F_4 \pm iF_5$ just as it is thought to be given in the $\Delta S = 0$ case by that of $F_1 \pm iF_2$ (the conserved current) and if the $\Delta S = 0$ and $|\Delta S| = 1$ currents are of equal strength, like the $e\nu$ and $\mu\nu$ currents. Then the leptonic $|\Delta S| = 1$ decays show renormalization factors that must be related to the vertex renormalization factors for the M meson, since the source currents are assumed to be the same. The experimental evidence on the decay $K \to \pi$ + leptons then indicates a renormalization factor, in the square of the amplitude, of the order of 1/20. In the decays $\Lambda \to p$ + leptons and $\Sigma^- \to n$ + leptons, both vector and axial vector currents appear to be renormalized by comparable factors.

The width for decay of M into $K + \pi$, if it is really about 15 MeV, indicates that the renormalized coupling constant $\gamma^2_{K\pi M}/4\pi$ is <u>not</u> much smaller than $\gamma^2_{\rho\pi\pi}/4\pi \approx 2/3$ and so there is at present no sign of these small factors in the coupling constants of M. It will be interesting, however, to see what the coupling constant $\gamma^2_{N\Lambda M}/4\pi$ comes out, as determined from the pole in $\pi^- + p \to \Lambda + K^0$.

We have seen that the prospect is rather gloomy for a quantitative test of unitary symmetry, or indeed of any proposed higher symmetry that is broken by mass differences or strong interactions. The best hope seems to lie in the possibility of direct study of the ratios of bare constants in experiments involving very high energies and momentum transfers, much larger than all masses.[24)] However, the theoretical work on this subject is restricted to renormalizable theories. At present, theories of the Yang-Mills type with a mass do not seem to be renormalizable,[25)] and no one knows how to improve the situation.

It is in any case an important challenge to theoreticians to construct a satisfactory theory of vector mesons. It may be useful to remark that the difficulty in Yang-Mills theories is caused by the mass. It is also the mass which spoils the gauge invariance of the first kind. Likewise, as in the μ-e case, it may be the mass that produces the violation of symmetry. Similarly, the nucleon and pion masses break the conservation of any axial vector current in the theory of weak interactions. It may be that a new approach to the rest masses of elementary particles can solve many of our present theoretical problems.

IX Acknowledgments

The author takes great pleasure in thanking Dr. S. L. Glashow and Professor R. P. Feynman for their enthusiastic help and encouragement and for numerous ideas, although they bear none of the blame for any errors or defects in the theory. Conversations with Professor R. Block about Lie algebras have been very enlightening.

TABLE I.

A Set of Matrices λ_i.

$$\lambda_1 = \begin{pmatrix} 0 & 1 & 0 \\ 1 & 0 & 0 \\ 0 & 0 & 0 \end{pmatrix} \qquad \lambda_2 = \begin{pmatrix} 0 & -i & 0 \\ i & 0 & 0 \\ 0 & 0 & 0 \end{pmatrix} \qquad \lambda_3 = \begin{pmatrix} 1 & 0 & 0 \\ 0 & -1 & 0 \\ 0 & 0 & 0 \end{pmatrix}$$

$$\lambda_4 = \begin{pmatrix} 0 & 0 & 1 \\ 0 & 0 & 0 \\ 1 & 0 & 0 \end{pmatrix} \qquad \lambda_5 = \begin{pmatrix} 0 & 0 & -i \\ 0 & 0 & 0 \\ i & 0 & 0 \end{pmatrix} \qquad \lambda_6 = \begin{pmatrix} 0 & 0 & 0 \\ 0 & 0 & 1 \\ 0 & 1 & 0 \end{pmatrix}$$

$$\lambda_7 = \begin{pmatrix} 0 & 0 & 0 \\ 0 & 0 & -i \\ 0 & i & 0 \end{pmatrix} \qquad \lambda_8 = \begin{pmatrix} \frac{1}{\sqrt{3}} & 0 & 0 \\ 0 & \frac{1}{\sqrt{3}} & 0 \\ 0 & 0 & \frac{-2}{\sqrt{3}} \end{pmatrix}$$

TABLE II.

Non-zero elements of f_{ijk} and d_{ijk}. The f_{ijk} are odd under permutations of any two indices while the d_{ijk} are even.

ijk	f_{ijk}	ijk	d_{ijk}
123	1	118	$1/\sqrt{3}$
147	1/2	146	1/2
156	-1/2	157	1/2
246	1/2	228	$1/\sqrt{3}$
257	1/2	247	-1/2
345	1/2	256	1/2
367	-1/2	338	$1/\sqrt{3}$
458	$\sqrt{3}/2$	344	1/2
678	$\sqrt{3}/2$	355	1/2
		366	-1/2
		377	-1/2
		448	$-1/(2\sqrt{3})$
		558	$-1/(2\sqrt{3})$
		668	$-1/(2\sqrt{3})$
		778	$-1/(2\sqrt{3})$
		888	$-1/\sqrt{3}$

TABLE III.

Yukawa interactions of pseudoscalar mesons with baryons,

assuming pure coupling through D.

$$L_{int}/ig_o = \pi^o \left\{\bar{p}\gamma_5 p - \bar{n}\gamma_5 n + \frac{2}{\sqrt{3}}\ \overline{\Sigma^o}\gamma_5\Lambda + \frac{2}{\sqrt{3}}\ \bar{\Lambda}\gamma_5\Sigma^o - \overline{\Xi^o}\gamma_5\Xi^o + \overline{\Xi^-}\gamma_5\Xi^-\right\}$$

$$+ \pi^+ \left\{\sqrt{2}\ \bar{p}\gamma_5 n + \frac{2}{\sqrt{3}}\ \overline{\Sigma^+}\gamma_5\Lambda + \frac{2}{\sqrt{3}}\ \bar{\Lambda}\gamma_5\Sigma^- - \sqrt{2}\ \overline{\Xi^o}\gamma_5\Xi^-\right\}$$

$$+ \text{h.c.}$$

$$+ K^+ \left\{- \frac{1}{\sqrt{3}}\ \bar{p}\gamma_5\Lambda + \bar{p}\gamma_5\Sigma^o + \sqrt{2}\ \bar{n}\gamma_5\Sigma^- - \frac{1}{\sqrt{3}}\ \bar{\Lambda}\gamma_5\Xi^- + \overline{\Sigma^o}\gamma_5\Xi^- + \sqrt{2}\ \overline{\Sigma^+}\gamma_5\Xi^o\right\}$$

$$+ \text{h.c.}$$

$$+ K^o \left\{- \frac{1}{\sqrt{3}}\ \bar{n}\gamma_5\Lambda - \bar{n}\gamma_5\Sigma^o + \sqrt{2}\ \bar{p}\gamma_5\Sigma^+ - \frac{1}{\sqrt{3}}\ \bar{\Lambda}\gamma_5\Xi^o - \overline{\Sigma^o}\gamma_5\Xi^o + \sqrt{2}\ \overline{\Sigma^-}\gamma_5\Xi^-\right\}$$

$$+ \text{h.c.}$$

$$+ \chi^o \left\{- \frac{1}{\sqrt{3}}\ \bar{p}\gamma_5 p - \frac{1}{\sqrt{3}}\ \bar{n}\gamma_5 n - \frac{2}{\sqrt{3}}\ \bar{\Lambda}\gamma_5\Lambda + \frac{2}{\sqrt{3}}\ \overline{\Sigma^+}\gamma_5\Sigma^+ + \frac{2}{\sqrt{3}}\ \overline{\Sigma^o}\gamma_5\Sigma^o + \frac{2}{\sqrt{3}}\ \overline{\Sigma^-}\gamma_5\Sigma^- - \frac{1}{\sqrt{3}}\ \overline{\Xi^o}\gamma_5\Xi^o - \frac{1}{\sqrt{3}}\ \overline{\Xi^-}\gamma_5\Xi^-\right\}$$

TABLE III (cont.)

Yukawa interactions of pseudoscalar mesons with baryons, assuming pure coupling through F.

$$L_{int}/ig_0 = \pi^0 (\bar{p}\gamma_5 p - \bar{n}\gamma_5 n + 2\,\overline{\Sigma^+}\gamma_5\Sigma^+ - 2\,\overline{\Sigma^-}\gamma_5\Sigma^- + \overline{\Xi^0}\gamma_5\Xi^0 - \overline{\Xi^-}\gamma_5\Xi^-)$$

$$+ \pi^+ (\sqrt{2}\ \bar{p}\gamma_5 n - \sqrt{2}\ \overline{\Xi^0}\gamma_5\Xi^- - 2\,\overline{\Sigma^+}\gamma_5\Sigma^0 + 2\,\overline{\Sigma^0}\gamma_5\Sigma^-)$$

$$+ \text{h.c.}$$

$$+ K^+ (-\sqrt{3}\ \bar{p}\gamma_5\Lambda + \sqrt{3}\ \bar{\Lambda}\gamma_5\Xi^- - \bar{p}\gamma_5\Sigma^0 - \sqrt{2}\ \bar{n}\gamma_5\Sigma^- + \overline{\Sigma^0}\gamma_5\Xi^- + \sqrt{2}\ \overline{\Sigma^+}\gamma_5\Xi^0)$$

$$+ \text{h.c.}$$

$$+ K^0 (-\sqrt{3}\ \bar{n}\gamma_5\Lambda + \sqrt{3}\ \bar{\Lambda}\gamma_5\Xi^0 + \bar{n}\gamma_5\Sigma^0 - \sqrt{2}\ \bar{p}\gamma_5\Sigma^+ - \overline{\Sigma^0}\gamma_5\Xi^0 + \sqrt{2}\ \overline{\Sigma^-}\gamma_5\Xi^-)$$

$$+ \text{h.c.}$$

$$+ \chi^0 (\sqrt{3}\ \bar{p}\gamma_5 p + \sqrt{3}\ \bar{n}\gamma_5 n - \sqrt{3}\ \overline{\Xi^0}\gamma_5\Xi^0 - \sqrt{3}\ \overline{\Xi^-}\gamma_5\Xi^-)$$

TABLE IV.

Trilinear couplings of ρ's to π's and N's.

$$L_{int}/i\gamma_0 = M^+_\alpha \{ -\sqrt{3}\ \bar{p}\gamma_\alpha\Lambda + \sqrt{3}\ \bar{\Lambda}\gamma_\alpha\Xi^- - \bar{p}\gamma_\alpha\Sigma^0 - \sqrt{2}\ \bar{n}\gamma_\alpha\Sigma^- + \overline{\Sigma^0}\gamma_\alpha\Xi^-$$

$$+\sqrt{2}\ \overline{\Sigma^+}\gamma_\alpha\Xi^0 - \sqrt{3}\ K^-\partial_\alpha\chi^0 + \sqrt{3}\ \chi^0\partial_\alpha K^- - K^-\partial_\alpha\pi^0$$

$$+ \pi^0\partial_\alpha K^- - \sqrt{2}\ \overline{K^0}\partial_\alpha\pi^- + \sqrt{2}\ \pi^-\partial_\alpha\overline{K^0}\}$$

$$+ \text{h.c.}$$

$$+ M^0_\alpha \{ -\sqrt{3}\ \bar{n}\gamma_\alpha\Lambda + \sqrt{3}\ \bar{\Lambda}\gamma_\alpha\Xi^0 + \bar{n}\gamma_\alpha\Sigma^0 - \sqrt{2}\ \bar{p}\gamma_\alpha\Sigma^+ - \overline{\Sigma^0}\gamma_\alpha\Xi^0$$

$$+ \sqrt{2}\ \overline{\Sigma^-}\gamma_\alpha\Xi^- - \sqrt{3}\ \overline{K^0}\partial_\alpha\chi^0 + \sqrt{3}\ \chi^0\partial_\alpha K^0 + \overline{K^0}\partial_\alpha\pi^0$$

$$- \pi^0\partial_\alpha\overline{K^0} - \sqrt{2}\ K^-\partial_\alpha\pi^+ + \sqrt{2}\ \pi^+\partial_\alpha K^-\}$$

$$+ \text{h.c.}$$

$$+ \rho^+_\alpha \{\sqrt{2}\ \bar{p}\gamma_\alpha n - \sqrt{2}\ \overline{\Xi^0}\gamma_\alpha\Xi^- - 2\ \overline{\Sigma^+}\gamma_\alpha\Sigma^0 + 2\ \overline{\Sigma^0}\gamma_\alpha\Sigma^- + \sqrt{2}\ K^-\partial_\alpha K^0$$

$$- \sqrt{2}\ K^0\partial_\alpha K^- - 2\ \pi^-\partial_\alpha\pi^0 + 2\ \pi^0\partial_\alpha\pi^-\}$$

$$+ \text{h.c.}$$

$$+ \rho^0_\alpha \{\bar{p}\gamma_\alpha p - \bar{n}\gamma_\alpha n + 2\ \overline{\Sigma^+}\gamma_\alpha\Sigma^+ - 2\ \overline{\Sigma^-}\gamma_\alpha\Sigma^- + \overline{\Xi^0}\gamma_\alpha\Xi^0 - \overline{\Xi^-}\gamma_\alpha\Xi^-$$

$$+ K^-\partial_\alpha K^+ - K^+\partial_\alpha K^- - \overline{K^0}\partial_\alpha K^0 + K^0\partial_\alpha\overline{K^0} + 2\ \pi^-\partial_\alpha\pi^+ - 2\ \pi^+\partial_\alpha\pi^-\}$$

$$+ \omega^0_\alpha \{\sqrt{3}\ \bar{p}\gamma_\alpha p + \sqrt{3}\ \bar{n}\gamma_\alpha n - \sqrt{3}\ \overline{\Xi^0}\gamma_\alpha\Xi^0 - \sqrt{3}\ \overline{\Xi^-}\gamma_\alpha\Xi^- + \sqrt{3}\ K^-\partial_\alpha K^+$$

$$- \sqrt{3}\ K^+\partial_\alpha K^- + \sqrt{3}\ \overline{K^0}\partial_\alpha K^0 - \sqrt{3}\ K^0\partial_\alpha\overline{K^0}\}$$

TABLE V.

Transformation properties of baryons and mesons, assuming pseudoscalar mesons coupled through D.

$$K^+ \sim \frac{\mu^+\nu + S^+\overline{D^0}}{\sqrt{2}}$$

$$K^0 \sim \frac{\mu^+e^- + S^+D^-}{\sqrt{2}}$$

$$\pi^+ \sim \frac{e^+\nu + D^+\overline{D^0}}{\sqrt{2}}$$

$$\pi^0 \sim \frac{\bar{\nu}\nu - e^+e^- + D^0\overline{D^0} - D^+D^-}{2}$$

$$\pi^- \sim \frac{\bar{\nu}e^- + D^0D^-}{\sqrt{2}}$$

$$\chi^0 \sim \frac{\bar{\nu}\nu + e^+e^- - 2\mu^+\mu^- + D^0\overline{D^0} + D^+D^- - 2S^+S^-}{\sqrt{12}}$$

$$\overline{K^0} \sim \frac{e^+\mu^- + D^+S^-}{\sqrt{2}}$$

$$K^- \sim \frac{\bar{\nu}\mu^- + D^0S^-}{\sqrt{2}}$$

$p \sim S^+\nu$ $\qquad$ $n \sim S^+e^-$

$\Sigma^+ \sim D^+\nu$ $\qquad$ $\Sigma^0 \sim \dfrac{D^0\nu - D^+e^-}{\sqrt{2}}$

$\Sigma^- \sim D^0e^-$ $\qquad$ $\Lambda \sim \dfrac{D^0\nu + D^+e^- - 2S^+\mu^-}{\sqrt{6}}$

$\Xi^0 \sim D^+\mu^-$ $\qquad$ $\Xi^- \sim D^0\mu^-$

TABLE V (cont.)

$$M^+ \sim \frac{\mu^+\nu - S^+\overline{D^o}}{\sqrt{2}}$$

$$M^o \sim \frac{\mu^+e^- - S^+D^-}{\sqrt{2}}$$

$$\rho^+ \sim \frac{e^+\nu - D^+\overline{D^o}}{\sqrt{2}}$$

$$\rho^o \sim \frac{\bar{\nu}\nu - e^+e^- - D^o\overline{D^o} + D^+D^-}{2}$$

$$\rho^- \sim \frac{\bar{\nu}e^- - D^oD^-}{\sqrt{2}}$$

$$\omega^o \sim \frac{\bar{\nu}\nu + e^+e^- - 2\mu^+\mu^- - D^o\overline{D^o} - D^+D^- + 2S^+S^-}{\sqrt{12}}$$

$$\overline{M^o} \sim \frac{e^+\mu^- - D^+S^-}{\sqrt{2}}$$

$$M^- \sim \frac{\bar{\nu}\mu^- - D^oS^-}{\sqrt{2}}$$

REFERENCES

1. M. Gell-Mann, Phys. Rev. 106, 1296 (1957).
2. J. Schwinger, Ann. Phys. 2, 407 (1957).
3. J. J. Sakurai, Ann. Phys. 11, 1 (1960).
4. J. J. Sakurai, "Vector Theory of Strong Interactions", unpublished.
5. C. N. Yang and R. Mills, Phys. Rev. 96, 191 (1954). Also, R. Shaw, unpublished.
6. After the circulation of the preliminary version of this work (January 1961) the author has learned of a similar theory put forward independently and simultaneously by Y. Ne'eman (Nuclear Phys., to be published). Earlier uses of the 3-dimensional unitary group in connection with the Sakata model are reported by Y. Ohnuki at the 1960 Rochester Conference on High Energy Physics. A. Salam and J. Ward (Nuovo Cimento, to be published) have considered related questions. The author would like to thank Dr. Ne'eman and Professor Salam for communicating their results to him.
7. W. R. Frazer and J. R. Fulco, Phys. Rev. 117, 1609 (1960). See also J. Bowcock, W. N. Cottingham, and D. Lurie, Phys. Rev. Letters 5, 386 (1960).
8. Y. Nambu, Phys. Rev. 106, 1366 (1957).
9. G. F. Chew, Phys. Rev. Letters 4, 142 (1960).
10. M. Alston et al., to be published.
11. E. Teller, Proceedings of the Rochester Conference, 1956.
12. C. N. Yang and T. D. Lee, Phys. Rev. 98, 1501 (1955).
13. S. L. Glashow, Nuclear Phys. 10, 107 (1959).

REFERENCES (cont.)

14. A. Salam and J. C. Ward, Nuovo Cimento 11, 568 (1959).

15. E. Fermi and C. N. Yang, Phys. Rev. 76, 1739 (1949).

16. S. L. Glashow and M. Gell-Mann, to be published.

17. R. P. Feynman and M. Gell-Mann, Phys. Rev. 109, 193 (1958).

18. S. Bludman, Nuovo Cimento 9, 433 (1958).

19. M. Gell-Mann and M. Levy, Nuovo Cimento 16, 705 (1960).

20. M. Gell-Mann, talk at Rochester Conference on High Energy Physics, 1960.

21. Earlier attempts to draw a parallel between leptons and baryons in the weak interactions have been made by A. Gamba, R. E. Marshak, and S. Okubo, Proc. Nat. Acad. Sci. 45, 881 (1959), and Y. Yamaguchi, unpublished. Dr. S. L. Glashow reports that Yamaguchi's scheme has much in common with the one discussed in this paper.

22. M. Alston, L. W. Alvarez, P. Eberhard, M. L. Good, W. Graziano, H. K. Ticho, and S. G. Wojcicki, Phys. Rev. Letters 5, 518 (1960).

23. M. Gell-Mann and F. Zachariasen, "Form Factors and Vector Mesons", to be published.

24. M. Gell-Mann and F. Zachariasen, "Broken Symmetries and Bare Coupling Constants", to be published.

25. Kamefuchi and Umezawa, to be published. Salam and Kumar, to be published.

DERIVATION OF STRONG INTERACTIONS FROM A GAUGE INVARIANCE

Y. NE'EMAN
Department of Physics, Imperial College, London

Received 13 February 1961

Abstract: A representation for the baryons and bosons is suggested, based on the Lie algebra of the 3-dimensional traceless matrices. This enables us to generate the strong interactions from a gauge invariance principle, involving 8 vector bosons. Some connections with the electromagnetic and weak interactions are further discussed.

1. Introduction

Following Yang and Mills [1]), two new theories deriving the strong interactions from a gauge invariance principle have been published lately, by Sakurai [2]) and by Salam and Ward [3]). Sakurai's treatment is based on three separate gauges — isospin, hypercharge and baryonic charge — unrelated from the point of view of group-theory; Salam and Ward postulate one unified gauge, an 8-dimensional rotation gauge, combining isospin and hypercharge through Tiomno's [4]) representation.

One important advantage of the latter theory is the emergence of Yukawa-like terms, allowing for the production of single π or K mesons. Such terms do not arise normally from the boson-currents, and it is through the reintroduction of the σ scalar isoscalar meson [5]), and the assumption that it has a non vanishing vacuum expectation value, that they now appear in ref. [3]). On the other hand, boson-current terms with no σ factor then lead to weak interactions, as it is the creation and re-absorption of these σ mesons that generates the strong coupling. A 9-dimensional version, with a gauge based on restricted rotations, involves 13 vector bosons, of which only seven mediate the strong interactions; the remainder would generate weak interactions — though no way has been found to induce parity non conservation into these without affecting the strong interactions as well. The seven vector bosons of the strong interactions look like a K set and a π set; in Sakurai's theory they are replaced by a π set and two singlets.

The following treatment is an attempt to formulate a unified gauge, while reducing the number of vector bosons. It does, indeed, generate a set of 8 mediating fields, seven of which are similar to the above seven, the eighth is rather like Sakurai's B_y singlet. Still, one important factor is missed, namely, there is no room for the σ meson, and thus there are no single-pion terms.

To minimise the number of parameters of the gauge, and thus the number of vector bosons it will generate, we have adopted the following method: we abandoned the usual procedure of describing fields as vector components in a Euclidean isospace, and replace it by a matrix-algebra manifold. Fields still form vectorial sets only in the space of the group operators themselves, invariance of the Lagrangians being achieved by taking the traces of product matrices.

We have also abandoned rotations and use a group first investigated by Ikeda, Ogawa and Ohnuki [6]) in connection with the construction of bound states in the Sakata model. Our present use of this group is in an entirely different context, as our assumptions with regard to the representation of the fermions do not follow the prescriptions of the model.

2. Matrix Formalism

We use an 8-dimensional linear vector space P spanned by the semisimple Lie algebra of the 3×3 matrices X_{ij} of ref. [6]). We have excluded the identity transformation and use as basis the 8 linearly independent $\mathbf{u}^i \in \mathbf{U}$ given by the following formulae:

$$\mathbf{U}\begin{cases} \mathbf{u}^1 = \frac{1}{2}\sqrt{2}(X_{(31)}-iX_{[31]}), & \mathbf{u}^4 = \frac{1}{2}\sqrt{2}(X_{(31)}+iX_{[31]}),\\ \mathbf{u}^2 = \frac{1}{2}\sqrt{2}(X_{(23)}-iX_{[23]}), & \mathbf{u}^3 = \frac{1}{2}\sqrt{2}(X_{(23)}+iX_{[23]}),\\ \mathbf{u}^5 = \frac{1}{2}\sqrt{2}(X_{(12)}+iX_{[12]}), & \mathbf{u}^6 = \frac{1}{2}\sqrt{2}(X_{(12)}-iX_{[12]}),\\ \mathbf{u}^7 = \frac{1}{2}(X_{11}-X_{22}), & \mathbf{u}^8 = \frac{1}{6}\sqrt{3}(X_{11}+X_{22}-2X_{33}),\end{cases} \tag{1}$$

$$X_{ij}^{\alpha\beta} = \tfrac{1}{2}\delta_{i\alpha}\delta_{j\beta}(1-i)+\tfrac{1}{2}\delta_{i\beta}\delta_{j\alpha}(1+i),$$

$$X_{(ij)} = \tfrac{1}{2}(X_{ij}+X_{ji}), \qquad X_{[ij]} = \tfrac{1}{2}(X_{ij}-X_{ji}),$$

the indices α and β denoting the matrix elements. The X_{ij} are hermitian, whereas the basis matrices $\mathbf{u}_i$ are not, with the exception of $\mathbf{u}^7$ and $\mathbf{u}^8$, both diagonal. $\mathbf{U}$ can contain only two linearly independent diagonal elements, and the 2-dimensional sub-space $P_\mathrm{d} \subset P$ spanned by the set of all diagonal elements can be represented by a real Euclidean 2-space. In this 2-space, $\mathbf{u}^7$ and $\mathbf{u}^8$ are orthogonal: not only do they commute with each other, as any $[\mathbf{u}'_\mathrm{d}, \mathbf{u}''_\mathrm{d}]=0$ for $\mathbf{u}'_\mathrm{d}, \mathbf{u}''_\mathrm{d} \subset P_\mathrm{d}$; each also commutes with a 3-rotation constructed by taking the other as an M_z. In the set (1), U_a ($\mathbf{u}^5$, $\mathbf{u}^6$, $\mathbf{u}^7$) forms such a 3-rotation, and

$$[\mathbf{u}^8, \mathbf{u}^a] = 0. \tag{2}$$

We also use a basis $\mathbf{U}'$ differing from $\mathbf{U}$ only in P_d,

$$\mathbf{U}'\begin{cases} \mathbf{u}^{i\prime} = \mathbf{u}^i, & i = 1, 2, \ldots, 6,\\ \mathbf{u}^{7\prime} = -\frac{1}{2}(\mathbf{u}^7-\sqrt{3}\,\mathbf{u}^8), & \\ \mathbf{u}^{8\prime} = \frac{1}{2}\sqrt{3}(\mathbf{u}^7+\frac{1}{3}\sqrt{3}\,\mathbf{u}^8), & \end{cases} \tag{3}$$

where again $\mathbf{u}^{7'}$ and $\mathbf{u}^{8'}$ are orthogonal, $\mathbf{u}^{8'}$ commuting with the 3-rotation U_b $(\mathbf{u}^2, \mathbf{u}^3, \mathbf{u}^{7'})$:

$$[\mathbf{u}^{8'}, \mathbf{u}^b] = 0. \tag{4}$$

We now define a metric g_{ij} in P space,

$$g_{ij} = \begin{vmatrix} & & & 1 & & & & \\ & & 1 & & & & & \\ & 1 & & & & & & \\ 1 & & & & & & & \\ & & & & & 1 & & \\ & & & & 1 & & & \\ & & & & & & 1 & \\ & & & & & & & 1 \end{vmatrix}, \tag{5}$$

such that

$$\sum_{i=1}^{8} g_{ij}\mathbf{u}^i = \mathbf{u}_j = \tilde{\mathbf{u}}^j. \tag{6}$$

Note that

$$2\mathrm{Tr}\{\tilde{\mathbf{u}}^i\, \mathbf{u}^i\} = 2\mathrm{Tr}\{\mathbf{u}^i\, \tilde{\mathbf{u}}^i\} = 1. \tag{7}$$

Thus

$$\mathbf{A} \cdot \mathbf{B} = 2\mathrm{Tr}\{ \sum_{i,j=1}^{8} g_{ij} A^i \mathbf{u}^i B^j \mathbf{u}^j\} \equiv \sum_{i,j} g_{ij} A^i B^j \tag{8}$$

is a scalar product in P.

When using our algebra for unitary transformations, we shall take the hermitian set V as a basis for the infinitesimal operators,

$$\mathbf{V} \begin{cases} \mathbf{v}^{14} = \frac{1}{2}\sqrt{2}(\mathbf{u}^1+\mathbf{u}^4), & \mathbf{v}^{41} = \frac{1}{2}i\sqrt{2}(\mathbf{u}^1-\mathbf{u}^4), \\ \mathbf{v}^{23} = \frac{1}{2}\sqrt{2}(\mathbf{u}^2+\mathbf{u}^3), & \mathbf{v}^{32} = -\frac{1}{2}i\sqrt{2}(\mathbf{u}^2-\mathbf{u}^3), \\ \mathbf{v}^{56} = \frac{1}{2}\sqrt{2}(\mathbf{u}^5+\mathbf{u}^6), & \mathbf{v}^{65} = -\frac{1}{2}i\sqrt{2}(\mathbf{u}^5-\mathbf{u}^6), \\ \mathbf{v}^7 = \mathbf{u}^7, & \mathbf{v}^8 = \mathbf{u}^8, \end{cases} \tag{9}$$

so that

$$\sum_{k=1}^{8} A_V{}^k B_V{}^k = \sum_{i=1}^{8} A_U{}^i B_{U,i}, \tag{10}$$

i.e. the scalar product (8) is Euclidean in the V system.

Under a unitary transformation $E^{(m_V)} = \exp(i\varepsilon^{m_V}\mathbf{v}^{m_V})$ (m_V is the single or double index in $\mathbf{V}$), the component $A^k\mathbf{u}^k$ transforms like

$$\sum_{l=1}^{8} \delta_{(m_V)} A^l \mathbf{u}^l = i\varepsilon^{m_V} A^k [\mathbf{v}^{m_V}, \mathbf{u}^k] = i\varepsilon^{m_V} A^k \sum_{l=1}^{8} f^l_{m_V, k}\mathbf{u}^l$$

and for

$$E = \exp(i \sum_{m_V} \varepsilon^{m_V} \mathbf{v}^{m_V}) \tag{11}$$

we get variations

$$\delta A^l = i \sum_{m_V} \varepsilon^{m_V} \sum_{k=1}^{8} f^l_{m_V, k} A^k. \tag{12}$$

The $f^l_{m_V, k}$ define an 8×8 representation of our algebra in P space,

$$C^{l, k}_{m_V} = f^l_{m_V, k}, \tag{13}$$

so that (12) becomes in P

$$\delta A^l = i \sum_{m_V=1}^{8} \varepsilon^{m_V} \sum_{k=1}^{8} C^{l, k}_{m_V} A^k,$$

or

$$\delta \mathbf{A} = i \sum_{m_V} \varepsilon^{m_V} C_{m_V} \mathbf{A} = i \sum_{i, j} g_{ij} \varepsilon^i C^j \mathbf{A}, \tag{14}$$

where we have returned to the basis $\mathbf{U}$ or $\mathbf{U}'$.

3. Fields and Interactions

We define the quantum operators

$$\mathbf{I}(C_5, C_6, C_7), \qquad I_Z = C^7, \qquad Q = \tfrac{2}{3}\sqrt{3}\, C^{8'}, \qquad Y = \tfrac{2}{3}\sqrt{3}\, C^8, \tag{15}$$

and write the fields as vectors in P space

$$\begin{aligned} &\psi(\mathrm{p}, \mathrm{n}, \Xi^0, \Xi^-, \Sigma^+, \Sigma^-, \Sigma^0, \Lambda), \qquad \overline{\psi}(\overline{\Xi^-}, \overline{\Xi^0}, \overline{\mathrm{n}}, \overline{\mathrm{p}}, \overline{\Sigma^-}, \overline{\Sigma^+}, \overline{\Sigma^0}, \overline{\Lambda}) \\ &\varphi(\mathrm{K}^+, \mathrm{K}^0, \overline{\mathrm{K}^0}, \overline{\mathrm{K}^-}, \pi^+, \pi^-, \pi^0, \pi^{0'}), \qquad \overline{\varphi} = \varphi, \end{aligned} \tag{16}$$

or in matrix form

$$\psi = \tfrac{1}{2}\sqrt{2} \left| \begin{array}{c:c:c} \frac{1}{2}\sqrt{2}\,\Sigma^0 + \frac{1}{6}\sqrt{6}\,\Lambda & \Sigma^+ & \mathrm{p} \\ \hdashline \Sigma^- & -\frac{1}{2}\sqrt{2}\,\Sigma^0 + \frac{1}{6}\sqrt{6}\,\Lambda & \mathrm{n} \\ \hdashline \Xi^- & \Xi^0 & -\sqrt{\frac{2}{3}}\,\Lambda \end{array} \right|,$$

$$\overline{\psi} = \tfrac{1}{2}\sqrt{2} \left| \begin{array}{c:c:c} \frac{1}{2}\sqrt{2}\,\overline{\Sigma^0} + \frac{1}{6}\sqrt{6}\,\overline{\Lambda} & \overline{\Sigma^-} & \overline{\Xi^-} \\ \hdashline \overline{\Sigma^+} & -\frac{1}{2}\sqrt{2}\,\overline{\Sigma^0} + \frac{1}{6}\sqrt{6}\,\overline{\Lambda} & \overline{\Xi^0} \\ \hdashline \overline{\mathrm{p}} & \overline{\mathrm{n}} & -\sqrt{\frac{2}{3}}\,\overline{\Lambda} \end{array} \right|, \tag{17}$$

$$\varphi = \tfrac{1}{2}\sqrt{2} \left| \begin{array}{c:c:c} \frac{1}{2}\sqrt{2}\,\pi^0 + \frac{1}{6}\sqrt{6}\,\pi^{0'} & \pi^+ & \mathrm{K}^+ \\ \hdashline \pi^- & -\frac{1}{2}\sqrt{2}\,\pi^0 + \frac{1}{6}\sqrt{6}\,\pi^{0'} & \mathrm{K}^0 \\ \hdashline \overline{\mathrm{K}^-} & \overline{\mathrm{K}^0} & -\sqrt{\frac{2}{3}}\,\pi^{0'} \end{array} \right|,$$

The free field Lagrangians are

$$\mathscr{L}_{\psi}{}^{0} = -\overline{\psi}\cdot(\gamma^{\mu}\partial_{\mu}+m_{\psi})\psi, \qquad \mathscr{L}_{\varphi}{}^{0} = -\tfrac{1}{2}(\partial^{\mu}\boldsymbol{\varphi}\cdot\partial_{\mu}\boldsymbol{\varphi}+m_{\varphi}{}^{2}\boldsymbol{\varphi}\cdot\boldsymbol{\varphi}). \quad (18)$$

We postulate the invariance of these Lagrangians under the unitary gauge transformation

$$E = \exp(i\sum_{m_V}\varepsilon^{m_V}(x)C^{m_V}) \quad (19)$$

and follow the now standardized technique of Yang and Mills [1]) and Utiyama [7]), recombining the C set in terms of the basis **U** of (1).

The total Lagrangian becomes

$$\mathscr{L}_{\text{total}} = \mathscr{L}_{\psi}{}^{0}+\mathscr{L}_{\varphi}{}^{0}+\mathscr{L}_{\psi}{}^{\mathscr{I}}+\mathscr{L}_{\varphi}{}^{\mathscr{I}}+\mathscr{L}_{B}{}^{0}, \quad (20)$$

$$\mathscr{L}_{\psi}{}^{\mathscr{I}} = -\sum_{i=1}^{8}\sum_{n=1}^{8}\bar{\psi}^{n}\gamma^{\mu}C_{i}\psi_{n}B_{\mu}{}^{i}, \quad (21)$$

$$\mathscr{L}_{\varphi}{}^{\mathscr{I}} = -\sum_{\substack{i=1\\ j=1}}^{8}\sum_{n=1}^{8}(\partial_{\mu}\bar{\varphi}_{n}+C_{i}\bar{\varphi}_{n}B_{\mu}{}^{i})C_{j}\varphi^{n}B^{j\mu}. \quad (22)$$

The $B_{\mu}{}^{i}$ is a set of 8 vector bosons, with the following isobaric and strangeness qualities:

$$\begin{array}{llll} B_{\mu}{}^{1}\to K^{+}, & B_{\mu}{}^{2}\to K^{0}, & B_{\mu}{}^{3}\to \overline{K^{0}}, & B_{\mu}{}^{4}\to \overline{K^{-}}, \\ B_{\mu}{}^{5}\to \pi^{+}, & B_{\mu}{}^{6}\to \pi^{-}, & B_{\mu}{}^{7}\to \pi^{0}, & B_{\mu}{}^{8}\to \pi^{0\prime}. \end{array} \quad (23)$$

We have here the same set of vector bosons Salam and Ward got out of the 8-dimensional rotation gauge — with an additional $\pi^{0\prime}$—like interaction. Denoting as in ref. [3]) the K-like set by Z_{μ},

$$Z_{\mu} = (Z_{\mu}{}^{+}, Z_{\mu}{}^{0}), \qquad \bar{Z}_{\mu} = (\bar{Z}_{\mu}{}^{-}, \bar{Z}_{\mu}{}^{0}),$$

the π-like one by $V_{\mu}{}^{i}$ with

$$V_{\mu}{}^{\pm} = \tfrac{1}{2}\sqrt{2}(V_{\mu}{}^{1}\pm iV_{\mu}{}^{2}), \qquad V_{\mu}{}^{0} = V_{\mu}{}^{3}$$

and the $\pi^{0\prime}$-like by $B_{\mu}{}^{8} = X_{\mu}{}^{0}$ we get

$$\mathscr{L}_{B}{}^{0} = -\tfrac{1}{4}(\mathbf{F}_{\mu\nu}\cdot\mathbf{F}^{\mu\nu}), \quad (24)$$

$$\mathbf{F}_{\mu\nu} = \mathbf{H}_{\mu\nu}+\mathbf{G}_{\mu\nu}, \quad (24')$$

$$\mathbf{H}_{\mu\nu} = \partial_{\mu}\mathbf{B}_{\nu}-\partial_{\nu}\mathbf{B}_{\mu}, \quad (24'')$$

$$\begin{aligned} G^{Z}_{\mu\nu} &= \tfrac{1}{2}\{Z_{\mu}(\mathbf{V}_{\nu}\cdot\boldsymbol{\tau}+\sqrt{3}X_{\nu}{}^{0})-(\mathbf{V}_{\mu}\cdot\boldsymbol{\tau}+\sqrt{3}X_{\mu}{}^{0})^{\sim}Z_{\nu}\}, \\ G^{\bar{Z}}_{\mu\nu} &= -\tfrac{1}{2}\{\bar{Z}_{\mu}(\mathbf{V}_{\nu}\cdot\boldsymbol{\tau}+\sqrt{3}X_{\nu}{}^{0})^{\sim}-(\mathbf{V}_{\mu}\cdot\boldsymbol{\tau}+\sqrt{3}X_{\mu}{}^{0})\bar{Z}_{\nu}\}, \\ G^{V}_{\mu\nu} &= i\mathbf{V}_{\mu}\wedge\mathbf{V}_{\nu}+\tfrac{1}{2}\{\bar{Z}_{\mu}\boldsymbol{\tau}Z_{\nu}-Z_{\mu}\boldsymbol{\tau}\bar{Z}_{\nu}\}, \\ G^{X}_{\mu\nu} &= \tfrac{1}{2}\sqrt{3}\{\bar{Z}_{\mu}Z_{\nu}-Z_{\mu}\bar{Z}_{\nu}\}. \end{aligned} \quad (24''')$$

In the 3-space of the matrix elements of **U**, we have the set

$$\mathbf{B}_\mu = \tfrac{1}{2}\sqrt{2} \left| \begin{array}{c|c|c} \tfrac{1}{2}\sqrt{2}\,V_\mu{}^0+\tfrac{1}{6}\sqrt{6}\,X^0 & V_\mu{}^+ & Z_\mu{}^+ \\ \hline V_\mu{}^- & -\tfrac{1}{2}\sqrt{2}\,V_\mu{}^0+\tfrac{1}{6}\sqrt{6}\,X_\mu{}^0 & Z_\mu{}^0 \\ \hline \bar{Z}_\mu{}^- & \bar{Z}_\mu{}^0 & -\sqrt{\tfrac{2}{3}}\,X_\mu{}^0 \end{array} \right|. \qquad (25)$$

4. Discussion

The fermion and boson interaction Lagrangians provide us with the full set of known strong interactions (plus the $\pi^{0\prime}$ set) through the current-current-like 2nd order terms — but with no Yukawa-like simple processes for π or K.

In its general features, our Lagrangian reflects a certain similarity with Sakurai's theory [2]). The V_μ is similar to the $B_T{}^\mu$ (isospin-current boson) of the latter, and the $X_\mu{}^0$ is similarly related to its $B_Y{}^\mu$ (the hypercharge-current boson singlet). On the other hand we have no $B_B{}^\mu$ (baryon-current singlet) and do have a Z_μ set which has no place in ref. [2]).

We note that we do get directly from our group structure a ratio between the couplings; for the V and X fields, this is $f_X = f_V\sqrt{3}$, a value that fits Sakurai's phenomenological conclusion (from KN and $\overline{\text{K}}$N at low energies) that $(1/4\pi m_X{}^2)f_X{}^2 \approx (3/4\pi m_V{}^2)f_V{}^2$ if we assume the masses of be similar. Our X^0 field does not interact with the (Σ, Λ) set, and V does not interact with Λ, so that we get a split (N Ξ), Σ, Λ but though the interactions of X and Z with N and Ξ have opposite signs, lacking $B_B{}^\mu$ we cannot repeat here Sakurai's simple interpretation of the origin of the N-Ξ mass split. The arguments explaining the π-N S-wave scattering exist in our gauge. We also note that $\mathscr{L}_B{}^0$ in (24) with its $G^i_{\mu\nu}G^{\mu\nu}_i$ provides us with effective mass terms (in the sense of the mass of $A_\mu{}^\pm$ in a former work [8]) of Salam and Ward) for V_μ, Z_μ and $X_\mu{}^0$ (from $G^Z_{\mu\nu}$ and $G^{\bar{Z}}_{\mu\nu}$), whereas ref. [2]) lacks such terms for the singlets B_Y and B_B. From $\mathscr{L}_\varphi{}^{\mathscr{I}}$ we see that provided the masses are sufficient, there exist fast decays

$$Z \to \text{K}+\pi,$$

$$V \to 2\pi \quad \text{or} \quad V \to \text{K}+\overline{\text{K}} \qquad \text{(the even } G \text{ combination)},$$

$$X \to \text{K}+\overline{\text{K}} \qquad \text{(odd } G \text{ combination)}.$$

Note the possibility that $m_X \approx m_V$.

From the point of view of the Lagrangian formalism, it seems preferable to us to have what is in fact one conservation law for a "charge" that behaves like a 2nd rank tensor in three dimensions, than three separate unrelated conservation laws; this is even more important in view of the necessity to bring in at some further stage the electromagnetic and the weak interactions. In Sakurai's theory, these seem to imply two new independently conserved quantities Q and l, though the relation $Q = I_Z + \frac{1}{2}Y$ seems to indicate that the interactions are not wholly independent (and so does the $|\Delta \mathbf{I}| = \frac{1}{2}$ rule of the weak interactions). We think that the "aesthetic' value of Sakurai's theory and the "Urschmiere" approach would be enhanced if there proved to be only one kind of "Urschmiere" instead of five.

Our gauge has not given us directly an additional electromagnetic and weak Lagrangian. Still, it is interesting to check the connection it may have with these. We can do that by using (3) and rewriting our gauge invariance and the vector bosons in that basis. We get

$$B_\mu^{8'} \to A_\mu, \tag{26}$$

$$\mathbf{B}'_\mu = \tfrac{1}{2}\sqrt{2} \begin{vmatrix} \sqrt{\frac{2}{3}}\,A_\mu & V_\mu^+ & Z_\mu^+ \\ V_\mu^- & \frac{1}{2}\sqrt{2}\,B_\mu^{7'} - \frac{1}{6}\sqrt{6}\,A_\mu & Z_\mu^0 \\ Z_\mu^- & \bar{Z}_\mu^0 & -\frac{1}{2}\sqrt{2}\,B_\mu^{7'} - \frac{1}{6}\sqrt{6}\,A_\mu \end{vmatrix}. \tag{26'}$$

The interaction Lagrangian corresponding to (26) will be identical with the electromagnetic Lagrangian. It leads to the conditions

$$|\Delta \mathbf{I}| = 0{,}1 \qquad |\Delta I_Z| = 0, \qquad |\Delta Y| = 0, \tag{27}$$

though, of course, all quantum numbers are fully conserved when one adds the remaining interactions of P_d, mediated by $B_\mu^{7'}$. This last, with quantum numbers similar to (27), belongs with B_μ^2 and B_μ^3 (the K^0, $\overline{K}^0$-like vector bosons) to the subspace defined by $\mathbf{U}_b$ in (4). We note (4), from which we can see that $B_\mu^{8'}$ is the only matrix in P_d orthogonal to the $\mathbf{U}_b$ set. If, following Salam and Ward's [3]) treatment of weak interactions, we assume that the $|\Delta \mathbf{I}| = \frac{1}{2}$ law results from a non zero vacuum expectation value for the field K_1^0, our system could mediate the weak interactions through this B_μ^2, B_μ^3 subset; thus the $\mathbf{U}_b$ subset is apparently responsible for the generation of weak interactions as a secondary effect, with non conservation of parity — but whatever the mechanism involved, it cannot affect the electromagnetic interactions, generated by an orthogonal gauge. Still, we have no suggestion to explain why the strong interactions, mediated by a $\mathbf{U}$ gauge, should be accompanied by weaker interactions involving a change of basis into $\mathbf{U}'$.

One last remark, concerning the $\pi^{0'}$. From the group-structure aspect, it is related to the spinor-like subgroups (i.e. the K and not the π). From (2) we note that its matrix representation commutes with that of the pion. It has no direct interaction with the pions, while it does interact with all the kaons. If we assume a single parity for all the components of the $\boldsymbol{\varphi}$ vector, it is a pseudoscalar particle, with a fast decay into $K^0 + \overline{K}^- + \pi^+$ (mediated by Z_μ) or $K^+ + \overline{K}^- + \pi^0$ (through Z_μ again) etc., provided it has sufficient mass. On the other hand, if it were scalar, it could do for the medium strong interactions what the vacuum decay of the σ meson does for all strong interactions in ref. [3]). In fact it would then be identical with the σ' particle suggested in a variant of ref. [3]).

I am indebted to Prof. A. Salam for discussions on this problem. In fact, when I presented this paper to him, he showed me a study he had done on the unitary theory of the Sakata model, treated as a gauge, and thus producing a similar set of vector bosons [9]).

Shortly after the present paper was written, a further version, utilizing the 8-representation for baryons, as in this paper, reached us in a preprint by Prof. M. Gell Mann.

References

1) C. N. Yang and H. Mills, Phys. Rev. **96** (1954) 192
2) J. J. Sakurai, Ann. of Phys. **11** (1960) 1
3) A. Salam and J. C. Ward, Nuovo Cim. **19** (1961) 167
4) J. Tiomno, Nuovo Cim. **6** (1957) 1
5) J. Schwinger, Ann. of Phys. **2** (1957) 407
6) M. Ikeda, S. Ogawa, Y. Ohnuki, Progr. Theor. Phys. **22** (1959) 5, 719
7) R. Utiyama, Phys. Rev. **101** (1956) 1597
8) A. Salam and J. Ward, Nuovo Cim. **11** (1959) 4, 569
9) A. Salam and J. Ward, Nuovo Cim., to be published

Note on Unitary Symmetry in Strong Interactions*)

Susumu OKUBO

Department of Physics, University of Tokyo, Tokyo
and
Department of Physics, University of Rochester
Rochester, N.Y., U.S.A

(Received December 6, 1961)

Assuming invariance of theory under three-dimensional unitary group, various consequences have been investigated. Both Sakata's and Gell-Mann's scheme can be treated in the same fashion and in a simpler way. Mass formula for particles belonging to the same irreducible representation has been derived and compared with experiments.

§ 1. Introduction

The purpose of this note is to investigate consequences of the three-dimensional unitary group (denoted as U_3 hereafter), which is a certain generalization of the usual isotopic space group. Though many authors[1),2),3)] have examined this problem, our procedure is simpler and some new results have been obtained. Also, we can treat different schemes of U_3 such as Sakata's[1),2)] or Gell-Mann's[3)] on the same footing by our method.

First of all, we shall give some motivations for introducing U_3. All known interactions obey certain symmetries, i.e. they are subject to the corresponding transformation groups. We can classify all known groups appearing in the studies of elementary particles into the following three categories.

(I) *Space-group*
  (i) Lorentz group (ii) Charge conjugation
(II) *Isotopic-groups*
  (i) Isotopic spin rotation $R_3^{(I)}$
  (ii) Baryon gauge transformation $R_2^{(B)}$
  (iii) Charge gauge transformation $R_2^{(Q)}$
  (iv) Strangeness gauge transformation $R_2^{(S)}$
  (v) Leptonic gauge transformation $R_2^{(L)}$
(III) *Gauge-transformation of the 2nd kind*
  (i) Electro-magnetic field
  (ii) Yang-Mills field

*) A part of this paper has been presented at the La-Jolla Conference held at La-Jolla, California, June 12, 1961.

In this list, we have included the charge conjugation into the space-group, because of the TCP theorem. These three groups of transformations are correlated with each other in some degree, but here we do not go into details. Furthermore, we restrict ourselves only in the study of the iso-space groups (II), in this paper. Moreover, we do not take account of leptons also, though they might be treated on the same footing.[4] Then, the groups (II) consist of 4 groups. However, by virtue of the Nakano-Nishijima-Gell-Mann formula, we have one following relation:

$$Q=I_3+1/2\cdot(N+S). \qquad (1)$$

Thus, only 3 out of the 4 groups are independent. So, the known strong interactions have to be invariant under the following group G:

$$G=R_3^{(I)}\times R_2^{(B)}\times R_2^{(Q)}.$$

Now, for the moment, let us suppose that the nature obeys some higher symmetry than this. Then, the invariant group U of this higher symmetry must include G as a sub-group. One of them including G is U_3, which is relatively uncomplicated. This is one motivation for adopting U_3. Besides, we may note that the 3-dimension is the minimum dimension for non-trivial representation of the group G. This may be taken as another motivation for U_3.[5]

In the next section, we shall give the classification of particles belonging to a given irreducible representation by means of restricting U_3 into U_2 (two-dimensional unitary group). In § 3 we shall give applications of U_3. Furthermore, the following mass formula will be proved:

$$M=a+b\cdot S+c\cdot[I(I+1)-1/4\cdot S^2]. \qquad (2)$$

This relation holds for particles belonging to a given irreducible representation of U_3, and S and I stand for the strangeness and isospin of particles contained in the representation, respectively. This formula has been proved in the lowest order perturbation violating U_3-symmetry of the type $\bar{\Lambda}\Lambda$, but in any orders for the strong U_3-invariant interactions. The proof of Eq. (2) will be given in the Appendix. As an application of Eq. (2), we note that if N, Λ, Σ and Ξ belong to an irreducible representation as in the Gell-Mann scheme, we have

$$1/2\cdot[M_N+M_\Xi]=3/4\cdot M_\Lambda+1/4\cdot M_\Sigma,$$

which is satisfied in good accuracy. Another application of our formula Eq. (2) is that the mass of a neutral-isoscalar meson π_0' would be given by

$$M(\pi_0')=4/3\cdot M(K)-1/3\cdot M(\pi)\simeq 600\text{ Mev},$$

where π_0' is the meson belonging to the same representation as π, K and $\bar{K}$ mesons. Similarly, we should have

$$M(K^*)=3/4\cdot M(\omega)+1/4\cdot M(\rho)$$

where ρ, ω and K^* are bosons representing resonant states of (π-π), (π-π-π) and (π-K) system, respectively. We note that this relation is satisfied within an error of 12%.

§ 2. Classification of particles in U_3

The three-dimensional unitary group U_3 is defined by the following transformation on a vector $\phi_\mu (\mu=1, 2, 3)$:

$$\phi_\mu \to \sum_{\lambda=1,2,3} a_\mu{}^\lambda \phi_\lambda \quad (\mu=1, 2, 3) \tag{3}$$

where $a_\mu{}^\lambda$ satisfies

$$\sum_{\mu=1,2,3} (a_\mu{}^\lambda)^* a_\mu{}^\nu = \delta_\lambda{}^\nu \quad (\nu, \lambda=1, 2, 3). \tag{4}$$

In the Sakata model,[6] we identify ϕ_1, ϕ_2 and ϕ_3 with the proton, the neutron and the Λ, respectively. However, this is not the only way. We shall assume that ϕ_1, and ϕ_2 form an isotopic doublet and ϕ_3 an isotopic singlet. As for other quantum numbers, we can assign according to the following cases:

(a) ϕ_1, ϕ_2 and ϕ_3 have the baryon number $N=1$. ϕ_1 and ϕ_2 have the strangeness quantum number $S=0$, ϕ_3 has the strangeness $S=-1$.

(b) We do not assign any baryon numbers to ϕ_1, ϕ_2 and ϕ_3, but assign $Y=0$ for ϕ_1 and ϕ_2, and $Y=-1$ for ϕ_3 where Y stands for the hypercharge $Y=N+S$.

(c) We do not assign any baryon numbers to ϕ_1, ϕ_2 and ϕ_3, but assign a new quantum number $Z=N+3\cdot S$ as $Z=1$ for ϕ_1 and ϕ_2, and $Z=-2$ for ϕ_3.

The first assignment (a) corresponds to the usual Sakata model, and the second one (b) is practically the same as the Gell-Mann scheme,[3] and so we refer to it as "Gell-Mann scheme" for simplicity,*) though not exactly. The third scheme is actually convenient if we consider the unitary-unimodular group of 3 dimensions instead of U_3, and so refer to it as "the unitary-unimodular scheme". We may give possible schemes other than (a), (b) and (c), but it will not be so fruitful.

First, let us consider the case (a) (referred to as "Sakata scheme" hereafter). In this scheme, consider a special transformation:

$$\phi_1 \to \varepsilon_1 \phi_1, \qquad \phi_2 \to \varepsilon_2 \phi_2, \qquad \phi_3 \to \varepsilon_3 \phi_3$$
$$|\varepsilon_\mu| = 1 \quad (\mu=1, 2, 3). \tag{5}$$

This is a special transformation of Eqs. (3) and (4). Then, a component of every tensor $T^{\alpha_1 \cdots \alpha_n}_{\beta_1 \cdots \beta_m}$ would transform as

*) Note added in proof: Exactly the same scheme has been proposed by Y. Yamaguchi in 1960, so that we should call it as Yamaguchi-Gell-Mann scheme hereafter. Y. Yamaguchi: private communication.

$$T \rightarrow \varepsilon_1^{\alpha}\, \varepsilon_2^{\beta}\, \varepsilon_3^{\gamma}\, T.$$

In our case, the baryon number N and the strangeness S is obviously given by

$$N = \alpha + \beta + \gamma$$
$$S = -\gamma. \tag{6}$$

Now, all irreducible tensor representation of U_3 are characterized by three integers f_1, f_2 and f_3 satisfying a condition $f_1 \geq f_2 \geq f_3$. We shall denote it as $U_3(f_1, f_2, f_3)$, hereafter. The dimension of the representation is given[7] by

$$D = 1/2 \cdot (f_1 - f_2 + 1)(f_1 - f_3 + 2)(f_2 - f_3 + 1). \tag{7}$$

Also, comparing the character of $U_3(f_1, f_2, f_3)$ with Eq. (6), we find that the baryon number N of this representation is

$$N = f_1 + f_2 + f_3. \tag{8}$$

Now, to specify sub-quantum numbers S and the isospin I in $U_3(f_1, f_2, f_3)$, we fix the direction of the 3rd component ϕ_3. So, we restrict ourselves within the two-dimensional unitary group U_2, whose irreducible representations are specified by two integers f_1', and f_2' satisfying $f_1' \geq f_2'$ and will be referred to as $U_2(f_1', f_2')$. Then, the branching rule[8] for this decomposition tells us that U_3 can be decomposed according as

$$U_3(f_1, f_2, f_3) \rightarrow \sum_{(f_1', f_2')} U_2(f_1', f_2'), \tag{9}$$

where we sum over all possible integer pairs (f_1', f_2') satisfying the following conditions:

$$f_1 \geq f_1' \geq f_2 \geq f_2' \geq f_3. \tag{10}$$

The decomposition Eq. (9) is an analogue of the well-known decomposition of R_4 into R_3 (R_n being the n-dimensional rotation group).

$$R_4(l, l') \rightarrow \sum_{L=|l-l'|}^{(l+l')} R_3(L).$$

Now, two-dimensional unitary group is a product of two-dimensional unitary-unimodular group (which we can identify as the usual isotopic rotation group) and a gauge group, which defines the nucleon charge. Then, the isospin I is immediately given by

$$I = 1/2 \cdot (f_1' - f_2') \tag{11}$$

and also, comparing the character of $U_2(f_1', f_2')$ with Eq. (6), we get

$$S = (f_1' + f_2') - (f_1 + f_2 + f_3). \tag{12}$$

In this way, we could specify sub-quantum numbers S and I. Furthermore, we note[9] that two representations $U_3(f_1, f_2, f_3)$ and $U_3(-f_3, -f_2, -f_1)$ are contragradient to each other, i.e. they are charge-conjugate of each other in our case.

This remark does not apply to the cases (b) and (c), since the nucleon number is not defined in these cases.

In order to explain our procedure, consider various cases:

(i) $(f_1, f_2, f_3) = (1, 0, 0)$

This is a 3-dimensional representation by Eq. (7) and the decomposition Eqs. (9) and (10) tells us two choices $(f_1', f_2') = (1, 0)$ or $(0, 0)$. By Eqs. (8), (11) and (12), $N=1$ and the former belongs to $(I=1/2, S=0)$, and the latter to $(I=0, S=-1)$. So the natural identification would be the triplet (p, n, Λ).

(ii) $(f_1, f_2, f_3) = (1, 0, -1)$

By Eqs. (7) and (8), this is a boson representation with 8 components. Also, by the remark given after Eq. (12), it must be self-conjugate, i.e. it must contain a particle and its anti-particle together. Now, the decomposition Eqs. (9) and (10) gives us the choice $(f_1', f_2') = (1, 0)$, $(0, -1)$, $(1, -1)$ and $(0, 0)$, and by Eqs. (11) and (12) they have $(I=1/2, S=1)$, $(I=1/2, S=-1)$, $(I=1, S=0)$, and $(I=0, S=0)$. respectively. By the remark given in the beginning, the first two must be charge conjugate of each other and the last two must be self-conjugate under charge conjugation operation. Natural identification would be (K_+, K_0), $(\overline{K}_+, \overline{K}_0)$, (π_+, π_0, π_-) and π_0', where the last one is a new pseudoscalar boson. We may identify the newly found states K^*, $\overline{K}^*$, ρ and ω mesons under the same category.

(iii) $(f_1, f_2, f_3) = (2, 0, -1)$

This is a fermion state with 15 components by Eqs. (7) and (8), and they contain the following particles by Eqs. (10), (11) and (12).

$$(I=1/2, S=-2),\ (I=1, S=-1),\ (I=0, S=-1),$$

$$(I=1/2, S=0),\quad (I=1, S=+1),\ (I=3/2, S=0).$$

We might identify the first four as Ξ, Σ, Λ and N, respectively, but then we have two other unwanted particles. This interpretation is originally due to Yamaguchi,[2] but as we will see in a later section this identification seems to give small masses for $(I=1, S=1)$ and $(I=3/2, S=0)$ particles so as to make them stable, and so it would be more natural to adopt the case (i) as representing Λ and N. Furthermore, if we take the viewpoint (ii) for bosons, then $(I=1/2, S=-2)$ has to be identified still as Ξ particles. This is because the transition $\Xi \to \Lambda + \overline{K}$ must be possible and therefore Ξ (and also Σ since $\Sigma \to \Lambda + \pi$) has to be in a product representation $U_3(1, 0, 0) \times U_3(1, 0, -1)$. However,[10] we have

$$U_3(1, 0, 0) \times U_3(1, 0, -1) = U_3(2, 0, -1) + U_3(1, 1, -1) + U_3(1, 0, 0)$$

but $U_3(1, 1, -1)$ and $U_3(1, 0, 0)$ do not contain a particle with $(I=1/2, S=-2)$. As for Σ, the same argument shows that it must belong either to $U_3(2, 0, -1)$ or to $U_3(1, 1, -1)$. Ikeda et al.[1] identify $(I=3/2, S=0)$ in $U_3(2, 0, -1)$ as

N^* (the first π-N scattering resonance), then the spin of Ξ has to be 3/2, since N^* has the space-spin 3/2. Similarly, $(I=1, S=-1)$ and $(I=0, S=-1)$ states in $U_3(2, 0, -1)$ may be interpreted as Y_1^*(π-Λ scattering resonance) and Y_0^* (π-Σ scattering resonance), respectively. Then, they must have spin 3/2 also. In this case, we have to assign $U_3(1, 1, -1)$ for Σ.

(iv) $(f_1, f_2, f_3) = (1, 1, -1)$

This is a fermion state with six components. We have $(I=1/2, S=0)$, $(I=0, S=+1)$ and $(I=1, S=-1)$, and the last one may be interpreted as Σ. However, we have a new state with $(I=0, S=+1)$, so, we should observe a resonance for the reaction $K_+ + n$ scattering, which has not so far been found experimentally.

Up to now, we have investigated the case (a), i.e. the Sakata-scheme. Now, let us consider the case (b). In this case, we cannot assign any baryon numbers to ϕ_μ, so that Eq. (8) has no meaning as to indicate the baryon number. Eq. (11) is unchanged as before, but in Eq. (12), S has to be replaced by Y, so that in our scheme (b), we have

$$I = 1/2 \cdot (f_1' - f_2')$$
$$Y = (f_1' + f_2') - (f_1 + f_2 + f_3). \tag{13}$$

In this case, the representation $(1, 0, -1)$ gives four states; $(I=1/2, Y=1)$, $(I=1/2, Y=-1)$, $(I=1, Y=0)$ ahd $(I=0, Y=0)$. As for bosons, our assignment is unchanged, since S and Y are the same for bosons. So, we can assign $(\pi, K, \overline{K}, \pi_0')$ and $(\rho, K^*, \overline{K}^*, \omega)$ to $U_3(1, 0, -1)$. A new phenomenon is that we can also assign $(N, \Xi, \Sigma, \Lambda)$ to $U_3(1, 0, -1)$ since the nucleon number is no longer defined and the corresponding quantum numbers Y and I can be given correctly. This is exactly the same as in Gell-Mann's scheme, though the starting points are quite different. As we shall see in the next section, our scheme is essentially the same as Gell-Mann's as for all practical purposes, and so we can call our scheme (b) as Gell-Mann's. We may note the following decomposition:[10)]

$$U_3(1, 0, -1) \times U_3(1, 0, -1) = 2U_3(1, 0, -1) + U_3(0, 0, 0) + U_3(2, 0, -2)$$
$$+ U_3(2, -1, -1) + U_3(1, 1, -2)$$

so that Y_1^*, Y_0^* and N^* in the Gell-Mann scheme have to be included in one of the right-hand side, since they decay into one-boson and one-fermion state. This will be treated in a forthcoming paper.

Finally, we may study the consequence of our scheme (c). This was given, since it is more natural when we think of the unitary-unimodular group of 3-dimension (we refer to it as $SL(3)$) rather than U_3. In $SL(3)$, there is no distinction between covariant and contravariant tensors. This is because a constant totally anti-symmetric tensor $\epsilon^{\lambda\mu\nu}$ is invariant under $SL(3)$, so that ϕ^λ behaves like $\epsilon^{\lambda\mu\nu}T_{\mu\nu}$ where $T_{\mu\nu}$ is a tensor. More generally, we have that the representation (f_1, f_2, f_3), which we have written[11] as $U_3(f_1, f_2, f_3)$ up to now, is the same representation as (f_1+e, f_2+e, f_3+e) where e is an arbitrary integer. Then, obviously Eqs. (12) or (13) is not invariant under $SL(3)$, since it is not invariant under $f_\mu \to f_\mu + e(\mu=1, 2, 3)$ and $f_\mu' \to f_\mu' + e(\mu=1, 2)$. Invariant quantum numbers under $SL(3)$ under our decomposition Eq. (9) are given by

$$Z=3(f_1'+f_2')-2(f_1+f_2+f_3),$$
$$I=1/2(f_1'-f_2') \tag{14}$$

where $Z=N+3\cdot S$. We omit the details for these derivations. In this case, we can repeat the same procedures as before, but it gives almost the same results as in the case (a), so we will not go too far. Here we may note also that if we give up additivity of quantum numbers, we may assign $Z=3\cdot Y+N(N-1)$ for Eq. (14). In this case, we can assign $(1, 0, -1)$ both for bosons and fermion, and we have the same result as Gell-Mann's again. We shall not consider our case (c) any longer in this paper, and restrict ourselves only in discussions of the cases (a) and (b).

§ 3. Tensor representation and applications

First, let us consider the Sakata scheme (a), and we take the representations $U_3(1, 0, 0)$ and $U_3(1, 0, -1)$ for (Λ, n, p) and $(\pi, \pi_0', K, \overline{K})$ systems, respectively. Then, p, n and Λ can be represented by a vector ψ_μ.

$$\psi_1=p,\ \psi_2=n,\ \psi_3=\Lambda \tag{15}$$

and $(\pi, \pi_0', K, \overline{K})$ can be represented by a traceless tensor f_ν^μ, so that $f_\mu^\mu=0$. The identification is

$$\pi_+=f_1^2,\ \pi_-=f_2^1,\ \pi_0=\frac{1}{\sqrt{2}}(f_1^1-f_2^2),\ \pi_0'=-\frac{3}{\sqrt{6}}f_3^3,$$
$$K_+=f_1^3,\ K_0=f_2^3,\ \overline{K}_+=f_3^1,\ \overline{K}_0=f_3^2 \tag{16}$$

and also $(\rho, \omega, K^*, \overline{K}^*)$ can be represented by a traceless tensor F_ν^μ exactly in the same fashion as Eq. (16) by replacing $\pi\to\rho$, $\pi_0'\to\omega$, $K\to K^*$, $\overline{K}\to\overline{K}^*$. Actually, F_ν^μ has a vector suffix due to space-spin, but we omit it for simplicity.

The invariant interactions among baryon-boson and among boson-boson would be given by

$$H_1=ig\,\overline{\psi}_\mu\gamma_5\psi_\nu f_\mu^\nu, \tag{17}$$

$$H_2=ig\,F_\nu^\mu\cdot(f_\lambda^\nu\cdot\partial f_\mu^\lambda-\partial f_\lambda^\nu\cdot f_\mu^\lambda) \tag{18}$$

where the repeated indices mean summations over 1, 2 and 3. In Eq. (17), we note that $\overline{\psi}_\mu$ behaves as a contra-variant vector ϕ^μ. Using the representations Eqs. (15) and (16), these Hamiltonians can be written as

$$H_1=ig\frac{1}{\sqrt{2}}\overline{N}\gamma_5(\boldsymbol{\tau}\cdot\boldsymbol{\pi})N+ig\overline{N}\gamma_5\Lambda K+ig\overline{\Lambda}\gamma_5N\overline{K}$$

$$+ig\frac{1}{\sqrt{6}}(\bar{N}\gamma_5 N - 2\bar{\Lambda}\gamma_5\Lambda)\pi_0', \tag{17$'$}$$

$$H_2 = \frac{ig}{\sqrt{2}}\rho(\bar{K}\boldsymbol{\tau}\partial K - \partial\bar{K}\boldsymbol{\tau}K) + \sqrt{2}\cdot g\cdot\boldsymbol{\rho}(\boldsymbol{\pi}\times\partial\boldsymbol{\pi})$$

$$+\frac{ig}{\sqrt{2}}\bar{K}^*\boldsymbol{\tau}[K\partial\boldsymbol{\pi} - (\partial K)\boldsymbol{\pi}] + \frac{3}{\sqrt{6}}ig\bar{K}^*[K\partial\pi_0' - \partial K\pi_0']$$

$$+\frac{ig}{\sqrt{2}}[\boldsymbol{\pi}(\partial\bar{K}) - \partial\boldsymbol{\pi}\bar{K}]\cdot\boldsymbol{\tau}K^* + \frac{3}{\sqrt{6}}ig[\pi_0'\partial\bar{K} - \partial\pi_0'\bar{K}]K^*$$

$$+\frac{3}{\sqrt{6}}ig\omega[\bar{K}\partial K - \partial\bar{K}K]. \tag{18$'$}$$

We note that Eq. (18)′ agrees with that given by Gell-Mann.[3]

Now, let us consider the Gell-Mann scheme (b). Here, as for bosons, Eq. (16) is unchanged. For baryons, we introduce two traceless tensors N_ν^μ and M_ν^μ (so that $M_\mu^\mu = N_\mu^\mu = 0$) as representing

$$\Sigma_+ = N_1^2,\ \Sigma_- = N_2^1,\ \Sigma_0 = \frac{1}{\sqrt{2}}(N_1^1 - N_2^2),\ \Lambda = -\frac{3}{\sqrt{6}}N_3^3, \tag{19a}$$

$$p = N_1^3,\ n = N_2^3,\ \Xi_- = N_3^1,\ \Xi_0 = N_3^2,$$

$$\bar{\Sigma}_- = M_1^2,\ \bar{\Sigma}_+ = M_2^1,\ \bar{\Sigma}_0 = \frac{1}{\sqrt{2}}(M_1^1 - M_2^2),\ \bar{\Lambda} = -\frac{3}{\sqrt{6}}M_3^3, \tag{19b}$$

$$\bar{\Xi}_- = M_1^3,\ \bar{\Xi}_0 = M_2^3,\ \bar{p} = M_3^1,\ \bar{n} = M_3^2.$$

Then, we have two invariant forms for baryon-boson interactions.

$$H_3 = ig\,M_\nu^\mu\,\gamma_5\,N_\lambda^\nu\,f_\mu^\lambda, \tag{20a}$$

$$H_4 = ig\,M_\nu^\mu\,\gamma_5 f_\lambda^\nu\,N_\mu^\lambda. \tag{20b}$$

Explicit calculation gives

$$H_3 = \frac{ig}{\sqrt{2}}\bar{N}\gamma_5(\boldsymbol{\tau}\cdot\boldsymbol{\pi})N + \frac{g}{\sqrt{2}}(\bar{\boldsymbol{\Sigma}}\gamma_5\times\boldsymbol{\Sigma})\boldsymbol{\pi} + \frac{g}{\sqrt{6}}[i\bar{\boldsymbol{\Sigma}}\boldsymbol{\pi}\gamma_5\Lambda + \text{c.c.}]$$

$$+\frac{g}{\sqrt{6}}[\bar{\Lambda}\gamma_5\Xi\tau_2 K + \text{c.c.}] - \frac{\sqrt{6}}{3}g[i\bar{N}\gamma_5 K\Lambda + \text{c.c.}]$$

$$-\frac{g}{\sqrt{2}}[\bar{K}\boldsymbol{\tau}\tau_2\bar{\Xi}\gamma_5\boldsymbol{\Sigma} + \text{c.c.}]$$

$$-\frac{ig}{\sqrt{6}}\pi_0'[2(\bar{\Xi}\gamma_5\Xi) + \bar{\Lambda}\gamma_5\Lambda - \bar{\boldsymbol{\Sigma}}\gamma_5\boldsymbol{\Sigma} - \bar{N}\gamma_5 N], \tag{21a}$$

$$H_4 = \frac{g}{\sqrt{6}}[i\bar{\Lambda}\gamma_5\boldsymbol{\Sigma}\boldsymbol{\pi} + \text{c.c.}] - \frac{g}{\sqrt{2}}(\bar{\boldsymbol{\Sigma}}\gamma_5\times\boldsymbol{\Sigma})\boldsymbol{\pi}$$

$$-\frac{ig}{\sqrt{2}}\bar{\Xi}(\boldsymbol{\tau}\cdot\boldsymbol{\pi})\gamma_5\Xi$$

$$+\frac{g}{\sqrt{6}}[i\bar{N}\gamma_5 K\Lambda+\text{c.c.}]$$

$$-\frac{2g}{\sqrt{6}}[\bar{\Lambda}\gamma_5\Xi\tau_2 K+\text{c.c.}]$$

$$+\frac{g}{\sqrt{2}}[i\bar{N}\gamma_5\boldsymbol{\tau}K\boldsymbol{\Sigma}+\text{c.c.}]$$

$$+\frac{ig}{\sqrt{6}}\pi_0'[\bar{\boldsymbol{\Sigma}}\gamma_5\boldsymbol{\Sigma}+\bar{\Xi}\gamma_5\Xi-\bar{\Lambda}\gamma_5\Lambda-2\bar{N}\gamma_5 N] \tag{21b}$$

where we have put

$$N=\begin{pmatrix}p\\n\end{pmatrix},\quad \boldsymbol{\Sigma}=\begin{pmatrix}\Sigma_1\\\Sigma_2\\\Sigma_3\end{pmatrix},\quad \Xi=\begin{pmatrix}-\Xi_0\\\Xi_-\end{pmatrix},\quad K=\begin{pmatrix}K_+\\K_0\end{pmatrix},\quad \boldsymbol{\pi}=\begin{pmatrix}\pi_1\\\pi_2\\\pi_3\end{pmatrix},$$

and Eqs. (21a) and (21b) are connected with L_D and L_F of Gell-Mann[3] by

$$H_3=\frac{1}{2\sqrt{2}}[L_D+L_F],$$

$$H_4=\frac{1}{2\sqrt{2}}[L_D-L_F],$$

when we take the same coupling constants.

As applications of our formalism, we may think of the boson-baryon scattering in the case of the Sakata scheme. In this case, we can form the following invariants of which the S-matrix element is a linear combination:

$$T_\nu^{\ \mu}f_\lambda^{\ \nu}\tilde{f}_\mu^{\ \lambda},\ T_\nu^{\ \mu}\tilde{f}_\lambda^{\ \nu}f_\mu^{\ \lambda},\ T_\mu^{\ \mu}f_\beta^{\ \alpha}\tilde{f}_\alpha^{\ \beta}$$

where we have put $T_\nu^{\ \mu}=\bar{\psi}_\mu\psi_\nu$, and f, and $\tilde{f}$ represent for incoming and outgoing bosons. From this, we can prove the following identities among total cross-sections.

$$\sigma(\pi_++p)=\sigma(K_++p),\quad \sigma(K_-+n)=\sigma(\pi_++\Lambda),$$

$$\sigma(\pi_-+p)=\sigma(K_-+p)=\sigma(K_++\Lambda),\ \text{etc.}$$

$$\sigma(\pi_0'+p)=1/3\cdot\sigma(\pi_0+p)+2/3\cdot\sigma(K_0+p).$$

These have been derived also by Hara and Singh.[12] They are also investigating similar identities in the case of Gell-Mann scheme. We can get similar identities among magnetic moments of baryons. In the case of Sakata-scheme, let us assume that the electromagnetic current j_μ has a transformation property as $T_1^{\ 1}$ component of a tensor $T_\nu^{\ \mu}$. This can be taken, since the usual current $ie\bar{p}\gamma_\mu p$ has such form. Then, the method mentioned in the above immediately gives

$\mu(\Lambda)=\mu(n)$ and also we can prove that K_0 and $\overline{K_0}$ have no electromagnetic structures. This is because we can prove $\langle K_0|j_\mu|K_0\rangle=\langle\overline{K_0}|j_\mu|\overline{K_0}\rangle$ similarly, but j_μ changes its sign under charge conjugation, and therefore $\langle K_0|j_\mu|K_0\rangle$ has to be identically zero.

In the case of Gell-Mann scheme (b), we can give some relations among magnetic moments of baryons. By the same reason as in the above, let us assume that the electromagnetic current j_μ behaves as T_1^1 of a tensor T_ν^μ, with respect to U_3. We have to take the expectation value of j_μ, i.e. T_1^1. From invariance, we have

$$\langle T_\nu^\mu\rangle=aM_\lambda^\mu N_\nu^\lambda+bM_\nu^\lambda N_\lambda^\mu+c\cdot\delta_\nu^\mu\cdot(M_\beta^\alpha N_\alpha^\beta)$$

where M and N represent baryons as in Eq. (19) and we have omitted spinor indices. By putting $\mu=\nu=1$, and comparing with Eq. (19), we have $\mu(p)=a+c$, $\mu(n)=c$, etc. Then, we have the following relations:

$$\begin{aligned}&\mu(p)=\mu(\Sigma_+),\\&\mu(\Xi_0)=\mu(n),\\&\mu(\Xi_-)=\mu(\Sigma_-),\\&\mu(\Lambda)=1/6\cdot[\mu(p)+\mu(\Sigma_-)+4\mu(n)],\\&\mu(\Sigma_0)=1/2\cdot[\mu(\Sigma_+)+\mu(\Sigma_-)].\end{aligned}\tag{22}$$

Furthermore, if we demand that T_ν^μ is traceless, i.e. $T_\mu^\mu=0$, then we should have $a+b+3c=0$ and then this condition gives one more relation:

$$\mu(\Lambda)=(1/2)\mu(n).\tag{23}$$

Relations Eqs. (22) and (23) have been given also by Coleman and Glaschow[13] by somewhat more direct method. We note that they used $T_\beta^\alpha=M_\lambda^\alpha N_\beta^\lambda-M_\beta^\lambda N_\lambda^\alpha$, so that obviously $T_\alpha^\alpha=0$ is satisfied. From our derivation, however, it is clear that the explicit form for T_1^1 is unnecessary.

We can give other applications of our method for the weak leptonic decays of bosons and fermions. In case of the strangeness-violating leptonic decays, the interaction Hamiltonian would be given by

$$H_1=G\,\mathfrak{J}_\mu[\bar{\nu}\gamma_\mu(1+\gamma_5)e+\bar{\nu}\gamma_\mu(1+\gamma_5)\mu]+\text{c.c.}\tag{24}$$

where $\mathfrak{J}_\mu$ is the strangeness-violating current. Let us consider the case of Gell-Mann scheme, and assume that $\mathfrak{J}_\mu$ has the transformation property as T_1^3 component of a tensor T_ν^μ, so that it has the same character as K_+. Then, we may construct two tensors $M_\lambda^3N_1^\lambda$ and $M_1^\lambda N_\lambda^3$ out of M and N, and it would be natural to take

$$\begin{aligned}\mathfrak{J}_\mu&=aM_\lambda^3N_1^\lambda+bM_1^\lambda N_\lambda^3\\&=a\left[\frac{1}{\sqrt{6}}(\overline{\Xi}_-\cdot\Lambda)+(\overline{\Xi}_0\cdot\Sigma_++\frac{1}{\sqrt{2}}\overline{\Xi}_-\cdot\Sigma_0)-\frac{\sqrt{6}}{3}(\bar{\Lambda}\cdot p)\right]\cdot\end{aligned}\tag{25}$$

$$+b\left[\frac{1}{\sqrt{6}}(\bar{\Lambda}\cdot p)+(\bar{\Sigma}_{-}\cdot n+\frac{1}{\sqrt{2}}\bar{\Sigma}_0\cdot p)-\frac{\sqrt{6}}{3}(\bar{\Xi}_{-}\cdot\Lambda)\right]$$

where we omitted γ-matrices. Of course, this behaves as a component of an isotopic spinor[14] in the usual isospin assignment.

§ 4. Applications of mass formula

If there are no interactions violating U_3 symmetry, all particles belonging to the same irreducible representation have to have the same mass, the same spin and parity. So we should have the same mass for pion and kaon, which is not true. We must therefore have some interactions violating U_3. According to Yamaguchi,[2] we may suppose that such interactions may be moderately strong, as compared with the very strong U_3-conserving interactions. Our purpose in this note is to investigate the result of mass-splitting among particles in a given irreducible representation due to this moderately strong U_3-violating interaction. In the Appendix, we shall prove that the mass splitting is given by the following formula.*)

$$M=a+b\cdot S+c\cdot[1/4\cdot S^2-I(I+1)]. \qquad (26)$$

Eq. (26) has been proved in the lowest order perturbation for such U_3-violating interaction with the transformation property T_3^3 of a tensor T_ν^μ but in any orders for U_3-conserving very strong interactions. In Eq. (26), a, b and c are constants which do not depend upon such sub-quantum numbers as the strangeness S and isospin I, but may depend upon the nature of the interaction and upon the irreducible representation to be considered. Eq. (26) may be rewritten as

$$M=a'+b'Y+c'[1/4\cdot Y^2-I(I+1)] \qquad (27)$$

if we use the hypercharge $Y=N+S$ instead of S. Formula Eqs. (26) or (27) holds for both the Sakata and the Gell-Mann scheme. For the details, the reader may consult the Appendix.

Now, in this section, we shall investigate the result of Eqs. (26) or (27). First, let us consider boson system (π, π_0', K and $\bar{K}$). An application of (26) or (27) immediately gives that we have a relation

$$M(K)=1/2\cdot[M(K)+M(\bar{K})]=3/4\cdot M(\pi_0')+1/4\cdot M(\pi). \qquad (28)$$

From this, we can calculate the mass of π_0' with $M(\pi_0')\simeq 600$ Mev. It is interesting to note that a similar value has been predicted by other methods.[15] The same formula as Eqs. (28) holds for the (ω, ρ, K^*, $\bar{K}^*$) system.

$$M(K^*)=1/2\cdot[M(K^*)+M(\bar{K}^*)]=3/4\cdot M(\omega)+1/4\cdot M(\rho). \qquad (29)$$

---

*) A similar formula has already been suggested by R. P. Feynman at Gatlingburg Conference held in 1958.

The calculated value for $M(K^*)$ by using $M(\omega)$ and $M(\rho)$ is 780 Mev, compared to the experimental value 885 Mev. This relation Eq. (29) holds as long as (ρ, ω, K^*, $\bar{K}^*$) belongs to the same irreducible representation. Previously we have assigned (1, 0, −1) for these, but another possibility is that these may belong to 27-dimensional representation (2, 0, −2) instead of the 8-dimensional $U_3(1, 0, -1)$ representation. Then, the method of § 2 tells us that we have 5 more states ($I=1, S=\pm 2$), ($I=3/2, S=\pm 1$) and ($I=2, S=0$) in addition to (ρ, ω, K^*, $\bar{K}^*$). Then, we can use our formula Eq. (26) and we can calculate the mass of these states in terms of $M(\rho)$ and $M(\omega)$, to get

$$M(I=1, S=\pm 2) \simeq 770 \text{ Mev},$$

$$M(I=3/2, S=\pm 1) \simeq 720 \text{ Mev}, \quad M(I=2, S=0) \simeq 700 \text{ Mev}.$$

However, we do not observe $I=3/2$ resonance for the K-π system, and so this value for $M(I=3/2, S=\pm 1)$ contradicts the experiment. Accordingly, it seems that our assignment of (1, 0, −1) for (ρ, ω, K^*, $\bar{K}^*$) is more reasonable than that of (2, 0, −2). The above argument equally applies both to the Sakata and the Gell-Mann schemes.

As for baryons, let us first consider the Gell-Mann scheme; then (Λ, Σ, N, Ξ) belongs to $U_3(1, 0, -1)$ representation. Then, by using Eq. (27), we have a relation

$$1/2[M(N)+M(\Xi)] = 3/4 \cdot M(\Lambda) + 1/4 \cdot M(\Sigma) \tag{30}$$

which is satisfied with good accuracy.

In the case of Sakata scheme, we do not have such relation unless we include (N, Ξ, Λ, Σ) in $U_3(2, 0, -1)$ representation as we mentioned in § 2. Then, we have Eq. (30) still. However, $U_3(2, 0, -1)$ representation contains two other states with ($I=3/2, S=0$) and ($I=0, S=+1$). We can calculate the masses of these particles by Eq. (26) and by using the experimental masses of N, Λ, and Σ. Then, we get

$$M(I=3/2, S=0) \simeq 1050 \text{ Mev} \quad (<M(N)+M(\pi)),$$

$$M(I=0, S=+1) \simeq 770 \text{ Mev} \quad (<M(N))$$

which seems to have too small masses not to be detected experimentally. Thus, this assignment originally due to Yamaguchi would not be so good. Therefore, we take the view that $U_3(2, 0, -1)$ represents Ξ, N^*, Y_0^*, Y_1^*, etc., as has been mentioned in § 2. In this case, we have the following relations:

$$\begin{aligned} &M(Y_1^*) = 1/2 \cdot [M(\Xi) + M(N^*)], \\ &M(I=1/2, S=0) = 1/2 \cdot [M(Y_0^*) + M(I=1, S=+1)], \\ &M(I=1, S=+1) = M(Y_1^*) + 2[M(Y_0^*) - M(\Xi)]. \end{aligned} \tag{31}$$

The first relation gives us $M(Y_1^*) \simeq 1280$ Mev by using the experimental values for $M(\Xi)$ and $M(N^*)$ and it should be compared to the experimental value of $M(Y_1^*) \simeq 1385$ Mev. Similarly, the last two equations give us

$$M(I=1, S=+1) \simeq 1560 \text{ Mev},$$
$$M(I=1/2, S=0) \simeq 1480 \text{ Mev}$$

where we have used the experimental masses for Y_0^* and Y_1^*. Consequently, we may identify the $(I=1/2, S=0)$ state as the 2nd pion-nucleon resonance, if it corresponds to the $p_{3/2}$ resonance instead of the usual $d_{3/2}$ resonance. As for $(I=1, S=+1)$, resonance for $K_+ + n$ or $K_+ + p$ scattering has not been discovered yet, and this gives a trouble to this scheme.

Acknowledgements

A part of this work was carried out when the author was at CERN. He is grateful for the hospitality at CERN. He would like to express his thanks to Professors R.E.Marshak and Y. Yamaguchi for their encouragements and discussions. He is also grateful to Professors H. Miyazawa and H. Umezawa and to Drs. M. Fukui, Y. Hara and T. Okabayashi, for the hospitality extended to him at the University of Tokyo.

Appendix

Derivation of Mass Formula

Here, we shall prove the mass formula Eq. (26).

Let us consider infinitesimal U_3 transformation. Then, the infinitesimal generator $A_\nu{}^\mu$ of U_3 satisfies the Lie equation:

$$[A_\beta{}^\alpha, A_\nu{}^\mu] = \delta_\beta{}^\mu \cdot A_\nu{}^\alpha - \delta_\nu{}^\alpha \cdot A_\beta{}^\mu. \tag{A·1}$$

This relation holds actually for general linear transformation of arbitrary dimension. The unitary restriction gives

$$(A_\beta{}^\alpha)^\dagger = A_\alpha{}^\beta \tag{A·2}$$

where $Q^\dagger$ means the hermitian conjugate of Q. For comparison's sake, our $A_\nu{}^\mu$ is related to Ikeda et al.[1]'s $X_{\mu\nu}$ by

$$A_\nu{}^\mu = -1/2 \cdot [(1+i) X_{\nu\mu} + (1-i) X_{\mu\nu}],$$
$$X_{\mu\nu} = -1/2 \cdot [(1+i) A_\nu{}^\mu + (1-i) A_\mu{}^\nu]. \tag{A·3}$$

However, their notation $X_{\mu\nu}$ makes the mixed tensor character of $A_\nu{}^\mu$ obscure.

For an arbitrary mixed tensor $T_\nu{}^\mu$, the commutation relation is given by

$$[A_\beta{}^\alpha, T_\nu{}^\mu] = \delta_\beta{}^\mu \cdot T_\nu{}^\alpha - \delta_\nu{}^\alpha \cdot T_\beta{}^\mu. \tag{A·4}$$

Comparing this with Eq. (A·1), we see that $A_\nu{}^\mu$ has the property of a mixed tensor.

Generalized Casimir operators of our Lie algebra can be given by

$$M_1=A_\mu{}^\mu=\langle A\rangle,$$
$$M_2=A_\nu{}^\mu\cdot A_\mu{}^\nu=\langle A\cdot A\rangle, \tag{A·5}$$
$$M_3=A_\nu{}^\mu\cdot A_\lambda{}^\nu\cdot A_\mu{}^\lambda=\langle A\cdot A\cdot A\rangle$$

where the repeated indices mean summation over 1, 2 and 3, and we used the notations $\langle Q\rangle$ and defined product tensor $Q\cdot R$ of two tensor $Q_\nu{}^\mu$ and $R_\nu{}^\mu$ by

$$\langle Q\rangle=Q_\mu{}^\mu,$$
$$(Q\cdot R)_\nu{}^\mu=Q_\lambda{}^\mu\cdot R_\nu{}^\lambda. \tag{A·6}$$

It is easy to see that M_1, M_2 and M_3 commute with all $A_\nu{}^\mu$ and therefore they commute with each other. Thus, they are constants in a given irreducible representation. Again, we will give a relation between our M_i and N, M, M' of Ikeda et al.[1)]

$$N=-M_1,$$
$$M=1/2\cdot M_2,$$
$$M'=-1/2\cdot M_3+3/4\cdot M_2-1/4\cdot (M_1)^2,$$

and so the relation between eigenvalues of M_i and f_1, f_2, f_3 of $U_3(f_1,f_2,f_3)$ is given[1)] by

$$M_1=-(f_1+f_2+f_3),$$
$$M_2=(f_1^2+f_2^2+f_3^2)+2(f_1-f_3),$$
$$M_3=-(f_1^3+f_2^3+f_3^3)+[-3/2\cdot f_1^2+3/2\cdot f_2^2+9/2\cdot f_3^2] \tag{A·7}$$
$$-1/2\cdot(f_1+f_2+f_3)^2+(2f_1+2f_2-4f_3).$$

Note that $M_4=\langle A\cdot A\cdot A\cdot A\rangle$, etc., are unnecessary. They are given as functions of M_1, M_2 and M_3 as will be seen shortly.

Now, we will prove the following theorem.

[Theorem I]

In any irreducible representations of U_3, any mixed tensors $T_\nu{}^\mu$ can be regarded as a linear combination:

$$T_\nu{}^\mu=a\cdot\delta_\nu{}^\mu+b\cdot A_\nu{}^\mu+c(A\cdot A)_\nu{}^\mu. \tag{A·8}$$

Eq. (A·8) means that it holds good when we take matrix elements of both sides in a given irreducible representation. Constants a, b and c are independent of tensor suffices μ and ν and of sub-quantum numbers S and I of the representation, but may depend upon f_1, f_2 and f_3 and upon the nature of the tensor $T_\nu{}^\mu$. Eq. (A·8) is an analogue of the so-called vector algebra in R_3, i.e.

$$\langle J,m|V_\mu|J,m'\rangle=\langle J||V||J\rangle\langle J,m|J_\mu|J,m'\rangle$$

where $V_\mu(\mu=1, 2, 3)$ is a vector in R_3, and J_μ means the angular momentum operator in R_3.

Before proving our theorem, we will show that this equation will give the desired mass formula Eq. (26).

First, let us consider the case of Sakata scheme. In that case, the nucleon number N, the strangeness quantum number S, and the isotopic spin operator $\boldsymbol{I}$ are defined[1] by

$$
\begin{aligned}
&N=-\langle A\rangle,\\
&S=A_3^3,\\
&I_+=(I_1+iI_2)=-A_1^2,\ I_-=(I_1-iI_2)=-A_2^1, \qquad\qquad (A\cdot 9)\\
&I_3=1/2(A_2^2-A_1^1).
\end{aligned}
$$

Now, let us suppose that the mass-splitting interaction is given by T_3^3 which has the same property as $\bar{\Lambda}\cdot\Lambda$ in the case of Sakata model. Then, the mass splitting is given by diagonal matrix element of T_3^3.

$$\Delta M=\langle i|T_3^3|i\rangle.$$

Then, noting (A9) and

$$(A\cdot A)_3^3=1/2\cdot\langle A\cdot A\rangle+1/2\cdot S^2+1/2(3\cdot S-\langle A\rangle)-(\boldsymbol{I})^2-1/4\cdot(S-\langle A\rangle)^2,$$

we find that our theorem I (Eq. (A·8)) gives the desired mass formula Eq. (26).

In the case of Gell-Mann scheme, we have only to replace S by Y, hence we get Eq. (27). In this case, N is simply a parameter to distinguish representations.

Now, let us prove our theorem Eq. (A·8). First, we will show the following lemma.

[Lemma I]

In the three-dimensional space, suppose that a tensor $S_{\mu\nu}^{\alpha\beta}$ is anti-symmetric with respect to exchanges of α and β and of μ and ν and furthermore $S_{\mu\nu}^{\mu\beta}=0$, i.e. traceless; then $S_{\mu\nu}^{\alpha\beta}$ is identically zero. Schematically, this means that $S_{\mu\nu}^{\alpha\beta}=-S_{\mu\nu}^{\beta\alpha}=-S_{\nu\mu}^{\alpha\beta}$ and $S_{\mu\nu}^{\mu\beta}=0\rightarrow S_{\mu\nu}^{\alpha\beta}\equiv 0$.

[Proof]

Let us consider a tensor

$$
\begin{aligned}
T_{\mu\nu\lambda}^{\alpha\beta\gamma}=&S_{\mu\nu}^{\alpha\beta}\cdot\delta_\lambda{}^\gamma-S_{\mu\nu}^{\alpha\gamma}\cdot\delta_\lambda{}^\beta-S_{\mu\nu}^{\gamma\beta}\cdot\delta_\lambda{}^\alpha\\
&+S_{\lambda\nu}^{\gamma\beta}\cdot\delta_\mu{}^\alpha-S_{\lambda\nu}^{\gamma\alpha}\cdot\delta_\mu{}^\beta-S_{\lambda\nu}^{\alpha\beta}\cdot\delta_\mu{}^\gamma.
\end{aligned}
$$

Then, $T_{\mu\nu\lambda}^{\alpha\beta\gamma}$ is totally anti-symmetric for any two exchanges of α, β and γ and satisfies traceless condition $T_{\alpha\nu\lambda}^{\alpha\beta\gamma}=0$. However, such tensor must be identically zero in the three-dimensional space, since only non-zero independent component must be $T_{\mu\nu\lambda}^{123}$ and by traceless-condition, this has to be identically zero, (for example, consider the case $\mu=1$). Thus, we have $T_{\mu\nu\lambda}^{\alpha\beta\gamma}\equiv 0$. Then, by putting $\gamma=\nu$ and summing over ν, we find

$$T^{\alpha\beta\nu}_{\mu\nu\lambda}=S^{\alpha\beta}_{\mu\lambda}-S^{\alpha\beta}_{\lambda\mu}=2S^{\alpha\beta}_{\mu\lambda}\equiv 0. \quad (Q.E.D.)$$

Our lemma I is not surprising at all, since such tensor $S^{\alpha\beta}_{\mu\nu}$ must be an irreducible representation in U_n but such type of irreducible representation is not possible in U_3. (However, it is possible in $U_n (n\geq 4)$ and has signature $(1, 1, -1, -1)$ in U_4.)

[Lemma II]

In U_3, for any two arbitrary tensors $M_\nu{}^\mu$ and $N_\nu{}^\mu$, we have the following identities:

$$\begin{aligned}
&(M_\nu{}^\mu \cdot N_\beta{}^\alpha + M_\beta{}^\alpha \cdot N_\nu{}^\mu)-(M_\nu{}^\alpha \cdot N_\beta{}^\mu + M_\beta{}^\mu \cdot N_\nu{}^\alpha)\\
&=\delta_\nu{}^\mu[\langle M\rangle N_\beta{}^\alpha + M_\beta{}^\alpha\langle N\rangle-(M\cdot N)_\beta{}^\alpha - M_\beta{}^\lambda\cdot N_\lambda{}^\alpha]\\
&-\delta_\nu{}^\alpha[\langle M\rangle N_\beta{}^\mu + M_\beta{}^\mu\langle N\rangle-(M\cdot N)_\beta{}^\mu - M_\beta{}^\lambda\cdot N_\lambda{}^\mu]\\
&-\delta_\beta{}^\mu[\langle M\rangle\cdot N_\nu{}^\alpha + M_\nu{}^\alpha\cdot\langle N\rangle-(M\cdot N)_\nu{}^\alpha - M_\nu{}^\lambda\cdot N_\lambda{}^\alpha]\\
&+\delta_\beta{}^\alpha[\langle M\rangle\cdot N_\nu{}^\mu + M_\nu{}^\mu\cdot\langle N\rangle-(M\cdot N)_\nu{}^\mu - M_\nu{}^\lambda\cdot N_\lambda{}^\mu]\\
&-(\delta_\nu{}^\mu\cdot\delta_\beta{}^\alpha-\delta_\nu{}^\alpha\cdot\delta_\beta{}^\mu)\cdot[\langle M\rangle\cdot\langle N\rangle-\langle M\cdot N\rangle].
\end{aligned}$$

[Proof]

Define a tensor $Q^{\alpha\beta}_{\mu\nu}$ by

$$Q^{\alpha\beta}_{\mu\nu}=(M_\mu{}^\alpha\cdot N_\nu{}^\beta - M_\mu{}^\beta\cdot N_\nu{}^\alpha)-(M_\nu{}^\alpha\cdot N_\mu{}^\beta - M_\nu{}^\beta\cdot N_\mu{}^\alpha)\cdot$$

Then, $Q^{\alpha\beta}_{\mu\nu}$ is anti-symmetric for exchanges of α and β and of μ and ν. Furtheremore, construct a new tensor $S^{\alpha\beta}_{\mu\nu}$ by

$$\begin{aligned}
S^{\alpha\beta}_{\mu\nu}=Q^{\alpha\beta}_{\mu\nu}&-(\delta_\mu{}^\alpha\cdot Q^{\lambda\beta}_{\lambda\nu}+\delta_\mu{}^\beta\cdot Q^{\alpha\lambda}_{\lambda\nu}+\delta_\nu{}^\alpha\cdot Q^{\lambda\beta}_{\mu\lambda}+\delta_\nu{}^\beta\cdot Q^{\alpha\lambda}_{\mu\lambda})\\
&+1/2\cdot(\delta_\mu{}^\alpha\cdot\delta_\nu{}^\beta-\delta_\nu{}^\alpha\cdot\delta_\mu{}^\beta)Q^{\lambda\theta}_{\lambda\theta}.
\end{aligned}$$

We can see that $S^{\alpha\beta}_{\mu\nu}$ satisfies the conditions of lemma I, and must be identically zero. This gives the desired identity. (Q.E.D.)

[Theorem II]

In U_3, for any tensor $T_\nu{}^\mu$ and for infinitesimal operator $A_\nu{}^\mu$, which satisfy the commutation relations Eqs. (A·1) and (A·4), we have the following identity.

$$\begin{aligned}
&2\cdot[(A\cdot T\cdot A)_\nu{}^\mu+(T\cdot A\cdot A)_\nu{}^\mu+(A\cdot A\cdot T)_\nu{}^\mu]-(2\langle A\rangle+9)\cdot[(A\cdot T)_\nu{}^\mu+(T\cdot A)_\nu{}^\mu]\\
&-2\cdot\langle T\rangle(A\cdot A)_\nu{}^\mu+[6\langle A\rangle+12+(\langle A\rangle)^2]T_\nu{}^\mu\\
&-1/2\cdot[\langle A\cdot A\rangle T_\nu{}^\mu+T_\nu{}^\mu\cdot\langle A\cdot A\rangle]+[6\langle T\rangle+2\langle A\rangle\langle T\rangle-2\langle A\cdot T\rangle]A_\nu{}^\mu\\
&+\delta_\nu{}^\mu\cdot(-\langle T\rangle\cdot[(\langle A\rangle)^2-\langle A\cdot A\rangle+4\langle A\rangle+4]+(2\langle A\rangle+6)\langle A\cdot T\rangle\\
&\qquad -2\langle A\cdot A\cdot T\rangle)\equiv 0.
\end{aligned}$$

Note that $[\langle A\rangle, T_\nu{}^\mu]=0$, $[\langle T\rangle, A_\nu{}^\mu]=0$ but $[\langle A\cdot A\rangle, T_\nu{}^\mu]\neq 0$.

[Theorem III]

$$6(A\cdot A\cdot A)_\nu{}^\mu-[6\langle A\rangle+18]\cdot(A\cdot A)_\nu{}^\mu+[3\cdot(\langle A\rangle)^2-3\cdot\langle A\cdot A\rangle+$$

$$+12\cdot\langle A\rangle+12]\cdot A_\nu{}^\mu$$
$$-[(\langle A\rangle)^3+4(\langle A\rangle)^2+4\langle A\rangle-3\langle A\rangle\cdot\langle A\cdot A\rangle+2\langle A\cdot A\cdot A\rangle$$
$$-6\langle A\cdot A\rangle]\delta_\nu{}^\mu=0.$$

Theorem III can be obtained from theorem II by putting $T=A$. From this, we see that $(A\cdot A\cdot A\cdot A)_\nu{}^\mu$ can be expressed as a linear combination of $\delta_\nu{}^\mu$, $A_\nu{}^\mu$, $(A\cdot A)_\nu{}^\mu$ and $(A\cdot A\cdot A)_\nu{}^\mu$, and so $\langle A\cdot A\cdot A\cdot A\rangle$ is a function of $\langle A\rangle$, $\langle A\cdot A\rangle$ and $\langle A\cdot A\cdot A\rangle$. So are $\langle A^n\rangle$ $(n\geq 4)$, as has already been mentioned.

To prove theorem II, we put $M_\nu{}^\mu=N_\nu{}^\mu=A_\nu{}^\mu$ in lemma II, and multiply $T_\alpha{}^\nu$ from the left, and using commutation relations Eqs. (A·1) and (A·4), we find our theorem II, when we change the indices suitably. We may give another direct proof of theorem II as follows. Any tensor $Q^{\alpha\beta\gamma\theta}_{\pi\lambda\mu\nu}$ which is anti-symmetric with respect to any exchanges of two variables among α, β, γ and θ must be identically zero in U_3. Therefore, we have

$$\sum_P(-1)^P T_\pi{}^\alpha\cdot A_\lambda{}^\beta\cdot A_\nu{}^\gamma\cdot\delta_\mu{}^\theta=0$$

where P means permutations among α, β, γ and θ. Then putting $\pi=\beta$, $\gamma=\lambda$, $\theta=\nu$ and taking traces, we find our theorem II again after somewhat long calculations.

Now, we shall prove our theorem I, Eq. (A·8). Using the commutation relations

$$[M_3, T_\nu{}^\mu]=3(A\cdot A\cdot T)_\nu{}^\mu-3(T\cdot A\cdot A)_\nu{}^\mu-3[M_2, T_\nu{}^\mu],$$
$$[M_2, S_\nu{}^\mu]=2(A\cdot S)_\nu{}^\mu-2(S\cdot A)_\nu{}^\mu,$$

we can rewrite theorem II as follows.

$$3(T\cdot A\cdot A)_\nu{}^\mu-(T\cdot A)_\nu{}^\mu\cdot(2\langle A\rangle+9)+T_\nu{}^\mu\cdot[1/2\cdot(\langle A\rangle)^2$$
$$-1/2\cdot\langle A\cdot A\rangle+3\langle A\rangle+6]$$
$$=-1/2\cdot[M_2, (TA)_\nu{}^\mu-(\langle A\rangle+3)T_\nu{}^\mu]-1/3[M_3, T_\nu{}^\mu]$$
$$+(A\cdot A)_\nu{}^\mu\cdot\langle T\rangle-A_\nu{}^\mu\cdot[(\langle A\rangle+3)\cdot\langle T\rangle-\langle T\cdot A\rangle]$$
$$-\delta_\nu{}^\mu([\langle A\rangle+3]\langle T\cdot A\rangle-\langle T\cdot A\cdot A\rangle-1/2\cdot\langle T\rangle\cdot[(\langle A\rangle)^2-\langle A\cdot A\rangle+4\langle A\rangle+4]). \quad (A\cdot 9)$$

Now, in a given irreducible representation, M_2 and M_3 are constants, so that matrix elements $\langle\alpha|[M_2, Q]|\beta\rangle=0$ and $\langle\alpha|[M_3, Q]|\beta\rangle=0$, hence we can omit the first and second terms in the right-hand side of Eq. (A·9) in our case. Thus, we have

$$3(T\cdot A\cdot A)_\nu{}^\mu-(T\cdot A)_\nu{}^\mu(2\langle A\rangle+9)+T_\nu{}^\mu\cdot[1/2\cdot(\langle A\rangle)^2-1/2\cdot\langle A\cdot A\rangle+3\langle A\rangle+6]$$
$$=(A\cdot A)_\nu{}^\mu\cdot\langle T\rangle-A_\nu{}^\mu\cdot[(\langle A\rangle+3)\langle T\rangle-\langle T\cdot A\rangle]$$
$$-\delta_\nu{}^\mu([\langle A\rangle+3]\langle T\cdot A\rangle-\langle T\cdot A\cdot A\rangle-$$

$$-1/2\cdot\langle T\rangle\cdot[(\langle A\rangle)^2-\langle A\cdot A\rangle+4\langle A\rangle+4]). \qquad (A\cdot 10)$$

Eq. (A·10) is true when we take any matrix elements in a given irreducible representation. Now, T_ν^μ is arbitrary, as long as it satisfies the commutation relation Eq. (A·4), and so we can replace T by $T\cdot A$ and $T\cdot A\cdot A$ in Eq. (A·10). For quantities like $T\cdot A\cdot A\cdot A$ or $T\cdot(A\cdot A\cdot A\cdot A)$, we use our theorem III and we can reduce them to a linear combination of T, $T\cdot A$ and $T\cdot A\cdot A$. Then, Eq. (A·10) gives three equations of the form

$$a_{1i}(T\cdot A\cdot A)_\nu^\mu+a_{2i}(T\cdot A)_\nu^\mu+a_{3i}(T)_\nu^\mu$$
$$=b_{1i}(A\cdot A)_\nu^\mu+b_{2i}(A)_\nu^\mu+b_{3i}\cdot\delta_\nu^\mu. \qquad (i=1,2,3) \qquad (A\cdot 10)$$

We can give an explicit form for a_{ij} and b_{ij}, but as it is a little complicated, here we simply remark that a_{ij} are functions of only $\langle A\rangle$, $\langle A\cdot A\rangle$ and $\langle A\cdot A\cdot A\rangle$, i.e. a_{ij} depend only upon f_1, f_2 and f_3 by Eq. (A·7), b_{ij} depend upon f_1, f_2 and f_3, and also upon $\langle T\rangle$, $\langle T\cdot A\rangle$ and $\langle T\cdot A\cdot A\rangle$, which are constants in the irreducible representation which we are considering. We can solve Eq. (A·10), since the determinant $\det(a_{ij})$ is, in general, not identically zero; thus we get

$$T_\nu^\mu=a\cdot\delta_\nu^\mu+b\cdot A_\nu^\mu+c(A\cdot A)_\nu^\mu$$

and two other equations for $(T\cdot A)_\nu^\mu$ and $(T\cdot A\cdot A)_\nu^\mu$. This is the desired formula theorem I.

References

1) M. Ikeda, S. Ogawa and Y. Ohnuki, Prog. Theor. Phys. **22** (1959), 715; **23** (1960), 1073.
2) Y. Yamaguchi, Prog. Theor. Phys. Suppl. **11** (1960), 1 and 37.
3) M. Gell-Mann, preprint: A Theory of Strong Interaction Symmetry, California Institute of Technology, March 15, 1961; Phys. Rev. **125** (1962), 1067.
4) A. Gamba, R. E. Marshak and S. Okubo, Proc. Nat. Acad. Sci. **45** (1959), 881.
R. E. Marshak and S. Okubo, Nuovo Cimento **19** (1961), 1226.
5) A similar analysis has been done by W. E. Thirring, Nuclear Phys. **10** (1959), 97.
6) S. Sakata, Prog. Theor. Phys. **16** (1956), 686.
7) H. Weyl, *The Classical Groups* (Princeton University Press, 1939).
8) H. Weyl, *The Theory of Groups and Quantum Mechanics* (Dover Publications. Inc., 1931), see the last chapter.
9) H. Weyl, reference 7), see Theorem (4. 5. A), p. 135.
10) H. Weyl, reference 7), use Theorem (7. 10. A), pp. 231~232.
11) H. Weyl, reference 7), pp. 132 and 208.
12) Y. Hara and G. P. Singh, private communication.
13) S. Coleman and S. L. Glaschow, Phys. Rev. Letters **6** (1961), 423.
14) S. Okubo, R. E. Marshak, E. G. Sudarshan, W. Teutsch and S. Weinbers, Phys. Rev. **112** (1958), 665.
15) K. Matumoto, Prog. Theor. Phys. **16** (1956), 583.
S. Sawada and M. Yonezawa, Prog. Theor. Phyg. **23** (1960), 662.

II

THE DECIMET AND THE OMEGA-MINUS

The classification of the known $J = \frac{1}{2}^+$ baryons and pseudoscalar mesons as two SU(3) octets raises the question of how to classify the baryon states that can decay by the strong interaction into one $J = \frac{1}{2}^+$ baryon and one $J = 0^-$ meson. These unstable particles are encountered as meson-baryon scattering resonances.

Without violating SU(3) invariance, there are only a few representations that can be made by combining two octets, as pointed out above in "The Eightfold Way" paper: the singlet representation **1**, the octet **8** (occurring twice), and three more representations called **10**, $\overline{\mathbf{10}}$, and **27**. A resonance of the type we have mentioned, if its formation and decay are allowed by SU(3), must belong to one of these representations.

The most familiar resonance, discovered at Chicago in 1952, is the one at 1238 MeV, with spin and parity $\frac{3}{2}^+$, decaying into $\pi + N$; we may call it $\Delta(1238)$. It has $I = \frac{3}{2}$ and $Y = +1$; a multiplet with these quantum numbers appears in **27** and in **10**. These two assignments are discussed by Glashow and Sakurai [Nuovo Cimento, **25**, 337 (1962)].

In the excerpt from the Proceedings of the 1962 Geneva Conference that appears below, Gell-Mann points out that the **27** assignment is ruled out by the absence of any nearby $I = 1$, $Y = +2$ resonance in KN scattering. (We may refer to this striking experimental phenomenon as the "Goldhaber gap.") That leaves only the **10** supermultiplet, which contains the quartet with $Y = +1$ along with a triplet with $Y = 0$, a doublet with $Y = -1$, and a singlet with $Y = -2$. The first-order mass rule in this case is just a rule of equal spacing. (That can be seen, for example, from the Okubo mass formula; the result can also be easily derived directly.) It is further pointed out that a resonance at 1385 MeV with $I = 1$ and $Y = 0$, which we may call $\Sigma(1385)$, is suspected of having $J = \frac{3}{2}^+$. Thus, the 10 assignment, together with the mass rule, predicts a particle with $I = \frac{1}{2}$ and $Y = -1$ at about 1532 MeV, which we may call $\Xi(1532)$ and one with $I = 0$ and $Y = -2$ at about 1679 MeV, which may be called Ω^- or $\Omega(1679)$. Both of these would have $J = \frac{3}{2}^+$. The Ω^- would have the striking property of metastability, decaying only by weak interactions and yielding $\pi^0 + \Xi^-$, $\pi^- + \Xi^0$, and $K^- + \Lambda$.

Now a resonance just like $\Xi(1532)$ was reported at the Geneva meeting just before Gell-Mann's remarks (and evidence has since appeared for the assignment $J = \frac{3}{2}^+$). Thus, an experimental search for Ω^- at about 1679 MeV was clearly indicated as a crucial test of the broken symmetry.

The search was successful, as the second paper makes clear. The observation of two artificially produced Ω hyperons, in the full splendor of their cascading decays, has provided a striking confirmation of the broken eightfold way. Furthermore, it has made us more conscious than ever that we do not understand why the first-order mass formula is so good. Except for the special case of ϕ-ω mixing, the formula is accurate to a few MeV for each supermultiplet, while the mass splittings are hundreds of MeV. From now on, any fundamental theory of strongly interacting particles will have to account for this remarkable circumstance.

The confirmation of the Ω prediction has also laid to rest any lingering notion that there is necessarily anything sacred or "elementary" about a baryon or meson that happens to be stable under strong interactions, such as the eight $J = \frac{1}{2}^+$ baryons and the eight $J = 0^-$ mesons. The Ω^- hyperon is metastable, but it belongs to the same supermultiplet as the very broad resonance $\Delta(1238)$, which has always been considered obviously composite. Mere stability is no more a criterion of elementary character for baryon number 0 or 1 than it is for nuclei, which are strongly interacting particles of higher baryon number.

From: *Proc Intern Conf High Energy Phys* (CERN, 1962), p. 805

Strange Particle Physics. Strong Interactions

M. Gell-Mann

SNOW: I know that at 1.5 GeV in the CERN K^-p experiment in the fourbody reaction $\Sigma^\pm\pi^\mp\pi^-\pi^+$ one looks for $T=2$ resonances, but when one looks at these reactions the dominant things that one sees are $T=0$ resonances. I think the Berkeley data which has more statistical weight shows the same thing: they do not see any evidence for this.

SAKURAI: It is now plausible that the 1405 MeV Y_0^* but not the 1385 MeV Y_1^* is an *s*-wave $\bar{K}N$ bound state of the Dalitz-Tuan type. So it is worth asking what kind of dynamical mechanism is responsible for binding a $\bar{K}$ and an N in the $T=0$ state. Along this line a student of mine, Richard Arnold at Chicago, has performed a very crude N/D calculation to show that for reasonable values of coupling constants the forces due to the exchanges of ϱ and ω are sufficiently attractive to bind the $T=0$, $\bar{K}N$ system. On this calculation the signs of the ϱ and ω exchange force are fixed by the universality principle of the vector theory of strong interactions. Another interesting point is that the same mechanism predicts that the $T=1$, KN system is strongly repulsive, in agreement with observation. I want to make another remark which is somewhat more general. It is interesting to conjecture that the isospin dependent force for any low energy scattering is dominated by the exchange of the ϱ meson coupled universally to the isospin current. The ϱ exchange force is then attractive whenever the isospins are antiparallel and repulsive whenever the isospin are parallel. This rule works remarkably well in five cases examined so far. In the case of *s*-wave πN scattering essentially the entire isospin dependence is due to the ϱ exchange as conjectured by Cini and Fubini and by myself independently and as proved by Hamilton *et al.* In the *s*-wave $\pi\pi$ scattering case the $T=0$ seems more attractive than the $T=2$ state. In low energy $K\bar{K}$ scattering there is some evidence for a strong attractive interaction in the $T=0$ state as we have just heard. In the $\bar{K}N$ case the currently accepted idea, that the 1405 MeV Y_0^* but not the 1385 MeV Y_1^* is likely to be a $\bar{K}N$ bound state resonance, shows that the $T=0$ state is more attractive. In the $S=1$, KN case, the $T=1$ state is definitely more repulsive than the $T=0$ state. In general bound states and resonances are more likely for states with lower isospins as you can see from the table of elementary particles and resonances., So there seems to be a correlation between the simplicity of quantum numbers and the possibility of bound or resonant states.

ROSENFELD: (in reply to Good): Good asked how carefully we have looked for doubly charged ($T=2$) resonances. 1.51 GeV/c $K^-+p\to\Sigma^\pm\pi^\mp\pi^+\pi^-$ ($E^*_{CM}=2025$ MeV, 400 events) allowed Alston *et al.*, to explore doubly charged $\Sigma\pi$ combination fairly well up to $\simeq$1600 MeV. Actually, we saw one bump of perhaps 2 standard deviations at 1560 MeV, but nothing that looked interesting. But I repeat for negative strangeness Berkeley has not looked above 1600 MeV.

TICHO: In reply to Good, we are sensitive to $\Xi\pi$ $T=3/2$ resonance up to Q values of 170 MeV. Within our statistics we see no evidence for resonances other than the reported Ξ^* of $T=1/2$.

SANDWEISS: In a paper submitted to this conference, results of a π^+p study at 2 GeV are reported. In about 70 events of $\Sigma^+\pi^+K^0$, $\Sigma^0\pi^+K^+$ and $\Sigma^+\pi^0K^+$ no evidence for a $T=2$ $\Sigma\pi$ resonance was found.

GELL-MANN: If we take the unitary symmetry model with baryon and meson octets, with first order violation giving rise to mass differences, we obtain some rules for supermultiplets. The broken symmetry picture is hard to interpret on any fundamental theoretical basis, but I hope that such a justification may be forthcoming on the basis of analytic continuation of resonant states in isotopic spin and strangeness. Instead of constructing just the inverse Regge function E (J), we can consider surfaces E (J, I, Y, etc.). Certainly the dynamical equations are as smooth in I and Y as they are in J.

Anyway, we may look at the success of the mass rules:

$$\frac{m_N+m_\Xi}{2}=\frac{3m_\Lambda}{4}+\frac{m_\Sigma}{4}$$

and

$$m_K^2=\frac{3m_\chi^2}{4}+\frac{m_\pi^2}{4}$$

work fine, while

$$m_M^2=\frac{3m_\omega^2}{4}+\frac{m_\rho^2}{4}$$

does not work quite so well if $M=K^*$.

Suppose, now we try to incorporate the 3/2-3/2 nucleon resonance into the scheme. The only supermultiplet that does not lead to non-existent resonances in the $K-N$ channels is the *10* representation, which gives 4 states:

$$I=3/2,\quad S=0$$
$$I=1\ ,\quad S=-1$$
$$I=1/2,\quad S=-2$$
$$I=0\ ,\quad S=-3$$

The mass rule is stronger here and yields *equal spacing* of these states. Starting with the resonance at 1238 MeV, we may conjecture that the Y_1^*, at 1385 MeV and the Ξ^* at 1535 MeV might belong to this supermultiplet. Certainly they fulfil the requirement of equal spacing. If $J=3/2^+$ is really right for these two cases, then our speculation might have some value and we should look for the last particle, called, say, Ω^- with $S=-3$, $I=0$. At 1685 MeV, it would be metastable and should decay by the weak interactions into $K^-+\Lambda$, $\pi^-+\Xi^0$, or $\pi^0+\Xi^-$. Perhaps it would explain the old Eisenberg event. A beam of K^- with momentum $\geqslant$3,5 GeV/c could yield Ω^- by means of $K^-+p\to K^++K^0+\Omega^-$.

ADAIR: I would like to clarify the statements attributed to me concerning the $\Sigma-K$ parity. The conclusions of Tripp, Ferro-Luzzi, and Watson, and, implicitly, those of Capps, that the $\Sigma-K$ parity is odd, actually refer to a particular model of the K-nucleon interaction, a model which at best can be but a first approximation to the real world, and which does not exhibit certain important features of their data. In particular, this model assumes that the energy dependence of background amplitudes and resonance widths may be neglected, that charge dependent effects may be neglected, and most important, that there are no background amplitudes in the states with the j and parity of the resonance. Elementary considerations show that for the high angular momentum states of interest such approximations are quite inadequate. For example, the magnitude of background amplitudes and widths must vary with energy perhaps by a factor of two, over the resonance region. Their model results in equal resonant cross-sections for the

OBSERVATION OF A HYPERON WITH STRANGENESS MINUS THREE*

V. E. Barnes, P. L. Connolly, D. J. Crennell, B. B. Culwick, W. C. Delaney, W. B. Fowler, P. E. Hagerty,† E. L. Hart, N. Horwitz,† P. V. C. Hough, J. E. Jensen, J. K. Kopp, K. W. Lai, J. Leitner,† J. L. Lloyd, G. W. London,‡ T. W. Morris, Y. Oren, R. B. Palmer, A. G. Prodell, D. Radojičić, D. C. Rahm, C. R. Richardson, N. P. Samios, J. R. Sanford, R. P. Shutt, J. R. Smith, D. L. Stonehill, R. C. Strand, A. M. Thorndike, M. S. Webster, W. J. Willis, and S. S. Yamamoto
Brookhaven National Laboratory, Upton, New York
(Received 11 February 1964)

It has been pointed out[1] that among the multitude of resonances which have been discovered recently, the $N_{3/2}^*(1238)$, $Y_1^*(1385)$, and $\Xi_{1/2}^*(1532)$ can be arranged as a decuplet with one member still missing. Figure 1 illustrates the position of the nine known resonant states and the postulated tenth particle plotted as a function of mass and the third component of isotopic spin. As can be seen from Fig. 1, this particle (which we call Ω^-, following Gell-Mann[1]) is predicted to be a negatively charged isotopic singlet with strangeness minus three.[2] The spin and parity should be the same as those of the $N_{3/2}^*$, namely, $3/2^+$. The 10-dimensional representation of the group SU_3 can be identified with just such a decuplet. Consequently, the existence of the Ω^- has been cited as a crucial test of the theory of unitary symmetry of strong interactions.[3,4] The mass is predicted[5] by the Gell-Mann–Okubo mass formula to be about 1680 MeV/c^2. We wish to report the observation of an event which we believe to be an example of the production and decay of such a particle.

The BNL 80-in. hydrogen bubble chamber was exposed to a mass-separated beam of 5.0-BeV/c K^- mesons at the Brookhaven AGS. About 100 000 pictures were taken containing a total K^- track length of ~10^6 feet. These pictures have been partially analyzed to search for the more characteristic decay modes of the Ω^-.

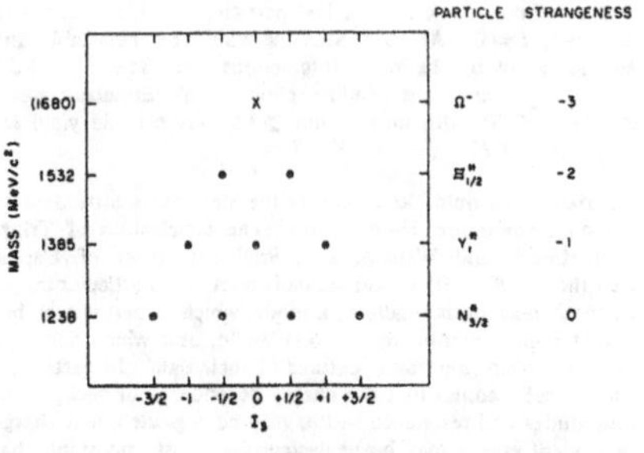

FIG. 1. Decuplet of $\frac{3}{2}^+$ particles plotted as a function of mass versus third component of isotopic spin.

The event in question is shown in Fig. 2, and the pertinent measured quantities are given in Table I. Our interpretation of this event is

$$
\begin{aligned}
& K^- + p \to \Omega^- + K^+ + K^0 \\
& \quad \Omega^- \to \Xi^0 + \pi^- \\
& \quad\quad \Xi^0 \to \Lambda^0 + \pi^0 \\
& \quad\quad\quad \pi^0 \to \gamma_1 + \gamma_2 \\
& \quad\quad\quad\quad \gamma_2 \to e^+ + e^- \\
& \quad\quad\quad\quad \gamma_1 \to e^+ + e^- \\
& \quad\quad\quad \Lambda^0 \to \pi^- + p.
\end{aligned}
\qquad (1)
$$

From the momentum and gap length measurements, track 2 is identified as a K^+. (A bubble density of 1.9 times minimum was expected for this track while the measured value was 1.7 ± 0.2.) Tracks 5 and 6 are in good agreement with the decay of a Λ^0, but the Λ^0 cannot come from the primary interaction. The Λ^0 mass as calculated from the measured proton and π^- kinematic quantities is 1116 ± 2 MeV/c^2. Since the bubble density from gap length measurement of track 6 is 1.52 ± 0.17, compared to 1.0 expected for a π^+ and 1.4 for a proton, the interpretation of the V as a K^0 is unlikely. In any case, from kinematical considerations such a K^0 could not come from the production vertex. The Λ^0 appears six decay lengths from the wall of the bubble chamber, and there is no other visible origin in the chamber.

The event is unusual in that two gamma rays, apparently associated with it, convert to electron-positron pairs in the liquid hydrogen. From measurements of the electron momenta and angles, we determine that the effective mass of the two gamma rays is 135.1 ± 1.5 MeV/c^2, consistent with a π^0 decay. In a similar manner, we have used the calculated π^0 momentum and angles, and the values from the fitted Λ^0 to deter-

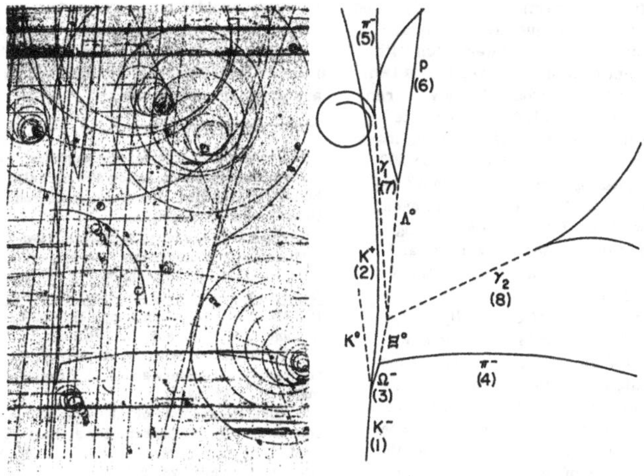

FIG. 2. Photograph and line diagram of event showing decay of Ω^-.

mine the mass of the neutral decaying hyperon to be 1316 ± 4 MeV/c^2 in excellent agreement with that of the Ξ^0. The projections of the lines of flight of the two gammas and the Λ^0 onto the XY plane (parallel to the film) intersect within 1 mm and in the XZ plane within 3 mm. The calculated momentum vector of the Ξ^0 points back to the decay point of track 3 within 1 mm and misses the production vertex by 5 mm in the XY plane. The length of the Ξ^0 flight path is 3 cm with a calculated momentum of 1906 ± 20 MeV/c. The transverse momenta of the Ξ^0 and of track 4 balance within the errors, indicating that no other particle is emitted in the decay of particle 3.

We will now discuss the decay of particle 3. From the momentum and gap length measurements on track 4, we conclude that its mass is less than that of a K. Using the Ξ^0 momentum and assuming particle 4 to be a π^-, the mass of particle 3 is computed to be 1686 ± 12 MeV/c^2 and its momentum to be 2015 ± 20 MeV/c. Note that the measured transverse momentum of track 4, 248 ± 5 MeV/c, is greater than the maximum momentum for the possible decay modes of the known particles (given in Table II), except for $\Xi^- \to e^- + n + \nu$. We reject this hypothesis not only because it involves $\Delta S = 2$, but also because it disregards the previously established associations of the Λ and two gammas with the event.

Table I. Measured quantities.

Track	Azimuth (deg)	Dip (deg)	Momentum (MeV/c)
1	4.2 ± 0.1	1.1 ± 0.1	4890 ± 100
2	6.9 ± 0.1	3.3 ± 0.1	501 ± 5.5
3	14.5 ± 0.5	-1.5 ± 0.6	...
4	79.5 ± 0.1	-2.7 ± 0.1	281 ± 6
5	344.5 ± 0.1	-12.0 ± 0.2	256 ± 3
6	9.6 ± 0.1	-2.5 ± 0.1	1500 ± 15
7	357.0 ± 0.3	3.9 ± 0.4	82 ± 2
8	63.3 ± 0.3	-2.4 ± 0.2	177 ± 2

Table II. Maximum transverse momentum of the negative decay product for various particle decays.

Decay modes	Maximum transverse momentum (MeV/c)
$\pi^- \to \mu^- + \nu$	30
$K^- \to \mu^- + \nu$	236
$K^- \to \pi^- + \pi$	205
$K^- \to e^- + \pi^0 + \nu$	229
$\Sigma^- \to \pi^- + n$	192
$\Sigma^- \to e^- + \Lambda^0 + \nu$	78
$\Sigma^- \to e^- + n + \nu$	229
$\Xi^- \to \pi^- + \Lambda^0$	139
$\Xi^- \to e^- + \Lambda^0 + \nu$	190
$\Xi^- \to e^- + n + \nu$	327

The proper lifetime of particle 3 was calculated to be 0.7×10^{-10} sec; consequently we may assume that it decayed by a weak interaction with $\Delta S=1$ into a system with strangeness minus two. Since a particle with $S=-1$ would decay very rapidly into $Y+\pi$, we may conclude that particle 3 has strangeness minus three. The missing mass at the production vertex is calculated to be 500 $\pm$ 25 MeV/c^2, in good agreement with the K^0 assumed in Reaction (1). Production of the event by an incoming π^- is excluded by the missing mass calculated at the production vertex, and would not alter the interpretation of the decay chain starting with track 3.

In view of the properties of charge ($Q=-1$), strangeness ($S=-3$), and mass ($M=1686\pm12$ MeV/c^2) established for particle 3, we feel justified in identifying it with the sought-for Ω^-. Of course, it is expected that the Ω^- will have other observable decay modes, and we are continuing to search for them. We defer a detailed discussion of the mass of the Ω^- until we have analyzed further examples and have a better understanding of the systematic errors.

The observation of a particle with this mass and strangeness eliminates the possibility which has been put forward[6] that interactions with $\Delta S=4$ proceed with the rates typical of the strong interactions, since in that case the Ω^- would decay very rapidly into $n+K^0+\pi^-$.

We wish to acknowledge the excellent cooperation of the staff of the AGS and the untiring efforts of the 80-in. bubble chamber and scanning and programming staffs.

---

*Work performed under the auspices of the U. S. Atomic Energy Commission and partially supported by the U. S. Office of National Research and the National Science Foundation.

†Syracuse University, Syracuse, New York.

‡University of Rochester, Rochester, New York.

[1]M. Gell-Mann, Proceedings of the International Conference on High-Energy Nuclear Physics, Geneva, 1962 (CERN Scientific Information Service, Geneva, Switzerland, 1962), p. 805; R. Behrends, J. Dreitlein, C. Fronsdal, and W. Lee, Rev. Mod. Phys. 34, 1 (1962); S. L. Glashow and J. J. Sakurai, Nuovo Cimento 25, 337 (1962).

[2]A possible example of the decay of this particle was observed by Y. Eisenberg, Phys. Rev. 96, 541 (1954).

[3]M. Gell-Mann, Phys. Rev. 125, 1067 (1962); Y. Ne'eman, Nucl. Phys. 26, 222 (1961).

[4]See, however, R. J. Oakes and C. N. Yang, Phys. Rev. Letters 11, 174 (1963).

[5]M. Gell-Mann, Synchrotron Laboratory, California Institute of Technology, Internal Report No. CTSL-20, 1961 (unpublished); S. Okubo, Progr. Theoret. Phys. (Kyoto) 27, 949 (1962).

[6]G. Racah, Nucl. Phys. 1, 302 (1956); H. J. Lipkin, Phys. Letters 1, 68 (1962).

III

FURTHER ASSIGNMENTS OF SUPERMULTIPLETS

Identifying the SU(3) transformation properties of further hadron states is a matter of comparison with experiment. Any set of assignments can be checked for predictions as to missing states, intensity rules, masses, widths, and so on.

The situation is extremely vague once we leave the multiplets discussed in the previous chapters. Experimental evidence is rather meager, and theoretical attempts can be no more than a learned guess. Glashow and Rosenfeld study the hypothetical recurrence of the known multiplets in higher rotational levels; they also attempt to identify a $J = \frac{1}{2}^-$ singlet at 1405 MeV and a rather uncertain $J = \frac{3}{2}^-$ octet. Since their work was done, recent experimental and theoretical developments with respect to N-π-π resonances have suggested the possible existence of a $J = \frac{5}{2}^+$, **35** baryon supermultiplet, perhaps even including some further metastable states as in the case of the decimet.

Gell-Mann's report treats some possible assignments with respect to the $\kappa(725)$, a K-π resonance which has been occasionally reported in various experiments. One possibility is specially emphasized because it is interesting, although not any more likely to be right than the others. It is based on an "abnormal" charge-conjugation behavior of the third and eighth components of a scalar octet; this could explain by unitary parity conservation [that is, charge conjugation combined with SU(3)] the small width of $\kappa(725)$. Somewhat similar suggestions were made by A. Katz and H. J. Lipkin [Phys. Letters,

7, 44 (1963)], and unitary parity has also been studied by Y. Dothan [Nuovo Cimento, **30**, 399 (1963)].

Oakes treats SU(3) assignments and their implications for nuclei and hypernuclei.

All of the papers in this part are presented as examples of the method of SU(3) and its promise; the content is in each case speculative and the results will take a long time to check.

EIGHTFOLD-WAY ASSIGNMENTS FOR $Y_1^*(1660)$ AND OTHER BARYONS†

Sheldon L. Glashow* and Arthur H. Rosenfeld

Department of Physics and Lawrence Radiation Laboratory, University of California, Berkeley, California
(Received 17 December 1962)

In the preceding papers establishing the existence of a new $T=1$, $Y=0$ hyperon resonance at 1660 MeV, $Y_1^*(1660)$,[1] it is noted that the baryons and the many low-lying baryon-meson resonances may be included within just four unitary-symmetry multiplets[2]: the $j=\frac{1}{2}^+$ α octet, the $j=\frac{1}{2}^-$ β singlet, the $j=\frac{3}{2}^-$ γ octet, and the $j=\frac{3}{2}^+$ δ decuplet.[3] All members of these unitary multiplets are seen except Ω_δ^- (sometimes called Z^-) and Ξ_γ, whose discoveries would give crucial tests of the eightfold way. Each of the four unitary multiplets may be viewed as the "ground" (i.e., lowest angular momentum) state of a different Regge trajectory, and each such trajectory may have several manifestations with different physical values of angular momentum. These "Regge recurrences" (should they occur at all) must characterize all the members of a given unitary multiplet, and must be spaced by two units of angular momentum. Identifying the $N_{3/2}^*(1920)$ resonance with Δ_δ $(j=\frac{7}{2}^+)$, the $Y_0^*(1815)$ with Λ_α $(j=\frac{5}{2}^+)$, and the $N_{1/2}^*(1688)$ resonance with N_α $(j=\frac{5}{2}^+)$, we find that all presently known baryon states are described with only four unitary-multiplet Regge trajectories. Because the first Regge recurrences of some

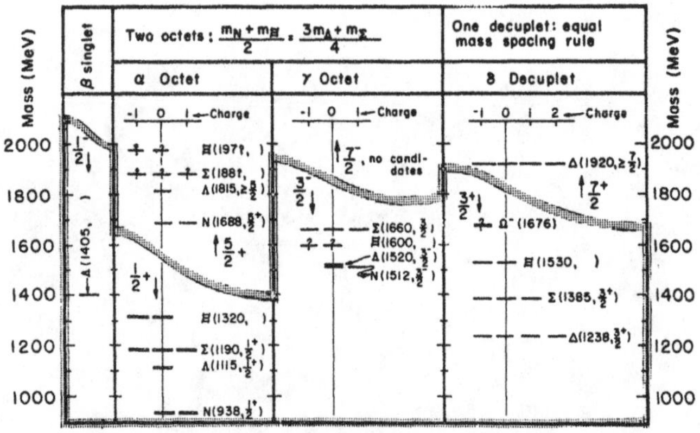

FIG. 1. Baryons: the four unitary multiplets and their Regge recurrences. Spin and parity assignments J^P are written beside each particle if they are supported by any experimental evidence; if not, J^P have been conjectured by assigning one known resonance to each set of quantum numbers. The notation was introduced in the Proceedings of the International Conference on High-Energy Nuclear Physics, Geneva, 1962 (CERN Scientific Information Service, Geneva, Switzerland, 1962), pp. 783 and 325. Observe that the families so defined coincide with the unitary multiplets of the eightfold way. Heavy bars show stable or metastable particles; light lines show resonances. States predicted by the eightfold way but not yet seen are indicated by question marks. The masses of Ξ_γ and Ω_δ^- follow from the mass formulas alone; those of the $\frac{5}{2}^+$ Σ_α and Ξ_α also require the assumption of nearly parallel Regge trajectories.

members of the α and δ unitary multiplets have appeared, one predicts the existence of Σ_α and Ξ_α states of $j = \frac{5}{2}^+$ and of the remaining members of the $j = \frac{7}{2}^+$ δ decuplets. These four unitary multiplets, and their first recurrences, are displayed in Fig. 1. The analogous meson display appears in Fig. 2.

Beyond the mere classification of particles and resonances, there must be more quantitative, dynamical verifications of the "broken eightfold way."[4] In this note, we show that the partial widths for the various two-body decay modes of the γ octet and of the δ decuplet are compatible with unitary symmetry of strong interactions. In Table I the experimental partial widths for decay into meson plus baryon are summarized. Two of these are used as input variables determining the eightfold-way D and F decay-coupling constants for the γ octet; the remaining five partial widths are calculated after adjustment of a radius of interaction.[5] With the same form factor, the calculation is repeated for the δ decuplet, for which there is a unique invariant coupling to meson plus baryon. In each case, agreement with experiment is excellent, in general, to within the accuracy with which the partial widths are known.

Some of the agreement may be due merely to our use of a reasonable form factor. On the other hand, the theoretical result for $\Gamma(\Sigma_\gamma \to \bar{K}N):\Gamma(\Sigma_\gamma \to \pi\Lambda):\Gamma(N_\gamma \to \pi N)$ of about 1:4:22 (compared with the experimental values of about 1:4:27) refers to modes of roughly the same momenta. Hence we see that the eightfold way supplies nontrivial coefficients. Moreover, the δ coupling strengths are consistently greater by an order of magnitude than the γ coupling strengths [e.g., $\Gamma(\Sigma_\delta \to \pi\Lambda) \approx 4\Gamma(\Sigma_\gamma \to \pi\Lambda)$, even though the γ mode has three times as much available energy as the δ mode].[6] These considerations preclude any purely kinematical explanation for the agreement between theory and experiment.[7]

Three-body decay modes of baryon resonances are less of a test of the eightfold way than two-body modes, because there are many possible

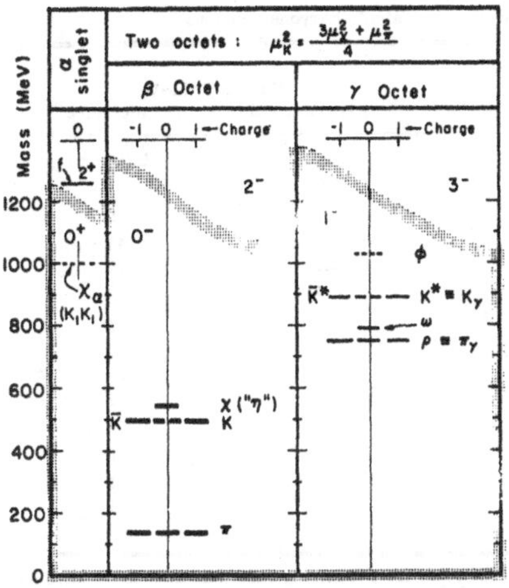

FIG. 2. Mesons: the meson unitary multiplets include a β (pseudoscalar) octet and a γ (vector) octet. There are two observed χ_γ states (i.e., vector isotopic singlets) called ω and ϕ; one linear combination of these is presumably the eighth member of the γ octet, and the orthogonal linear combination is assigned to a unitary singlet. ϕ is seen as a K_1K_2 enhancement at 1030 MeV [see L. Bertanza et al., Phys. Rev. Letters 9, 180 (1962), and J. J. Sakurai, Phys. Rev. Letters 9, 472 (1962)]. In addition there appear to be two α singlets: a $J^{PG} = 2^{++}$ pion-pion resonance at 1250 MeV called f, and a 0^{++} K_1K_1 interaction near $\bar{K}K$ threshold. For more complete references, see W. H. Barkas and A. H. Rosenfeld, Lawrence Radiation Laboratory Report UCRL-8030, February 1963 (unpublished). All the meson states have charge-conjugation properties such that they may couple to baryon-antibaryon states, e.g., π^0 and χ have $C = +1$ (and decay into two photons), while ρ^0 and ω have $C = -1$ (and couple to a single photon).

Table I. Two-body partial widths for the γ octet and δ decuplet.

Resonance and total width Γ	Decay mode	Momentum (MeV/c)	Width, Γ (MeV) Experimental[a]	Calculated[b]
		γ octet		
$\Xi(1600?)$	$\Xi\pi$	220	?	0.6[c]
$\Sigma(1600)$	$\bar{K}N$	406	3[d]	3
$\Gamma = 40$ MeV	$\Lambda\pi$	441	11	Input = 11
	$\Sigma\pi$	386	13	Input = 13
$\Lambda(1520)$	$\bar{K}N$	244	5	6
$\Gamma = 16$ MeV	$\Sigma\pi$	267	9	Input = 8
$N(1512)$	$N\pi$	450	80	67
$\Gamma = 100$ MeV				
		δ decuplet		
$\Omega^-(1676?)$		Decays weakly into $\Xi\pi$, $\Lambda\bar{K}$, or leptonically		
$\Xi(1530)$	$\Xi\pi$	148	<7	12
$\Gamma < 7$ MeV				
$\Sigma(1385)$	$\Lambda\pi$	210	50	35
$\Gamma = 50$ MeV	$\Sigma\pi$	119	$\lesssim 4$	5
$\Delta(1238)$	$N\pi$	233	100	Input = 100
$\Gamma = 100$ MeV				

[a] For references to the data, see W. H. Barkas and A. H. Rosenfeld, Lawrence Radiation Laboratory Report UCRL-8030, February 1963 (unpublished).

[b] The $D-F$ mixing ratio giving best fit between experiment and theory is $\alpha = 0.655$ (in Gell-Mann's notation) or $\theta = 35°$ [in the notation of R. Cutkosky, Carnegie Institute of Technology (unpublished)]. This value is in good agreement with dynamical considerations given in preprints by Capps, by Cutkosky, and by Martin and Wali. Moreover, Cutkosky shows that a value of θ near 33° is probably demanded for the bootstrap appearance of the γ octet and the δ decuplet. We thank Professor Cutkosky for telling us of his result after our calculation was completed.

[c] The very low predicted partial width for $\Xi_\gamma \to \Xi\pi$ suggests that an alternative decay mode may predominate (such as $\Xi + \gamma$, $\Xi\pi\pi$, or ΛK, if $\Xi\gamma$ is sufficiently heavy).

[d] There is some discrepancy in the measurement of $\Gamma(\Sigma_\gamma \to \bar{K}N)$. According to Alvarez et al., it is $\lesssim 2$ Mev; according to Bastien and Berge, it is $\gtrsim 4$ MeV.

Table II. Two-body decays of the α recurrences into stable baryons + mesons.

Resonance, total width, and threshold	Decay mode	Momentum (MeV/c)	Width, Γ (MeV) Experimental	Calculated[a]
$N_\alpha(1688)$	$N\pi$	572	60	Input = 80
$\Gamma = 100$ MeV	$N\chi$("η")	387	<20	0.5
$P_\pi = 1.00$ GeV/c	ΛK	235	< 2	1
$\Lambda_\alpha(1815)$	$N\bar{K}$	538	70	41
$\Gamma = 120$ MeV	$\Sigma\pi$	504	<40	29
$P_{K^-} = 1.05$ GeV/c	$\Lambda\chi$("η")	345	< 1.3	3
$\Sigma_\alpha(1875)$	$N\bar{K}$	586		5
	$\Lambda\pi$	595		15
$P_{K^-} = 1.20$ GeV/c	$\Sigma\pi$	548		20
	ΞK	208		1
	$\Sigma\chi$("η")	322		2
$\Xi_\alpha(1972)$	$\Xi\pi$	531		5
	$\Xi\chi$("η")	290		2
$P_{K^-} = 1.40$ GeV/c	$\Lambda\bar{K}$	540		1
	$\Sigma\bar{K}$	479		41

[a]There are again two invariant coupling constants. We use the same $D-F$ coupling constant ratio as we have determined from the decays of the γ octet, and we adjust the strength to fit the $N\pi$ decay mode of $N_\alpha(1688)$ (theoretical arguments of Cutkosky suggest that the same "self-consistent" value of θ should apply to all couplings of mesons to baryon octets). The same form factor is used as earlier (reference 6).

forms of symmetrical interaction involving two final-state mesons. However, it is instructive to consider the possible two-body decays of one baryon resonance into another baryon resonance plus a meson (giving, eventually, a three-body mode), e.g., $N_\gamma \to N_\delta\pi$, $\Sigma_\gamma \to \Sigma_\delta\pi$, $\Sigma_\gamma \to \Lambda_\beta\pi$.

Only inequalities for these partial widths have thus far been experimentally determined:

$$\Gamma(N_\gamma \to N_\delta\pi) \lesssim \Gamma(N_\gamma \to N\pi\pi) = 30 \text{ MeV}, \quad (1)$$

$$\Gamma(\Sigma_\gamma \to \Sigma_\delta\pi) \lesssim \Gamma(\Sigma_\gamma \to \Lambda\pi\pi) = 8 \text{ MeV}, \quad (2)$$

$$\Gamma(\Sigma_\gamma \to \Lambda_\beta\pi) \lesssim \Gamma(\Sigma_\gamma \to \Sigma\pi\pi) = 8 \text{ MeV}. \quad (3)$$

There is again only a single invariant coupling strength for modes (1) and (2) so that their ratio is determined in the eightfold way:

$$\Gamma(N_\gamma \to N_\delta\pi)/\Gamma(\Sigma_\gamma \to \Sigma_\delta\pi) = 6.5.$$

The existence of the mode (3), $\Sigma_\gamma \to \Lambda_\beta\pi$, is a test of the correctness of the assignments of resonances to unitary multiplets: If Λ_β is a unitary singlet, then it can be shown that unitary symmetry allows this mode if Σ_γ is a member of an octet, but forbids it if it is a member of either the 10- or 27-plet.

Finally, we consider decays of the two higher energy resonances supposed to be first Regge recurrences of the α octet. The experimental situation is summarized in Table II: $N_\alpha(1688)$ and $\Lambda_\alpha(1815)$ are known to be mainly "elastic" resonances; i.e., for $N_\alpha(1688)$ we know $\Gamma(N\pi)/\Gamma \approx 80\%$,[8] for $\Lambda_\alpha(1815)$, $\Gamma(N\bar{K})/\Gamma \approx 60\%$[9]; perhaps the reason that $\Sigma_\alpha(187?)$ has not yet been found is because its coupling to $N\bar{K}$ is small; in any case it is not strongly formed by 1.2-GeV/c K^- on protons. Also in Table II are given the calculated partial widths for all decays of the Regge recurrences of the α octet into two-body states of meson plus stable baryon, according to the eightfold way. With satisfaction and relief we find that the calculated results are completely compatible with experiment.

---

†Work done under the auspices of the U. S. Atomic Energy Commission.

*Alfred P. Sloan Foundation Fellow.

[1]L. W. Alvarez et al., second preceding Letter [Phys. Rev. Letters 10, 184 (1963)]; J. P. Berge, preceding Letter [Phys. Rev. Letters 10, 188 (1963)].

[2]Reference 1 contains the appropriate citations; for further detail, see J. J. Sakurai, in Proceedings of the International Summer School at Varenna, 1962 (to be

published); S. L. Glashow, in Proceedings of the International Summer School in Theoretical Physics, Istanbul, 1962 (Gordon and Breach, London, to be published).

[3]Words signifying sets of similar particles find their origins in musical terminology. Thus, a trio, quartet, ···, octet, nonet, decimet,··· is a composition for 3, 4, ···, 8, 9, 10, ··· voices or instruments; but a triplet, quadruplet, ···, octuplet, nonuplet, decuplet, ··· refers to 3, 4, ···, 8, 9, 10, ··· notes played in one beat. After triplet (of pions) and quadruplet (of Δ isobars), we use "decuplet" for the 10. Because of an unfortunate earlier misuse, "octet" has become commonplace for the 8, rather than the more appropriate "octuplet." [See P. Scholes, Oxford Companion to Music (Oxford University Press, London, 1955), 9th ed.]

[4]The success of Gell-Mann's mass formula and its generalizations is certainly one such verification. More important is to understand why an octet resonates at $J=\frac{3}{2}^-$ and a decuplet resonates at $J=\frac{3}{2}^+$, and why there seem to be no resonances in the 27-plet and other decuplet channels. Some progress in this direction has been reported by R. Cutkosky, J. Kalcar, and P. Tarjanne, Phys. Letters 1, 93 (1962), and by R. H. Capps (to be published).

[5]The momentum dependence used included a form factor and p/M, i.e.,

$$\Gamma \propto \left|\frac{p^2}{p^2+X^2}\right|^l \frac{p}{M},$$

where p = momentum of decay products of a resonance of mass M, and X is related to the size of the interaction. The two coupling constants D and F were adjusted along with X^2 to fit the three input data shown in Table I. We found $X = 350$ MeV.

[6]In this connection, there also seems to be evidence that the δ resonances are produced more copiously than the γ resonances.

[7]Also relevant are the two-body decay modes of the vector particles. Using coupling constants of the eightfold way, and with the same form factor as above, we find $\Gamma(\rho\to 2\pi)/\Gamma(K^*\to K\pi) = 2.4$, in rough agreement with experiment.

[8]From summary prepared by R. Omnès and G. Valladas, at the Aix-en-Provence Conference on Elementary Particles, 1961 (C. E. N. Saclay, France, 1961), p. 472.

[9]W. F. Beall et al., in Proceedings of the International Conference on High-Energy Nuclear Physics, Geneva, 1962 (CERN Scientific Information Service, Geneva, Switzerland, 1962), p. 368.

The Meson $\kappa(725)$ and the Eightfold Way*

MURRAY GELL-MANN

California Institute of Technology, Pasadena, California

(October 1963)

*Work supported in part by the U. S. Atomic Energy Commission.

(California Institute of Technology Internal Report)

Recent experiments[1,2] seem to have established the existence of a meson with a mass of about 725 MeV, a width $\Gamma \lesssim 15$ MeV, and quantum numbers $Y = +1$, $I = 1/2$, along with the corresponding antiparticle having $Y = -1$, $I = 1/2$.

In accordance with a new notation,[3] we shall denote a meson with $Y = +1$, $I = 1/2$, spin and parity J^P, and mass M in MeV by $\kappa(J^P, M)$ or $\kappa(J^P)$ or $\kappa(M)$. Thus the pseudoscalar K meson has the alternative designations $\kappa(0^-, 495)$ or $\kappa(0^-)$ or $\kappa(495)$. In a similar way, a meson with $Y = 0$, $I = 1$, spin and parity and G-value J^{PG}, and mass M is called $\pi(J^{PG}, M)$ or $\pi(J^{PG})$ or $\pi(M)$; and a meson with $Y = 0$, $I = 0$, spin and parity and G-value J^{PG}, and mass M is called $\eta(J^{PG}, M)$ or $\eta(J^{PG})$ or $\eta(M)$. The plain symbols κ, π, and η are used as optional names for the ground states $\kappa(0^-, 495)$, $\pi(0^{--}, 140)$, and $\eta(0^{-+}, 550)$, respectively.

If we assume the general correctness of the broken eightfold way[4,5] scheme of approximate higher symmetry, it is interesting to speculate about the relation to the scheme of the new meson $\kappa(725)$.

Nambu and Sakurai[6] have discussed the possibility that $\kappa(725)$ does not belong to a supermultiplet in the scheme; instead, its existence is supposed to be a manifestation of the mechanism that breaks the degeneracy of supermultiplets. This idea requires that $\kappa(725)$ have $J^P = 0^+$, but the same spin and parity assignments can, of course, obtain also under other conditions.

A different suggestion[7] is that the isotopic doublet $\kappa(725)$ forms part of a unitary triplet in the eightfold way, along with a neutral isotopic singlet. These three particles would transform like the representation $\underset{\sim}{3}$ of SU(3) in the limit of unitary symmetry and their three antiparticles would transform like $\underset{\sim}{\overline{3}}$. The violation of unitary symmetry by part of the strong interactions would then transform like $\underset{\sim}{3}$ and $\underset{\sim}{\overline{3}}$ in first order, leading to decays like $\kappa(725) \rightarrow \kappa + \pi$ and production processes like $\kappa + N \rightarrow \kappa(725) + N$. In second order, the violation would lead to the usual mass sum rules for octets and decuplets.

A simpler possibility is that $\kappa(725)$ forms part of a unitary octet with the usual pattern κ, $\bar{\kappa}$, η, and π. Minami[8] and others have considered, for example, the case of an ordinary vector octet $\kappa(1^-)$, $\bar{\kappa}(1^-)$, $\eta(1^{--})$, $\pi(1^{-+})$ similar to the octet $\kappa(1^-, 885)$, $\bar{\kappa}(1^-, 885)$, $\eta(1^{--}, 1020)$, $\pi(1^{-+}, 750)$. The new $\pi(1^{-+})$ should decay conspicuously into 2π and has not so far been observed.

In the same spirit, the case of a conventional scalar octet $\kappa(0^+)$, $\bar{\kappa}(0^+)$, $\eta(0^{++})$, $\pi(0^{+-})$ has been widely discussed. In such a model, it is somewhat mysterious that $\kappa(725)$ has such a small width for decay into $\kappa + \pi$. The meson $\eta(0^{++})$ should decay readily into 2π; however, unitary symmetry, together with the small width of $\kappa(0^+)$, would give η also a mysteriously low decay rate. The meson $\pi(0^{+-})$ has an allowed decay into $\pi+\eta$ or 5π if it is heavy enough; otherwise, electromagnetically induced decays should be the most important. To order e in the decay amplitude, we have $\pi^0(0^{+-}) \rightarrow \pi^+ + \pi^- + \gamma$, $\pi^{\pm}(0^{+-}) \rightarrow \pi^{\pm} + \pi^0 + \gamma$ and to order e^2 we have $\pi(0^{+-}) \rightarrow 2\pi$ and $\pi^0(0^{+-}) \rightarrow 2\gamma$.

In this Letter, we call attention to another possibility, that of an abnormal scalar octet of the form $\kappa(0^+)$, $\bar{\kappa}(0^+)$, $\eta(0^{+-})$, and $\pi(0^{++})$, with opposite transformation properties under charge conjugation to those of the normal scalar octet. The mesons $\eta(0^{+-})$ and $\pi(0^{++})$ are incapable of virtual transitions to

$N + \bar{N}$ configurations; that is why these assignments are infrequently discussed. Similarly, virtual or real transitions $\eta(0^{+-}) \to \kappa + \bar{\kappa}$, 2π, 2η and $\pi(0^{++}) \to \kappa + \bar{\kappa}$, 2π, $\pi + \eta$ are forbidden by G or I conservation. In the limit of unitary symmetry, then, the transition $\kappa(0^{+}) \to \kappa + \pi$ is likewise forbidden if $\kappa(0^{+})$ is part of such an abnormal scalar octet.[9] In the same way, the virtual transitions $\kappa(0^{+}) \to N + \bar{\Lambda}$, $N + \bar{\Sigma}$ are forbidden in the limit of unitary symmetry.

Violation of unitary symmetry by part of the strong interactions induces the decay $\kappa(0^{+}) \to \kappa + \pi$ and virtual transitions $\kappa(0^{+}) \to N + \bar{\Lambda}$, $N + \bar{\Sigma}$, etc. Violation of I and G conservation by electromagnetism permits the virtual or real transitions $\pi^{\pm}(0^{++}) \to 2\pi$, etc. The forbidden transitions mentioned for $\pi^{0}(0^{++})$ and $\eta(0^{+-})$ involve violation of C conservation and are not permitted even by electromagnetism.

The small width of $\kappa(725)$ is then explained; its low production rates in many processes can be attributed, at least partially, to the absence or smallness of many pole terms that are fully allowed in the case of $\kappa(1^{-}, 885)$.

The meson $\eta(0^{+-})$ behaves very differently from the corresponding member $\eta(0^{++})$ of a normal scalar octet. The decays of $\eta(0^{+-})$ into 2π, 3π, $\pi + \gamma$, 2γ, and $\pi + 2\gamma$ are all strictly forbidden. The meson has an allowed decay into 5π if it is heavy enough; the symmetry pattern of the spatial and isospin wave functions is that of the Young diagram 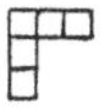, giving a charge ratio 2 : 1 for $\pi^{+}\pi^{-}\pi^{+}\pi^{-}\pi^{0} : \pi^{+}\pi^{-}\pi^{0}\pi^{0}\pi^{0}$ and appreciable centrifugal barriers on account of antisymmetrizing with respect to three mesons out of the five. To order e in the amplitude, there is the decay into $2\pi + \gamma$, with the pions having $I = 0$ and a relative orbital angular momentum L of 2 or higher. To order e^2, the decay into 4π comes into play, where the pions have $I = 1$, $C = -1$, with symmetry patterns 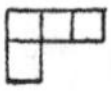 and 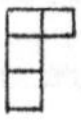 ; the former pattern gives fewer centrifugal barriers and charge ratio 2 : 1 for $\pi^+ \pi^- \pi^+ \pi^- : \pi^+ \pi^- \pi^0 \pi^0$.

The particle $\pi(0^{++})$ is allowed to decay into 4π if it is heavy enough; the symmetry pattern of the final state is the same as in the forbidden decay $\eta(0^{+-}) \to 4\pi$. To order e, we have $\pi(0^{++}) \to 2\pi + \gamma$; in the neutral case, the pions have $I = 0$ and $L = 2$ or higher, while in the charged case they have $I = 1$ and $L = 1$ or higher. To order e^2, there is the decay of the charged $\pi(0^{++})$ into 2π, although such a decay is impossible for the neutral $\pi(0^{++})$.

Glashow[10] has pointed out that the pseudoscalar and vector octets exhibit about equal values of $m_\kappa^2 - m_\pi^2 \sim 0.22\ (\text{BeV})^2$ and that if the same applies to $\kappa(725)$, the corresponding π mass is around 550 MeV. The mass rule for octets[4] then would predict a mass of around 760 MeV for the corresponding η. Under these conditions, we might expect the most important decays to be $\eta(0^{+-}) \to 2\pi + \gamma$, $\pi(0^{++}) \to 2\pi + \gamma$, and $\pi^{\pm}(0^{++}) \to 2\pi$, all electromagnetically induced.

REFERENCES

1. G. Alexander, G. R. Kalbfleisch, D. H. Miller, and D. A. Smith, Phys. Rev. Letters 8, 447 (1962).

2. S. Wojcicki, G. R. Kalbfleisch, and M. H. Alston, Bull. Am. Phys. Soc. 8, 341 (1962).

3. G. F. Chew, M. Gell-Mann, and A. H. Rosenfeld, to be published.

4. M. Gell-Mann, California Institute of Technology Synchrotron Laboratory Report CTSL-20 (1961); Phys. Rev. 125, 1067 (1962).

5. Y. Ne'eman, Nuclear Phys. 26, 222 (1961).

6. Y. Nambu and J. J. Sakurai, Phys. Rev. Letters 11, 42 (1963).

7. This suggestion forms part of the scheme of P. Tarjanne and V. L. Teplitz, to be published.

8. S. Minami, to be published.

9. It should be noted that an abnormal vector octet is also a possibility.

10. S. Glashow, private communication.

Baryon-Baryon Interactions and the Eightfold Way*

R. J. OAKES†
Institute of Theoretical Physics, Department of Physics, Stanford University, Stanford, California
(Received 22 April 1963)

The deuteron is shown to belong to a ten-dimensional irreducible representation of the group $SU(3)$ in the symmetry scheme called the eightfold way. The baryon-baryon states, which together with the deuteron comprise this supermultiplet, are studied. Data are discussed which favor the existence of some of these states, and the masses of the remaining states are estimated.

I. INTRODUCTION

RECENTLY, Gell-Mann[1] and Ne'eman[2] have independently proposed a scheme, called the eightfold way, in which the eight baryons (N, Λ, Σ, and Ξ) are assumed to form a supermultiplet, degenerate in the limit of a certain symmetry, and split into the familiar isotopic multiplets by a dynamical interaction. The symmetry is called unitary symmetry and the relevant group is the group of unitary unimodular transformations in three dimensions, which is denoted by $SU(3)$. The eight baryons belong to the regular (eight-dimensional) representation of $SU(3)$. All the known pseudoscalar mesons, vector mesons, and meson-baryon resonances have been tentatively identified with various representations of $SU(3)$.[3]

In this note, we examine the role of the deuteron in the eightfold way and find it must belong to a ten-dimensional representation of $SU(3)$. In the limit of exact unitary symmetry, the implication is that all ten of the baryon-baryon states which comprise the supermultiplet are similarly bound. Since it is clear that unitary symmetry is not exact in the physical world, some of these might not occur as actual bound states. However, unless the symmetry is broken so badly that it is meaningless, one might expect these states to be nearly bound, or resonant.[4] As will be discussed below, there is some evidence to support this point of view.

Since the symmetry of the eightfold way seems to be broken in a particularly simple and definite way, the mass splittings in the supermultiplets are related. Using this fact and certain existing data, the masses of the heretofore unobserved constituents of this deuteron decuplet are estimated. It is suggested that these be looked for as a test of the applicability of unitary symmetry and the eightfold way to the study of the hyperon and cascade interactions.

II. BARYON-BARYON STATES

In the eightfold way, the eight baryons comprise an eight-dimensional representation of the special unitary group $SU(3)$, which corresponds to a tensor having one upper and one lower index. These indexes range from 1 to 3, and their contraction vanishes. This baryon supermultiplet can be represented by a 3×3 traceless matrix as follows:

$$\psi_\alpha{}^\beta=\begin{bmatrix} -(\frac{2}{3})^{1/2}\Lambda & p & n \\ \Xi^- & (1/\sqrt{6})\Lambda+(1/\sqrt{2})\Sigma^0 & \Sigma^- \\ \Xi^0 & \Sigma^+ & (1/\sqrt{6})\Lambda-(1/\sqrt{2})\Sigma^0 \end{bmatrix}. \tag{1}$$

The possible baryon-baryon states are given by the Kronecker product of this representation with itself, i.e., by the product $\psi_\alpha{}^\beta\otimes\psi_\mu{}^\nu$. If we denote this product by $\mathbf{8}\otimes\mathbf{8}$, we find it can be decomposed into irreducible representations of $SU(3)$, as follows:

$$\mathbf{8}\otimes\mathbf{8}=\mathbf{1}\oplus\mathbf{8}\oplus\mathbf{8}\oplus\mathbf{10}\oplus\overline{\mathbf{10}}\oplus\mathbf{27}. \tag{2}$$

Here we have used the dimension to denote the representation and the $\overline{\mathbf{10}}$ is the adjoint of the $\mathbf{10}$. The $\mathbf{8}$ appears twice and it is conventional to choose linear combinations such that one is symmetric and the other is antisymmetric. The hypercharges (Y) and isotopic spins (I) of the states comprising the irreducible representations appearing in the decomposition of the product $\mathbf{8}\otimes\mathbf{8}$ are given in Table I.

Inspection of Table I shows the deuteron, which has $Y=2$ and $I=0$, must belong to the $\overline{\mathbf{10}}$, since no other representation contained in the product $\mathbf{8}\otimes\mathbf{8}$ admits such a state. The baryon-baryon states comprising the $\overline{\mathbf{10}}$ can be expanded in terms of the single baryon states $\psi_\alpha{}^\beta$ using the Wigner coefficients for $SU(3)$. We effect this expansion by noting that the $\overline{\mathbf{10}}$ corresponds to a

* This work was supported in part by the National Science Foundation and by the U. S. Air Force Office of Scientific Research under Contract AF 49(638)-388.
† National Science Foundation Postdoctoral Fellow.

[1] M. Gell-Mann, Phys. Rev. **125**, 1067 (1962) and California Institute of Technology Synchrotron Laboratory Report, CTSL-20 (unpublished).

[2] Y. Ne'eman, Nucl. Phys. **26**, 222 (1961).

[3] For a summary see S. L. Glashow and A. H. Rosenfeld, Phys. Rev. Letters **10**, 192 (1963).

[4] We use "resonance" here in its loosest sense to denote a relative enhancement of an interaction at a reasonably well-defined energy.

completely symmetric tensor having no upper indexes and three lower indexes.

One can construct only one such tensor that is bilinear in $\psi_\alpha{}^\beta$. We denote this tensor by $\chi_{\alpha\beta\gamma}$ and find it to be

$$\chi_{\alpha\beta\gamma}=(\psi_\alpha{}^\mu\psi_\beta{}^\nu+\psi_\beta{}^\mu\psi_\alpha{}^\nu)\epsilon_{\mu\nu\gamma}+(\psi_\alpha{}^\mu\psi_\gamma{}^\nu+\psi_\gamma{}^\mu\psi_\alpha{}^\nu)\epsilon_{\mu\nu\beta} +(\psi_\beta{}^\mu\psi_\gamma{}^\nu+\psi_\gamma{}^\mu\psi_\beta{}^\nu)\epsilon_{\mu\nu\alpha}, \quad (3)$$

where $\epsilon_{\alpha\beta\gamma}$ is the completely antisymmetric tensor density with $\epsilon_{123}=+1$, and repeated indexes are summed from 1 to 3. If one puts in the baryon states $\psi_\alpha{}^\beta$, performs the summations, and normalizes the resulting wave functions, one obtains

$$\chi_{111}=|Y=2,I=0,I_3=0\rangle=\frac{1}{\sqrt{2}}(pn-np),$$

$$\chi_{113}=|Y=1,I=\tfrac{1}{2},I_3=\tfrac{1}{2}\rangle =\frac{1}{(12)^{1/2}}[\sqrt{3}(p\Lambda-\Lambda p)+\Sigma^0p-p\Sigma^0+\sqrt{2}(\Sigma^+n-n\Sigma^+)],$$

$$\chi_{112}=|Y=1,I=\tfrac{1}{2},I_3=-\tfrac{1}{2}\rangle =\frac{1}{(12)^{1/2}}[\sqrt{3}(\Lambda n-n\Lambda)+\Sigma^0n-n\Sigma^0+\sqrt{2}(p\Sigma^--\Sigma^-p)],$$

$$\chi_{133}=|Y=0,I=1,I_3=1\rangle =\frac{1}{(12)^{1/2}}[\sqrt{3}(\Sigma^+\Lambda-\Lambda\Sigma^+)+\Sigma^0\Sigma^+-\Sigma^+\Sigma^0 +\sqrt{2}(\Xi^0p-p\Xi^0)],$$

$$\chi_{123}=|Y=0,I=1,I_3=0\rangle =\frac{1}{(12)^{1/2}}[\sqrt{3}(\Sigma^0\Lambda-\Lambda\Sigma^0)+\Sigma^+\Sigma^--\Sigma^-\Sigma^+ +n\Xi^0-\Xi^0n+\Xi^-p-p\Xi^-],$$

$$\chi_{122}=|Y=0,I=1,I_3=-1\rangle =\frac{1}{(12)^{1/2}}[\sqrt{3}(\Lambda\Sigma^--\Sigma^-\Lambda)+\Sigma^0\Sigma^--\Sigma^-\Sigma^0 +\sqrt{2}(n\Xi^--\Xi^-n)],$$

$$\chi_{333}=|Y=-1,I=\tfrac{3}{2},I_3=\tfrac{3}{2}\rangle=\frac{1}{\sqrt{2}}(\Xi^0\Sigma^+-\Sigma^+\Xi^0),$$

$$\chi_{233}=|Y=-1,I=\tfrac{3}{2},I_3=\tfrac{1}{2}\rangle =\frac{1}{\sqrt{6}}[\sqrt{2}(\Xi^0\Sigma^0-\Sigma^0\Xi^0)+\Xi^-\Sigma^+-\Sigma^+\Xi^-],$$

$$\chi_{223}=|Y=-1,I=\tfrac{3}{2},I_3=-\tfrac{1}{2}\rangle =\frac{1}{\sqrt{6}}[\sqrt{2}(\Xi^-\Sigma^0-\Sigma^0\Xi^-)+\Sigma^-\Xi^0-\Xi^0\Sigma^-],$$

$$\chi_{222}=|Y=-1,I=\tfrac{3}{2},I_3=-\tfrac{3}{2}\rangle=\frac{1}{\sqrt{2}}(\Sigma^-\Xi^--\Xi^-\Sigma^-). \quad (4)$$

TABLE I. Hypercharges and isotopic spins of the constituents of the irreducible representations contained in the Kronecker product $8\otimes8$.

Representation	(Y,I)
1	(0,0)
8	$(1,\frac{1}{2})$ (0,0) (0,1) $(-1,\frac{1}{2})$
10	$(1,\frac{3}{2})$ (0,1) $(-1,\frac{1}{2})$ (−2,0)
$\overline{10}$	(2,0) $(1,\frac{1}{2})$ (0,1) $(-1,\frac{3}{2})$
27	(2,1) $(1,\frac{1}{2})$ $(1,\frac{3}{2})$ (0,0) (0,1) (0,2) $(-1,\frac{3}{2})$ $(-1,\frac{1}{2})$ (−2,1)

The first of these (χ_{111}) clearly corresponds to the channel in which the deuteron occurs as a bound state. In the limit of exact unitary symmetry, in which the eight baryons are degenerate, the forces that bind the deuteron are the same in each of the ten baryon-baryon states $\chi_{\alpha\beta\gamma}$. In this limit they are all bound states with the same spin and parity (1^+) and with the same binding energy (2.2 MeV) as the deuteron. Clearly, unitary symmetry is not exact in the physical world. However, if the symmetry is not broken so badly that the classification is no longer useful, one might expect nearly bound, or resonant, states to occur in these baryon-baryon channels. The experimental data bearing on this point of view will be discussed next.

III. COMPARISON WITH EXPERIMENT

The experimental information concerning the interactions in the states $\chi_{\alpha\beta\gamma}$ decreases rapidly with the strangenesses of the baryons involved, being virtually nonexistent for the $Y=0$ triplet and the $Y=-1$ quadruplet. The $Y=1$ doublet does not seem to be bound.[5] However, there is some evidence that the hyperon-nucleon interaction is quite strong near the ΣN threshold in the state with $I=\frac{1}{2}$ and spin and parity 1^+. This can be interpreted as a manifestation of a resonant hyperon-nucleon interaction in the states χ_{112} and χ_{113} around 2130 MeV.

The evidence comes from K^- mesons stopping in deuterium[6] and Σ^- capture on protons.[7] In the former experiments one finds that the kinetic-energy spectrum of the final pion in the reaction $K^-+d\to\Lambda+p+\pi^-$ shows two peaks, one at 147 MeV and one at 92 MeV. The higher peak has been interpreted as being caused by the K^- meson interacting directly with the neutron in the deuteron to produce the Λ and π^-, while the proton is a spectator. The lower peak at 97 MeV corresponds to the energy expected if the K^- meson interacted directly with one nucleon to produce a Σ and π^-,

[5] R. H. Dalitz, in *Proceedings of the Rutherford Jubilee International Conference* (Heywood and Company Ltd., London, 1961), p. 103.

[6] L. Alvarez, in *Proceedings of the Ninth International Annual Conference on High Energy Physics* (Academy of Sciences, U.S.S.R., Moscow, 1960), Vol. I, p. 471; O. I. Dahl, N. Horowitz, D. H. Miller, J. J. Murray, and P. G. White, Phys. Rev. Letters **6**, 142 (1961); and D. H. Miller (private communication).

[7] R. R. Ross, Bull. Am. Phys. Soc. **3**, 335 (1958).

while the other nucleon was merely a spectator. Since the observed final state contains a Λ rather than a Σ, a two-step process is suggested in which the K^- is first absorbed by one of the nucleons to produce a Σ and π^-, after which the rather low-energy Σ interacts with the other nucleon and converts into a Λ. Analyses[8,9] based on this model and phenomenological interactions are in agreement with the data in this region of the pion-energy spectrum. These calculations suggest the conversion reaction $\Sigma+N\rightarrow\Lambda+N$ occurs primarily in the S state and is quite strong. This result agrees qualitatively with the prediction of the eightfold way that the hyperon-nucleon interaction in the states χ_{112} and χ_{113} should be strong in analogy to the nucleon-nucleon forces which bind the deuteron.

In the Σ^-p capture data, one also finds a considerable amount of $\Sigma^0\Lambda$ conversion in the $^3S_1$ state. In capture at rest one observes the reactions $\Sigma^-+p\rightarrow\Sigma^0+n$ and $\Sigma^-+p\rightarrow\Lambda+n$ with a branching ratio[7] $R=\Lambda/(\Lambda+\Sigma^0)=0.67\pm0.05$. To estimate this branching ratio it is reasonable to assume the conversion takes place from the S state, since the capture occurs at rest. However, there is still a contribution from the $^1S_0$ state which is not related by the eightfold way to the hyperon-nucleon interaction in the $\overline{\mathbf{10}}$. Consequently, let us neglect this contribution and estimate the branching ratio, R, assuming the $\Sigma\Lambda$ conversion occurs predominantly from $^3S_1$ state represented by χ_{112}. Including phase-space factors, one then finds $R=0.70$, which compares favorably with the experimental value. Although this crude estimate cannot be taken too seriously, it is a qualitative indication that the hyperon-nucleon interaction in the $^3S_1$ state χ_{112} of the baryon-baryon $\overline{\mathbf{10}}$ is relatively strong in analogy to the interaction responsible for the deuteron's binding energy, as required by the eightfold way. This conclusion is extended to the state χ_{113} simply by isotopic spin invariance.

Unfortunately, there is insufficient data regarding the hyperon-hyperon, nucleon-cascade, and hyperon-cascade interactions to discuss the $Y=0$ triplet and the $Y=-1$ quadruplet occurring in baryon-baryon $\overline{\mathbf{10}}$. In the next section we will attempt to estimate where resonant interactions in these states are most likely to occur.

[8] R. Karplus and L. Rodberg, Phys. Rev. **115**, 1058 (1959).
[9] T. Kotani and M. Ross, Nuovo Cimento **14**, 1282 (1959).

IV. PREDICTIONS

Since the symmetry of the eightfold way seems to be broken in a particularly simple manner, there exists a relation among the masses within a given supermultiplet. In particular, the squares of the masses[10] of the baryon-baryon $\overline{\mathbf{10}}$ are equally spaced, according to the Gell-Mann–Okubo[11] mass formula. Therefore, one has the relation

$$M^2=M_0^2(1+\alpha Y)\,, \qquad (5)$$

where M_0 and α are parameters that one can evaluate once two of the masses are known. Using the deuteron mass of 1876 MeV for the $Y=2$ state and 2130 MeV for the $Y=1$ state, as suggested by the K^-d and Σ^-p capture data, one finds these parameters to be $M_0=2357$ MeV and $\alpha=-0.1832$. Using these values, the masses of the $Y=0$, $I=1$ and $Y=-1$, $I=\frac{3}{2}$ baryon-baryon resonances are predicted to be 2357 MeV and 2564 MeV, respectively. Comparing these masses with the masses of their constituents, one finds they are far from being bound states. Nevertheless, they can manifest themselves as important final-state interactions in certain production reactions. For example, the production reactions $p+p\rightarrow\Sigma^++\Lambda+K+K^0$, $p+p\rightarrow p+\Xi^-+K^++K^+$, and $\Sigma^-+p\rightarrow\Lambda+\Sigma^-+K^+$ should show an enhancement of events with the final two hyperons in the 1^+ state when their center-of-mass energy is near 2357 MeV.

Confirmation of the existence of the $Y=0$ triplet and the $Y=1$ quadruplet would support the use of the eightfold way as a theoretical means of relating the virtually unknown hyperon and cascade interactions to the relatively well-known nucleon interactions.

ACKNOWLEDGMENTS

It is a pleasure to thank Professor S. L. Glashow, Professor C. K. Iddings, and Professor D. H. Miller for interesting discussions.

[10] It is not clear that the masses themselves rather than their squares should not be equally spaced, even though the states are formally bosons. However, the numerical results are not much different if the masses rather than their squares are used in Eq. (6). In this case one finds $M_0=2384$ MeV and $\alpha=-0.1065$. Then the $Y=0$ triplet is at 2384 MeV and the $Y=-1$ quadruplet is at 2638 MeV.
[11] Ref. 1 and S. Okubo, Progr. Theoret. Phys. (Kyoto) **27**, 949 (1962).

IV

ϕ-ω MIXING

In the following papers, attempts are made to deal quantitatively with the suggestion discussed above, that the mass formula fails for the vector mesons because an accidental degeneracy of a singlet and the eighth member of an octet leads to a large mixing and repulsion of levels.

Sakurai's work has an additional feature: he emphasizes his belief that all nine of the vector mesons are somehow elementary. He then goes on to suppose that in ϕ-ω mixing, he is not only discussing an important manifestation of the violation of the eightfold way, but also that he is perhaps identifying the basic mechanism of the violation.

In a later part we discuss other views of the violation mechanism, including Ne'eman's proposal of a "fifth interaction," which is different from Sakurai's idea but has some features in common with it.

POSSIBLE EXISTENCE OF A $T = 0$ VECTOR MESON AT 1020 MeV†

J. J. Sakurai*
The Enrico Fermi Institute and the Department of Physics, The University of Chicago, Chicago, Illinois
(Received September 7, 1962)

The existence of two $T = 0$, negative-G vector mesons—one coupled to the baryon current (B_B meson), the other coupled to the hypercharge current (B_Y meson)—was first discussed within the framework of the vector theory of strong interactions[1] (VTSI). Subsequently Gell-Mann[2] and Ne'eman[3] have constructed a symmetry model based on the group SU(3) (unitary unimodular group in three dimensions) which can accommodate (but does not necessarily require) the two $T = 0$ vector mesons of VTSI—the B_Y meson as a member of a unitary octet and the B_B meson as a unitary singlet. Although we have failed in our earlier attempt[4] to identify the 550-MeV meson of Pevsner et al.[5] as the conjectured second $T = 0, J = 1^{--}$ ($\text{spin}^{\text{parity},\ G\text{-parity}}$) meson, there now appears new (rather weak) evidence for a $K\bar{K}$ resonance[6] which, as we shall show below, might be interpreted as the decay of a $T = 0, J = 1^{--}$ meson. We wish to discuss experimental and theoretical consequences of having a $T = 0, J = 1^{--}$ meson with mass $> 2m_K$ in addition to the well established $T = 0, J = 1^{--}$ ω meson.

In a recent issue of this journal a Brookhaven-Syracuse group[6] reports the possible presence of a narrow resonance ($\Gamma \approx 20$ MeV) in the $K\bar{K}$ system with mass ≈ 1020 MeV in the reactions

$$K^- + p \to K^+ + K^- + \Lambda, \quad K^0 + \bar{K}^0 + \Lambda.$$

In discussing the quantum numbers of the resonating $K\bar{K}$ pair, which we tentatively refer to as the φ meson, let us first observe that the G-conjugation parity of a $K\bar{K}$ system is given by

$$G = (-1)^{l + T}.$$

If the G-parity of the φ meson is even, the G-allowed two-pion mode is likely to give a very large width. For instance, if the φ meson were a $T = 0, J = 2^{++}$ particle, the two-pion mode would be expected to be more frequent than the $K\bar{K}$ mode by a factor of several hundred. So it is natural to assign a negative G-parity to the φ meson, which leaves the following possibilities open:

$$T = 0;\ 1^{--}, 3^{--}, 5^{--}, \cdots,$$

$$T = 1;\ 0^{+-}, 2^{+-}, 4^{+-}, \cdots,$$

Fortunately the two sets of possibilities can readily be distinguished in bubble-chamber experiments. Let us recall[7]

$$|T = 0, T_3 = 0, G \text{ odd}\rangle$$

$$= \tfrac{1}{2}[-|K^+K^-\rangle + |K^-K^+\rangle + i|K_1^0K_2^0\rangle + i|K_2^0K_1^0\rangle],$$

$$|T = 1, T_3 = 0, G \text{ odd}\rangle$$

$$= \tfrac{1}{2}[-|K^+K^-\rangle - |K^-K^+\rangle + |K_1^0K_1^0\rangle + |K_2^0K_2^0\rangle].$$

We note that a $T = 0$, negative-G meson is forbidden to decay into $K_1^0 + K_1^0$, which means that the φ peak should not show up for "double-V^0" K-pair events. (This is evident from Bose statistics alone.) On the other hand, a $T = 1$, negative-G meson is allowed to decay into "double V^0's." We understand that the preliminary data[8] of the Brookhaven-Syracuse group[6] are, within limited statistics, consistent with the absence of the decay mode $\varphi \to K_1^0 + K_1^0$. Of course, the isospin of the φ meson can be directly established by studying

$$K^- + d \to K^- + K^0 + \Lambda + p$$

at the same incident K^--beam momentum as that of reference 6.

It appears that if future experiments conclusively confirm the existence of peaks in the K^+K^- and $K_1^0K_2^0$ mass distributions but not in the $K_1^0K_1^0$, nor in the $K^+\bar{K}^0$ nor K^-K^0 distribution, then the only reasonable quantum number assignment for the φ meson is $T = 0, J = 1^{--}$. For $T = 0$, $J \geq 3$, and $G = -1$, the $K\bar{K}$ mode would be very much suppressed in comparison to the $\rho + \pi$ ($3\pi, \cdots$) mode because of the f- (h-, $\cdots$) wave centrifugal barrier. We have assumed throughout that $\varphi \to K + \bar{K}$ is a G-allowed transition; otherwise multipion modes would certainly predominate.

With the proposed $T = 0, 1^{--}$ assignment, the φ meson, which is forbidden to decay into $K_1^0 + K_1^0$, cannot be studied easily in the reaction $\pi^- + p \to K^0 + \bar{K}^0 + n$. Note also that G-conjugation invariance forbids $2\pi \rightleftarrows \varphi$ so that the φ meson cannot be produced via the well known one-pion-exchange mechanism in the reaction $\pi + N \to K + \bar{K} + N$. For these reasons, the low-energy $K\bar{K}$ interactions recently discussed by Erwin et al.[9] have no relations whatsoever to the conjectured $T = 0, 1^{--}$ φ meson.

Work of the CERN-Paris group[10] on $p+\bar{p}\to K+\bar{K}+\omega$ is not too relevant either, because in $p+\bar{p}\to\varphi+\omega$ the $\varphi\omega$ system must be in a relative p state for s-state capture.[11]

Because the $K\bar{K}$ threshold is near the mass of the φ meson (Q value of 32 MeV for K^++K^-, 24 MeV for $K^0+\bar{K}^0$), we expect a large apparent violation of charge independence in the decay $\varphi\to K+\bar{K}$. We obtain on the basis of the $T=0,1^{--}$ hypothesis

$$N(K^0+\bar{K}^0)/N(K^++K^-)=N(K_1^0+K_2^0)/N(K^++K^-)=0.65$$

(instead of unity), at 1020 MeV. It is also worth noting that because of the rapidly varying p^3 dependence that multiplies the usual Breit-Wigner form, the shapes of the K^+K^- and $K_1^0K_2^0$ bumps are expected to be very asymmetric.

The decay rate for $\varphi\to K+\bar{K}$ can be calculated if the φ meson is the B_Y meson of VTSI coupled to the hypercharge current with strength f_φ (where f_φ is equal to f_Y of reference 1). We obtain

$$\Gamma(\varphi\to K+\bar{K};\text{ both charge modes})=\tfrac{4}{3}(f_\varphi{}^2/4\pi)p_K{}^3/m_\varphi{}^2,$$

where p_K stands for the momentum of either K meson in the rest system of the decaying φ meson. Meanwhile, if the φ meson is the $T=0$ partner of the vector meson octet (together with ρ and K^*) in the unitary symmetry model,[2,3] we must have

$$(1/\sqrt{3})f_\varphi=\tfrac{1}{2}f_\rho=\gamma,.$$

where f_ρ ($=f_T$ in reference 1) is the coupling constant of the ρ meson to the isospin current, and γ is the vector-meson coupling constant in Gell-Mann's notation.[2] Since f_ρ can be determined from the observed width of the ρ meson (taken to be 100 MeV), we obtain

$$\Gamma(\varphi\to K+\bar{K})=\tfrac{3}{2}[(p_K{}^3/m_\varphi{}^2)/(p_\pi{}^3/m_\rho{}^2)]\Gamma(\rho\to\pi+\pi)$$
$$\approx 3.4\text{ MeV}.$$

(A similar calculation leads to a K^* width of ≈30 MeV.[2]) On the other hand, if the φ meson were a unitary-singlet vector meson, its decay into $K+\bar{K}$ would be forbidden to the extent that unitary symmetry is exact. It would then be impossible to estimate the decay rate for the $K+\bar{K}$ mode on the basis of unitary symmetry.

Because the φ mass is above the $\rho\pi$ threshold, the proposed 1^{--} assignment fully allows the decay modes

$$\varphi\to\rho^{\pm,0}+\pi^{\mp,0}.$$

If the matrix element for this process is given (apart from the usual kinematical factors) by

$$(f_{\rho\pi\varphi}/m)\epsilon_{\mu\nu\lambda\sigma}\epsilon_\mu^{(\varphi)}k_\nu^{(\varphi)}\epsilon_\lambda^{(\rho)}k_\sigma^{(\rho)}$$

for each charge mode, we obtain[12]

$$\Gamma(\varphi\to\rho+\pi;\text{ three charge modes})=(f_{\rho\pi\varphi}{}^2/4\pi)p^3/m^2.$$

The constant $f_{\rho\pi\varphi}/m$ (where m has been inserted to make the coupling constant dimensionless) cannot be determined without detailed dynamical considerations. This constant, however, can be deduced if we assume that π^0 decay is completely dominated by $\pi^0\to\rho+\varphi\to 2\gamma$. (This model is not expected to represent reality since we completely ignore the ω meson, but if both ω and φ contributed, we could not arrive at a definite value for $f_{\rho\pi\varphi}/m$; hence we proceed.) Straightforward calculations based on the pole-dominance approximation outlined in Gell-Mann and Zachariasen[13] and in Gell-Mann, Sharp, and Wagner[14] lead to the results

$$f_{\rho\pi\varphi}{}^2/(4\pi m^2)\approx 0.02/m_\pi{}^2,$$

$$\Gamma(\varphi\to\rho+\pi;\text{ three charge modes})\approx 9\text{ MeV},$$

for a π^0 lifetime of 2.2×10^{-16} sec.

We emphasize again that our calculation on the $\rho\pi$ mode is useful only as an order-of-magnitude estimate. All we wish to claim is that we should not be too much surprised if, despite its small Q value, the $K\bar{K}$ mode turns out to be almost as frequent as the $\rho\pi$ mode. Note also that if there are no cancellations between the $\rho+\varphi$ contribution and the $\rho+\omega$ contribution in π^0 decay,[15] the actual decay rate for the $\rho+\pi$ mode may well be smaller than the value estimated above. In any case, an experimental attempt should be made to detect $\varphi\to\rho+\pi$ by studying the $\pi^+\pi^-\pi^0$ mass distribution in various reactions. (It may be difficult to look for the $\rho+\pi$ mode in high-energy K^-p collisions if $K^-+p\to Y_1{}^*+2\pi$ turns out to be frequent.) The Dalitz plot for $\varphi\to\pi^++\pi^-+\pi^0$ should reveal three very striking bands (similar to the kind observed in $p+\bar{p}\to\pi^++\pi^-+\pi^0$ by the Oxford-Padua group[16]) corresponding to the three charge states of the ρ meson.

Let us now return to unitary symmetry. It has been suggested that a mass formula of the form

$$m=m_0\{1+aY+b[T(T+1)-\tfrac{1}{4}Y^2]\}$$

(where Y and T stand for hypercharge and isospin) be applicable to the members of a given unitary symmetry multiplet.[2,17] This formula holds remarkably well for the baryon octet[2] $(N, \Lambda, \Sigma, \Xi)$ and for the pseudoscalar meson octet[2] (π, K, η) and possibly for the $J = \frac{3}{2}^+$ isobars[18] $(N_{3/2}^*, Y_1^*, \Xi_{1/2}^*$, and a "stable" $Y = -2, T = 0$ hyperon[19] at 1680 MeV). Whatever the deep reason for these numerologies may be, it is somewhat disturbing that the formula does not work for the vector mesons. With the observed ρ and the K^* mass of 750 MeV and 885 MeV, respectively, the $T = 0$ partner of the vector meson octet is predicted to be at 930 MeV, which is 150 MeV above the ω mass and 90 MeV below the φ mass.

We would like to suggest that this discrepancy is due to mixing between ω and φ arising from the fact that the ω and the φ have the same quantum numbers as far as spin-parity, isospin, and G-parity are concerned. Specifically, we conjecture that without ω-φ mixing, the φ meson would be the $T = 0$ member of a "pure" unitary octet with mass ≈ 930 MeV, and the ω meson, which we assume to be a unitary singlet, would lie somewhat lower. The φ mass and the ω mass are supposed to be rather close to start with, but the two states tend to repel each other as the mixing effect becomes operative.[20] Assuming that the two states repel symmetrically (in analogy with perturbation theory in nonrelativistic quantum mechanics), we can "predict" the mass of the physical ω meson to be lower than 840 MeV, but not very much lower than that (if there is to be mixing to start with). This is in agreement with observation.

Mixing between ω and φ would imply violation of R invariance[2,21] as well as violation of unitary symmetry. In fact, if the vector meson coupled to the baryon current turns itself into the kind coupled to the hypercharge current, the N-Ξ level would split even in lowest order.[22] Thus the observed N-Ξ mass difference (which corresponds to the term linear in Y in the mass formula for the baryon octet) may emerge as a result of ω-φ mixing. Conversely, if there is a small mass difference between N and Ξ, the ω and the φ are expected to mix with each other; we can forbid the transition $\omega \rightleftarrows \varphi$ only in a model in which the N-Ξ degeneracy is exact.

We may argue that if we start with a theory that is completely invariant under unitary symmetry transformations, it should stay invariant, hence no ω-φ mixing, etc. However, it may well be that unitary symmetry is dynamically unstable with respect to ω-φ mixing. Once a slight violation of unitary symmetry is triggered, ω and φ get mixed; this leads to a small N-Ξ mass difference, which in turn results in a larger amount of ω-φ mixing. This cycle goes on until the ω and the φ get separated so well that there is no further mixing. Here we may have, so to speak, a "bootstrap mechanism" for unitary symmetry violation.

All this, of course, is extremely speculative. In any case, it appears that major progress in strong interaction physics will be made when we understand the detailed dynamical mechanism responsible for the observed breakdown of unitary symmetry.

Most of the ideas contained in this paper were developed while the author was on leave at the California Institute of Technology. He wishes to thank Professor M. Gell-Mann for many enlightening discussions[23] as well as for the hospitality extended to him at California Institute of Technology. He also wishes to acknowledge the value of an interesting communication from Professor S. L. Glashow who independently noted the importance of the K^+K^0 mass difference and the possible $\rho + \pi$ decay mode of the φ meson. Thanks are also due to Dr. N. P. Samios for informative discussions on the experimental status of the $K\bar{K}$ interaction.

---

†Work supported by the U. S. Atomic Energy Commission.

*Alfred P. Sloan Foundation Fellow.

[1] J. J. Sakurai, Ann. Phys. 11, 1 (1960).

[2] M. Gell-Mann, Phys. Rev. 125, 1067 (1962), see especially Sec. VIII; see also California Institute of Technology Report, CTSL-20, 1961 (unpublished).

[3] Y. Ne'eman, Nuclear Phys. 26, 222 (1961).

[4] J. J. Sakurai, Phys. Rev. Letters 7, 355 (1961).

[5] A. Pevsner et al., Phys. Rev. Letters 7, 421 (1961). For evidence in favor of the 0^{-+} assignment, see, e.g., P. L. Bastien et al., Phys. Rev. Letters 8, 114 (1962); M. Chrétien et al., Phys. Rev. Letters 9, 127 (1962); reports by J. Steinberger and by A. H. Rosenfeld in the Proceedings of the International Conference on High-Energy Nuclear Physics, Geneva, 1962 (CERN, Geneva, Switzerland, to be published).

[6] L. Bertanza et al., Phys. Rev. Letters 9, 180 (1962).

[7] M. Goldhaber, T. D. Lee, and C. N. Yang, Phys. Rev. 112, 1796 (1958).

[8] Remark made by N. Samios in the Proceedings of the International Conference on High-Energy Nuclear Physics, Geneva, 1962 (CERN, Geneva, Switzerland, to be published). To the extent that there is no single "double-V" K-pair event in the $K\bar{K}$ bump region, the $K\bar{K}$ bump of reference 6 cannot be interpreted as a low-energy s-wave scattering length effect.

[9] A. R. Erwin, G. A. Hoyer, R. A. March, W. D.

Walker, and T. P. Wangler, Phys. Rev. Letters 9, 34 (1962). Preliminary data of the Berkeley hydrogen-bubble-chamber group also reveal the existence of a strong J = even, $K\bar{K}$ interaction in the reaction $\pi^- + p \rightarrow K^0 + \bar{K}^0 + n$ [G. Alexander and D. Miller (private communication)].

[10]R. Armenteros et al. (to be published).

[11]G. A. Snow, Phys. Letters 1, 213 (1962).

[12]We neglect possible interference effects among the three charge modes arising from the fact that the three ρ bands in the Dalitz plot for $\varphi \rightarrow \pi^+ + \pi^- + \pi^0$ overlap each other. (This is essentially a zero-width approximation for ρ.)

[13]M. Gell-Mann and F. Zachariasen, Phys. Rev. 124, 953 (1961).

[14]M. Gell-Mann, D. Sharp, and W. D. Wagner, Phys. Rev. Letters 8, 261 (1962).

[15]See, however, D. Geffen, Phys. Rev. 128, 374 (1962).

[16]G. B. Chadwick et al. (to be published).

[17]S. Okubo, Progr. Theoret. Phys. (Kyoto) 27, 949 (1962). R. P. Feynman (private communication) suggests that (mass)^2 be used in the mass formula, especially for bosons.

[18]S. L. Glashow and J. J. Sakurai, Nuovo cimento 25, 337 (1962) (see especially Appendix); Nuovo cimento (to be published). See also a remark made by M. Gell-Mann in the Proceedings of the International Conference on High-Energy Nuclear Physics, Geneva, 1962 (CERN, Geneva, Switzerland, to be published).

[19]This may correspond to Eisenberg's "new hyperon" if the decay mode $K^- + \Sigma^0$ is assumed for his peculiar event[18] [Y. Eisenberg, Phys. Rev. 96, 541 (1954)]. The possible existence of such a metastable $Y = -2$ (strangeness = -3), $T = 0$ hyperon decaying via weak interactions into $\pi + \Xi$, $\bar{K} + \Lambda$, or $\bar{K} + \Sigma$ was discussed by Gell-Mann many years ago [M. Gell-Mann, Suppl. Nuovo cimento 4, 848 (1956)].

[20]A similar mixing mechanism has been discussed by Gell-Mann[2] in connection with the mass formula for the vector meson octet.

[21]J. J. Sakurai, Phys. Rev. Letters 7, 426 (1961).

[22]A symmetry-breaking mechanism of this kind was considered already in reference 1.

[23]This does not necessarily mean that Professor Gell-Mann agrees with the content of this paper.

$\varphi\omega$ MIXING*

S. L. Glashow†
Department of Physics, University of California, Berkeley, California
(Received 10 May 1963)

We consider the possibility that the nine known strongly interacting vector mesons comprise a unitary octet and a unitary singlet. Symmetry breakdown is assumed to result from a mixing of the singlet with the $T=Y=0$ member of the octet.[1,2] We assume that the couplings of the vector mesons are otherwise invariant under the transformations of the eightfold way.[3,4] Because of the mixing, the particle eigenstates, ω and φ, are linear combinations of ω_1 (the unitary singlet) and ω_8 (the $T=Y=0$ member of the octet):

$$\omega_1=\omega\cos\theta-\varphi\sin\theta,$$
$$\omega_8=\varphi\cos\theta+\omega\sin\theta. \quad (1)$$

In this note, we relate the mixing angle θ to more experimentally accessible parameters.

The SU_3-invariant vecton-vecton-meson couplings (we use the word "vecton" to mean "vector meson") are[5]

$$g\epsilon^{\mu\nu\lambda\sigma}\omega_1{}^{\mu\nu}\,\mathrm{Tr}(V^{\lambda\sigma}P)=g\omega_1(\rho^+\pi^-+K^{*+}K^-)+\cdots,$$
$$f\epsilon^{\mu\nu\lambda\sigma}\,\mathrm{Tr}(V^{\mu\nu}V^{\lambda\sigma}P)=f\omega_8(\rho^+\pi^--\tfrac{1}{2}K^{*+}K^-)+\cdots, \quad (2)$$

where V and P are 3×3 matrices representing the vector and pseudoscalar meson octets, and space-time indices are omitted from the right-hand sides. The first interaction couples the singlet vecton and the octet vectons, the second interaction involves only the octet vectons and is D type. The octet of vectons, but not the singlet, also participates in vecton-meson-meson couplings of the F type,

$$h\,\mathrm{Tr}(V^\mu P\partial_\mu P)=h\omega_8K^+K^-+\cdots. \quad (3)$$

For the couplings of the particles, ω and φ, we obtain

$$G_{\rho\omega\pi}=g\cos\theta+f\sin\theta,$$
$$G_{\omega K^*\bar{K}}=g\cos\theta-\tfrac{1}{2}f\sin\theta,$$
$$G_{\omega\bar{K}K}=h\sin\theta, \quad (4)$$

and

$$G_{\rho\varphi\pi}=-g\sin\theta+f\cos\theta,$$
$$G_{\varphi K^*\bar{K}}=-g\sin\theta-\tfrac{1}{2}f\cos\theta,$$
$$G_{\varphi\bar{K}K}=h\cos\theta. \quad (5)$$

One method of determining the mixing angle θ is to measure the one-K-exchange (OKE) contribution to $K^-p\to\Lambda$ (ω or ϕ). The ratio of the OKE cross sections is proportional to $\cot^2\theta$ and certain kinematical factors. Unfortunately, these factors involve unknown meson-baryon and meson-meson form factors. Comparison must be made at identical momentum transfer, and even then the $\bar{K}K\phi$ and $\bar{K}K\omega$ form factors may be different. Far less sensitive to these form factors would be the comparison of one-K^* exchange with one-K exchange in both ϕ production and ω production. This determines the coupling constant ratio

$$R = (G_{\phi K^*\bar{K}}/G_{\phi\bar{K}K})(G_{\omega K^*\bar{K}}/G_{\omega\bar{K}K})^{-1},$$

or, in terms of θ and $\epsilon = G_{\rho\phi\pi}/G_{\rho\omega\pi}$, we obtain

$$R = -\left[\frac{\frac{3}{4}\sin\theta\cos\theta - \epsilon(1-\frac{3}{4}\cos^2\theta)}{1-\frac{3}{4}\sin^2\theta - \frac{3}{4}\epsilon\sin\theta\cos\theta}\right]\tan\theta.$$

Experimentally, $|\epsilon|$ seems to be small because the partial decay width for $\varphi\to\rho\pi$ is less than 1 MeV,[6,7] an order of magnitude smaller than Sakurai[2] and others have anticipated.[8] Assuming $\epsilon \sim 0$, we find that $|R|\leq 1$ requires $\sin^2\theta \leq \frac{1}{3}$. With the weaker assumption that $|\epsilon|\leq 1$, we find that $|R|\leq 1$ requires $\sin^2\theta \leq \frac{1}{2}$.

There are preliminary indications that (i) $K^-p\to\Lambda\omega$ occurs peripherally with contributions from both one-K exchange and one-K^* exchange,[9] and (ii) $K^-p\to\Lambda\phi$ is consistent with pure K exchange.[10] No stronger conclusion than $|R|\leq 1$ may be made at present, but this fact is already sufficient to require that the $\phi\omega$ mixing is such that ω is mostly singlet and ϕ is mostly octet.[11]

The hypothesis of eightfold symmetry broken principally by $\omega\phi$ mixing is compatible with the present experimental situation. In particular, the theory can accomodate any value of ϵ, so that a great disparity between ω and ϕ production in pion experiments is possible.

Also interesting from the point of view of $\omega\varphi$ mixing are the various electromagnetic decay modes of ρ, ω, and φ. These depend upon the couplings of the photon to the vector mesons: $G_{\rho\gamma}$, $G_{\omega_1\gamma}$, $G_{\omega_8\gamma}$. In the eightfold-way limit, we have $G_{\rho\gamma}=\sqrt{3}G_{\omega_8\gamma}$ and $G_{\omega_1\gamma}=0$, and for the matrix elements of the electromagnetic decay modes we obtain

$$M(\omega\to\pi^0\gamma)\sim\sqrt{3}(g\cos\theta+f\sin\theta),$$

$$M(\varphi\to\pi^0\gamma)\sim\sqrt{3}(-g\sin\theta+f\cos\theta),$$

$$M(\rho\to\pi^0\gamma)\sim f,$$

$$M(\omega\to\eta\gamma)\sim g\cos\theta - f\sin\theta,$$

$$M(\varphi\to\eta\gamma)\sim-(g\sin\theta+f\cos\theta),$$

$$M(\rho\to\eta\gamma)\sim\sqrt{3}f.$$

Comparison of these results to experiment can give an independent determination of f, g, and θ. The decay mode $\varphi\to\pi^0\gamma$ is suppressed because $G_{\rho\varphi\pi}\sim 0$, but we anticipate $\Gamma(\varphi\to\eta\gamma)\cong\Gamma(\omega\to\pi^0\gamma)$ [the available momenta are nearly equal, and $M^2(\omega\to\pi^0\gamma)\cong M^2(\varphi\to\eta\gamma)$ for $\sin^2\theta\cong\frac{1}{3}$].

We gratefully acknowledge helpful conversations with Professor G. Goldhaber, Professor S. Goldhaber, Professor L. Stevenson, and Professor H. Ticho. We thank Professor C. Levinson, Professor H. Lipkin, and Professor S. Meshkov for a copy of their work prior to publication.

---

*This research was supported in part by the U. S. Air Force Office of Scientific Research Grant USAF-OSR-62-373.

†Alfred P. Sloan Foundation Fellow.

[1]M. Gell-Mann, Phys. Rev. 125, 1067 (1962); see also S. L. Glashow, in Istanbul Summer School in Theoretical Physics, 1962 (Gordon and Breach, to be published).

[2]J. J. Sakurai, Phys. Rev. Letters 9, 472 (1962).

[3]M. Gell-Mann, California Institute of Technology Synchrotron Report No. 20, 1961 (unpublished).

[4]Y. Ne'eman, Nucl. Phys. 26, 222 (1961).

[5]Related considerations of C. Levinson, H. Lipkin, and S. Meshkov (unpublished) omit the possibility of vecton-vecton-meson couplings involving ω_1.

[6]P. L. Connolly et al., Phys. Rev. Letters 10, 371 (1963).

[7]P. Schlein, W. E. Slater, L. T. Smith, D. H. Stork, and H. K. Ticho, Phys. Rev. Letters 10, 368 (1963).

[8]Moreover, in π^+p collisions from 2.3-2.9 BeV/c, ω production is copious and forward peaked [C. Alff et al., Phys. Rev. Letters 9, 322 (1962)], while ϕ production is rare [as reported by N. Gelfand and D. Berley, American Physical Society Washington Meeting, 1963, postdeadline paper (unpublished)].

[9]L. Stevenson (private communication).

[10]H. Ticho, Proceedings of the Athens Conference on Resonant Particles, Ohio University, 1963 (to be published). It should be kept in mind that this result is based on a small number of observed φ's.

[11]This result agrees with Sakurai's estimate of $\phi\omega$ mixing obtained in his attempt to understand the vector mass spectrum with the hypothesis that all breaking of eightfold symmetry is due to $\phi\omega$ mixing [J. Sakurai (to be published)].

Experimental Consequences of ϕ–ω Mixing

ROGER F. DASHEN*
California Institute of Technology, Pasadena, California

AND

DAVID H. SHARP†
Palmer Physical Laboratory, Princeton University, Princeton, New Jersey

(Received 8 August 1963; revised manuscript received 2 December 1963)

In this paper a simple model is proposed in which the observed ϕ and ω resonant states are considered as mixtures of "pure" states $|Y\rangle$ and $|B\rangle$ corresponding to hypercharge and baryonic mesons. The implications of this model for the isoscalar nucleon form factor; the decays of the ϕ, ω, and π^0 mesons; the role of the ω and ϕ mesons in nuclear forces; the mass distribution of Dalitz pairs in the decay $\pi^0 \rightarrow \gamma+e^++e^-$; and the photoproduction of η mesons are briefly considered.

I. INTRODUCTION

WE shall consider two $T=0$, $J=1^-$, $G=-1$ vector mesons, B and Y. The B meson we shall assume is coupled universally to the conserved baryon current and in the unitary symmetry scheme[1–3] belongs to the singlet representation, while Y is coupled universally to the conserved hypercharge current and is a member of the unitary symmetry octet representation.

The states $|B\rangle$ and $|Y\rangle$ are eigenstates of a Hamiltonian H_0 which describes their interactions when unitary symmetry is not violated. We shall suppose here that unitary symmetry is *broken* by adding a "small" perturbing potential V to H_0. The complete Hamiltonian $H=H_0+V$ then has eigenstates $|\phi\rangle$ and $|\omega\rangle$ which can be written as

$$|\phi\rangle=a|Y\rangle+b|B\rangle,$$
$$|\omega\rangle=a|B\rangle-b|Y\rangle, \tag{1}$$

and

$$a^2+b^2=1, \tag{2}$$

where the phases are chosen so that a and b are real and positive. We shall identify $|\omega\rangle$ with the observed[4] $T=0$, $J=1^-$, $G=-1$ 3π resonance at 780 MeV, and $|\phi\rangle$ with the observed[5,6] $T=0$, $J=1^-$, $G=-1$ $K\bar{K}$ resonance at 1020 MeV.

In this note we shall study some experimental consequences of a simple model in which the $|\phi\rangle$ and $|\omega\rangle$ states are described as mixtures of pure $|Y\rangle$ and $|B\rangle$ states. We retain the universality hypothesis and neglect those violations of unitary symmetry which are not directly implied by Y, B (or ϕ, ω) mixing.

To determine the mixing parameters, a and b, we write the Hamiltonian H in the form

$$H=H_0+V=\begin{pmatrix} m_{0Y}+V_{YY} & V_{YB} \\ V_{BY} & m_{0B}+V_{BB}\end{pmatrix},\quad V_{YB}=V_{BY}. \tag{3}$$

Presumably the Okubo mass formula[7] fails to work well for the vector-meson octet because it takes into account only the diagonal elements of the matrix (3). However, we assume that the diagonal element $m_{0Y}+V_{YY}$ is the mass predicted for the Y meson by the Okubo relation

$$m_Y\approx m_{0Y}+V_{YY}\approx\left(\frac{4m_{K^*}{}^2-m_\rho{}^2}{3}\right)^{1/2}\approx 925 \text{ MeV}.$$

Then the fact that $|\phi\rangle$ and $|\omega\rangle$ are eigenstates of H with known eigenvalues m_ϕ and m_ω allows us to determine the mixing parameters a and b. We find:

$$a\approx 0.78 \quad\text{and}\quad b\approx 0.62. \tag{4}$$

These values are in good agreement with those recently found by several other workers each using somewhat different methods and each concerned with somewhat different consequences of ϕ-ω mixing from those considered here.[8–11]

In the introduction and throughout this paper we use freely the vector-meson description of the ϕ and ω resonances. This is not to imply that we necessarily take very seriously the usual field theoretic formulations which underlie such a description. In particular, all of the important results obtained or used here can also be obtained as approximations in a purely dispersion theoretic calculation, a point emphasized by Gell-Mann and Zachariasen.[3,12] We regard the vector-meson

* National Science Foundation Predoctoral Fellow.
† National Science Foundation Postdoctoral Fellow, 1963–64.
1 M. Gell-Mann, California Institute of Technology Synchrotron Laboratory Report CTSL-20, 1961 (unpublished).
2 Y. Ne'eman, Nucl. Phys. 26, 222 (1961).
3 M. Gell-Mann, Phys. Rev. **125**, 1067 (1962).
4 B. C. Maglić, L. W. Alvarez, A. H. Rosenfeld, and M. L. Stevenson, Phys. Rev. Letters **7**, 178 (1961).
5 P. Schlein, W. E. Slater, L. T. Smith, D. H. Stork, and H. K. Ticho, Phys. Rev. Letters **10**, 368 (1963).
6 P. L. Connolly, E. L. Hart, K. W. Lai, G. London, C. G. Moneti *et al.*, Phys. Rev. Letters **10**, 371 (1963).
7 S. Okubo, Progr. Theoret. Phys. (Kyoto) 27, 949 (1962).
8 S. L. Glashow, Phys. Rev. Letters **11**, 48 (1963).
9 J. J. Sakurai, University of Chicago preprint EF INS 63-28 (unpublished).
10 S. Okubo, Phys. Letters **5**, 165 (1963).
11 V. Singh (private communication).
12 M. Gell-Mann and F. Zachariasen, Phys. Rev. **124**, 965 (1961).

language as a convenient and intuitive way of presenting an approximate dispersion theoretic calculation.

II. INFLUENCE OF THE ϕ AND ω MESONS ON THE ISOSCALAR CHARGE FORM FACTOR OF THE NUCLEON. APPLICATIONS.

First, we must determine the couplings of the ϕ and ω mesons to the photon and to nucleons.

Following Sakurai,[13] we introduce the coupling strength f_Y of the Y meson to the hypercharge current j_Y and the coupling strength f_B of the B meson to the baryon current j_B, both defined in analogy with the electric charge e. The photon is coupled to the hypercharge current with strength $e/2$ and the Y meson is coupled to this current with strength f_Y. Then, according to Gell-Mann and Zachariasen,[12] the coupling of a Y meson with mass m_Y to a photon is given by $\gamma_{Y\gamma}=-em_Y{}^2/2f_Y$. On the other hand, the B is coupled to the independent baryon current and can therefore not couple directly to the photon. Intuitively, one might think of ω as a superposition of Y and B mesons *with masses* m_ω, and the ϕ as a superposition of Y and B mesons *with masses* m_ϕ. Then Eq. (1) implies that

$$\gamma_{\phi\gamma}=-(aem_\phi{}^2/2f_Y),\quad \gamma_{\omega\gamma}=+bem_\omega{}^2/2f_Y. \tag{5}$$

As these relations play an important role in extracting information about coupling constants from processes like $\pi^0\to 2\gamma$, it may be of interest to note that if the bare mass of the Y meson is zero they can be derived from field theory.[14]

It would seem likely that the general features of Eq. (5) are correct. But whether the masses $m_\phi{}^2$ and $m_\omega{}^2$, rather than some average mass such as $m_Y{}^2$, should appear in these equations seems to us a delicate point, which we shall not try to settle here. Instead we note that a direct experimental test of Eq. (5) is obtained by observing the decays $\omega\to e^++e^-$ and $\phi\to e^++e^-$. Neglecting the mass of the electron and using the couplings of Eq. (5), one finds[15,16]

$$\Gamma(\omega\to e^++e^-)\approx b^2[\alpha^2/12][f_Y{}^2/4\pi]^{-1}m_\omega\approx 0.89\text{ keV},$$

and

$$\Gamma(\phi\to e^++e^-)\approx a^2[\alpha^2/12][f_Y{}^2/4\pi]^{-1}m_\phi\approx 1.8\text{ keV}. \tag{6}$$

In arriving at the numerical values quoted above, we have taken the coupling[17] $f_Y\approx(\frac{3}{4})^{1/2}f_\rho$ from unitary symmetry, and estimated f_ρ from the measured width $\Gamma(\rho\to 2\pi)$. For a ρ width of 100 MeV, one finds $f_\rho{}^2/4\pi\approx 2$ and $f_Y{}^2/4\pi\approx 1.5$. An accurate experimental determination of the ratio $\Gamma(\omega\to e^++e^-)/\Gamma(\phi\to e^++e^-)\approx(b^2/a^2)(m_\omega/m_\phi)\approx 0.48$ would also provide a critical test of this model and in particular of Eq. (5).

An experimental study of the decays $\omega\to\pi^++\pi^-$ and $\phi\to\pi^++\pi^-$ could provide interesting information about the pion form factor $F_\pi(t)$. In this model the partial widths are

$$\Gamma(\omega\to\pi^++\pi^-)=b^2[\alpha^2/48]\times[f_Y{}^2/4\pi]^{-1}m_\omega[1-(4m_\pi{}^2/m_\omega{}^2)]^{3/2}|F_\pi(m_\omega{}^2)|^2,$$

and (7)

$$\Gamma(\phi\to\pi^++\pi^-)=a^2[\alpha^2/48]\times[f_Y{}^2/4\pi]^{-1}m_\phi[1-(4m_\pi{}^2/m_\phi{}^2)]^{3/2}|F_\pi(m_\phi{}^2)|^2.$$

Now let us consider the coupling of the ϕ and ω mesons to nucleons. These are given by Eq. (1) as

$$f_{\phi NN}=af_{YNN}+bf_{BNN},$$

and (8)

$$f_{\omega NN}=af_{BNN}-bf_{YNN}.$$

In accord with our universality hypothesis we will set $f_{YNN}\approx f_Y$ and $f_{BNN}\approx f_B$.

We suppose that the isoscalar charge form factor of the nucleon is dominated by the ϕ and ω resonant states. Then one writes:

$$F_{1S}(t)\approx\left(1-\frac{f_{YNN}}{f_Y}\right)+\frac{a}{f_Y}\frac{f_{\phi NN}m_\phi{}^2}{m_\phi{}^2-t}-\frac{b}{f_Y}\frac{f_{\omega NN}m_\omega{}^2}{m_\omega{}^2-t}. \tag{9}$$

Note that since $F_{1S}(0)=1$, *complete* ϕ-ω dominance would require that $f_{YNN}=f_Y$, in close analogy with the case of no mixing.

Since there is only one free parameter in (9) one could, in principle, determine $f_{\phi NN}$ and $f_{\omega NN}$ by fitting the experimental form factors. However, given the present uncertainty in the experimental data and the theoretical uncertainty as to the effect of the higher mass states on Eq. (9), we have concluded that it would only be misleading to attempt to use the form factors to estimate $f_{\phi NN}$ and $f_{\omega NN}$.

On the other hand, some rather definite qualitative statements can be made. At low t, F_{1S} appears to fall off faster than either the ϕ or ω terms separately.[18] This

[13] J. J. Sakurai, Ann. Phys. (N. Y.) 11, 1 (1960).

[14] The derivation is as follows. Let $Y(x)$ be the renormalized Heisenberg field operator of a Y meson of bare mass zero. It satisfies the field equation

$$\Box^2 Y(x)=f_Yj_Y(x).$$

Then we have:

(a) $$\Box^2\langle 0|Y(x)|\phi\rangle=\Box^2\langle 0|Y(0)|\phi\rangle e^{-i(P_\phi\cdot x)}=-m_\phi{}^2\langle 0|Y(0)|\phi\rangle e^{-i(P_\phi\cdot x)}$$

and

(b) $$\Box^2\langle 0|Y(x)|\phi\rangle=f_Y\langle 0|j_Y(x)|\phi\rangle=f_Y\langle 0|j_Y(0)|\phi\rangle e^{-i(P_\phi\cdot x)}.$$

By definition, $\gamma_{\phi\gamma}=(e/2)\langle 0|j_Y(0)|\phi\rangle$ and we also note that $\langle 0|Y(0)|\phi\rangle$ is simply the mixing coefficient a. Hence, (a) and (b) together give

$$\gamma_{\phi\gamma}=-aem_\phi{}^2/2f_Y.$$

[15] Y. Nambu and J. J. Sakurai, Phys. Rev. Letters 8, 79 (1962).

[16] M. Gell-Mann, D. Sharp, and W. Wagner, Phys. Rev. Letters 8, 261 (1962).

[17] Our coupling is $f_\rho=2\gamma_\rho$; γ_ρ being the coupling of Gell-Mann *et al.* (see Refs. 3, 12, and 16).

[18] L. Hand, D. Miller, and R. Wilson, Rev. Mod. Phys. 35, 335 (1963).

more rapid decrease can only be obtained (i), if $f_{\omega NN}$ and $f_{\phi NN}$ are of the same sign and (ii), if $|f_{\omega NN}|$ is considerably larger than $|f_{\phi NN}|$. Preliminary data[19] from the Cambridge Electron Accelerator indicates that a fit to $F_{1\,\mathrm{proton}}$ can be obtained with expressions of the form $m_\omega^2/(m_\omega^2-t)$ or $m_\rho^2/(m_\rho^2-t)$, for t ranging from ≈ 1 to 5 (BeV)^2. This would seem to indicate that the neutron form factor is negligible compared to the proton form factor even at high-momentum transfers and that the isovector and isoscalar charge form factors are, to a fair approximation, dominated by the ρ and ω poles. In particular, the high-energy data suggest that F_{1S} is at most 40% ϕ.

Moreover, if one assumes that $F_{1\,\mathrm{neutron}}$ is always small compared to $F_{1\,\mathrm{proton}}$, the experimental data indicate that

$$-1 \lesssim (a/f_Y) f_{\phi NN} \lesssim 0, \quad -2 \lesssim (b/f_Y) f_{\omega NN} \lesssim -1. \tag{10}$$

If as before we take $f_Y^2/4\pi \sim 1.5$, then we find

$$0 \lesssim f_{\phi NN}^2/4\pi \lesssim 2.5, \quad 4 \lesssim f_{\omega NN}^2/4\pi \lesssim 16. \tag{11}$$

Sakurai[13] has pointed out that an isoscalar vector meson could account for the hard core and spin-orbit interactions which are essential features of the nuclear forces. Judging from Eqs. (10) or (11) it seems that the ω meson must be a major contributor to the nuclear forces while the ϕ meson with its higher mass and weaker coupling should be less important.

A number of authors[20–22] have attempted to determine the vector meson-nucleon couplings from nucleon-nucleon scattering data. The details of the results vary considerably, but all the fits, whether they are to the low-energy phase shifts,[20] the phenomenological potentials,[21] or the high-energy pp cross sections,[22] seem to require that $f_{\omega NN}^2/4\pi$ be considerably larger than $f_{\rho NN}^2/4\pi$, in agreement with Eqs. (10) and (11).

Arnold and Sakurai[23] have pointed out that a vector meson coupled to the hypercharge current, which has the property that the product of its couplings to nucleons and kaons is $f_{\omega KK} f_{\omega NN}/4\pi \sim 5$ could explain in a rough way the low-energy Kp scattering data including the Y_0^*. It is interesting to observe that Eqs. (1), (4), and (11) and the assumed universality of the Y-meson couplings give $1.5 \lesssim f_{\omega KK} f_{\omega NN}/4\pi \lesssim 3.0$, which may perhaps be considered reasonably good agreement in view of the crude approximations involved.

III. DECAY RATES OF THE ϕ, ω, AND π^0 MESONS

In this section we shall try to give a coherent description of the decays of the ϕ, ω and π^0 mesons on the basis of the present model.

[19] Private communication from N. F. Ramsey to M. Gell-Mann. We wish to thank Professor Gell-Mann for bringing these results to our attention.

[20] R. S. McKean, Phys. Rev. 125, 1399 (1962).

[21] D. Amati, E. Leader, and B. Vitale, Phys. Rev. 130, 750 (1963).

[22] Riazuddin and Fayyazuddin, Phys. Rev. 132, 873 (1963).

[23] R. Arnold and J. J. Sakurai, Phys. Rev. 128, 2808 (1963).

The decay $\phi \to K\bar{K}$ will proceed through the coupling of the Y-component of the ϕ to the hypercharge carried by the K mesons. The width is obtained simply by multiplying Sakurai's result[24] by a^2, which gives

$$\Gamma(\phi \to K\bar{K};\ \text{both charge modes}) \approx 2.1\ \text{MeV}. \tag{12}$$

Since the experimental value[25] of the branching ratio $\Gamma(\phi \to \rho+\pi)/\Gamma(\phi \to K\bar{K})$ now appears to be $\approx 0.10 \pm 0.10$, this estimate for $\Gamma(\phi \to K\bar{K})$ indicates a total ϕ width of $\approx 2.3 \pm 0.2$ MeV, which is comparable with the experimental width[6,25,26] of $\approx 3.1 \pm 0.8$ MeV.

The ω width has recently been measured[27] to be 9.5 ± 2.1 MeV. For an experimental branching ratio[27] $\Gamma(\omega \to \text{neutrals})/\Gamma(\omega \to 3\pi) \approx 12\%$ one finds $\Gamma(\omega \to 3\pi) \approx 8.5 \pm 1.9$ MeV. With the assumption that this decay is dominated[16] by $\omega \to \rho + \pi \to 3\pi$, one finds[28]

$$f^2_{\omega\rho\pi}/4\pi \approx (0.41 \pm 0.09)/m_\pi^2. \tag{13}$$

The width $\Gamma(\phi \to \rho+\pi)$ and the branching ratio $\Gamma(\phi \to \rho+\pi)/\Gamma(\phi \to K\bar{K})$ can be predicted once $f_{\phi\rho\pi}^2/4\pi$ is known. Lacking this information, we may note instead that the present experimental value of the branching ratio $\Gamma(\phi \to \rho+\pi)/\Gamma(\phi \to K\bar{K})$ ($\approx 0.10 \pm 0.10$), our estimate for $\Gamma(\phi \to K\bar{K})$ and Eq. (13) suggest that $(f_{\omega\rho\pi}^2/4\pi)/(f_{\phi\rho\pi}^2/4\pi) \approx 850$. On the other hand, a preliminary analysis[29] of ϕ and ω production in the reaction $\pi + N \to N + 4\pi$ indicates that $(f_{\omega\rho\pi}^2/4\pi)/(f_{\phi\rho\pi}^2/4\pi) \approx 100$. This corresponds to a branching ratio $\Gamma(\phi \to \rho+\pi)/\Gamma(\phi \to K\bar{K}) \approx 0.8 \pm 0.2$, which is <1 and in somewhat better agreement with the older determinations[6] of $\Gamma(\phi \to \rho+\pi)/\Gamma(\phi \to K\bar{K})$ which gave 0.35 ± 0.20. Further experimental results bearing on the ratio $f_{\omega\rho\pi}^2/f_{\phi\rho\pi}^2$ would clearly be of high interest.

Assuming that the decay $\pi^0 \to 2\gamma$ is dominated by $\pi^0 \to \rho + \phi \to 2\gamma$ plus $\pi^0 \to \omega + \rho \to 2\gamma$, we find for a coupling $f_{\omega\rho\pi}^2/4\pi \approx (0.41 \pm 0.09)/m_\pi^2$

$$\Gamma(\pi^0 \to 2\gamma) \approx [1-(a f_{\phi\rho\pi}/b f_{\omega\rho\pi})]^2 (24.8 \pm 5.3\ \text{eV}). \tag{14}$$

If $f_{\phi\rho\pi}/f_{\omega\rho\pi} \approx 1/10$, Eq. (14) predicts $\Gamma(\pi^0 \to 2\gamma) \approx 19 \pm 4.0$ eV, a value which is somewhat greater than the latest experimental value[30] of $\approx 6.3 \pm 1.0$ eV. For a smaller or negative value of $f_{\phi\rho\pi}/f_{\omega\rho\pi}$, the discrepancy would be larger. Thus, for $f_{\phi\rho\pi}/f_{\omega\rho\pi} \approx 1/(850)^{1/2}$, the predicted width would be $\approx 22.6 \pm 4.9$ eV. Considering the crudeness of the calculations and the uncertainty in the data, a factor of 3 or so difference between theory and experiment is perhaps not alarming. For example,

[24] J. J. Sakurai, Phys. Rev. Letters 9, 472 (1962).

[25] P. L. Connolly, E. L. Hart, K. W. Lai, G. London, G. C. Moneti *et al.* (to be published).

[26] N. Gelfand, D. Miller, M. Nussbaum, J. Ratau, J. Schultz *et al.*, Phys. Rev. Letters 11, 438 (1963).

[27] N. Gelfand, D. Miller, M. Nussbaum, J. Ratau, J. Schultz *et al.*, Phys. Rev. Letters 11, 436 (1963).

[28] At an $\omega\rho\pi$ vertex we write $f_{\omega\rho\pi}\epsilon_{\mu\nu\sigma\tau}k_\mu{}^\omega\epsilon_\nu{}^\omega k_\sigma{}^\rho\epsilon_\tau{}^\rho$.

[29] N. Xuong (private communication).

[30] G. Von Dardel, D. Dekkers, R. Mermod, J. Van Putten, M. Vivargent, G. Weber, and K. Winter, Phys. Rev. Letters 4, 51 (1963).

the π^0 width is proportional to[16] f_ρ^{-4}, assuming $f_Y = (\frac{3}{4})^{1/2} f_\rho$. If we took the ρ-width to be 120 MeV instead of 100 MeV, then our prediction for the width would be reduced by a factor of 1.5.

Previous calculations[12,16] of the branching ratio $\Gamma(\pi^0 \to 2\gamma)/\Gamma(\omega \to \pi^0+\gamma)$ can easily be adapted to include the ϕ meson. One finds

$$\frac{\Gamma(\pi^0 \to 2\gamma)}{\Gamma(\omega \to \pi^0+\gamma)} = (1.7\times 10^{-5})\left[1-\frac{a}{b}\frac{f_{\phi\rho\pi}}{f_{\omega\rho\pi}}\right]^2,$$
$$\approx 1.3\times 10^{-5} \text{ if } f_{\phi\rho\pi}^2/f_{\omega\rho\pi}^2 = 1/100,$$
$$\approx 1.6\times 10^{-5} \text{ if } f_{\phi\rho\pi}^2/f_{\omega\rho\pi}^2 = 1/850. \quad (15)$$

The experimental π^0 width, the experimental ω width and the experimental branching ratio $\Gamma(\omega \to \text{neutrals})/\Gamma(\omega \to 3\pi)$ of $\approx 12\%$ suggest $\Gamma(\pi^0 \to 2\gamma)/\Gamma(\omega \to \pi^0+\gamma) \approx 0.63\times 10^{-5}$.

The distribution in mass of the Dalitz pairs in the decay $\pi^0 \to \gamma + e^+ + e^-$ is of considerable interest in connection with models of π^0 decay of the type discussed here. Applying the present model to this decay gives the form factor

$$\Gamma(t) = \frac{1}{2}\left\{\frac{m_\rho^2}{m_\rho^2-t} + [f_{Y\rho\pi}]^{-1}\right.$$
$$\left.\times\left[af_{\phi\rho\pi}\frac{m_\phi^2}{m_\phi^2-t} - bf_{\omega\rho\pi}\frac{m_\omega^2}{m_\omega^2-t}\right]\right\}. \quad (16)$$

Neglecting $f_{\phi\rho\pi}$ compared to $f_{\omega\rho\pi}$ in the derivative one finds $[d\Gamma(t)/dt]_{t=0} \approx +0.03/m_\pi^2$. Experiments indicate[31,32] a negative value for $\Gamma'(0)$, but do not rule out a small positive value.[33] It is of interest to check this point experimentally, because if it is definitely established that $\Gamma'(0)$ is negative, it would seem very difficult indeed to escape the conclusion that the ϕ, ω and ρ resonances do *not* dominate the form factor entering in π^0 decay.

[31] N. Samios, Phys. Rev. **121**, 275 (1961).
[32] H. Kobrak, Nuovo Cimento **20**, 1115 (1961).
[33] H. Kobrak (private communication).

IV. APPLICATIONS TO PHOTOPRODUCTION OF π^0 AND η MESONS

We shall estimate some coupling constants which may be of interest in photoproduction. Following the usual "pole dominance" method we set $f_{\omega\pi\gamma} \approx ef_{\omega\rho\pi}/f_\rho$, $f_{\phi\pi\gamma} \approx ef_{\phi\rho\pi}/f_\rho$, and $f_{\rho\pi\gamma} \approx ef_{Y\rho\pi}/2f_Y = e(af_{\phi\rho\pi} - bf_{\omega\rho\pi})/2f_Y$. Since $|f_{\omega\rho\pi}|^2 \gg |f_{\phi\rho\pi}|^2$, one can probably neglect ϕ exchange in analyzing π^0 photoproduction. We have also seen that $f_{\omega NN}^2$ is probably large compared to $f_{\rho NN}^2$ and at any rate $f_{\rho\pi\gamma}^2/f_{\omega\rho\gamma}^2 \approx 0.12$, so the major vector-meson effects in π^0 photoproduction should come from ω exchange. The analogous couplings in the case of η-meson production can be obtained from unitary symmetry and we find [using $f_{\phi\rho\pi}/f_{\omega\rho\pi} \approx 1/(850)^{1/2}$] that

$$\frac{f_{\omega\eta\gamma}}{f_{\omega\pi\gamma}} \approx \left(\frac{f_\rho}{2f_Y}\right)\left[a^2 - \frac{b^2}{2} + \frac{3}{2}ab\frac{f_{\phi\rho\pi}}{f_{\omega\rho\pi}}\right] \approx 0.26,$$
$$\frac{f_{\phi\eta\gamma}}{f_{\omega\pi\gamma}} \approx \left(\frac{f_\rho}{2f_Y}\right)\left[\frac{3}{2}ab + \left(b^2 - \frac{a^2}{2}\right)\frac{f_{\phi\rho\pi}}{f_{\omega\rho\pi}}\right] \approx 0.41,$$

and

$$\frac{f_{\rho\eta\gamma}}{f_{\omega\pi\gamma}} \approx \frac{1}{2}\left[a\frac{f_{\phi\rho\pi}}{f_{\omega\rho\pi}} - b\right] \approx -0.30. \quad (17)$$

The ϕ pole is considerably further from the physical region than the ω pole and we expect the ρ to be less strongly coupled to nucleons than the ω. These facts, plus the results of Eq. (17), mean that the vector-meson effects which appear to be present in π^0 photoproduction[34] at high energies will be considerably suppressed in η photoproduction.

ACKNOWLEDGMENTS

It is a pleasure for the authors to express their appreciation to Professor M. Gell-Mann, Professor J. Pines, and Professor A. V. Tollestrup for many helpful conversations.

[34] R. Talman, C. Clinesmith, R. Gomez, and A. V. Tollestrup, Phys. Rev. Letters **9**, 177 (1963).

V

INTENSITY RULES

The most spectacular results of the eightfold way have been obtained in connection with supermultiplet assignments and mass rules, but there are a great many other consequences of approximate SU(3) symmetry for the strong interactions.

SU(3) assignments provide a direct means of predicting intensity rules for decays, scatterings, and so on. Some of the simpler cases were included in the first papers, mainly covering the decay of vector mesons and baryon resonances. The more complicated cases require explicit calculation of Clebsch-Gordan or Racah coefficients, described in the article by de Swart. Complete tables of reduction coefficients exist for $\mathbf{8} \times \mathbf{8}$ (H. Goldberg, Israel AEC Report IA-834) and $\mathbf{8} \times \mathbf{8} \times \mathbf{8}$ (Y. Dothan and H. Harari, Israel AEC Report IA-777).

Some simplifications can be attained through the use of Weyl reflections, as explained by Meshkov, Levinson, and Lipkin. Alternatively, Macfarlane, Sudarshan, and Dullemond have applied Shmushkevich's method [Doklady Akad. Nauk SSR, **103**, 235 (1955)] of symmetrizing probability amplitudes.

Actual computation of cross sections, branching ratios, and so on should include the effects of the symmetry-breaking part of the strong interactions. This involves thresholds and vertices in a complicated way. Meshkov, Snow, and Yodh have suggested an approximate treatment: they compare cross sections at equal distances from the corresponding thresholds.

A recent paper by V. Gupta and V. Singh [Phys. Rev., **135**, B1442 (1964)] treats the amplitudes for decay of the $J = \frac{3}{2}^+$ baryon decimet into the $J = \frac{1}{2}^+$ baryon octet and the pseudoscalar meson octet, including first order violations of SU(3). One of the sum rules they derive is experimentally verifiable and checks quite well.

The Octet Model and its Clebsch-Gordan Coefficients

J. J. DE SWART*
CERN, Geneva

1. INTRODUCTION

IN trying to understand the structure of the strong interactions, several higher symmetry schemes have been proposed.[1,2] These higher symmetries should conserve the isospin I and the hypercharge Y. Especially interesting in this respect is the octet model (unitary symmetry) proposed independently by Gell-Mann[3] and Ne'eman.[4] In this model one assumes the strongest interactions to be invariant under transformations belonging to $SU(3)$, i.e., under unimodular unitary transformations in some three-dimensional complex linear vector space ("unitary spin space"). The symmetry of these strong interactions is broken by some unknown weaker mechanism, but in such a way that the isospin and the hypercharge are still conserved. A still weaker interaction, the electromagnetic interaction, breaks this lower symmetry in such a way that only the hypercharge and the third component of isospin are conserved. In this unitary symmetry model one assigns groups of strongly interacting particles with the same quantum numbers (not the same are I, Y, I_3, and directly related ones as strangeness, charge, G parity, etc.), to irreducible representations (IR's) of the group $SU(3)$. The lowest nontrivial IR in the octet model, which is physically possible (i.e., has integer quantum numbers for the hypercharge), is the IR $\{8\}$. The eight well-known baryons N,Λ,Σ, and Ξ, as well as the eight pseudoscalar mesons, K,η,π, and $\bar{K}$, are assigned to IR's $\{8\}$. One assumes, moreover, the existence of eight vector mesons which belong to such a representation. Perhaps the mesons ρ,ω,K^*, and $\bar{K}^*$ constitute this octet. A difficulty here is which K^* to take. There seem to be two $(K\pi)$ resonances, one[5] at 730 MeV and the other[6] at 888 MeV. One favors the 888-MeV resonance because it seems to have all the correct quantum numbers. The next higher IR can contain 10 particles. It is suggested[7] that the familiar (3,3) pion-nucleon resonance, the Y_1^* (1385 MeV), the recently discovered[8,9] $I = \frac{1}{2}$, $\Xi\pi$ resonance at 1532 MeV and a still unknown baryon $\Omega^-(Y = -2,\ I = 0, \pm 1685$ MeV) belong to this IR $\{10\}$. A discovery of this Ω^- would be a great triumph for this octet model. Okubo[10] has derived a mass formula for the different members belonging to the same IR. For the octets (IR $\{8\}$), this formula reduces to a mass relation between the different members. This mass relation is very well satisfied for the baryons and for the pseudoscalar mesons. However, for the vector mesons, neither the 888-MeV nor the 730-MeV $(K\pi)$ resonance fulfills this relation. For the IR $\{10\}$ this mass formula is again very well satisfied. Coleman and Glashow[11] have given a relation connecting the electromagnetic mass differences within the baryon octet. This relation is also very well satisfied.

The main purpose of this paper is to derive the

---

* On leave from the University of Nijmegen, Nijmegen, The Netherlands.

[1] A very nice survey of the different higher symmetry schemes in strong interactions is given by R. E. Behrends, J. Dreitlein, C. Fronsdal, and B. W. Lee, Rev. Mod. Phys. **34**, 1 (1962). The reader is referred there to the large existing literature about this subject.

[2] D. R. Speiser and J. Tarski, Math. Phys. **4**, 588 (1963).

[3] M. Gell-Mann, California Institute of Technology, Report CTSL-20, March, 1961 (unpublished); Phys. Rev. **125**, 1067 (1962).

[4] Y. Ne'eman, Nucl. Phys. **26**, 222 (1961).

[5] G. Alexander, G. R. Kalbfleisch, D. H. Miller, and G. A. Smith, Phys. Rev. Letters **8**, 447 (1962).

[6] For extensive references, see, *Proceedings of the 1962 Annual International Conference on High-Energy Physics, at CERN* (CERN, Geneva, 1962), p. 781.

[7] M. Gell-Mann, *Proceedings of the 1962 Annual International Conference on High-Energy Physics, at CERN* (CERN, Geneva, 1962), p. 805.

[8] G. M. Pjerrou, D. J. Prowse, P. Schlein, W. E. Slater, D. H. Stork, and H. K. Ticho, *Proceedings of the 1962 Annual International Conference on High-Energy Physics, at CERN* (CERN, Geneva, 1962), p. 289.

[9] L. Bertanza, V. Brisson, P. L. Connolly, E. L. Hart, I. S. Mittra, G. C. Moneti, R. R. Rau, N. P. Samios, S. S. Yamamoto, M. Goldberg, L. Gray, J. Leitner, S. Lichtman, and J. Westgard, *Proceedings of the 1962 Annual International Conference on High-Energy Physics, at CERN* (CERN, Geneva, 1962), p. 279.

[10] S. Okubo, Progr. Theoret. Phys. (Kyoto) **27**, 949 (1962).

[11] S. Coleman and S. L. Glashow, Phys. Rev. Letters **6**, 423 (1961).

Clebsch–Gordan coefficients[12–14] (CG coefficients) of $SU(3)$ for the products of the most important irreducible representations (Secs. 10, 11, and 18). Special care is taken to define properly all the relevant phase factors (Secs. 7, 10). Some useful symmetry relations for the CG coefficients are derived (Sec. 14). The Wigner–Eckart theorem[15] for this group is given (Sec. 15) and applied to derive a general mass formula for the octets (Sec. 16). A special case gives the Gell-Mann–Okubo mass relation (16.3). Another special case, however, gives a mass relation (16.15) for the octets which is very well satisfied by the vector mesons if one takes as the K^* the 730-MeV $(K-\pi)$ resonance. For completeness and to demonstrate how to handle some special phase assignments, we have considered in Sec. 17 the Yukawa couplings between the baryons and the mesons. To be able to show clearly how the results in the later sections are derived, some additional sections were necessary to define properly the different symbols and concepts used. This leads us to the alternate purpose of this paper; only slight extensions of the existing sections and a few additional ones were necessary to give a rather complete insight in the mathematical framework of this special model for the strong interactions. The treatment is as much as possible "physical"[16] and tries not to rely too heavily on results obtained by purely abstract group theoretic methods. However, where necessary, results only easily obtained (to the author's knowledge) by such methods are stated and used. A very good example is Sec. 12 where the beautifully simple method of Speiser[17] for reducing the direct product of two IR's is explained, but not proved.

2. TENSORS[18]

The group $SU(3)$ consists of all the unitary unimodular transformations in the three-dimensional vector space C_3 over the complex numbers. Let us denote a vector in this space by x^i and its complex conjugate by x_i; thus $x_i = (x^i)^*$. Under a transformation of the group, the vectors x^i and x_i get transformed into the vectors $\bar{x}^i$ and $\bar{x}_i$ according to[19]

$$\bar{x}^i = \alpha_{ij} x^j \,, \tag{2.1a}$$

$$\bar{x}_j = \alpha^*_{ij} x_j = \alpha^{-1}_{ji} x_j \,, \tag{2.1b}$$

because unitarity of α implies[20]

$$\alpha^+ = \alpha^{-1} \,, \quad \text{or} \quad \alpha^*_{ij} = \alpha^{-1}_{ji} \,.$$

In this vector space C_3 we can define mixed tensors $A^{\alpha\beta\cdots\gamma}_{ij\cdots k}$ which transform according to

$$\bar{A}^{\alpha\beta\cdots\gamma}_{ij\cdots k} = \alpha_{\alpha\lambda}\alpha_{\beta\mu}\cdots\alpha_{\gamma\nu}\alpha^{-1}_{li}\alpha^{-1}_{mj}\cdots\alpha^{-1}_{nk}A^{\lambda\mu\cdots\nu}_{lm\cdots n} \,. \tag{2.2}$$

Very special tensors are δ^i_j, ϵ^{ijk}, and ϵ_{ijk}; they are unchanged under a transformation of the group. We have

$$\bar{\delta}^i_j = \alpha_{ik}\alpha^{-1}_{lj}\delta^k_l = \alpha_{ik}\alpha^{-1}_{kj} = \delta^i_j \,,$$

and

$$\bar{\epsilon}^{ijk} = \alpha_{il}\alpha_{jm}\alpha_{kn}\epsilon^{lmn} = \det \alpha\, \epsilon^{ijk} = \epsilon^{ijk} \,,$$

because of the restriction to unimodular transformations ($\det \alpha = 1$).

The monomials $M(p,q)$

$$x^\alpha y^\beta \cdots z^\gamma u_i v_j \cdots w_k$$

with p upper indices and q lower indices are transformed into each other by transformations of the group. These monomials could, therefore, conveniently be used as a basis to construct representations of the group. These representations will, in general, be reducible, as will be shown below, because of the existence of the tensors δ^i_j, ϵ^{ijk}, and ϵ_{ijk}.

With the help of these special tensors, we can construct, from the general mixed tensor $A^{\alpha\beta\cdots\delta}_{ij\cdots l}$ with p upper and q lower indices, the mixed tensors B, C, and D. Where

$$B^{\beta\cdots\delta}_{j\cdots l} = \delta^i_\alpha A^{\alpha\beta\cdots\delta}_{ij\cdots l}$$

is a tensor with $(p-1)$ upper indices and $(q-1)$ lower indices,

$$C^{\gamma\cdots\delta}_{\mu ij\cdots l} = \epsilon_{\mu\alpha\beta}A^{\alpha\beta\gamma\cdots\delta}_{ij\cdots l}$$

is a tensor with $(p-2)$ upper and $(q+1)$ lower indices, and

$$D^{m\alpha\beta\cdots\delta}_{k\cdots l} = \epsilon^{mij}A^{\alpha\beta\cdots\delta}_{ijk\cdots l}$$

is a tensor with $(p+1)$ upper and $(q-2)$ lower indices. The tensors B, C, and D are linear combina-

---

[12] In several other papers [e.g., Refs. 13 and 14] tables of CG coefficients can be found. Special care has to be taken in using these tables in combination with some of the theorems of this paper. The phase definitions for these CG coefficients are not the same and mostly not stated.

[13] A. R. Edmonds, Proc. Roy. Soc. (London) **A268**, 567 (1962).

[14] M. A. Rashid, Nuovo Cimento **26**, 118 (1962).

[15] C. Eckart, Rev. Mod. Phys. **2**, 302 (1930); E. P. Wigner, Gruppentheorie, Vieweg (1931).

[16] With "physical" we mean in this context: "Along the lines familiar to most physicists with some knowledge of the theory of angular momentum." If in some place in the following sections, jumps in the reasoning are made which are a little large for the reader, he is advised to look at the analogous situation in the theory of angular momentum and the point will almost always become clear (at least this was our experience).

[17] D. R. Speiser, in Proceedings of the Istanbul International Summer School 1962 (to be published).

[18] R. H. Dalitz, Lectures, University of Chicago, summer 1962.

[19] We will use the Einstein summation convention.

[20] α^+ denotes the Hermitian conjugate of α, α^T the transpose, and α^* the complex conjugate.

tions of the elements of the tensor A with p upper and q lower indices. The transformation properties of B, C, and D are, however, different from a tensor with p upper and q lower indices. The tensor A is therefore reducible, unless B, C, and D are identically zero.

We find that $B = 0$ when $A^{i\beta\cdots\gamma}_{ij\cdots k} = 0$, thus when the trace of A with respect to the indices α and i is zero; $C = 0$ when A is symmetric in the indices α and β; and $D = 0$ when A is symmetric in the indices i and j.

It is now clear how to construct bases for irreducible representations of $SU(3)$. We take such linear combinations $P(p,q)$ of the monomials $M(p,q)$ that these polynomials $P(p,q)$ are

(1) totally symmetric in all p upper indices,
(2) totally symmetric in all q lower indices,
(3) traceless.[21]

These polynomials $P^{\alpha\beta\cdots\gamma}_{ij\cdots k}$ form a basis for the IR $D(p,q)$ of $SU(3)$. The dimension N of $D(p,q)$, i.e., the number of basis vectors is

$$N = (1 + p)(1 + q)[1 + \tfrac{1}{2}(p + q)] \,. \qquad (2.3)$$

Proof: A tensor, with only upper indices symmetric in these p indices, has $\frac{1}{2}(p + 1)(p + 2)$ linearly independent components. This can be seen in the following way. Due to the symmetry requirement, the order of the indices is irrelevant. We could, therefore, arrange the indices in such a way that we have first all the ones, then all the twos, and finally all the threes. Let us assume that we have α indices equal to one, then α could run from zero to p. For the twos and threes are so left $(p - \alpha)$ indices. We could make up, therefore, $(p - \alpha + 1)$ different combinations, with α ones and the rest of the indices twos and/or threes. In total there are thus

$$\sum_{\alpha=0}^{p} (p - \alpha + 1) = \tfrac{1}{2}(p + 1)(p + 2)$$

different components. A tensor with only lower indices and symmetric in these q indices has $\frac{1}{2}(q + 1)(q + 2)$ linearly independent components. A mixed tensor totally symmetric in its p upper and q lower indices has therefore $N_1 = \frac{1}{4}(p + 1)(p + 2)(q + 1)(q + 2)$ linearly independent components. The requirement that the trace should be zero gives further restrictions. The trace is a tensor with $(p - 1)$ upper and $(q - 1)$ lower indices. The trace of a mixed tensor, totally symmetric in its p upper and totally symmetric in its q lower indices, has N_2 linearly independent components, where $N_2 = \frac{1}{4}p(p + 1)q(q + 1)$. All these components should be identically zero. A traceless tensor symmetric in its p upper and symmetric in its q lower indices has therefore $N = N_1 - N_2$ linearly independent components, *q.e.d.*

A way to denote an IR is to write $\{N\}$, e.g., $D(1,1) = \{8\}$, $D(2,2) = \{27\}$, $D(3,0) = \{10\}$, etc. When more than one IR has the same dimension we could distinguish them by stars, primes, etc. For example, $D(p,q)$ and $D(q,p)$ $(p > q)$ have the same dimension. We denote then $D(p,q) = \{N\}$ and $D(q,p) = \{N^*\}$.

3. GENERATORS OF THE GROUP

In Sec. 2 we have shown that a suitable basis for the IR $D(p,q)$ of $SU(3)$ is formed by the N polynomials $P(p,q)$. There polynomials span a linear vector space V_N. A transformation α of $SU(3)$ in the space C_3 corresponds to a transformation U in the space V_N. These transformations U form the IR $D(p,q)$ of $SU(3)$, they are unitary[22] $(U^+ = U^{-1})$ and unimodular[23] $(\det U = 1)$.

Any unitary transformation U can be written as

$$U = e^{iH} \,, \qquad (3.1)$$

where H is Hermitian; $H = H^+$. Also the inverse is true; for any Hermitian H the U defined by (3.1) is unitary. The requirement of unimodularity implies

$$\operatorname{Tr} H = 0 \,. \qquad (3.2)$$

Also here the inverse is true; (3.2) ensures that U defined by (3.1) is unimodular.

We can also write the transformations α in the space C_3 in the form (3.1) with the condition (3.2). In a three-dimensional space, there exist nine linearly independent Hermitian operators but only eight traceless ones. To these eight operators in C_3 (the generators of the group) correspond eight Hermitian traceless operators F_i in V_N. We write, therefore,[3]

$$H = \sum_{i=1}^{8} \alpha_i F_i \,, \qquad (3.3)$$

where $F_i = F_i^\dagger$, $\operatorname{Tr} F_i = 0$, and α_i is real. To obtain the commutation relations for F_i, it is more con-

[21] Due to the symmetry requirements 1 and 2, every polynomial has only one trace. By trace we mean here only the contraction of one upper and one lower index and not the contraction of two upper or two lower indices.

[22] We can always choose the basis such that the matrices are unitary. See for example L. S. Pontrjagin, *Topologische Gruppen* (B. G. Teubner Verlagsgesellschaft, Leipzig, 1957), Vol. 1, Sec. 32.

[23] The representations have to be unimodular, because otherwise the unimodular matrices would form an invariant subgroup. The group $SU(3)$ does not have an invariant subgroup, the representations are therefore unimodular.

venient to express (3.3) slightly differently. We introduce a set of nine traceless operators[24] A_k^i, which are defined such that their representation in C_3 is given by

$$(A_k^i)_{\mu\nu} = \delta_{i\nu}\delta_{k\mu} - \tfrac{1}{3}\,\delta_{ik}\delta_{\mu\nu} \tag{3.4}$$

(i,k,μ,ν = 1, 2, or 3). These operators satisfy

$$A_k^i = (A_i^k)^+ . \tag{3.5}$$

Moreover, they are not totally independent but

$$A_1^1 + A_2^2 + A_3^3 = 0 . \tag{3.6}$$

We can now write (3.3) in terms of these A_k^i; we get

$$H = \sum \beta_i^k A_k^i , \tag{3.7}$$

where the hermiticity of H requires

$$\beta_k^i = (\beta_i^k)^* . \tag{3.8}$$

It is easy to see that these matrices A_k^i satisfy the commutation relations

$$[A_k^i, A_l^j] = \delta_l^i A_k^j - \delta_k^j A_l^i . \tag{3.9}$$

To these 9 operators in C_3 correspond 9 operators in every space V_N satisfying the relations (3.5) to (3.9). Then they are, of course, not any more 3 by 3 but N by N matrices.

We will introduce here still another notation for the generators of the group, which we will use throughout the rest of this paper. With the help of this new notation the connection between the F_i (e.g., Gell-Mann's notation[8]) and the A_k^i (e.g., Okubo's notation[10]) is readily made.

We denote

$$\begin{aligned} &F_1 = I_1 , \quad F_4 = K_1 , \quad F_6 = L_1 , \quad F_8 = M , \\ &F_2 = I_2 , \quad F_5 = K_2 , \quad F_7 = L_2 , \\ &F_3 = I_3 . \end{aligned} \tag{3.10}$$

We can form then the operators

$$\begin{aligned} I_\pm &= I_1 \pm iI_2 , \\ K_\pm &= K_1 \pm iK_2 , \\ L_\pm &= L_1 \pm iL_2 . \end{aligned} \tag{3.11}$$

The operators A_k^i are then expressed in terms of these operators

$$\begin{aligned} &A_1^1 = I_3 + \tfrac{1}{3}\sqrt{3}M , \quad A_1^2 = I_+ , \quad A_2^1 = I_- , \\ &A_2^2 = -I_3 + \tfrac{1}{3}\sqrt{3}M , \quad A_1^3 = K_+ , \quad A_3^1 = K_- , \\ &A_3^3 = -\tfrac{2}{3}\sqrt{3}M , \quad A_2^3 = L_+ , \quad A_3^2 = L_- . \end{aligned} \tag{3.12}$$

[24] These operators A_k^i differ from the ones given by Okubo (Ref. 10) by an over-all minus sign.

For completeness we will also give the relation between our set of generators and the set used by Behrends *et al.*[1]

$$\begin{aligned} &H_1 = (1/\sqrt{3})I_3 \quad E_1 = (1/\sqrt{3})I_+ \quad E_{-1} = (1/\sqrt{3})I_- \\ &H_2 = (1/\sqrt{3})M \quad E_2 = (1/\sqrt{3})K_+ \quad E_{-2} = (1/\sqrt{3})K_- \\ &\qquad\qquad\qquad E_3 = (1/\sqrt{3})L_+ \quad E_{-3} = (1/\sqrt{3})L_- \end{aligned} \tag{3.13}$$

4. COMMUTATION RELATIONS

We note from the commutation relations (3.9) that I_3 and M are two commuting operators. Because the rank of the group is two, there exist no other linear operators commuting with these two. We will combine them into a vector $\mathbf{E} = (I_3, M)$. The commutation relations can then be written as

$$\begin{aligned} [\mathbf{E}, I_\pm] &= \pm\mathbf{i}I_\pm , \\ [\mathbf{E}, K_\pm] &= \pm\mathbf{k}K_\pm , \\ [\mathbf{E}, L_\pm] &= \pm\mathbf{l}L_\pm , \end{aligned} \tag{4.1a}$$

and

$$\begin{aligned} [I_+, I_-] &= 2\mathbf{i}\cdot\mathbf{E} , \\ [K_+, K_-] &= 2\mathbf{k}\cdot\mathbf{E} , \\ [L_+, L_-] &= 2\mathbf{l}\cdot\mathbf{E} , \end{aligned} \tag{4.1b}$$

where the unit vectors $\mathbf{i}$, $\mathbf{k}$, and $\mathbf{l}$ are defined by

$$\mathbf{i} = (1,0) , \quad \mathbf{k} = (\tfrac{1}{2}, \tfrac{1}{2}\sqrt{3}) , \quad \mathbf{l} = (-\tfrac{1}{2}, \tfrac{1}{2}\sqrt{3}) .$$

The other commutation relations, less symmetric in form, are

$$\begin{aligned} &[I_-, K_+] = L_+ , \quad [K_-, I_+] = L_- , \\ &[I_+, L_+] = K_+ , \quad [L_-, I_-] = K_- , \\ &[K_+, L_-] = I_+ , \quad [L_+, K_-] = I_- , \end{aligned} \tag{4.1c}$$

and the rest is zero.

We will introduce the operators[24a]

$$P_i = e^{i\pi I_2} , \quad P_k = e^{i\pi K_2} , \quad P_l = e^{i\pi L_2} . \tag{4.2}$$

Then

$$\begin{aligned} &P_i^{-1} I_\pm P_i = -I_\mp , \quad P_i^{-1} K_\pm P_i = L_\pm , \\ &P_k^{-1} I_\pm P_k = L_\mp , \quad P_k^{-1} K_\pm P_k = -K_\mp , \\ &P_l^{-1} I_\pm P_l = K_\pm , \quad P_l^{-1} K_\pm P_l = -I_\pm , \\ &P_i^{-1} L_\pm P_i = -K_\pm , \\ &P_k^{-1} L_\pm P_k = -I_\mp , \\ &P_l^{-1} L_\pm P_l = -L_\mp . \end{aligned} \tag{4.3}$$

[24a] Quite extensive use of these operators is made by C. A. Levinson, H. J. Lipkin, and S. Meshkov, Phys. Letters **1**, 44, 125, and 307 (1962); Nuovo Cimento **23**, 236 (1962); Phys. Rev. Letters **10**, 361 (1962). See also A. J. Macfarlane, E. C. G. Sudarshan, and C. Dullemond, Nuovo Cimento (to be published).

These relations can easily be proved in the following way:

$$P_k^{-1}I_\pm P_k = I_\pm + i\pi[I_\pm,K_2] + ((i\pi)^2/2!)[[I_\pm,K_2],K_2] + \cdots$$
$$= I_\pm \cos \tfrac{1}{2}\pi + L_\mp \sin \tfrac{1}{2}\pi = L_\mp ,$$

and analogously for the other relations. From the relations (4.3) follows directly that

$$P_i^{-1}P_kP_i = P_l \ , \quad P_i^{-1}P_lP_i = P_k^{-1} ,$$
$$P_k^{-1}P_iP_k = P_l^{-1} , \quad P_k^{-1}P_lP_k = P_i \ ,$$
$$P_l^{-1}P_iP_l = P_k \ , \quad P_l^{-1}P_kP_l = P_i^{-1} . \qquad (4.4)$$

Moreover, we have the relations

$$P_i^{-1}\mathbf{E}P_i = \mathbf{E} - 2\mathbf{i}(\mathbf{i}\cdot\mathbf{E}) ,$$
$$P_k^{-1}\mathbf{E}P_k = \mathbf{E} - 2\mathbf{k}(\mathbf{k}\cdot\mathbf{E}) ,$$
$$P_l^{-1}\mathbf{E}P_l = \mathbf{E} - 2\mathbf{l}(\mathbf{l}\cdot\mathbf{E}) . \qquad (4.5)$$

The relations (4.5) can easily be proved as follows:

$$P_k^{-1}\mathbf{E}P_k = \mathbf{E} + i\pi[\mathbf{E},K_2] + ((i\pi)^2/2!)[[\mathbf{E},K_2],K_2] + \cdots ,$$
$$= \mathbf{E} + \mathbf{k}(\mathbf{k}\cdot\mathbf{E})(\cos\pi - 1) + \mathbf{k}K_1 \sin\pi ,$$
$$= \mathbf{E} - 2\mathbf{k}(\mathbf{k}\cdot\mathbf{E}) .$$

5. COMPLETE SET OF COMMUTING OPERATORS

To denote the different eigenstates, it is convenient to label them with the eigenvalues of a complete set of commuting operators which are linearly independent. The set consisting of I_3 and M can be extended with $I^2 = I_1^2 + I_2^2 + I_3^2$ and with two more operators F^2 and G^3, called Casimir operators.[25,26] These are[10]

$$F^2 = \tfrac{1}{2} \sum_{\mu\nu} A_\mu^\nu A_\nu^\mu = \sum_{i=1}^{8} F_i^2 ,$$
$$G^3 = \sum_{\mu\nu\lambda} A_\mu^\nu A_\nu^\lambda A_\lambda^\mu .$$

The operators F^2 and G^3 have the property that they commute with every F_i. According to Schur's lemma[27,28] these operators are constants for an irreducible representation. The IR's can, therefore, also conveniently be labeled by the eigenvalues f^2 and g^3 of these operators F^2 and G^3. Of course, the set (p,q) is equivalent with the set (f^2, g^3). The states within an IR can be labeled by the eigenvalues $T(T + 1)$, T_z, and m of I^2, I_3, and M. Also for the eigenvalues of $\mathbf{E} = (I_3, M)$ we will use the vector notation $\mathbf{e} = (T_z, m)$. We will denote by $|\mathbf{e},\gamma)$ an eigenstate of the operator $\mathbf{E}$ belonging to the eigenvalue $\mathbf{e}$. The label γ describes the unspecified other quantum numbers. For an IR, we can also define the highest eigenvalue $\mathbf{e}_H$ and the highest eigenstate $|p, q, T, e_H, \gamma)$. This highest eigenvalue is that eigenvalue $\mathbf{e}$ within the IR which has the largest T_z; the highest eigenstate is the eigenstate corresponding to the highest eigenvalue.

6. TWO THEOREMS

We are now in the position to state some useful theorems.

Theorem 1: Let $|\mathbf{e}, \gamma)$ be an eigenstate of $\mathbf{E}$. If $K_+|\mathbf{e}, \gamma)$ is different from zero, then $K_+|\mathbf{e}, \gamma)$ is also an eigenstate of $\mathbf{E}$ with the eigenvalue $\mathbf{e} + \mathbf{k}$.

Proof: From $[\mathbf{E}, K_+] = \mathbf{k}K_+$ follows

$$\mathbf{E}K_+|\mathbf{e},\gamma) = K_+(\mathbf{E} + \mathbf{k})|\mathbf{e},\gamma) = (\mathbf{e} + \mathbf{k})K_+|\mathbf{e},\gamma) ,$$

Q.E.D.

Analogously we have, if $K_-|\mathbf{e}, \gamma) \neq 0$, then $K_-|\mathbf{e}, \gamma)$ has the eigenvalue $\mathbf{e} - \mathbf{k}$; if $I_\pm|\mathbf{e}, \gamma) \neq 0$ then $I_\pm|\mathbf{e}, \gamma)$ has the eigenvalue $\mathbf{e} \pm \mathbf{i}$, and if $L_\pm|\mathbf{e}, \gamma) \neq 0$ then $L_\pm|\mathbf{e}, \gamma)$ has the eigenvalue $\mathbf{e} \pm \mathbf{l}$.

This theorem has a simple geometrical interpretation in a two-dimensional eigenvalue diagram (Fig. 1). In this eigenvalue diagram every eigenvalue is represented by a point $\mathbf{e} = (T_z, m)$. These points form a regular pattern; the distances between neighboring points being $\mathbf{i}$, $\mathbf{k}$, or $\mathbf{l}$.

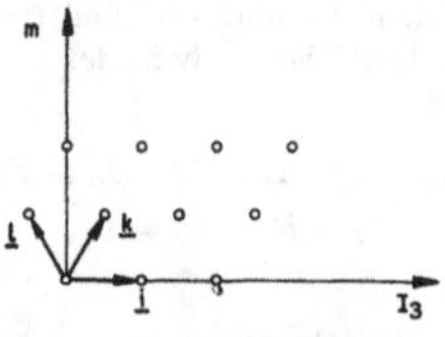

FIG. 1. Part of a two-dimensional eigenvalue diagram showing the regular pattern (Theorem 1).

Theorem 2: Let $|\mathbf{e}, \gamma)$ be an eigenstate of $\mathbf{E}$, then $P_i|\mathbf{e}, \gamma)$ is also an eigenstate of $\mathbf{E}$ with the eigenvalue $\mathbf{e} - 2\mathbf{i}(\mathbf{i}\cdot\mathbf{e})$. The degeneracy of the state $P_i|\mathbf{e}, \gamma)$ is the same as the degeneracy of the state $|\mathbf{e}, \gamma)$.

Proof: From $P_i^{-1}EP_i = \mathbf{E} - 2\mathbf{i}(\mathbf{i}\cdot\mathbf{E})$ follows

$$\mathbf{E}P_i|\mathbf{e},\gamma) = P_i\{\mathbf{E} - 2\mathbf{i}(\mathbf{i}\cdot\mathbf{E})\}|\mathbf{e},\gamma)$$
$$= \{\mathbf{e} - 2\mathbf{i}(\mathbf{i}\cdot\mathbf{e})\}P_i|\mathbf{e},\gamma)$$

[25] H. B. G. Casimir, Proc. Roy. Acad. Amsterdam **34**, 844 (1931).

[26] M. Hamermesh, *Group Theory and its Application to Physical Problems* (Addison-Wesley Publishing Company, Inc., New York, 1962), Secs. 8–13.

[27] I. Schur, Sitzber. Preuss. Akad. Wiss. Physik. math. Kl. **24**, 406 (1905).

[28] Ref. 26, Secs. 8–14, lemma II.

Moreover, P_i is a unitary operator; it conserves, therefore, the multiplicity of the state.

Analogously we find that $P_k|\mathbf{e}, \gamma)$ and $P_l|\mathbf{e}, \gamma)$ have the eigenvalues $\mathbf{e} - 2\mathbf{k}(\mathbf{k}\cdot\mathbf{e})$ and $\mathbf{e} - 2\mathbf{l}(\mathbf{l}\cdot\mathbf{e})$. The degeneracy of these states is again the same as the degeneracy of $|\mathbf{e}, \gamma)$.

Also this theorem has a simple geometrical interpretation in the two-dimensional eigenvalue diagram. If $\mathbf{e}_1$ is an eigenvalue, then also the values $\mathbf{e}_2$, $\mathbf{e}_3$, and $\mathbf{e}_4$, obtained from $\mathbf{e}_1$ by reflection with respect to lines through the origin perpendicular to the **i**, **k**, and **l** directions, are eigenvalues [see Fig. 2(a)]. Once more applying P_i gives two more eigenvalues $\mathbf{e}_5$ and $\mathbf{e}_6$. So, in general, the existence of one eigenvalue implies the existence of six eigenvalues[29,30] all with the same degeneracy.

There are two exceptions. The first exception is when the eigenvalue $\mathbf{e}_1$ lies on one of the reflection lines [see Fig. 2(b)]. In this case the existence of one eigenvalue implies only the existence of three eigenvalues. The second exception is when the eigenvalue $\mathbf{e}_1$ lies in the center; $\mathbf{e}_1 = (0,0)$. In this case no other eigenvalues are implied.

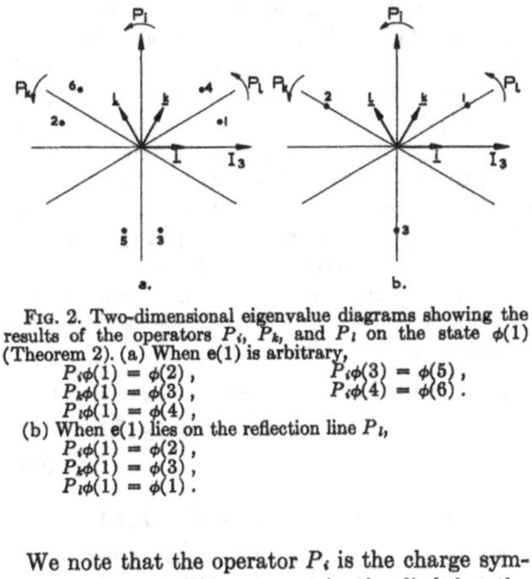

FIG. 2. Two-dimensional eigenvalue diagrams showing the results of the operators P_i, P_k, and P_l on the state $\phi(1)$ (Theorem 2). (a) When $\mathbf{e}(1)$ is arbitrary,

$$P_i\phi(1) = \phi(2)\,, \qquad P_i\phi(3) = \phi(5)\,,$$
$$P_k\phi(1) = \phi(3)\,, \qquad P_i\phi(4) = \phi(6)\,.$$
$$P_l\phi(1) = \phi(4)\,,$$

(b) When $\mathbf{e}(1)$ lies on the reflection line P_l,

$$P_i\phi(1) = \phi(2)\,,$$
$$P_k\phi(1) = \phi(3)\,,$$
$$P_l\phi(1) = \phi(1)\,.$$

We note that the operator P_i is the charge symmetry operator. The symmetries implied by the operators P_k and P_l are, therefore, generalizations of the principle of charge symmetry.

[29] In Racah's lecture notes (Ref. 30) is shown that not only $\mathbf{e}$ and $\mathbf{e} - \mathbf{i}(\mathbf{i}\cdot\mathbf{e})$ are eigenvalues, but the whole chain (isomultiplet) connecting these two eigenvalues.

[30] G. Racah, *Group Theory and Spectroscopy* (Institute for Advanced Study, Princeton, New Jersey, 1951).

7. IRREDUCIBLE REPRESENTATIONS I

In this section we consider the simplest irreducible representations which are of interest to us.

D(0,0) = {1}. This IR consists of only one state and because of theorem 2, the eigenvalue $\mathbf{e}$ belonging to this single state is $\mathbf{e} = (0,0)$. In the octet model, one identifies the hypercharge operator $Y = S + B$ with the operator

$$Y = (2/\sqrt{3})M\,. \tag{7.1}$$

This state is, therefore, an isosinglet state with $Y = 0$. The operators F_i we can represent by 1×1 matrices which are all identically zero (traceless).

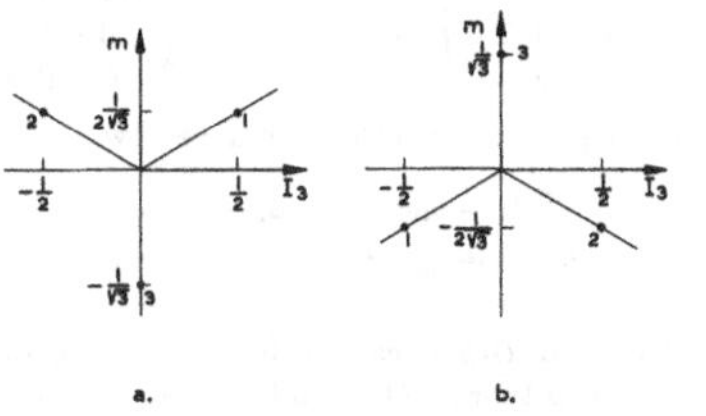

FIG. 3. Eigenvalue diagrams of the IR's with $N = 3$. (a) The irreducible representation $D(1,0) = \{3\}$. (b) The irreducible representation $D(0,1) = \{3^*\}$.

D(1,0) = {3}. There are three possibilities for this case according to theorem 2. The first possibility is three degenerate eigenvalues at the origin $I = Y = 0$. For this case, all the operators F_i are identically zero. Because they are 3×3 matrices, this case is reducible. The two other possibilities have eigenvalue diagrams as depicted in Fig. 3. Here we have to make a choice. We make the conventional choice [Fig. 3(a)]. We leave the eigenvalue diagram [Fig. 3(b)] for the contragredient representation $D(0,1) = \{3^*\}$. We are, therefore, able to write down

$$I_3 = \frac{1}{2}\begin{pmatrix} 1 & 0 & 0 \\ 0 & -1 & 0 \\ 0 & 0 & 0 \end{pmatrix}, \text{ and } M = \frac{1}{2\sqrt{3}}\begin{pmatrix} 1 & 0 & 0 \\ 0 & 1 & 0 \\ 0 & 0 & -2 \end{pmatrix}.$$

For the other matrices we have first to define the relative phases between these three states. We will do this the following way:

(1) Within an isomultiplet we use the Condon and Shortley phase convention.[31] This establishes that

$$I_+\phi_1 = 0\,, \quad I_+\phi_2 = \phi_1\,, \quad I_+\phi_3 = 0\,.$$

[31] E. U. Condon and G. H. Shortley, *The Theory of Atomic Spectra* (Cambridge University Press, Cambridge, England, 1935). See also Ref. 37.

(2) The relative phases between the different isomultiplets can also be defined rather easily. We require also for the operators $K_\pm$ the phase choice of Condon and Shortley. Therefore,

$$K_+\phi_1 = 0\,, \quad K_+\phi_2 = 0\,, \quad K_+\phi_3 = \phi_1\,.$$

The matrix elements of the operators $I_\pm$ and $K_\pm$ are now totally defined and, therefore, also of $L_+ = [I_-, K_+]$. We can identify $\phi_1 = x^1$, $\phi_2 = x^2$, and $\phi_3 = x^3$. These states ϕ_1, ϕ_2, and ϕ_3 in $SU(3)$ are equivalent to the states α (spin up) and β (spin down) in $SU(2)$.

D(0,1) = {3*}. The eigenvalue diagram is like Fig. 3(b). We can, therefore, write

$$I_3 = \frac{1}{2}\begin{pmatrix} -1 & 0 & 0 \\ 0 & 1 & 0 \\ 0 & 0 & 0 \end{pmatrix}, \text{ and } M = \frac{1}{2\sqrt{3}}\begin{pmatrix} -1 & 0 & 0 \\ 0 & -1 & 0 \\ 0 & 0 & 2 \end{pmatrix}.$$

With our phase conventions we have now

$$I_+\phi_i = \delta_{i1}\phi_2\,, \quad K_+\phi_i = \delta_{i1}\phi_3\,.$$

We can identify[32] $\phi_1(3^*) = -x_1$, $\phi_2(3^*) = x_2$, and $\phi_3(3^*) = x_3$.

All the IR $D(p,q)$ can be formed by the direct product of p times $D(1,0)$ and q times $D(0,1)$. This direct product contains of course more than the IR $D(p,q)$. However, it is clear that the highest eigenstate $\phi(e_H)$ of $D(p,q)$ can be formed in only one way

$$\phi(e_H) = \underbrace{x^1y^1\cdots z^1}_{p \text{ factors}}\ \underbrace{u_2v_2\cdots w_2}_{q \text{ factors}}\,. \tag{7.2}$$

The highest eigenvalue is, therefore,

$$\mathbf{e}_H = \left(\frac{p+q}{2},\ \frac{p-q}{2\sqrt{3}}\right). \tag{7.3}$$

Moreover, it is clear that this highest eigenvalue is nondegenerate. This highest eigenstate has hypercharge

$$Y = (p-q)/3\,. \tag{7.4}$$

Physically the hypercharge is an integer. If we want to restrict ourselves to the octet model, then not all the IR's of $SU(3)$ are interesting. The only IR's realizable in nature[33] are IR's for which $p - q = 3n$, where $n = 0, \pm1, \pm2, \pm3, \cdots$ etc.

These are, therefore, the representations

$$\{1\} = D(0,0),\quad \{10\} = D(3,0),\quad \{28\} = D(6,0),\ \text{etc.},$$
$$\{10^*\} = D(0,3),\quad \{28^*\} = D(0,6),\ \text{etc.},$$
$$\{8\} = D(1,1),\quad \{35\} = D(4,1),\ \text{etc.},$$
$$\{35^*\} = D(1,4),\ \text{etc.},$$
$$\{27\} = D(2,2),\quad \{81\} = D(5,2),\ \text{etc.},$$
$$\{81^*\} = D(2,5),\ \text{etc.},$$
$$\{64\} = D(3,3),\ \text{etc.}$$

To stay closer to the physics, we will work in the following solely with the hypercharge operator Y and its eigenvalues. The vertical scale in the eigenvalue diagrams we will, however, compress in such a way that the unit of length along this vertical axis still corresponds, as if we have plotted the eigenvalues of M. In this way we preserve the high symmetry in these eigenvalue diagrams.

To be able to refer simply to the different eigenstates of an IR we will enumerate them from 1 to N. We will choose the following ordering for the states:

(a) within an isomultiplet, the states are ordered so that I_z decreases;

(b) the isomultiplets belonging to the same Y are ordered so that I decreases;

(c) the groups for different Y are ordered such that Y decreases.

For the contragredient representation $\{N^*\}$, we adopt an opposite convention for (a) and (c); we order the states within a multiplet such that I_z increases and the groups of different Y we order such that Y increases. In this way the eigenvalue diagram for $\{N^*\}$ is the inverse with respect to the origin ($I_z = 0$, $Y = 0$) of the eigenvalue diagram for $\{N\}$.

To be able to define later uniquely the Clebsch-Gordan coefficients of $SU(3)$, it is necessary first of all to define precisely the relative phases within the IR's. We adopt, therefore, the following convention:

(1) The relative phases *within* a definite isomultiplet are determined by the Condon and Shortley phase convention.[31] Then

$$I_+\phi(I,I_z,Y) = [(I-I_z)(I+I_z+1)]^{\frac{1}{2}}\phi(I,I_z+1,Y),$$
$$I_-\phi(I,I_z,Y) = [(I+I_z)(I-I_z+1)]^{\frac{1}{2}}\phi(I,I_z-1,Y), \tag{7.5}$$

and

$$P_i\phi(I,I_z,Y) = (-)^{I+I_z}\phi(I,-I_z,Y)\,. \tag{7.6}$$

(2) The relative phases *between* the different isomultiplets we define then with the help of the operators $K_\pm$. Biedenharn[33a] has pointed out that

$$K_+\phi(I,I_z,Y) = b_+\phi(I+\tfrac{1}{2},I_z+\tfrac{1}{2},Y+1) + b_-\phi(I-\tfrac{1}{2},I_z+\tfrac{1}{2},Y+1), \tag{7.7}$$

---

[32] See Sec. 8 for the specific phases.
[33] This comes from our special choice (7.1) for the hypercharge operator.
[33a] L. C. Biedenharn, Phys. Letters 3, 69 and 254 (1962); J. Math. Phys. 4, 436 (1963). See also G. E. Baird and L. C. Biedenharn, J. Math. Phys. 4, 1449 (1963).

where

$$b_+ = \left\{\frac{(I + I_z + 1)[\frac{1}{3}(p - q) + I + \frac{1}{2}Y + 1][\frac{1}{3}(p + 2q) + I + \frac{1}{2}Y + 2][\frac{1}{3}(2p + q) - I - \frac{1}{2}Y]}{2(I + 1)(2I + 1)}\right\}^{\frac{1}{2}} \quad (7.8a)$$

and

$$b_- = \left\{\frac{(I - I_z)[\frac{1}{3}(q - p) + I - \frac{1}{2}Y][\frac{1}{3}(p + 2q) - I + \frac{1}{2}Y + 1][\frac{1}{3}(2p + q) + I - \frac{1}{2}Y + 1]}{2I(2I + 1)}\right\}^{\frac{1}{2}}. \quad (7.8b)$$

We require thus always $b_\pm$ to be real and positive.

This uniquely defines the relative phases between all the states of an IR.

8. CONTRAGREDIENT REPRESENTATIONS

The representation $D(p,q)$ is called contragredient to the representation $D(q,p)$. These representations are intimately connected.

If $U = e^{i\alpha_i F_i}$ is a representation of an element of $SU(3)$, then so is $U^* = (U^{-1})^T$. Now

$$(U^{-1})^T = e^{-i\alpha_i F_i^T} = e^{i\alpha_i F_i'}.$$

Therefore, we could choose [actually we do not, see Eq. (8.1)] the generators F_i' of the contragredient representation to be

$$F_i' = -F_i^T.$$

Then we have

$$I_3' = -I_3, \quad I_\pm' = -I_\mp,$$
$$Y' = -Y, \quad K_\pm' = -K_\mp,$$
$$L_\pm' = -L_\mp,$$

and the relation between the eigenstates is

$$\phi(\{N^*\},I,I_z,Y) = \eta\phi^*(\{N\},I, - I_z, - Y)$$

where η is an over-all phase factor.

This choice is certainly *inconvenient*, because it implies that not all of the elements of the matrices $I_\pm$ and $K_\pm$ are positive, which is required due to our phase conventions (1) and (2). We have to take the following choices for the matrices F_i'' of the contragredient representation $\{N^*\}$ if we have the matrices F_i of the representation $\{N\}$.

$$I_3'' = -I_3 \quad \text{and} \quad Y'' = -Y. \qquad (8.1a)$$

This is required because of the eigenvalue diagram and the specific ordering of the different states. Our phase conventions require the choice

$$I_\pm'' = I_\mp$$
$$K_\pm'' = K_\mp. \qquad (8.1b)$$

Then the commutation relations (4.1c) require

$$L_\pm'' = -L_\mp. \qquad (8.1c)$$

The relation between the eigenstates of the representations $\{N\}$ and $\{N^*\}$ is then

$$\phi(\{N^*\},I,I_z,Y) = \eta(-)^{I_z+\frac{1}{2}Y}\phi^*(\{N\},I, - I_z, - Y). \qquad (8.2)$$

Here η is an over-all phase which could conveniently be defined by the condition[34]

$$\phi(\{N^*\},I,0,0) = \phi^*(\{N\},I,0,0) \qquad (8.3)$$

then $\eta = 1$.

9. IRREDUCIBLE REPRESENTATIONS II

In this section we will discuss the IR's which we will use later on.

D(1,1) = {8}. One can immediately construct the eigenvalue diagram [Fig. 4(a)]. The highest eigenvalue $\mathbf{e}_3$ is $I_z = 1$, $Y = 0$, hence $I = 1$. Because of theorem 2 this eigenvalue ensures the existence of six eigenvalues $\mathbf{e}_1$, $\mathbf{e}_2$, $\mathbf{e}_3$, $\mathbf{e}_5$, $\mathbf{e}_7$, and $\mathbf{e}_8$ (all nondegenerate). The existence of $\mathbf{e}_4$ ($I = 1$, $I_z = 0$, $Y = 0$) is also implied by the existence of $\mathbf{e}_3$, because $\mathbf{e}_3$, $\mathbf{e}_4$, and $\mathbf{e}_5$ form the $I = 1$ isomultiplet. We count so in total 7 eigenstates. The missing eighth state $\mathbf{e}_6$ can be nothing else than a $I = 0$, $Y = 0$ state. The matrices $I_\pm$ are given by Eq. (7.5), the matrices K_+ by Eqs. (7.7) and (7.8).

D(3,0) = {10}. The eigenvalue diagram is given in Fig. 4(b). The highest eigenvalue $\mathbf{e}_1$ is $I = \frac{3}{2}$, $I_z = \frac{3}{2}$, $Y = 1$. Theorem 2 implies the existence of $\mathbf{e}_4$ and $\mathbf{e}_{10}$. The eigenvalues $\mathbf{e}_2$ and $\mathbf{e}_3$ have to exist to make up the $I = \frac{3}{2}$, $Y = 1$ isomultiplet. However, the existence of $\mathbf{e}_2$ and $\mathbf{e}_3$ implies the existence of $\mathbf{e}_5$, $\mathbf{e}_7$, $\mathbf{e}_8$, and $\mathbf{e}_9$ due to theorem 2. The last eigenvalue $\mathbf{e}_6$ is necessary to complete the $I = 1$, $Y = 0$ multiplet.

D(0,3) = {10*}. The eigenvalue diagram is given in Fig. 4(c). The matrix K_+ is easily determined from the corresponding matrix of the representation {10}. We have

$$K_+(10^*) = [K_+(10)]^\dagger.$$

[34] In the octet model every IR possesses a state with $I_z = Y = 0$. See also Sec. 14.

The representations

$$D(2,2) = \{27\}\,, \quad D(6,0) = \{28\}\,,$$

$$D(4,1) = \{35\}\,, \quad D(3,3) = \{64\}\,, \quad \text{etc.}\,,$$

and the corresponding conjugate representations $\{28^*\}$, $\{35^*\}$, etc., can be obtained along the same lines. The eigenvalue diagrams are given in Figs. 4(d)–4(g).

10. CLEBSCH-GORDAN COEFFICIENTS

When one forms the product representation of two IR's $D(p_1,q_1)$ and $D(p_2,q_2)$, then this product

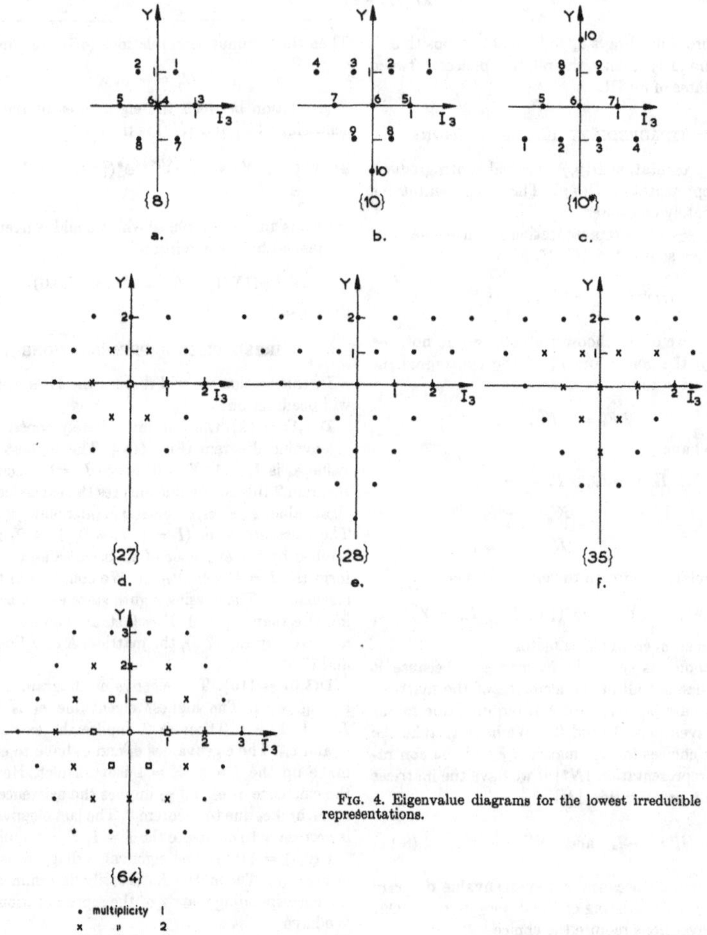

FIG. 4. Eigenvalue diagrams for the lowest irreducible representations.

representation is, in general, reducible. Symbolically one writes

$$D(p_1,q_1) \otimes D(p_2,q_2) = \sum_{P,Q} \oplus\, \sigma(P,Q)D(P,Q)\,, \quad (10.1)$$

where $\sigma(P,Q)$ is an integer. The series on the right-hand side of Eq. (10.1) is called the Clebsch–Gordan series (CG series). This symbolical equation states that the representation $D(P,Q)$ is contained $\sigma(P,Q)$ times in the direct product of the IR's $D(p_1,q_1)$ and $D(p_2,q_2)$.

We have seen in Sec. 5 that a complete set of operators necessary to specify uniquely the states of an IR is G^3, F^2, I^2, I_3, and Y. The states of the product representation $D(p_1,q_1) \otimes D(p_2,q_2)$ are, therefore, completely specified by the eigenvalues of the 10 linearly independent, commuting operators

$$G^3(1),\, G^3(2),\, F^2(1),\, F^2(2),\, I^2(1),\, I^2(2),\, I_3(1),\, I_3(2),\, Y(1), \text{ and } Y(2). \quad \text{(I)}$$

If we define the operators

$$F_i = F_i(1) + F_i(2) \quad (i = 1,\cdots,8) \quad (10.2)$$

then a (noncomplete) set of commuting operators is

$$G^3,\, G^3(1),\, G^3(2),\, F^2,\, F^2(1),\, F^2(2),\, I^2,\, I_3, \text{ and } Y.$$

However, this makes only 9 operators. We need another operator to make this set complete. This operator Γ is unfortunately not contained in the group. We have to go outside the group to find this operator[35] which is necessary to distinguish the different $D(P,Q)$ for the same P and Q, when $\sigma(P,Q) > 1$. A complete set of commuting operators is therefore

$$G^3,\, G^3(1),\, G^3(2),\, F^2,\, F^2(1),\, F^2(2),\, I^2,\, I_3,\, Y, \text{ and } \Gamma. \quad \text{(II)}$$

In the following we will use a shorthand notation when there is no chance for confusion. We denote then the eigenvalues g^3 and f^2 of G^3 and F^2 collectively by μ, the eigenvalues I, I_z, and Y by ν, and the eigenvalue of Γ by γ. We denote the eigenstates of the representation $D(p_1,q_1)$ by $\phi_{\nu_1}^{(\mu_1)}$, the eigenstates of $D(p_2,q_2)$ by $\phi_{\nu_2}^{(\mu_2)}$. We write the eigenstates of the set (II) in the product representation as

$$\psi\begin{pmatrix}\mu_1\,\mu_2\,\mu_\gamma\\ \nu\end{pmatrix}.$$

The dimension of the representation $D(p_i,q_i)$ is N_i.

[35] This is about all that can be said about this operator Γ. In practice, one uses symmetry properties of the wave function.

In the N_1N_2 dimensional product space, we can take as basis the vectors $\phi_{\nu_1}^{(\mu_1)}\,\phi_{\nu_2}^{(\mu_2)}$ of the product representation or the direct sum of the vectors

$$\psi\begin{pmatrix}\mu_1\,\mu_2\,\mu_\gamma\\ \nu\end{pmatrix}$$

of the different IR's contained in the product of $D(p_1,q_1) \otimes D(p_2,q_2)$. These two different sets of orthonormal basis vectors are connected by a unitary transformation[36]

$$\psi\begin{pmatrix}\mu_1\,\mu_2\,\mu_\gamma\\ \nu\end{pmatrix} = \sum_{\nu_1,\nu_2}\begin{pmatrix}\mu_1\,\mu_2\,\mu_\gamma\\ \nu_1\,\nu_2\,\nu\end{pmatrix}\phi_{\nu_1}^{(\mu_1)}\phi_{\nu_2}^{(\mu_2)}\,. \quad (10.3)$$

The coefficients $\left(\begin{smallmatrix}\mu_1\mu_2\mu_\alpha\\ \nu_1\nu_2\nu\end{smallmatrix}\right)$ are the Clebsch–Gordan coefficients (CG coefficients) of $SU(3)$.

We could have looked at the above problem in another way which is very useful. The product states $\phi_{\nu_1}^{(\mu_1)}\,\phi_{\nu_2}^{(\mu_2)}$ are eigenstates of the operators of the set (I). Therefore, they are also eigenstates of $I_z = I_{1z} + I_{2z}$ and $Y = Y_1 + Y_2$, but not of G^3, F^2, Γ, and I^2. With the help of the Clebsch–Gordan coefficients $C^{I_1\,I_2\,I}_{I_{1z}I_{2z}I_z}$ of $SU(2)$, we can construct the eigenfunctions χ of the operators $G^3(1)$, $G^3(2)$, $F^2(1)$, $F^2(2)$, I, I_z, and Y, but not yet of G^3, F^2, and Γ. Then

$$\chi\begin{pmatrix}\mu_1 & \mu_2 & \\ I_1Y_1 & I_2Y_2 & II_zY\end{pmatrix} = \sum_{I_{1z}I_{2z}} C^{I_1\,I_2\,I}_{I_{1z}I_{2z}I_z}\phi_{\nu_1}^{(\mu_1)}\phi_{\nu_2}^{(\mu_2)}\,. \quad (10.4)$$

Now we can combine the different χ to obtain eigenstates of the set (II)

$$\psi\begin{pmatrix}\mu_1\,\mu_2\,\mu_\gamma\\ \nu\end{pmatrix} = \sum_{\substack{I_1,Y_1\\ I_2,Y_2}}\left(\begin{matrix}\mu_1 & \mu_2\\ I_1Y_1 & I_2Y_2\end{matrix}\middle|\begin{matrix}\mu_\gamma\\ IY\end{matrix}\right) \times \chi\begin{pmatrix}\mu_1 & \mu_2 & \\ I_1Y_1 & I_2Y_2 & II_zY\end{pmatrix}. \quad (10.5)$$

The coefficients

$$\left(\begin{matrix}\mu_1 & \mu_2\\ I_1Y_1 & I_2Y_2\end{matrix}\middle|\begin{matrix}\mu_\gamma\\ IY\end{matrix}\right)$$

are called isoscalar factors.[13] Note the dependence of these isoscalar factors on the total isospin I.

Comparison of (10.4) and (10.5) with (10.3) shows

$$\begin{pmatrix}\mu_1\,\mu_2\,\mu_\gamma\\ \nu_1\,\nu_2\,\nu\end{pmatrix} = C^{I_1\,I_2\,I}_{I_{1z}I_{2z}I_z}\left(\begin{matrix}\mu_1 & \mu_2\\ I_1Y_1 & I_2Y_2\end{matrix}\middle|\begin{matrix}\mu_\gamma\\ IY\end{matrix}\right). \quad (10.6)$$

Because the CG coefficients of $SU(2)$ are well known, it is sufficient to give the isoscalar factors to specify the CG coefficients of $SU(3)$ uniquely. Care has to be taken, however, that the CG coefficients of $SU(2)$ have the correct phase factors.[31]

[36] It turns out that one can choose the different arbitrary phases in such a way that all the CG coefficients are real. The transformation matrix is then a real orthogonal matrix.

To define uniquely the isoscalar factors, one has to define the relative phase of the basis vectors of the IR $D(P,Q)$ in the CG series (10.1) with respect to the basis vectors of the product representation $D(p_1,q_1) \otimes D(p_2,q_2)$. We will take these phase factors always real; this will result then in real isoscalar factors. To decide on the sign of the phase, we consider the highest eigenstate $\phi^{(\mu)}_{\nu_H}$ of the IR $D(P,Q)$, then

$$\phi^{(\mu)}_{\nu_H} = \sum_{\nu_1,\nu_2} \begin{pmatrix} \mu_1 & \mu_2 & \mu_\gamma \\ \nu_1 & \nu_2 & \nu_H \end{pmatrix} \phi^{(\mu_1)}_{\nu_1} \phi^{(\mu_2)}_{\nu_2} .$$

Among the different CG coefficients, we choose the one with the largest possible I_1 to be positive. If this is not sufficient to decide, we take from the coefficients with the largest possible I_1 the one with the largest possible I_2 positive. This convention was sufficient to determine the phases in the cases met here. Perhaps this is not, in general, sufficient; however, this convention can easily be generalized. We have chosen this convention as the most direct generalization of the phase convention for the CG coefficients of $SU(2)$. There one requires

$$C^{j_1\ j_2\ j}_{j_1\ j-j_1\ j} > 0 .$$

11. ORTHOGONALITY RELATIONS

The CG coefficients of $SU(3)$ form a real orthogonal matrix. Therefore,

$$\phi^{(\mu_1)}_{\nu_1} \phi^{(\mu_2)}_{\nu_2} = \sum_{\mu\nu\gamma} \begin{pmatrix} \mu_1 & \mu_2 & \mu_\gamma \\ \nu_1 & \nu_2 & \nu \end{pmatrix} \psi \begin{pmatrix} \mu_1 & \mu_2 & \mu_\gamma \\ & & \nu \end{pmatrix}, \quad (11.1)$$

and

$$\sum_{\nu_1\nu_2} \begin{pmatrix} \mu_1 & \mu_2 & \mu_\gamma \\ \nu_1 & \nu_2 & \nu \end{pmatrix} \begin{pmatrix} \mu_1 & \mu_2 & \mu'_{\gamma'} \\ \nu_1 & \nu_2 & \nu' \end{pmatrix} = \delta_{\mu\mu'} \delta_{\gamma\gamma'} \delta_{\nu\nu'} , \quad (11.2a)$$

$$\sum_{\mu\gamma\nu} \begin{pmatrix} \mu_1 & \mu_2 & \mu_\gamma \\ \nu_1 & \nu_2 & \nu \end{pmatrix} \begin{pmatrix} \mu_1 & \mu_2 & \mu_\gamma \\ \nu'_1 & \nu'_2 & \nu \end{pmatrix} = \delta_{\nu_1\nu_1'} \delta_{\nu_2\nu_2'} . \quad (11.2b)$$

The orthogonality relations of the CG coefficients of $SU(2)$ are well known.[37] Therefore, the orthogonality relations for the isoscalar factors are

$$\sum_{\substack{I_1Y_1\\I_2Y_2}} \left(\begin{matrix} \mu_1 & \mu_2 \\ I_1Y_1 & I_2Y_2 \end{matrix} \middle| \begin{matrix} \mu_\gamma \\ IY \end{matrix} \right) \left(\begin{matrix} \mu_1 & \mu_2 \\ I_1Y_1 & I_2Y_2 \end{matrix} \middle| \begin{matrix} \mu'_{\gamma'} \\ IY' \end{matrix} \right) = \delta_{\mu\mu'} \delta_{\gamma\gamma'} \delta_{YY'} , \quad (11.3a)$$

and

$$\sum_{\mu\gamma Y} \left(\begin{matrix} \mu_1 & \mu_2 \\ I_1Y_1 & I_2Y_2 \end{matrix} \middle| \begin{matrix} \mu_\gamma \\ IY \end{matrix} \right) \left(\begin{matrix} \mu_1 & \mu_2 \\ I'_1Y'_1 & I'_2Y'_2 \end{matrix} \middle| \begin{matrix} \mu_\gamma \\ IY \end{matrix} \right) = \delta_{I_1I_1'} \delta_{I_2I_2'} \delta_{Y_1Y_1'} \delta_{Y_2Y_2'} . \quad (11.3b)$$

[37] M. E. Rose, *Theory of Angular Momentum* (John Wiley & Sons, Inc., New York, 1957).

12. CLEBSCH–GORDAN SERIES

The direct product $D \otimes D$ of two IR's of $SU(3)$ can be decomposed in several IR's of $SU(3)$. This is formally described by the Clebsch–Gordan series (10.1). To find the different $D(P,Q)$ with their multiplicities $\sigma(P,Q)$ we will follow the method of Speiser.[17,38]

We start by making a (p,q) coordinate system (Fig. 5). In this coordinate system we can represent every

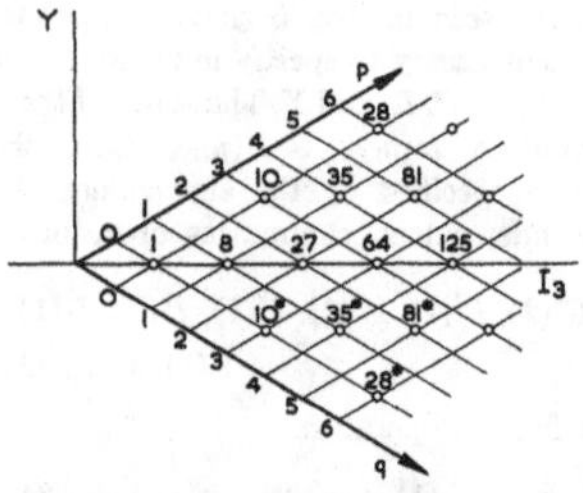

FIG. 5. Coordinate system in which we can represent every irreducible representation $\{N\} = D(p,q)$ by a point (p,q).

IR $D(p,q)$ by the point (p,q). We make this coordinate system oblique; the p and q axes make angles of 30° (see Fig. 5) with a horizontal line, which we will call the I_3 axis. The axis perpendicular to the I_3 axis is called the Y axis. The unit of length along the p and q axis we take to be $l/\sqrt{3}$.

Next we can reflect this figure about the p axis and about the q axis. Reflecting now about the Y axis, we obtain the "lattice" as shown in Fig. 6. We give a positive weight to the nonshaded sextants and to the shaded ones a negative weight.

In order to obtain the Clebsch–Gordan series of the direct product $D(p_1,q_1) \otimes D(p_2,q_2)$ we need the eigenvalue diagram of $D(p_1,q_1)$ or $D(p_2,q_2)$ with the multiplicities of the eigenvalues. We will assume that we have the eigenvalue diagram of $D(p_1,q_1)$ drawn on scale; the unit of length for I_3 is l, the unit of length for Y is $\frac{2}{3}\sqrt{3}l$. We place this eigenvalue diagram of $D(p_1,q_1)$ on top of Fig. 6 in such a way that the eigenvalue $(I_3,Y) = (0,0)$ coincides with the lattice point $\{N_2\} = D(p_2,q_2)$ in the first positive weight sextant and that the I_3 axis of the eigenvalue diagram coincides with the I_3 axis of the lattice. The eigenvalues of $D(p_1,q_1)$ coincide now all with points of the

[38] This method can be used for all the IR's of $SU(3)$ and not only for the subset of IR's used by us. Because we are only interested in the octet model, we give this restrictive version.

lattice representing IR's, except those eigenvalues which fall on the p, q, or Y axes. We can now state Speiser's theorem:

Every IR covered by an eigenvalue of $D(p_1,q_1)$, in the above described way, is contained in the direct product $D(p_1,q_1) \otimes D(p_2,q_2)$ as many times as the multiplicity of the eigenvalue which covers it and with a sign equal to the weight of the sextant. The contributions of the negative weight sextants have therefore to be subtracted from (instead of added to) the Clebsch–Gordan series. Eigenvalues covering the p, q, and Y axes can be neglected.

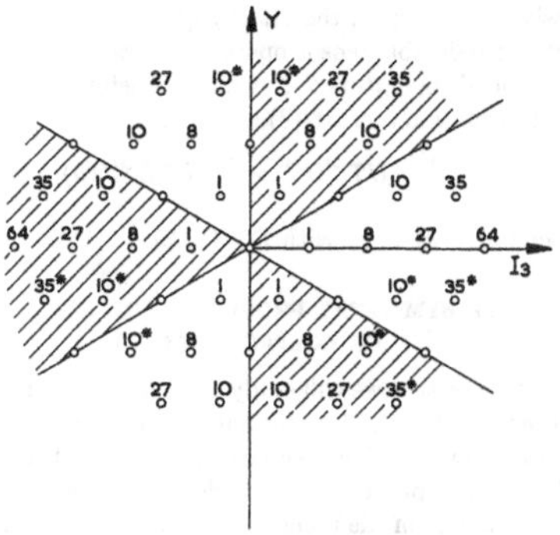

FIG. 6. The lattice. The nonshaded areas have positive weight, the shaded areas negative weight.

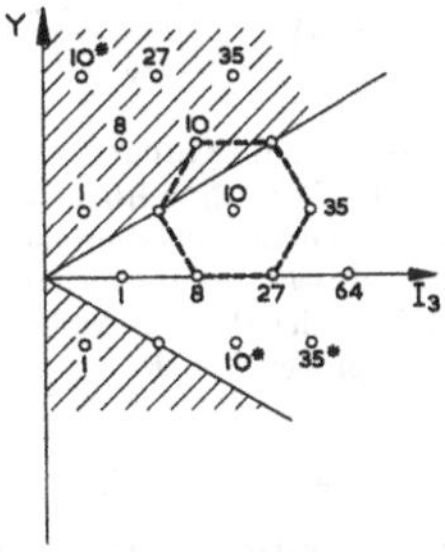

FIG. 7. Determination of the CG series for the direct product {8} ⊗ {10}.

To demonstrate the procedure we will obtain the CG series of {8} ⊗ {10}. We will do this in two ways.

First, we will place the center of the eigenvalue diagram [Fig. 4(a)] of {8} on top of the point {10} in a positive-weight section of the lattice (see Fig. 7). This point (IR) is covered by an eigenvalue of multiplicity two. This gives, therefore, a contribution to the CG series of ⊕ 2 × {10}. The points (IR's) {8}, {27}, and {35} in the positive-weight sextant, and the {10} in the negative-weight sextant are covered by eigenvalues of multiplicity one. They give, therefore, a contribution

$$\{8\} \oplus \{27\} \oplus \{35\} \ominus \{10\}$$

to the CG series. We can neglect the two eigenvalues falling on the p axis. The complete CG series becomes now

$$\begin{aligned}\{8\} \otimes \{10\} &= \{8\} \oplus \{27\} \oplus \{35\} \ominus \{10\} \oplus 2 \\ &\quad \times \{10\} \\ &= \{8\} \oplus \{10\} \oplus \{27\} \oplus \{35\} .\end{aligned}$$

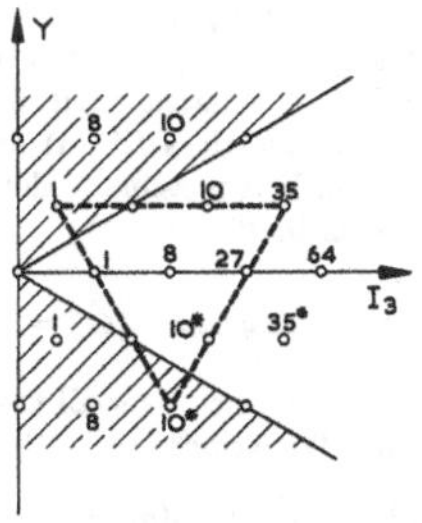

FIG. 8. Determination of the CG series for the direct product {10} ⊗ {8}.

To obtain the CG series the other way, we place the (0,0) eigenvalue of the eigenvalue diagram [Fig. 4(b)] of {10} on the point {8} in a positive-weight sextant of the lattice (see Fig. 8). All the eigenvalues of {10} have multiplicity one. The positive-weight sextant gives the contribution to the CG series

$$\{1\} \oplus \{8\} \oplus \{10\} \oplus \{10^*\} \oplus \{27\} \oplus \{35\} .$$

The negative-weight sextants give the contribution

$$\ominus \{1\} \ominus \{10^*\} .$$

We may neglect again the two eigenvalues of {10} falling on the p and q axes. The complete CG series is, therefore,

$$\{10\} \otimes \{8\} = \{8\} \oplus \{10\} \oplus \{27\} \oplus \{35\}$$

as we have seen before.

13. IRREDUCIBLE REPRESENTATIONS AND IRREDUCIBLE TENSOR OPERATORS

Under a transformation α of $SU(3)$ the vector x in the space C_3 transforms according to

$$x' = \alpha x \,. \tag{13.1}$$

The basis vectors $\psi_\nu^{(\mu)}$ of the IR $\{\mu\} = D(p,q) = \{N\}$ transform then as

$$\psi_\nu'^{(\mu)} = U\psi_\nu^{(\mu)} = \sum_{\nu'=1}^{N} \psi_{\nu'}^{(\mu)} D_{\nu'\nu}^{(\mu)*}(\alpha) \,. \tag{13.2}$$

Here $D^{(\mu)}$ is a unimodular unitary matrix.

Therefore,

$$\sum_{\nu} D_{\nu\nu'}^{(\mu)}(\alpha) D_{\nu\nu''}^{(\mu)*}(\alpha) = \delta_{\nu'\nu''} \tag{13.3}$$

and

$$\psi_\nu^{(\mu)} = \sum_{\nu'} D_{\nu\nu'}^{(\mu)}(\alpha)\psi_{\nu'}'^{(\mu)} \,. \tag{13.4}$$

As already pointed out in Sec. 3, the matrices $D^{(\mu)}$ form the IR $\{\mu\} = D(p,q)$ of the group $SU(3)$. In that same section, we have seen that every transformation of $SU(3)$ is characterized by eight real parameters α_i. The matrices $D^{(\mu)}(\alpha)$ are, therefore, functions of the α_i. It is possible to define a density function $\rho(\alpha_i)$ such that[39]

$$\int d\alpha_1 \cdots d\alpha_8 \rho(\alpha_1, \cdots, \alpha_8) = 1 \tag{13.5}$$

if the integral is performed over all the elements of the group $SU(3)$. We write[40]

$$d\Omega = d\alpha_1 \cdots d\alpha_8 \rho(\alpha_1, \cdots, \alpha_8) \,.$$

Then one can also show that[39,41]

$$\int d\Omega D_{ik}^{(\mu)}(\alpha) D_{jl}^{(\nu)*}(\alpha) = \frac{1}{N_\mu} \delta_{\mu\nu}\delta_{ij}\delta_{kl} \,. \tag{13.6}$$

From the transformation properties (13.2) and (13.4) of the basis vectors of the IR's and the definition (10.3) of the CG coefficients we find the relation

$$D_{\nu_1\lambda_1}^{(\mu_1)}(\alpha) D_{\nu_2\lambda_2}^{(\mu_2)}(\alpha) = \sum_{\mu\nu\lambda\gamma} \begin{pmatrix} \mu_1 & \mu_2 & \mu_\gamma \\ \nu_1 & \nu_2 & \nu \end{pmatrix} \begin{pmatrix} \mu_1 & \mu_2 & \mu_\gamma \\ \lambda_1 & \lambda_2 & \lambda \end{pmatrix} D_{\nu\lambda}^{(\mu)}(\alpha) \tag{13.7}$$

---

[39] Ref. 26, Chap. 8.

[40] For $SU(2)$ we have

$$\int d\Omega = \frac{1}{8\pi^2} \int_0^{2\pi} d\alpha \int_0^{\pi} \sin\beta d\beta \int_0^{2\pi} d\gamma \,,$$

where α, β, and γ are the three Euler angles.

[41] For $SU(2)$ this is the well-known relation

$$\int d\Omega D_{M_1M_1'}^{(J_1)*} D_{M_2M_2'}^{(J_2)*} = \frac{1}{2J_1+1} \delta_{J_1J_2}\delta_{M_1M_2}\delta_{M_1'M_2'} \,.$$

and the inverse relation

$$\sum_{\substack{\nu_1\nu_2 \\ \lambda_1\lambda_2}} \begin{pmatrix} \mu_1 & \mu_2 & \mu_\gamma \\ \nu_1 & \nu_2 & \nu \end{pmatrix} \begin{pmatrix} \mu_1 & \mu_2 & \mu_{\gamma'}' \\ \nu_1 & \nu_2 & \lambda \end{pmatrix} D_{\nu_1\lambda_1}^{(\mu_1)}(\alpha) D_{\nu_2\lambda_2}^{(\mu_2)}(\alpha) = \delta_{\mu\mu'}\delta_{\gamma\gamma'} D_{\nu\lambda}^{(\mu)}(\alpha) \,. \tag{13.8}$$

In (13.2) we defined the transformation of the basis vectors $\psi_\nu^{(\mu)}$ of the IR $D(p,q)$. We can now also define irreducible tensor operators of rank μ. This is a set of N_μ operators $T_\nu^{(\mu)}$ which transform under a transformation α of $SU(3)$ as

$$T_\nu'^{(\mu)} = UT_\nu^{(\mu)}U^{-1} = \sum_{\nu'} T_{\nu'}^{(\mu)} D_{\nu'\nu}^{(\mu)*}(\alpha) \tag{13.9}$$

in complete analogy with Eq. (13.2).

14. SYMMETRY PROPERTIES OF THE CG COEFFICIENTS

In this section we will study some of the symmetry properties of the CG coefficients of $SU(3)$ and of the isoscalar factors. Unfortunately, our lack of knowledge of the operator Γ will reflect itself here in the impossibility of defining rigorously some over-all phase factors. Fortunately, these phase factors are rather unimportant and can be fixed afterwards.

I

If the IR $\{\mu\}_\gamma$ appears in the CG series of $\{\mu_1\} \otimes \{\mu_2\}$ then it will also appear in the CG series $\{\mu_2\} \otimes \{\mu_1\}$, because these series are identical.

Therefore,

$$\begin{pmatrix} \mu_1 & \mu_2 & \mu_\gamma \\ \nu_1 & \nu_2 & \nu \end{pmatrix} = \xi_1 \begin{pmatrix} \mu_2 & \mu_1 & \mu_\gamma \\ \nu_2 & \nu_1 & \nu \end{pmatrix} \tag{14.1}$$

where the $\xi_1 = \pm 1$ according to our phase convention as given in Sec. 10. These $\xi_1 = \xi_1(\mu_1,\mu_2,\mu_\gamma)$ are independent of the "magnetic" quantum numbers ν_1, ν_2, and ν. Therefore, we can take the highest eigenvalue ν_H of $\{\mu\}_\gamma$ and determine there the value of ξ_1.

Because of the property of the CG coefficients of $SU(2)$

$$C_{I_{1z}I_{2z}I_z}^{I_1 I_2 I} = (-)^{I_1+I_2-I} C_{I_{2z}I_{1z}I_z}^{I_2 I_1 I} \,, \tag{14.2}$$

we obtain for the isoscalar factors the relation

$$\left(\begin{matrix} \mu_1 & \mu_2 \\ I_1Y_1 & I_2Y_2 \end{matrix}\middle|\begin{matrix} \mu_\gamma \\ IY \end{matrix}\right) = \xi_1(-)^{I_1+I_2-I} \left(\begin{matrix} \mu_2 & \mu_1 \\ I_2Y_2 & I_1Y_1 \end{matrix}\middle|\begin{matrix} \mu_\gamma \\ IY \end{matrix}\right) . \tag{14.3}$$

From this relation (14.3) we obtain directly that

$$\left(\begin{matrix} \mu_1 & \mu_1 \\ I_1Y_1 & I_1Y_1 \end{matrix}\middle|\begin{matrix} \mu_\gamma \\ I_2Y_1 \end{matrix}\right) = 0 \begin{matrix} \text{for } \xi_1 = 1 \quad \text{if } 2I_1 - I = \text{odd} \\ \text{for } \xi_1 = -1 \text{ if } 2I_1 - I = \text{even} \,. \end{matrix}$$

II

Consider the integral

$$I = \int d\Omega D^{(\mu_3)^*}_{\nu_3\lambda_3}(\alpha) D^{(\mu_2)}_{\nu_2\lambda_2}(\alpha) D^{(\mu_1)}_{\nu_1\lambda_1}(\alpha) . \qquad (14.4)$$

Using (13.7) and (13.6) we obtain

$$I = \frac{1}{N_3} \sum_{\gamma} \begin{pmatrix} \mu_1 & \mu_2 & \mu_{3\gamma} \\ \nu_1 & \nu_2 & \nu_3 \end{pmatrix} \begin{pmatrix} \mu_1 & \mu_2 & \mu_{3\gamma} \\ \lambda_1 & \lambda_2 & \lambda_3 \end{pmatrix} . \qquad (14.5)$$

From (8.2) we can write

$$\phi^{(\mu^*)}_{\nu} = \eta(-)^{\bar{\nu}} \phi^{(\mu)^*}_{-\nu} , \qquad (14.6)$$

where

$$\bar{\nu} = I_z + \tfrac{1}{2}Y \quad \text{and} \quad -\nu = (I, -I_z, -Y) .$$

The phase factor η is only dependent on μ but not on ν. Equation (13.4) together with (14.6) gives

$$D^{(\mu^*)^*}_{\nu'\nu}(\alpha) = (-)^{\bar{\nu}+\bar{\nu}'} D^{(\mu)}_{-\nu'-\nu}(\alpha) . \qquad (14.7)$$

Using (14.7) we can rewrite (14.4) as

$$I = (-)^{\bar{\nu}_3+\bar{\lambda}_3+\bar{\nu}_2+\bar{\lambda}_2} \int d\Omega D^{(\mu_2^*)^*}_{-\nu_2-\lambda_2}(\alpha) D^{(\mu_3^*)}_{-\nu_3-\lambda_3}(\alpha) D^{(\mu_1)}_{\nu_1\lambda_1}(\alpha)$$

$$= \frac{1}{N_2} \sum_{\gamma'} (-)^{\bar{\nu}_3+\bar{\nu}_2} \begin{pmatrix} \mu_1 & \mu_3^* & \mu_{2\gamma'}^* \\ \nu_1 & -\nu_3 & -\nu_2 \end{pmatrix} \times \begin{pmatrix} \mu_1 & \mu_3^* & \mu_{2\gamma'}^* \\ \lambda_1 & -\lambda_3 & -\lambda_2 \end{pmatrix} (-)^{\bar{\lambda}_3+\bar{\lambda}_2} .$$

Comparing this last expression for I with (14.5) and noting that this equation should hold independently of the values of the "magnetic" quantum numbers ν_1, ν_2, ν_3, λ_1, λ_2, and λ_3 gives in case of only one quantum number γ

$$\begin{pmatrix} \mu_1 & \mu_2 & \mu_{3\gamma} \\ \nu_1 & \nu_2 & \nu_3 \end{pmatrix} = \xi_2 (-)^{I_{1z}+Y_1/2} \left(\frac{N_3}{N_2}\right)^{\frac{1}{2}} \begin{pmatrix} \mu_1 & \mu_3^* & \mu_{2\gamma'}^* \\ \nu_1 & -\nu_3 & -\nu_2 \end{pmatrix} . \qquad (14.8)$$

It is expected that this formula holds, or can be made to hold by a suitable choice of Γ, in case there are different eigenvalues γ belonging to the same μ_3. For example, in the case $\{8\} \otimes \{8\} = \{8\}_1 \otimes \{8\}_2$ and $\{8\} \otimes \{27\} = \{27\}_1 \otimes \{27\}_2$, (14.8) does hold. The arbitrary phase $\xi_2 = \xi_2(\mu_1,\mu_2,\mu_3) = \pm 1$ is again independent of the magnetic quantum numbers. This phase can be determined in every specific case by considering the highest eigenvalue.

For the CG coefficients of $SU(2)$ we have the relation

$$C^{I_1\ I_2\ I_3}_{I_{1z} I_{2z} I_{3z}} = (-)^{I_1 - I_{1z}} \left(\frac{2I_3+1}{2I_2+1}\right)^{\frac{1}{2}} C^{I_1\ I_3\ I_2}_{I_{1z}-I_{3z}-I_{2z}} .$$

Therefore, the relation for the isoscalar factors is

$$\left(\begin{matrix} \mu_1 & \mu_2 \\ I_1Y_1 & I_2Y_2 \end{matrix}\middle|\begin{matrix} \mu_{3\gamma} \\ I_3Y_3 \end{matrix}\right) = \xi_2 (-)^{I_1+Y_1/2} \left(\frac{(2I_2+1)N_3}{(2I_3+1)N_2}\right)^{\frac{1}{2}} \times \left(\begin{matrix} \mu_1 & \mu_3^* \\ I_1Y_1 & I_3 -Y_3 \end{matrix}\middle|\begin{matrix} \mu_{2\gamma'}^* \\ I_2 -Y_2 \end{matrix}\right) . \qquad (14.9)$$

From (14.8) and $\left(\begin{smallmatrix} \mu & 1 & \mu \\ \nu & 0 & \nu \end{smallmatrix}\right) = 1$ follows

$$\begin{pmatrix} \mu & \mu^* & 1 \\ \nu & -\nu & 0 \end{pmatrix} = (-)^{I_z+I_H+\frac{1}{2}(Y+Y_H)} N^{-\frac{1}{2}} \qquad (14.10)$$

if the highest eigenvalue of $\{\mu\}$ has $I = I_H$ and $Y = Y_H$; and $\{\mu\} = \{N\}$. We find also

$$\left(\begin{matrix} \mu & \mu^* \\ IY & I-Y \end{matrix}\middle|\begin{matrix} 1 \\ 00 \end{matrix}\right) = (-)^{I+I_H+\frac{1}{2}(Y+Y_H)} \left(\frac{2I+1}{N}\right)^{\frac{1}{2}} . \qquad (14.11)$$

From (14.9) we obtain moreover that

$$\left(\begin{matrix} \mu_1 & \mu_2 \\ I_1 2Y_2 & I_2 -Y_2 \end{matrix}\middle|\begin{matrix} \mu_{2\gamma}^* \\ I_2Y_2 \end{matrix}\right) = 0 \quad \begin{matrix} \text{for } \xi_2 = 1 \\ \text{if } I_1 + Y_2 = \text{odd} , \\ \text{for } \xi_2 = -1 \\ \text{if } I_1 + Y_2 = \text{even} . \end{matrix}$$

III

From the definition (10.3) of the CG coefficients, from their reality property, and the relation (14.6) we deduce

$$\begin{pmatrix} \mu_1 & \mu_2 & \mu_{3\gamma} \\ \nu_1 & \nu_2 & \nu_3 \end{pmatrix} = \xi_3 \begin{pmatrix} \mu_1^* & \mu_2^* & \mu_{3\gamma}^* \\ -\nu_1 & -\nu_2 & -\nu_3 \end{pmatrix} , \qquad (14.12)$$

where ξ_3 is independent of the magnetic quantum numbers ν_1, ν_2, and ν_3, and $\xi_3 = \pm 1$. It is important to note here that the phase convention (8.3) cannot always be applied consistently at the same time to $\{\mu_1\}$, $\{\mu_2\}$, and $\{\mu_3\}$. Therefore, in certain cases $\xi_3 = -1$. Because of the relation

$$C^{I_1\ I_2\ I_3}_{I_{1z} I_{2z} I_{3z}} = (-)^{I_1+I_2-I_3} C^{\ I_1\ \ I_2\ \ I_3}_{-I_{1z}-I_{2z}-I_{3z}} ,$$

we get for the isoscalar factors

$$\left(\begin{matrix} \mu_1 & \mu_2 \\ I_1Y_1 & I_2Y_2 \end{matrix}\middle|\begin{matrix} \mu_{3\gamma} \\ I_3Y_3 \end{matrix}\right) = \xi_3 (-)^{I_1+I_2-I_3} \times \left(\begin{matrix} \mu_1^* & \mu_2^* \\ I_1 -Y_1 & I_2 -Y_2 \end{matrix}\middle|\begin{matrix} \mu_{3\gamma}^* \\ I_3 -Y_3 \end{matrix}\right) . \qquad (14.13)$$

Again the ξ_3 is easily determined in any particular case by considering the highest eigenstates.

15. WIGNER–ECKART THEOREM

The Wigner–Eckart[15] theorem can also be stated for $SU(3)$. This theorem concerns the matrix element $(\phi^{(\mu_3)}_{\nu_3}, T^{(\mu_2)}_{\nu_2} \phi^{(\mu_1)}_{\nu_1})$ of an irreducible tensor operator

$T^{(\mu_2)}_{\nu_2}$ between two basis states, $\phi^{(\mu_1)}_{\nu_1}$ and $\phi^{(\mu_3)}_{\nu_3}$ of IR's. It reads

$$(\phi^{(\mu_3)}_{\nu_3}, T^{(\mu_2)}_{\nu_2}\phi^{(\mu_1)}_{\nu_1}) = \sum_{\gamma} \begin{pmatrix} \mu_1 & \mu_2 & \mu_{3\gamma} \\ \nu_1 & \nu_2 & \nu_3 \end{pmatrix} (\mu_3||T^{(\mu_2)}||\mu_1)_\gamma . \tag{15.1}$$

We have to sum here over γ. The right-hand side contains, therefore, as many terms as the IR $\{\mu_3\}$ is contained in the product $\{\mu_1\} \otimes \{\mu_2\}$. The Eq. (15.1) is a theorem in as far as it predicts the dependence of the matrix element on ν_1, ν_2, and ν_3. At the same time (15.1) is a definition of the reduced matrix elements $(\mu_3||T^{(\mu_2)}||\mu_1)_\gamma$.

We can prove this theorem the following way. From the Eqs. (13.4) and (13.9) follows

$$(\phi^{(\mu_3)}_{\nu_3}, T^{(\mu_2)}_{\nu_2}\phi^{(\mu_1)}_{\nu_1}) = \sum_{\lambda_1\lambda_2\lambda_3} D^{(\mu_3)*}_{\nu_3\lambda_3} D^{(\mu_2)}_{\nu_2\lambda_2} D^{(\mu_1)}_{\nu_1\lambda_1} (\phi'^{(\mu_3)}_{\lambda_3}, T'^{(\mu_2)}_{\lambda_2}\phi'^{(\mu_1)}_{\lambda_1}).$$

Making use of the combination property (13.7) of the D functions one obtains

$$(\phi^{(\mu_3)}_{\nu_3}, T^{(\mu_2)}_{\nu_2}\phi^{(\mu_1)}_{\nu_1}) = \sum_{\substack{\lambda_1\lambda_2\lambda_3 \\ \mu\nu\lambda\gamma}} D^{(\mu_3)*}_{\nu_3\lambda_3} \begin{pmatrix} \mu_1 & \mu_2 & \mu_\gamma \\ \nu_1 & \nu_2 & \nu \end{pmatrix} \begin{pmatrix} \mu_1 & \mu_2 & \mu_\gamma \\ \lambda_1 & \lambda_2 & \lambda \end{pmatrix}$$
$$\times D^{(\mu)}_{\nu\lambda} (\phi'^{(\mu_3)}_{\lambda_3}, T'^{(\mu_2)}_{\lambda_2}\phi'^{(\mu_1)}_{\lambda_1}) .$$

The matrix elements are independent of our choice of coordinate system; they are, therefore, independent of the eight real transformation parameters $\alpha_1, \cdots, \alpha_8$. Thus we can perform the integration over $d\Omega$. After making use of (13.6) we obtain

$$(\phi^{(\mu_3)}_{\nu_3}, T^{(\mu_2)}_{\nu_2}\phi^{(\mu_1)}_{\nu_1}) = \sum_{\gamma} \begin{pmatrix} \mu_1 & \mu_2 & \mu_{3\gamma} \\ \nu_1 & \nu_2 & \nu_3 \end{pmatrix} \frac{1}{N_3} \sum_{\lambda_1\lambda_2\lambda_3} \begin{pmatrix} \mu_1 & \mu_2 & \mu_{3\gamma} \\ \lambda_1 & \lambda_2 & \lambda_3 \end{pmatrix}$$
$$\times (\phi^{(\mu_3)}_{\lambda_3}, T^{(\mu_2)}_{\lambda_2}\phi^{(\mu_1)}_{\lambda_1}) .$$

This gives us the form (15.1) of the Wigner–Eckart theorem if one defines

$$(\mu_3||T^{(\mu_2)}||\mu_1)_\gamma = \frac{1}{N_3} \sum_{\lambda_1\lambda_2\lambda_3} \begin{pmatrix} \mu_1 & \mu_2 & \mu_{3\gamma} \\ \lambda_1 & \lambda_2 & \lambda_3 \end{pmatrix} (\phi^{(\mu_3)}_{\lambda_3}, T^{(\mu_2)}_{\lambda_2}\phi^{(\mu_1)}_{\lambda_1}) .$$

The reduced matrix element is obviously independent of the quantum numbers ν_1, ν_2, and ν_3.

16. THE MASS FORMULA

We will consider the mass formula specifically here for the octet (the IR $\{8\}$). The discussion for every other IR can be done along the same lines. In the octet model one assumes that the strongest interactions are invariant under transformations belonging to the group $SU(3)$. In the absence of any other interactions the particles belonging to the same IR of $SU(3)$ should have, therefore, the same mass, the unitary multiplet mass. The symmetry of these strongest interactions (unitary symmetry) is broken by some unknown weaker mechanism but in such a way that the isospin I and the hypercharge Y are still conserved. Through the action of this unitary symmetry breaking mechanism, the mass degeneracy of the particles belonging to the same unitary multiplet will be removed. If one assumes the simplest form for the symmetry breaking interaction, then one can derive the Okubo mass formula[10,42] for fermions and for bosons[43]

$$M = M_0 + M_1 Y + M_2\{I(I+1) - \tfrac{1}{4}Y^2\} \tag{16.1}$$
$$m^2 = m_0^2 + m_2^2\{I(I+1) - \tfrac{1}{4}Y^2\} . \tag{16.2}$$

These formulas give the Gell-Mann–Okubo mass relations

$$M_N + M_\Xi = \tfrac{3}{2}M_\Lambda + \tfrac{1}{2}M_\Sigma , \tag{16.3a}$$
$$m_K^2 = \tfrac{3}{4}\, m_\eta^2 + \tfrac{1}{4}\, m_\pi^2 , \tag{16.3b}$$
$$m_{K^*}^2 = \tfrac{3}{4}\, m_\omega^2 + \tfrac{1}{4}\, m_\rho^2 . \tag{16.3c}$$

The relations (16.3a) and (16.3b) are very well satisfied; the relation (16.3c), however, is not so well satisfied.

If one introduces then the electromagnetic interaction, this has to be done in such a way that T_3 and Y are still conserved. The presence of this interaction results in mass differences between the different members of the same isomultiplet. If one introduces the electromagnetic interaction, but neglects the presence of the unknown unitary symmetry breaking mechanism, one can derive a relationship between the mass differences of the members of the isomultiplets which make up the same unitary multiplet. For the baryons one obtains[11]

$$M(\Xi^0) - M(\Xi^-) + M(p) - M(n)$$
$$= M(\Sigma^+) - M(\Sigma^-) . \tag{16.4}$$

This relation is also very well satisfied.

We have seen that when the weaker symmetry breaking interaction is switched on, the unitary multiplets split up in isomultiplets. As this breakdown of unitary symmetry has to be done in such a way as to preserve the selection rules for isospin and hypercharge, every such mass operator must have the form

$$0_1 = \sum_{\mu} T^{(\mu)}_{0,0,0} , \tag{16.5}$$

---

[42] The use of the mass for fermions and the (mass)^2 for bosons in formulas like (16.1) and (16.2) seems first to be suggested by R. P. Feynman. It is related to the fact that in the Lagrangian the mass term for bosons is $m^2\phi^+\phi$ and for fermions $M\bar{\psi}\psi$.

[43] We note that for bosons a term linear in Y cannot occur.

where the summation goes over all the physically allowable IR's of $SU(3)$, therefore[43a] $\mu = 1, 8, 27$, etc. The breakdown due to the electromagnetic interaction has to be such that T_3, as well as Y, are conserved. The most general operator[43b] achieving this is, therefore,

$$0_2 = \sum_{\mu,I} T^{(\mu)}_{I,0,0}\,, \tag{16.6}$$

where μ runs over all the possible IR's and I over all the values within an IR consistent with $Y = 0$. One notices that the operator 0_1 is a special case of 0_2. Hence, the general mass operator must have the form

$$\left.\begin{matrix} M \\ m^2 \end{matrix}\right\} = \sum_{\mu,I} T^{(\mu)}_{I,0,0} \tag{16.7}$$

where M is the mass operator for fermions and m^2 for bosons. For the IR $\{8\}$ only the irreducible tensor operators belonging to the IR's $\{1\}$, $\{8\}$, $\{10\}$, $\{10^*\}$, and $\{27\}$ give nonvanishing expectation values. Therefore, if we restrict ourselves only to these operators out of (16.7) which conserve the total I spin, then

$$M\,(\text{or } m^2) = T^{(1)}_{0,0,0} + T^{(8)}_{0,0,0} + T^{(27)}_{0,0,0}\,. \tag{16.8}$$

Using the Wigner–Eckart theorem (15.1) we obtain for the isomultiplet masses of the baryons

$$\begin{aligned}
M_N &= a_1 - (\sqrt{5}/10)a_{8_1} + \tfrac{1}{2}\,a_{8_2} + (1/3\sqrt{5})a_{27}\,,\\
M_\Xi &= a_1 - (\sqrt{5}/10)a_{8_1} - \tfrac{1}{2}\,a_{8_2} + (1/3\sqrt{5})a_{27}\,,\\
M_\Lambda &= a_1 - (\sqrt{5}/5)a_{8_1} \qquad\qquad - (1/\sqrt{5})a_{27}\,,\\
M_\Sigma &= a_1 + (\sqrt{5}/5)a_{8_1} \qquad\qquad - (1/9\sqrt{5})a_{27}\,,
\end{aligned} \tag{16.9}$$

where

$$a_{\mu_\gamma} = (8\,||\,T^{(\mu)}_{0,0,0}\,||\,8)_\gamma\,. \tag{16.10}$$

We have here four masses and four constants. Solving (16.9) for the constants gives

$$\begin{aligned}
a_1 &= \tfrac{1}{8}\,[2M_N + 2M_\Xi + M_\Lambda + 3M_\Sigma]\,,\\
a_{8_1} &= (1/\sqrt{5})[3M_\Sigma - M_\Lambda - M_N - M_\Xi]\\
a_{8_2} &= M_N - M_\Xi\,,\\
a_{27} &= -(9/8\sqrt{5})[3M_\Lambda + M_\Sigma - 2M_N - 2M_\Xi]\,.
\end{aligned} \tag{16.11}$$

The Gell-Mann–Okubo mass relation (16.3a) is based on the assumption that the mass differences transform as $T^{(8)}_{0,0,0}$, i.e., $a_{27} = 0$. In practice this condition is rather well satisfied because [44–46] $a_1 = 1150.84$ MeV/c^2, $a_{8_1} = 91.34$ MeV/c^2, $a_{8_2} = -379.54$ MeV/c^2, and $a_{27} = 11.9$ MeV/c^2.

The mass formula for the electromagnetic mass differences can be obtained by introducing those operators $T^{(\mu)}_{I,0,0}$, which exhibit explicitly the breakdown of isotopic spin conservation. Thus

$$T = T^{(8)}_{1,0,0} + T^{(10)}_{1,0,0} + T^{(10^*)}_{1,0,0} + T^{(27)}_{1,0,0} + T^{(27)}_{2,0,0}\,.$$

We get

$$\begin{aligned}
M_p &= M_N + (\sqrt{15}/90)(9b_{8_1} - 4b_{27})\\
&\qquad + (1/2\sqrt{3})b_{8_2} - (1/\sqrt{15})(b_{10} - b_{10^*})\,,\\
M_n &= M_N - (\sqrt{15}/90)(9b_{8_1} - 4b_{27})\\
&\qquad - (1/2\sqrt{3})b_{8_2} + (1/\sqrt{15})(b_{10} - b_{10^*})\,,\\
M_{\Xi^0} &= M_\Xi - (\sqrt{15}/90)(9b_{8_1} - 4b_{27})\\
&\qquad + (1/2\sqrt{3})b_{8_2} - (1/\sqrt{15})(b_{10} - b_{10^*})\,,\\
M_{\Xi^-} &= M_\Xi + (\sqrt{15}/90)(9b_{8_1} - 4b_{27})\\
&\qquad - (1/2\sqrt{3})b_{8_2} + (1/\sqrt{15})(b_{10} - b_{10^*})\,,\\
M_\Lambda &= M_\Lambda\,,\\
M_{\Sigma^0} &= M_\Sigma + \tfrac{4}{9}\,c_{27}\,,\\
M_{\Sigma^+} &= M_\Sigma - \tfrac{2}{9}\,c_{27} + (1/\sqrt{3})b_{8_2}\\
&\qquad + (1/\sqrt{15})(b_{10} - b_{10^*})\,,\\
M_{\Sigma^-} &= M_\Sigma - \tfrac{2}{9}\,c_{27} - (1/\sqrt{3})b_{8_2}\\
&\qquad - (1/\sqrt{15})(b_{10} - b_{10^*})\,,
\end{aligned} \tag{16.12}$$

where

$$b_{\mu_\gamma} = (8\,||\,T^{(\mu)}_{1,0,0}\,||\,8)_\gamma \tag{16.13}$$

and

$$c_{\mu_\gamma} = (8\,||\,T^{(\mu)}_{2,0,0}\,||\,8)_\gamma\,.$$

We notice from (16.12) that we have here essentially four arbitrary constants $(9b_{8_1} - 4b_{27})$, b_{8_2}, $(b_{10} - b_{10^*})$,

[43a] The IR's $\{10\}$ and $\{10^*\}$ do not have states with $I = I_z = Y = 0$.

[43b] Analogous results can be found in M. A. Rashid and I. I. Yamanaka, Phys. Rev. (to be published).

[44] We use here the following masses:

$$\begin{aligned}
M_p &= 938.21 \pm 0.01 \text{ MeV}/c^2,\\
M_n &= 939.51 \pm 0.01 \text{ MeV}/c^2,\\
M_\Lambda &= 1115.36 \pm 0.14 \text{ MeV}/c^2,\\
M_{\Sigma^+} &= 1189.40 \pm 0.20 \text{ MeV}/c^2,\\
M_{\Sigma^0} &= 1191.5 \pm 0.5 \text{ MeV}/c^2,\\
M_{\Sigma^-} &= 1195.96 \pm 0.30 \text{ MeV}/c^2,\\
M_{\Xi^0} &= 1315.8 \pm 0.8 \text{ MeV}/c^2,\\
M_{\Xi^-} &= 1321.0 \pm 0.5 \text{ MeV}/c^2.
\end{aligned}$$

The first six masses are from tables from W. H. Barkas and A. H. Rosenfeld, University of California, Berkeley (1960), UCRL–8050. The Ξ^- mass is from Ref. 45 and the Ξ^0 mass from Ref. 46.

[45] L. Bertanza, V. Brisson, P. L. Connolly, E. L. Hart, I. S. Mittra, G. C. Moneti, R. R. Rau, N. P. Samios, I. O. Skillicorn, S. S. Yamamoto, M. Goldberg, L. Gray, J. Leitner, S. Lichtman, and J. Westgard, *Proceedings of the 1962 Annual International Conference on High-Energy Physics at CERN* (CERN, Geneva, 1962), p. 437.

[46] D. H. Stork, Bull. Am. Phys. Soc. **8**, 46 (1963).

and c_{27}. However, there are also only four independent mass differences

$$(M_n - M_p)\,,\quad (M_{\Sigma^+} - M_{\Sigma^-})\,,\quad (M_{\Xi^0} - M_{\Xi^-})\,,$$

and

$$[M_{\Sigma^0} - \tfrac{1}{2}(M_{\Sigma^+} + M_{\Sigma^-})]\,.$$

The relation (16.4) between the electromagnetic mass differences can be obtained by assuming $b_{10} - b_{10^*} = 0$. From the experimentally observed masses[44] one obtains $b_{8_1} - \frac{4}{9} b_{27} = (2.52 \pm 0.7)$ MeV$/c^2$, $b_{8_2} = 22.6$ MeV$/c^2$, $b_{10} - b_{10^*} = (-0.12 \pm 0.6)$ MeV$/c^2$.

For the vector mesons, using[47]

$$m_\omega = 782 \text{ MeV}/c^2$$
$$m_\rho = 750 \text{ MeV}/c^2\,,$$

(16.3c) predicts $m_{K^*} = 774$ MeV$/c^2$. Now there exist two K^* resonances, one[6] at 888 MeV$/c^2$ and the other[5] at 730 MeV$/c^2$. None of these values is very close to 774 MeV$/c^2$. However, let us rewrite Eq. (16.9) for the vector mesons. We get

$$\begin{aligned} m_\omega^2 &= m_0^2 - 2m_1^2 - 9m_2^2 \\ m_\rho^2 &= m_0^2 + 2m_1^2 - m_2^2 \\ m_{K^*}^2 &= m_0^2 - m_1^2 + 3m_2^2 \end{aligned} \tag{16.14}$$

where

$$m_0^2 = a_1\,,\quad m_1^2 = (\sqrt{5}/10)a_{8_1}\,,\quad m_2^2 = -(1/9\sqrt{5})a_{27}\,, \quad \text{and} \quad a_{8_2} = 0\,.$$

The assumption made by Okubo[10] to obtain (16.2) was setting arbitrarily $m_2^2 = 0$. One could just as well set arbitrarily $m_1^2 = 0$. This leads to the mass relation

$$m_{K^*}^2 = \tfrac{3}{2} m_\rho^2 - \tfrac{1}{2} m_\omega^2\,. \tag{16.15}$$

This formula predicts the K^* at 733 MeV$/c^2$, surprisingly close to the observed 730 MeV$/c^2$ $K - \pi$ resonance. This might imply that the mass differences in the vector meson octet do not transform according to the representation $\{8\}$, but perhaps according to the representation $\{27\} = D(2,2)$.

Another explanation[48] is that the ω does not belong to the vector octet, but is a unitary singlet. The vector octet consists of the ρ, K^* (888 MeV$/c^2$), and another $I = 0$ vector meson ϕ. The mass relation (16.3c) predicts $m_\phi^2 = \frac{4}{3} m_{K^*}^2 - \frac{1}{3} m^2$, or $m_\phi = 927$ MeV$/c^2$. Due to the presence of the ω, with about the same mass and with the same quantum numbers as the ϕ, this level is pushed up and perhaps this ϕ meson is the 1020-MeV resonance[49] in the $K\bar{K}$ system.

17. YUKAWA COUPLINGS

In the unitary symmetry model of strong interactions very definite relations are predicted between the different meson baryon coupling constants. We shall restrict ourselves here to the Yukawa-type coupling between the baryon octet and the pseudoscalar meson octet. Generalizations to other couplings are obvious. In unitary spin space we shall denote the wave function of the baryons by B and of the mesons by M.

We make the following assignments[50]:

$$\begin{aligned} B_1 &= p\,, & M_1 &= K^+\,, \\ B_2 &= n\,, & M_2 &= K^0\,, \\ B_3 &= -\Sigma^+\,, & M_3 &= -\pi^+\,, \\ B_4 &= \Sigma^0\,, & M_4 &= \pi^0\,, \\ B_5 &= \Sigma^-\,, & M_5 &= \pi^-\,, \\ B_6 &= \Lambda\,, & M_6 &= \eta\,, \\ B_7 &= \Xi^0\,, & M_7 &= \bar{K}^0\,, \\ B_8 &= \Xi^-\,, & M_8 &= -K^-\,. \end{aligned} \tag{17.1}$$

The antiparticles belong to the conjugate representation. As the meson particle representation is equivalent to the antiparticle representation, one must be careful with the phases. We note that our above assignment for the mesons is consistent with the phase conventions (8.2) and (8.3) for the conjugate representation.

With the same conventions the antibaryon wave functions B$^+$ are

$$\begin{aligned} B_1^+ &= -(\Xi^-)^+\,, & B_3^+ &= -(\Sigma^-)^+\,, \\ B_2^+ &= (\Xi^0)^+\,, & B_4^+ &= (\Sigma^0)^+\,, \\ B_7^+ &= n^+\,, & B_5^+ &= (\Sigma^+)^+\,, \\ B_8^+ &= -p^+\,, & B_6^+ &= \Lambda^+\,. \end{aligned} \tag{17.2}$$

We assume an interaction Lagrangian of the Yukawa type

$$\mathcal{L}_{\text{int}} = -g(B^+B)M\,.$$

The Lagrangian should be a unitary singlet, i.e., belong to the IR $\{1\}$. The mesons transform as the

---

47 B. P. Gregory, *Proceedings of the 1962 Annual International Conference on High-Energy Physics at CERN* (CERN, Geneva, 1962), p. 779.

48 J. J. Sakurai, Phys. Rev. Letters 9, 472 (1962).

49 P. Schlein, W. E. Slater, L. T. Smith, D. H. Stork, and H. K. Ticho, Phys. Rev. Letters 10, 368 (1963); P. L. Connolly, E. L. Hart, K. W. Lai, G. London, G. C. Moneti, R. R. Rau, N. P. Samios, I. O. Skillicorn, S. S. Yamamoto, M. Goldberg, M. Gundzik, J. Leitner, and S. Lichtman, Phys. Rev. Letters 10, 371 (1963).

50 We assign $B_3 = -\Sigma^+$ in analogy with $M_3 = -\pi^+$. This gives $\Sigma \cdot \pi = \Sigma^+\pi^- + \Sigma^0\pi^0 + \Sigma^-\pi^+$.

IR {8}. To preserve unitary symmetry the invariant $\mathcal{G} = B^+B$ must also transform as the IR {8}. However, there are two ways to couple $\{8\} \otimes \{8\}$ to {8}. The two possible currents $\mathcal{G}^{(1)}$ and $\mathcal{G}^{(2)}$ are given by

$$\mathcal{G}_\nu^{(\gamma)} = \sum_{\nu_1\nu_2} \begin{pmatrix} 8 & 8 & 8_\gamma \\ \nu_1 & \nu_2 & \nu \end{pmatrix} B_{\nu_1}^+ B_{\nu_2} . \tag{17.3}$$

The interaction Lagrangian is, therefore,

$$\mathfrak{L}_{\text{int}} = -(g_1\mathcal{G}^{(1)} + g_2\mathcal{G}^{(2)})M = -\mathcal{G}M , \tag{17.4}$$

where

$$\mathcal{G}M = \sum_\nu \begin{pmatrix} 8 & 8 & 1 \\ \nu & -\nu & 0 \end{pmatrix} \mathcal{G}_\nu M_{-\nu} . \tag{17.5}$$

Introducing the constants

$$g_p = [(\sqrt{30}/40)g_1 + (\sqrt{6}/24)g_2] , \tag{17.6}$$

$$\alpha = (\sqrt{6}/24)(g_2/g_p) , \tag{17.7}$$

we write the interaction Lagrangian as

$$\begin{aligned}
\mathfrak{L}_{\text{int}} = {} & g_{NN\pi}(N_1^+\boldsymbol{\tau}N_1)\cdot\boldsymbol{\pi} + g_{\Xi\Xi\pi}(N_2^+\boldsymbol{\tau}N_2)\cdot\boldsymbol{\pi} \\
& + g_{\Lambda\Sigma\pi}(\Lambda^+\boldsymbol{\Sigma} + \boldsymbol{\Sigma}^+\Lambda)\cdot\boldsymbol{\pi} - ig_{\Sigma\Sigma\pi}(\boldsymbol{\Sigma}^+ \times \boldsymbol{\Sigma})\cdot\boldsymbol{\pi} \\
& + g_{NN\eta}(N_1^+N_1)\eta + g_{\Xi\Xi\eta}(N_2^+N_2)\eta \\
& + g_{\Lambda\Lambda\eta}(\Lambda^+\Lambda)\eta + g_{\Sigma\Sigma\eta}(\boldsymbol{\Sigma}^+\cdot\boldsymbol{\Sigma})\eta \\
& + g_{N\Lambda K}\{(N_1^+K)\Lambda + \Lambda^+(K^+N_1)\} \\
& + g_{\Xi\Lambda K}\{(N_2^+K_c)\Lambda + \Lambda^+(K_c^+N_2)\} \\
& + g_{N\Sigma K}\{\boldsymbol{\Sigma}^+\cdot(K^+\boldsymbol{\tau}N_1) + (N_1^+\boldsymbol{\tau}K)\cdot\boldsymbol{\Sigma}\} \\
& + g_{\Xi\Sigma K}\{\boldsymbol{\Sigma}^+(K_c^+\boldsymbol{\tau}N_2) + (N_2^+\boldsymbol{\tau}K_c)\cdot\boldsymbol{\Sigma}\} ,
\end{aligned} \tag{17.8}$$

where

$$\begin{aligned}
& g_{NN\pi} = g_p , \quad g_{\Xi\Xi\pi} = -g_p(1 - 2\alpha_p) , \\
& g_{\Lambda\Sigma\pi} = \tfrac{2}{3}\sqrt{3}g_p(1 - \alpha_p) , \quad g_{\Sigma\Sigma\pi} = 2g_p\alpha_p , \\
& g_{NN\eta} = \tfrac{1}{3}\sqrt{3}g_p(4\alpha_p - 1) , \\
& g_{\Xi\Xi\eta} = -\tfrac{1}{3}\sqrt{3}g_p(1 + 2\alpha_p) , \\
& g_{\Sigma\Sigma\eta} = \tfrac{2}{3}\sqrt{3}g_p(1 - \alpha_p) , \quad g_{\Lambda\Lambda\eta} = -\tfrac{2}{3}\sqrt{3}g_p(1 - \alpha_p) , \\
& g_{N\Lambda K} = -\tfrac{1}{3}\sqrt{3}g_p(1 + 2\alpha_p) , \\
& g_{\Xi\Lambda K} = \tfrac{1}{3}\sqrt{3}g_p(4\alpha_p - 1) , \quad g_{N\Sigma K} = g_p(1 - 2\alpha_p) , \\
& g_{\Xi\Sigma K} = -g_p .
\end{aligned} \tag{17.9}$$

We have used here the notation

$$N_1 = \begin{pmatrix} p \\ n \end{pmatrix} , \quad N_2 = \begin{pmatrix} \Xi^0 \\ \Xi^- \end{pmatrix} , \quad K = \begin{pmatrix} K^+ \\ K^0 \end{pmatrix} ,$$

$$K_c = \begin{pmatrix} \bar{K}^0 \\ -K^- \end{pmatrix} , \tag{17.10}$$

Λ, $\boldsymbol{\Sigma}$, $\boldsymbol{\pi}$, and η for the isospin wave functions of the particles. To illustrate the procedure followed we will consider in more detail the coupling of the η meson with the baryons. We have $\eta = M^6$. Equations (17.4), (17.5), and (14.10) give that

$$\mathfrak{L}_{\text{int}} = (1/2\sqrt{2})\mathcal{G}_6 M_6 .$$

Using the isoscalars from Table II, we obtain

$$\begin{aligned}
\mathcal{G}_6^{(1)} = {} & (1/2\sqrt{5})\{B_1^+B_8 + B_8^+B_1 - B_2^+B_7 - B_7^+B_2\} \\
& - (1/\sqrt{5})\{B_3^+B_5 - B_4^+B_4 + B_5^+B_3 \\
& - (1/\sqrt{5})B_6^+B_6 \\
= {} & (-1/2\sqrt{5})\{(\Xi^-)^+\Xi^- + p^+p + (\Xi^0)^+\Xi^0 \\
& + n^+n\} + (1/\sqrt{5})\{(\Sigma^-)^+\Sigma^- + (\Sigma^0)^+\Sigma^0 \\
& + (\Sigma^+)^+\Sigma^+\} - (1/\sqrt{5})\Lambda^+\Lambda \\
= {} & -(1/2\sqrt{5})\{N_1^+N_1 + N_2^+N_2\} \\
& + (1/\sqrt{5})\{\boldsymbol{\Sigma}^+\cdot\boldsymbol{\Sigma} - \Lambda^+\Lambda\}
\end{aligned}$$

and

$$\begin{aligned}
\mathcal{G}_6^{(2)} & = \tfrac{1}{2}\{B_1^+B_8 - B_8^+B_1 - B_2^+B_7 + B_7^+B_2\} \\
& = \tfrac{1}{2}\{N_1^+N_1 - N_2^+N_2\} .
\end{aligned}$$

Therefore, using (17.6), (17.7), and (17.9), one obtains

$$\begin{aligned}
\mathcal{G}_6 = g_1\mathcal{G}_6^{(1)} + g_2\mathcal{G}_6^{(2)} = {} & 2\sqrt{2}\Big[\Big(-\frac{g_1}{4\sqrt{10}} + \frac{g_2}{4\sqrt{2}}\Big)N_1^+N_1 \\
& + \Big(-\frac{g_1}{4\sqrt{10}} - \frac{g_2}{4\sqrt{2}}\Big)N_2^+N_2 \\
& + (1/2\sqrt{10})g_1(\boldsymbol{\Sigma}^+\cdot\boldsymbol{\Sigma} - \Lambda^+\Lambda)\Big] \\
= {} & 2\sqrt{2}[\tfrac{1}{3}\sqrt{3}g_p(4\alpha_p - 1)N_1^+N_1 \\
& - \tfrac{1}{3}\sqrt{3}g_p(1 + 2\alpha_p)N_2^+N_2 \\
& + \tfrac{2}{3}\sqrt{3}g_p(1 - \alpha_p)(\boldsymbol{\Sigma}^+\cdot\boldsymbol{\Sigma} - \Lambda^+\Lambda)] \\
= {} & 2\sqrt{2}[g_{NN\eta}N_1^+N_1 + g_{\Xi\Xi\eta}N_2^+N_2 + g_{\Sigma\Sigma\eta}\boldsymbol{\Sigma}^+\cdot\boldsymbol{\Sigma} \\
& + g_{\Lambda\Lambda\eta}\Lambda^+\Lambda] .
\end{aligned}$$

This leads directly to the interaction Lagrangian (17.8).

Experimentally one knows $g_{NN\pi}$ very well[51] and one has a pretty fair idea about $g_{\Lambda\Sigma\pi}$ and $g_{\Sigma\Sigma\pi}$. This establishes that α_p is small[52] and perhaps[53] $\alpha_p \sim \frac{1}{4}$.

Also the coupling of the vector mesons to the baryons is described by a formula like (17.8). We have only to make the replacement

$$\boldsymbol{\pi} \to \boldsymbol{\varrho} , \quad \eta \to \omega(\text{or } \phi) , \quad K \to K^* , \quad \text{and } K_c \to K_c^* .$$

---

[51] T. Spearman, Nucl. Phys. **16**, 402 (1960); G. Salzman and H. Schnitzer, Phys. Rev. **113**, 1153 (1959).

[52] J. J. de Swart and C. K. Iddings, Phys. Rev. **130**, 319 (1963).

[53] A. W. Martin and K. C. Wali, Phys. Rev. **130**, 2455 (1963).

However, we have now two coupling constants $g_v^{(e)}$ and $g_v^{(m)}$ and two factors $\alpha_v^{(e)}$ and $\alpha_v^{(m)}$ for the electric type and magnetic type of coupling.

In case there should exist a meson ϕ which is a unitary singlet, then the interaction Lagrangian of this meson with the baryon octet is

$$\mathcal{L}_{\text{int}} = g\{N_1^+N_1 + \Lambda^+\Lambda + \mathbf{\Sigma}^+\cdot\mathbf{\Sigma} + N_2^+N_2\}\phi\,. \quad (17.11)$$

A very nice special case of the vector meson octet-baryon octet coupling is obtained by setting $\alpha_v^{(e)} = 1$. We recover then the universal coupling of the $I = 1$ ρ meson to the isospin current and the universal coupling of the $I = 0$ octet vector meson ω (or ϕ) to the hypercharge current as proposed by Sakurai.[54] A unitary singlet vector meson is universally coupled to the baryonic current.

18. TABLES OF ISOSCALAR FACTORS

In this section, we present Tables I–VI of isoscalar factors for the following cases:

$$\{8\} \otimes \{8\} = \{27\} \oplus \{10\} \oplus \{10^*\} \oplus \{8\}_1 \oplus \{8\}_2 \oplus \{1\}$$

$$\{8\} \otimes \{10\} = \{35\} \oplus \{27\} \oplus \{10\} \oplus \{8\}$$

$$\{8\} \otimes \{27\} = \{64\} \oplus \{35\} \oplus \{35^*\} \oplus \{27\}_1 \oplus \{27\}_2 \oplus \{10\} \oplus \{10^*\} \oplus \{8\}$$

$$\{10\} \otimes \{10\} = \{35\} \oplus \{28\} \oplus \{27\} \oplus \{10^*\}$$

$$\{10\} \otimes \{10^*\} = \{64\} \oplus \{27\} \oplus \{8\} \oplus \{1\}\,.$$

With the help of the symmetry properties (14.3), (14.9), and (14.13), these tables can be extended quite a bit. In fact, already some of the entries in the tables are redundant, because they can be obtained with the help of Eqs. (14.3), (14.9), or (14.13) from other parts of the table. However, we feel that omitting these numbers would sometimes be confusing. In Table I we will give for some cases the phases ξ_1, ξ_2, and ξ_3, this will facilitate the extension of the tables of isoscalar factors for these cases.

ACKNOWLEDGMENTS

The author is indebted to Professor R. H. Dalitz for discussions and encouragement. He wishes to thank Dr. A. V. Morales, N. Brene, and Dr. D. R. Speiser for several discussions and helpful comments, and W. Klein for checking the orthogonality of the isoscalar factors in the tables.

[54] J. J. Sakurai, Ann. Phys. (N. Y.) 11, 1 (1960).

TABLE I. Phase factors to be used in Eqs. (14.1), (14.3), (14.8), (14.9), (14.12), and (14.13).

μ_1	μ_2	μ_3	ξ_1	ξ_2	ξ_3
8	8	27	1	−1	1
		10	−1	−1	1
		10*	−1	1	1
		8_1	1	1	1
		8_2	−1	−1	−1
		1	1	−1	1
8	10	35	1	−1	1
		27	−1	1	1
		10	−1	−1	−1
		8	1	1	−1
8	27	64	1	−1	1
		35	−1	−1	1
		35*	−1	1	1
		27_1	1	1	1
		27_2	−1	−1	−1
		10	1	−1	−1
		10*	1	1	−1
		8	1	−1	1

μ_1	μ_2	μ_3	ξ_1	ξ_2	ξ_3
8	10*	35*	1	−1	1
		27	−1	−1	1
		10*	−1	−1	−1
		8	1	−1	−1
10	10	28	1	1	1
		35	−1	−1	1
		27	1	1	1
		10*	−1	−1	1
10	10*	64	1	1	1
		27	−1	1	−1
		8	1	1	1
		1	−1	1	−1
10*	10*	28*	1	−1	1
		35*	−1	−1	1
		27	1	−1	1
		10	−1	−1	1

TABLE II. Isoscalar factors for {8} ⊗ {8}. Given are the isoscalar factors

$$\left(\begin{matrix} 8 & 8 \\ I_1Y_1 & I_2Y_2 \end{matrix}\middle|\begin{matrix} \mu_\gamma \\ I\,Y \end{matrix}\right)$$

for the CG series {8} ⊗ {8} = {27} ⊕ {10} ⊕ {10*} ⊕ {8}₁ ⊕ {8}₂ ⊕ {1}.

$Y = 2 \quad I = 1$

$I_1,\ Y_1;\ I_2,\ Y_2$	27	μ_γ
$\frac{1}{2}$, 1; $\frac{1}{2}$, 1	1	

$Y = 2 \quad I = 0$

$I_1,\ Y_1;\ I_2,\ Y_2$	10*	μ_γ
$\frac{1}{2}$, 1; $\frac{1}{2}$, 1	−1	

TABLE II. (*Continued*)

$Y = 1 \quad I = \frac{3}{2}$

$I_1, Y_1; I_2, Y_2$	27	10	μ_γ
$\frac{1}{2}, 1; 1, 0$	$\sqrt{2}/2$	$-\sqrt{2}/2$	
$1, 0; \frac{1}{2}, 1$	$\sqrt{2}/2$	$\sqrt{2}/2$	

$Y = 1 \quad I = \frac{1}{2}$

$I_1, Y_1; I_2, Y_2$	27	8_1	8_2	10^*	μ_γ
$\frac{1}{2}, 1; 1, 0$	$\sqrt{5}/10$	$3\sqrt{5}/10$	$1/2$	$-1/2$	
$1, 0; \frac{1}{2}, 1$	$-\sqrt{5}/10$	$-3\sqrt{5}/10$	$1/2$	$-1/2$	
$\frac{1}{2}, 1; 0, 0$	$3\sqrt{5}/10$	$-\sqrt{5}/10$	$1/2$	$1/2$	
$0, 0; \frac{1}{2}, 1$	$3\sqrt{5}/10$	$-\sqrt{5}/10$	$-1/2$	$-1/2$	

$Y = 0 \quad I = 2$

$I_1, Y_1; I_2, Y_2$	27	μ_γ
$1, 0; 1, 0$	1	

$Y = 0 \quad I = 1$

$I_1, Y_1; I_2, Y_2$	27	8_1	8_2	10	10^*	μ_γ
$\frac{1}{2}, 1; \frac{1}{2}, -1$	$\sqrt{5}/5$	$-\sqrt{30}/10$	$\sqrt{6}/6$	$-\sqrt{6}/6$	$\sqrt{6}/6$	
$\frac{1}{2}, -1; \frac{1}{2}, 1$	$\sqrt{5}/5$	$-\sqrt{30}/10$	$-\sqrt{6}/6$	$\sqrt{6}/6$	$-\sqrt{6}/6$	
$1, 0; 1, 0$	0	0	$\sqrt{6}/3$	$\sqrt{6}/6$	$-\sqrt{6}/6$	
$1, 0; 0, 0$	$\sqrt{30}/10$	$\sqrt{5}/5$	0	$1/2$	$1/2$	
$0, 0; 1, 0$	$\sqrt{30}/10$	$\sqrt{5}/5$	0	$-1/2$	$-1/2$	

$Y = 0 \quad I = 0$

$I_1, Y_1; I_2, Y_2$	27	8_1	1	8_2	μ_γ
$\frac{1}{2}, 1; \frac{1}{2}, -1$	$\sqrt{15}/10$	$\sqrt{10}/10$	$1/2$	$\sqrt{2}/2$	
$\frac{1}{2}, -1; \frac{1}{2}, 1$	$-\sqrt{15}/10$	$-\sqrt{10}/10$	$-1/2$	$\sqrt{2}/2$	
$1, 0; 1, 0$	$-\sqrt{10}/20$	$-\sqrt{15}/5$	$\sqrt{6}/4$	0	
$0, 0; 0, 0$	$3\sqrt{30}/20$	$-\sqrt{5}/5$	$-\sqrt{2}/4$	0	

$Y = -1 \quad I = \frac{3}{2}$

$I_1, Y_1; I_2, Y_2$	27	10^*	μ_γ
$\frac{1}{2}, -1; 1, 0$	$\sqrt{2}/2$	$-\sqrt{2}/2$	
$1, 0; \frac{1}{2}, -1$	$\sqrt{2}/2$	$\sqrt{2}/2$	

$Y = -1 \quad I = \frac{1}{2}$

$I_1, Y_1; I_2, Y_2$	27	8_1	8_2	10	μ_γ
$\frac{1}{2}, -1; 1, 0$	$-\sqrt{5}/10$	$-3\sqrt{5}/10$	$1/2$	$1/2$	
$1, 0; \frac{1}{2}, -1$	$\sqrt{5}/10$	$3\sqrt{5}/10$	$1/2$	$1/2$	
$\frac{1}{2}, -1; 0, 0$	$3\sqrt{5}/10$	$-\sqrt{5}/10$	$-1/2$	$1/2$	
$0, 0; \frac{1}{2}, -1$	$3\sqrt{5}/10$	$-\sqrt{5}/10$	$1/2$	$-1/2$	

$Y = -2 \quad I = 1$

$I_1, Y_1; I_2, Y_2$	27	μ_γ
$\frac{1}{2}, -1; \frac{1}{2}, -1$	1	

$Y = -2 \quad I = 0$

$I_1, Y_1; I_2, Y_2$	10	μ_γ
$\frac{1}{2}, -1; \frac{1}{2}, -1$	1	

TABLE III. Isoscalar factors for {8} ⊗ {10}. Given are the isoscalar factors

$$\left(\begin{matrix} 8 & 10 \\ I_1Y_1 & I_2Y_2 \end{matrix}\middle|\begin{matrix} \mu_\gamma \\ I\ Y \end{matrix}\right)$$

for the CG series {8} ⊗ {10} = {35} ⊕ {27} ⊕ {10} ⊕ {8}.

$Y = 2 \quad I = 2$

$I_1, Y_1; I_2, Y_2$	35	μ_γ
$\frac{1}{2}, 1; \frac{3}{2}, 1$	1	

$Y = 2 \quad I = 1$

$I_1, Y_1; I_2, Y_2$	27	μ_γ
$\frac{1}{2}, 1; \frac{3}{2}, 1$	-1	

$Y = 1 \quad I = \frac{5}{2}$

$I_1, Y_1; I_2, Y_2$	35	μ_γ
$1, 0; \frac{3}{2}, 1$	1	

$Y = 1 \quad I = \frac{3}{2}$

$I_1, Y_1; I_2, Y_2$	35	27	10	μ_γ
$1, 0; \frac{3}{2}, 1$	$-1/4$	$-\sqrt{5}/4$	$\sqrt{10}/4$	
$0, 0; \frac{3}{2}, 1$	$\sqrt{5}/4$	$-3/4$	$-\sqrt{2}/4$	
$\frac{1}{2}, 1; 1, 0$	$\sqrt{10}/4$	$\sqrt{2}/4$	$1/2$	

$Y = 1 \quad I = \frac{1}{2}$

$I_1, Y_1; I_2, Y_2$	27	8	μ_γ
$1, 0; \frac{3}{2}, 1$	$\sqrt{5}/5$	$-2\sqrt{5}/5$	
$\frac{1}{2}, 1; 1, 0$	$-2\sqrt{5}/5$	$-\sqrt{5}/5$	

$Y = 0 \quad I = 2$

$I_1, Y_1; I_2, Y_2$	35	27	μ_γ
$1, 0; 1, 0$	$\sqrt{3}/2$	$1/2$	
$\frac{1}{2}, -1; \frac{3}{2}, 1$	$1/2$	$-\sqrt{3}/2$	

$Y = 0 \quad I = 1$

$I_1, Y_1; I_2, Y_2$	35	27	10	8	μ_γ
$1, 0; 1, 0$	$-\sqrt{3}/6$	$-3\sqrt{5}/10$	$\sqrt{3}/3$	$\sqrt{30}/15$	
$0, 0; 1, 0$	$\sqrt{2}/2$	$-\sqrt{30}/10$	0	$-\sqrt{5}/5$	
$\frac{1}{2}, 1; \frac{1}{2}, -1$	$\sqrt{3}/3$	$\sqrt{5}/5$	$\sqrt{3}/3$	$\sqrt{30}/15$	
$\frac{1}{2}, -1; \frac{3}{2}, 1$	$-\sqrt{3}/6$	$\sqrt{5}/10$	$\sqrt{3}/3$	$-2\sqrt{30}/15$	

TABLE III. (*Continued*)

$Y = 0 \quad I = 0$

$I_1, Y_1; I_2, Y_2$	27	8	μ_γ
$1, 0; 1, 0$	$\sqrt{10}/5$	$-\sqrt{15}/5$	
$\frac{1}{2}, 1; \frac{1}{2}, -1$	$-\sqrt{15}/5$	$-\sqrt{10}/5$	

$Y = -1 \quad I = \frac{3}{2}$

$I_1, Y_1; I_2, Y_2$	35	27	μ_γ
$1, 0; \frac{1}{2}, -1$	$\sqrt{2}/2$	$\sqrt{2}/2$	
$\frac{1}{2}, -1; 1, 0$	$\sqrt{2}/2$	$-\sqrt{2}/2$	

$Y = -1 \quad I = \frac{1}{2}$

$I_1, Y_1; I_2, Y_2$	35	27	10	8	μ_γ
$1, 0; \frac{1}{2}, -1$	$-1/4$	$-7\sqrt{5}/20$	$\sqrt{2}/4$	$\sqrt{5}/5$	
$0, 0; \frac{1}{2}, -1$	$3/4$	$-3\sqrt{5}/20$	$\sqrt{2}/4$	$-\sqrt{5}/5$	
$\frac{1}{2}, 1; 0, -2$	$\sqrt{2}/4$	$3\sqrt{10}/20$	$1/2$	$\sqrt{10}/5$	
$\frac{1}{2}, -1; 1, 0$	$-1/2$	$\sqrt{5}/10$	$\sqrt{2}/2$	$-\sqrt{5}/5$	

$Y = -2 \quad I = 1$

$I_1, Y_1; I_2, Y_2$	35	27	μ_γ
$1, 0; 0, -2$	$1/2$	$\sqrt{3}/2$	
$\frac{1}{2}, -1; \frac{1}{2}, -1$	$\sqrt{3}/2$	$-1/2$	

$Y = -2 \quad I = 0$

$I_1, Y_1; I_2, Y_2$	35	10	μ_γ
$0, 0; 0, -2$	$\sqrt{2}/2$	$\sqrt{2}/2$	
$\frac{1}{2}, -1; \frac{1}{2}, -1$	$-\sqrt{2}/2$	$\sqrt{2}/2$	

$Y = -3 \quad I = \frac{1}{2}$

$I_1, Y_1; I_2, Y_2$	35	μ_γ
$\frac{1}{2}, -1; 0, -2$	1	

TABLE IV. Isoscalar factors for {8} ⊗ {27}. Given are the isoscalar factors

$$\left(\begin{matrix} 8 & 27 \\ I_1Y_1 & I_2Y_2 \end{matrix}\middle| \begin{matrix} \mu_\gamma \\ I\ Y \end{matrix}\right)$$

for the CG series $\{8\} \otimes \{27\} = \{64\} \oplus \{35\} \oplus \{35^*\} \oplus \{27\}_1 \oplus \{27\}_2 \oplus \{10\} \oplus \{10^*\} \oplus \{8\}$.

$Y = 3 \quad I = \frac{3}{2}$

$I_1, Y_1; I_2, Y_2$	64	μ_γ
$\frac{1}{2}, 1; 1, 2$	1	

$Y = 3 \quad I = \frac{1}{2}$

$I_1, Y_1; I_2, Y_2$	35*	μ_γ
$\frac{1}{2}, 1; 1, 2$	-1	

$Y = 2 \quad I = 2$

$I_1, Y_1; I_2, Y_2$	64	35	μ_γ
$\frac{1}{2}, 1; \frac{3}{2}, 1$	$\sqrt{6}/3$	$-\sqrt{3}/3$	
$1, 0; 1, 2$	$\sqrt{3}/3$	$\sqrt{6}/3$	

$Y = 2 \quad I = 1$

$I_1, Y_1; I_2, Y_2$	64	35*	27_1	27_2	μ_γ
$\frac{1}{2}, 1; \frac{3}{2}, 1$	$\sqrt{14}/21$	$-2/3$	$\sqrt{70}/14$	$\sqrt{6}/6$	
$\frac{1}{2}, 1; \frac{1}{2}, 1$	$2\sqrt{70}/21$	$\sqrt{5}/6$	$-\sqrt{14}/28$	$\sqrt{30}/12$	
$1, 0; 1, 2$	$-\sqrt{21}/21$	$-\sqrt{6}/6$	$-\sqrt{105}/14$	$1/2$	
$0, 0; 1, 2$	$\sqrt{14}/7$	$-1/2$	$-\sqrt{70}/28$	$-\sqrt{6}/4$	

$Y = 2 \quad I = 0$

$I_1, Y_1; I_2, Y_2$	35*	10*	μ_γ
$\frac{1}{2}, 1; \frac{1}{2}, 1$	$-\sqrt{30}/6$	$-\sqrt{6}/6$	
$1, 0; 1, 2$	$\sqrt{6}/6$	$-\sqrt{30}/6$	

$Y = 1 \quad I = \frac{5}{2}$

$I_1, Y_1; I_2, Y_2$	64	35	μ_γ
$\frac{1}{2}, 1; 2, 0$	$\sqrt{3}/3$	$-\sqrt{6}/3$	
$1, 0; \frac{3}{2}, 1$	$\sqrt{6}/3$	$\sqrt{3}/3$	

$Y = 1 \quad I = \frac{3}{2}$

$I_1, Y_1; I_2, Y_2$	64	35	35*	27_1	27_2	10	μ_γ
$\frac{1}{2}, 1; 2, 0$	$\sqrt{7}/21$	$-1/12$	$-\sqrt{5}/6$	$5\sqrt{42}/56$	$\sqrt{10}/8$	$-5\sqrt{2}/12$	
$\frac{1}{2}, 1; 1, 0$	$5\sqrt{7}/21$	$-5/12$	$\sqrt{5}/6$	$-3\sqrt{42}/56$	$\sqrt{10}/8$	$-\sqrt{2}/12$	
$1, 0; \frac{3}{2}, 1$	$-\sqrt{21}/63$	$7\sqrt{3}/36$	$-\sqrt{15}/9$	$-5\sqrt{14}/56$	$\sqrt{30}/8$	$5\sqrt{6}/36$	
$1, 0; \frac{1}{2}, 1$	$5\sqrt{42}/63$	$5\sqrt{6}/18$	$\sqrt{30}/18$	$\sqrt{7}/7$	0	$\sqrt{3}/9$	
$0, 0; \frac{3}{2}, 1$	$\sqrt{105}/21$	$-\sqrt{15}/12$	$-\sqrt{3}/3$	$\sqrt{70}/56$	$-\sqrt{6}/8$	$\sqrt{30}/12$	
$\frac{1}{2}, -1; 1, 2$	$\sqrt{35}/21$	$\sqrt{5}/6$	$-1/3$	$-\sqrt{210}/28$	$-\sqrt{2}/4$	$-\sqrt{10}/6$	

$Y = 1 \quad I = \frac{1}{2}$

$I_1, Y_1; I_2, Y_2$	64	35*	27_1	27_2	10*	8	μ_γ
$\frac{1}{2}, 1; 1, 0$	$\sqrt{35}/21$	$-\sqrt{10}/6$	$3\sqrt{105}/70$	$1/2$	$-1/3$	$2\sqrt{5}/15$	
$\frac{1}{2}, 1; 0, 0$	$2\sqrt{35}/21$	$\sqrt{10}/6$	$-\sqrt{105}/70$	$1/2$	$1/3$	$\sqrt{5}/15$	
$1, 0; \frac{3}{2}, 1$	$-\sqrt{42}/63$	$\sqrt{3}/9$	$-\sqrt{14}/7$	0	$-\sqrt{30}/9$	$2\sqrt{6}/9$	
$1, 0; \frac{1}{2}, 1$	$-\sqrt{210}/63$	$-5\sqrt{15}/36$	$-19\sqrt{70}/280$	$\sqrt{6}/8$	$7\sqrt{6}/36$	$\sqrt{30}/45$	
$0, 0; \frac{1}{2}, 1$	$\sqrt{210}/21$	$-\sqrt{15}/12$	$-13\sqrt{70}/280$	$-\sqrt{6}/8$	$-\sqrt{6}/12$	$-\sqrt{30}/15$	
$\frac{1}{2}, -1; 1, 2$	$-2\sqrt{7}/21$	$\sqrt{2}/12$	$-\sqrt{21}/28$	$\sqrt{5}/4$	$-\sqrt{5}/6$	$-2/3$	

TABLE IV. (*Continued*)

$Y = 0 \quad I = 3$

$I_1, Y_1; I_2, Y_2$	64	μ_γ
1, 0; 2, 0	1	

$Y = 0 \quad I = 2$

$I_1, Y_1; I_2, Y_2$	64	35	35*	27_1	27_2	μ_γ
$\frac{1}{2}$, 1; $\frac{3}{2}$, −1	$2\sqrt{21}/21$	$-\sqrt{3}/3$	$\sqrt{3}/6$	$-\sqrt{210}/28$	$\sqrt{2}/4$	
$\frac{1}{2}$, −1; $\frac{3}{2}$, 1	$2\sqrt{21}/21$	$\sqrt{3}/6$	$-\sqrt{3}/3$	$-\sqrt{210}/28$	$-\sqrt{2}/4$	
1, 0; 1, 0	$\sqrt{210}/21$	$\sqrt{30}/12$	$\sqrt{30}/12$	$\sqrt{21}/14$	0	
1, 0; 2, 0	0	$\sqrt{2}/4$	$-\sqrt{2}/4$	0	$\sqrt{3}/2$	
0, 0; 2, 0	$\sqrt{7}/7$	−1/2	−1/2	$\sqrt{70}/14$	0	

$Y = 0 \quad I = 1$

$I_1, Y_1; I_2, Y_2$	64	35	35*	27_1	27_2	10	10*	8	μ_γ
$\frac{1}{2}$, 1; $\frac{3}{2}$, −1	$2\sqrt{35}/63$	−1/9	−5/18	$3\sqrt{14}/28$	$\sqrt{30}/12$	$-2\sqrt{5}/9$	$-\sqrt{5}/9$	4/9	
$\frac{1}{2}$, −1; $\frac{3}{2}$, 1	$-2\sqrt{35}/63$	5/18	1/9	$-3\sqrt{14}/28$	$\sqrt{30}/12$	$-\sqrt{5}/9$	$-2\sqrt{5}/9$	−4/9	
$\frac{1}{2}$, 1; $\frac{1}{2}$, −1	$10\sqrt{7}/63$	$-\sqrt{5}/9$	$2\sqrt{5}/9$	$-3\sqrt{70}/70$	$\sqrt{6}/6$	−1/9	4/9	$2\sqrt{5}/45$	
$\frac{1}{2}$, −1; $\frac{1}{2}$, 1	$10\sqrt{7}/63$	$2\sqrt{5}/9$	$-\sqrt{5}/9$	$-3\sqrt{70}/70$	$-\sqrt{6}/6$	−4/9	1/9	$2\sqrt{5}/45$	
0, 0; 1, 0	$5\sqrt{7}/21$	$-\sqrt{5}/6$	$-\sqrt{5}/6$	$-\sqrt{70}/70$	0	1/3	−1/3	$-2\sqrt{5}/15$	
1, 0; 0, 0	$10\sqrt{7}/63$	$2\sqrt{5}/9$	$2\sqrt{5}/9$	$4\sqrt{70}/70$	0	2/9	−2/9	$-\sqrt{5}/45$	
1, 0; 1, 0	0	$\sqrt{30}/12$	$-\sqrt{30}/12$	0	1/2	$\sqrt{6}/6$	$\sqrt{6}/6$	0	
1, 0; 2, 0	$-\sqrt{14}/63$	$\sqrt{10}/36$	$\sqrt{10}/36$	$-\sqrt{35}/14$	0	$5\sqrt{2}/18$	$-5\sqrt{2}/18$	$2\sqrt{10}/9$	

$Y = 0 \quad I = 0$

$I_1, Y_1; I_2, Y_2$	64	27_1	27_2	8	μ_γ
$\frac{1}{2}$, 1; $\frac{1}{2}$, −1	$2\sqrt{21}/21$	$\sqrt{210}/70$	$\sqrt{2}/2$	$2\sqrt{15}/15$	
$\frac{1}{2}$, −1; $\frac{1}{2}$, 1	$-2\sqrt{21}/21$	$-\sqrt{210}/70$	$\sqrt{2}/2$	$-2\sqrt{15}/15$	
1, 0; 1, 0	$-\sqrt{21}/21$	$-4\sqrt{210}/70$	0	$2\sqrt{15}/15$	
0, 0; 0, 0	$2\sqrt{7}/7$	$-4\sqrt{70}/70$	0	$-\sqrt{5}/5$	

$Y = -1 \quad I = \frac{5}{2}$

$I_1, Y_1; I_2, Y_2$	64	35*	μ_γ
$\frac{1}{2}$, −1; 1, 0	$\sqrt{3}/3$	$-\sqrt{6}/3$	
1, 0; $\frac{3}{2}$, −1	$\sqrt{6}/3$	$\sqrt{3}/3$	

$Y = -1 \quad I = \frac{3}{2}$

$I_1, Y_1; I_2, Y_2$	64	35*	35	27_1	27_2	10*	μ_γ
$\frac{1}{2}$, −1; 2, 0	$-\sqrt{7}/21$	1/12	$\sqrt{5}/6$	$-5\sqrt{42}/56$	$\sqrt{10}/8$	$-5\sqrt{2}/12$	
$\frac{1}{2}$, −1; 1, 0	$5\sqrt{7}/21$	−5/12	$\sqrt{5}/6$	$-3\sqrt{42}/56$	$-\sqrt{10}/8$	$\sqrt{2}/12$	
1, 0; $\frac{3}{2}$, −1	$\sqrt{21}/63$	$-7\sqrt{3}/36$	$\sqrt{15}/9$	$5\sqrt{14}/56$	$\sqrt{30}/8$	$5\sqrt{6}/36$	
1, 0; $\frac{1}{2}$, −1	$5\sqrt{42}/63$	$5\sqrt{6}/18$	$\sqrt{30}/18$	$\sqrt{7}/7$	0	$-\sqrt{3}/9$	
0, 0; $\frac{3}{2}$, −1	$\sqrt{105}/21$	$-\sqrt{15}/21$	$-\sqrt{3}/3$	$\sqrt{70}/56$	$\sqrt{6}/8$	$-\sqrt{30}/12$	
$\frac{1}{2}$, 1; 1, −2	$\sqrt{35}/21$	$\sqrt{5}/6$	−1/3	$-\sqrt{210}/28$	$\sqrt{2}/4$	$\sqrt{10}/6$	

$Y = -1 \quad I = \frac{1}{2}$

$I_1, Y_1; I_2, Y_2$	64	35	27_1	27_2	10	8	μ_γ
$\frac{1}{2}$, −1; 1, 0	$-\sqrt{35}/21$	$\sqrt{10}/6$	$-3\sqrt{105}/70$	1/2	−1/3	$-2\sqrt{5}/15$	
$\frac{1}{2}$, −1; 0, 0	$2\sqrt{35}/21$	$\sqrt{10}/6$	$-\sqrt{105}/70$	−1/2	−1/3	$\sqrt{5}/15$	
1, 0; $\frac{3}{2}$, −1	$-\sqrt{42}/63$	$\sqrt{3}/9$	$-\sqrt{14}/7$	0	$\sqrt{30}/9$	$2\sqrt{6}/9$	
1, 0; $\frac{1}{2}$, −1	$\sqrt{210}/63$	$5\sqrt{15}/36$	$-19\sqrt{70}/280$	$\sqrt{6}/8$	$7\sqrt{6}/36$	$-\sqrt{30}/45$	
0, 0; $\frac{1}{2}$, −1	$\sqrt{210}/21$	$-\sqrt{15}/12$	$-13\sqrt{70}/280$	$\sqrt{6}/8$	$\sqrt{6}/12$	$-\sqrt{30}/15$	
$\frac{1}{2}$, 1; 1, −2	$2\sqrt{7}/21$	$-\sqrt{2}/12$	$\sqrt{21}/28$	$\sqrt{5}/4$	$-\sqrt{5}/6$	2/3	

$Y = -2 \quad I = 2$

$I_1, Y_1; I_2, Y_2$	64	35*	μ_γ
$\frac{1}{2}$, −1; $\frac{3}{2}$, −1	$\sqrt{6}/3$	$-\sqrt{3}/3$	
1, 0; 1, −2	$\sqrt{3}/3$	$\sqrt{6}/3$	

$Y = -2 \quad I = 1$

$I_1, Y_1; I_2, Y_2$	64	35	27_1	27_2	μ_γ
$\frac{1}{2}$, −1; $\frac{3}{2}$, −1	$-\sqrt{14}/21$	2/3	$-\sqrt{70}/14$	$\sqrt{6}/6$	
$\frac{1}{2}$, −1; $\frac{1}{2}$, −1	$2\sqrt{70}/21$	$\sqrt{5}/6$	$-\sqrt{14}/28$	$-\sqrt{30}/12$	
1, 0; 1, −2	$\sqrt{21}/21$	$\sqrt{6}/6$	$\sqrt{105}/14$	1/2	
0, 0; 1, −2	$\sqrt{14}/7$	−1/2	$-\sqrt{70}/28$	$\sqrt{6}/4$	

$Y = -2 \quad I = 0$

$I_1, Y_1; I_2, Y_2$	35	10	μ_γ
$\frac{1}{2}$, −1; $\frac{1}{2}$, −1	$\sqrt{30}/6$	$-\sqrt{6}/6$	
1, 0; 1, −2	$\sqrt{6}/6$	$\sqrt{30}/6$	

$Y = -3 \quad I = \frac{3}{2}$

$I_1, Y_1; I_2, Y_2$	64	μ_γ
$\frac{1}{2}$, −1; 1, −2	1	

$Y = -3 \quad I = \frac{1}{2}$

$I_1, Y_1; I_2, Y_2$	35	μ_γ
$\frac{1}{2}$, −1; 1, −2	1	

TABLE V. Isoscalar factors for {10} ⊗ {10}. Given are the isoscalar factors

$$\left(\begin{matrix} 10 & 10 \\ I_1Y_1 & I_2Y_2 \end{matrix}\middle| \begin{matrix} \mu_\gamma \\ I\ Y \end{matrix}\right)$$

for the CG series {10} ⊗ {10} = {35} ⊕ {28} ⊕ {27} ⊕ {10*}.

$Y = 2 \quad I = 3$

$I_1, Y_1; I_2, Y_2$	28	μ_γ
$\frac{3}{2}, 1; \frac{3}{2}, 1$	1	

$Y = 2 \quad I = 2$

$I_1, Y_1; I_2, Y_2$	35	μ_γ
$\frac{3}{2}, 1; \frac{3}{2}, 1$	-1	

$Y = 2 \quad I = 1$

$I_1, Y_1; I_2, Y_2$	27	μ_γ
$\frac{3}{2}, 1; \frac{3}{2}, 1$	1	

$Y = 2 \quad I = 0$

$I_1, Y_1; I_2, Y_2$	10*	μ_γ
$\frac{3}{2}, 1; \frac{3}{2}, 1$	-1	

$Y = 1 \quad I = \frac{5}{2}$

$I_1, Y_1; I_2, Y_2$	28	35	μ_γ
$\frac{3}{2}, 1; 1, 0$	$\sqrt{2}/2$	$\sqrt{2}/2$	
$1, 0; \frac{3}{2}, 1$	$\sqrt{2}/2$	$-\sqrt{2}/2$	

$Y = 1 \quad I = \frac{3}{2}$

$I_1, Y_1; I_2, Y_2$	35	27	μ_γ
$\frac{3}{2}, 1; 1, 0$	$-\sqrt{2}/2$	$-\sqrt{2}/2$	
$1, 0; \frac{3}{2}, 1$	$-\sqrt{2}/2$	$\sqrt{2}/2$	

$Y = 1 \quad I = \frac{1}{2}$

$I_1, Y_1; I_2, Y_2$	27	10*	μ_γ
$\frac{3}{2}, 1; 1, 0$	$\sqrt{2}/2$	$\sqrt{2}/2$	
$1, 0; \frac{3}{2}, 1$	$\sqrt{2}/2$	$-\sqrt{2}/2$	

$Y = 0 \quad I = 2$

$I_1, Y_1; I_2, Y_2$	28	35	27	μ_γ
$\frac{3}{2}, 1; \frac{1}{2}, -1$	$\sqrt{5}/5$	$\sqrt{2}/2$	$\sqrt{30}/10$	
$\frac{1}{2}, -1; \frac{3}{2}, 1$	$\sqrt{5}/5$	$-\sqrt{2}/2$	$\sqrt{30}/10$	
$1, 0; 1, 0$	$\sqrt{15}/5$	0	$-\sqrt{10}/5$	

$Y = 0 \quad I = 1$

$I_1, Y_1; I_2, Y_2$	35	27	10*	μ_γ
$\frac{3}{2}, 1; \frac{1}{2}, -1$	$-\sqrt{6}/6$	$-\sqrt{2}/2$	$-\sqrt{3}/3$	
$\frac{1}{2}, -1; \frac{3}{2}, 1$	$-\sqrt{6}/6$	$\sqrt{2}/2$	$-\sqrt{3}/3$	
$1, 0; 1, 0$	$-\sqrt{6}/3$	0	$\sqrt{3}/3$	

$Y = 0 \quad I = 0$

$I_1, Y_1; I_2, Y_2$	27	μ_γ
$1, 0; 1, 0$	1	

$Y = -1 \quad I = \frac{3}{2}$

$I_1, Y_1; I_2, Y_2$	28	35	27	10*	μ_γ
$\frac{3}{2}, 1; 0, -2$	$\sqrt{5}/10$	$1/2$	$3\sqrt{5}/10$	$1/2$	
$0, -2; \frac{3}{2}, 1$	$\sqrt{5}/10$	$-1/2$	$3\sqrt{5}/10$	$-1/2$	
$1, 0; \frac{1}{2}, -1$	$3\sqrt{5}/10$	$1/2$	$-\sqrt{5}/10$	$-1/2$	
$\frac{1}{2}, -1; 1, 0$	$3\sqrt{5}/10$	$-1/2$	$-\sqrt{5}/10$	$1/2$	

$Y = -1 \quad I = \frac{1}{2}$

$I_1, Y_1; I_2, Y_2$	35	27	μ_γ
$1, 0; \frac{1}{2}, -1$	$-\sqrt{2}/2$	$-\sqrt{2}/2$	
$\frac{1}{2}, -1; 1, 0$	$-\sqrt{2}/2$	$\sqrt{2}/2$	

$Y = -2 \quad I = 1$

$I_1, Y_1; I_2, Y_2$	28	35	27	μ_γ
$1, 0; 0, -2$	$\sqrt{5}/5$	$\sqrt{2}/2$	$\sqrt{30}/10$	
$0, -2; 1, 0$	$\sqrt{5}/5$	$-\sqrt{2}/2$	$\sqrt{30}/10$	
$\frac{1}{2}, -1; \frac{1}{2}, -1$	$\sqrt{15}/5$	0	$-\sqrt{10}/5$	

$Y = -2 \quad I = 0$

$I_1, Y_1; I_2, Y_2$	35	μ_γ
$\frac{1}{2}, -1; \frac{1}{2}, -1$	-1	

$Y = -3 \quad I = \frac{1}{2}$

$I_1, Y_1; I_2, Y_2$	28	35	μ_γ
$\frac{1}{2}, -1; 0, -2$	$\sqrt{2}/2$	$\sqrt{2}/2$	
$0, -2; \frac{1}{2}, -1$	$\sqrt{2}/2$	$-\sqrt{2}/2$	

$Y = -4 \quad I = 0$

$I_1, Y_1; I_2, Y_2$	28	μ_γ
$0, -2; 0, -2$	1	

TABLE VI. Isoscalar factors for {10} ⊗ {10*}. Given are the isoscalar factors

$$\left(\begin{matrix} 10 & 10^* \\ I_1Y_1 & I_2Y_2 \end{matrix}\middle|\begin{matrix} \mu_\gamma \\ I\,Y \end{matrix}\right)$$

for the CG series $\{10\} \otimes \{10\}^* = \{64\} \oplus \{27\} \oplus \{8\} \oplus \{1\}$.

$Y = 3 \quad I = \frac{3}{2}$

$I_1, Y_1; I_2, Y_2$	64	μ_γ
$\frac{3}{2}$, 1; 0, 2	1	

$Y = 2 \quad I = 2$

$I_1, Y_1; I_2, Y_2$	64	μ_γ
$\frac{3}{2}$, 1; $\frac{1}{2}$, 1	1	

$Y = 2 \quad I = 1$

$I_1, Y_1; I_2, Y_2$	64	27	μ_γ
$\frac{3}{2}$, 1; $\frac{1}{2}$, 1	$\sqrt{21}/7$	$2\sqrt{7}/7$	
1, 0; 0, 2	$2\sqrt{7}/7$	$-\sqrt{21}/7$	

$Y = 1 \quad I = \frac{5}{2}$

$I_1, Y_1; I_2, Y_2$	64	μ_γ
$\frac{5}{2}$, 1; 1, 0	1	

$Y = 1 \quad I = \frac{3}{2}$

$I_1, Y_1; I_2, Y_2$	64	27	μ_γ
$\frac{3}{2}$, 1; 1, 0	$\sqrt{14}/7$	$\sqrt{35}/7$	
1, 0; $\frac{1}{2}$, 1	$\sqrt{35}/7$	$-\sqrt{14}/7$	

$Y = 1 \quad I = \frac{1}{2}$

$I_1, Y_1; I_2, Y_2$	64	27	8	μ_γ
$\frac{3}{2}$, 1; 1, 0	$\sqrt{7}/7$	$4\sqrt{35}/35$	$\sqrt{10}/5$	
1, 0; $\frac{1}{2}$, 1	$2\sqrt{7}/7$	$\sqrt{35}/35$	$-\sqrt{10}/5$	
$\frac{1}{2}$, −1; 0, 2	$\sqrt{14}/7$	$-3\sqrt{70}/35$	$\sqrt{5}/5$	

$Y = 0 \quad I = 3$

$I_1, Y_1; I_2, Y_2$	64	μ_γ
$\frac{3}{2}$, 1; $\frac{3}{2}$, −1	1	

$Y = 0 \quad I = 2$

$I_1, Y_1; I_2, Y_2$	64	27	μ_γ
$\frac{3}{2}$, 1; $\frac{3}{2}$, −1	$\sqrt{7}/7$	$\sqrt{42}/7$	
1, 0; 1, 0	$\sqrt{42}/7$	$-\sqrt{7}/7$	

$Y = 0 \quad I = 1$

$I_1, Y_1; I_2, Y_2$	64	27	8	μ_γ
$\frac{3}{2}$, 1; $\frac{3}{2}$, −1	$\sqrt{21}/21$	$\sqrt{14}/7$	$\sqrt{6}/3$	
1, 0; 1, 0	$\sqrt{210}/21$	$3\sqrt{35}/35$	$-2\sqrt{15}/15$	
$\frac{1}{2}$, −1; $\frac{1}{2}$, +1	$\sqrt{210}/21$	$-4\sqrt{35}/35$	$\sqrt{15}/15$	

$Y = 0 \quad I = 0$

$I_1, Y_1; I_2, Y_2$	64	27	8	1	μ_γ
$\frac{3}{2}$, 1; $\frac{3}{2}$, −1	$\sqrt{35}/35$	$\sqrt{210}/35$	$\sqrt{10}/5$	$\sqrt{10}/5$	
1, 0; 1, 0	$2\sqrt{105}/35$	$\sqrt{70}/14$	0	$-\sqrt{30}/10$	
$\frac{1}{2}$, −1; $\frac{1}{2}$, 1	$3\sqrt{70}/35$	$-\sqrt{105}/35$	$-\sqrt{5}/5$	$\sqrt{5}/5$	
0, −2; 0, 2	$2\sqrt{35}/35$	$-3\sqrt{210}/70$	$\sqrt{10}/5$	$-\sqrt{10}/10$	

$Y = -1 \quad I = \frac{5}{2}$

$I_1, Y_1; I_2, Y_2$	64	μ_γ
1, 0; $\frac{3}{2}$, −1	1	

$Y = -1 \quad I = \frac{3}{2}$

$I_1, Y_1; I_2, Y_2$	64	27	μ_γ
1, 0; $\frac{3}{2}$, −1	$\sqrt{14}/7$	$\sqrt{35}/7$	
$\frac{1}{2}$, −1; 1, 0	$\sqrt{35}/7$	$-\sqrt{14}/7$	

$Y = -1 \quad I = \frac{1}{2}$

$I_1, Y_1; I_2, Y_2$	64	27	8	μ_γ
1, 0; $\frac{3}{2}$, −1	$\sqrt{7}/7$	$4\sqrt{35}/35$	$\sqrt{10}/5$	
$\frac{1}{2}$, −1; 1, 0	$2\sqrt{7}/7$	$\sqrt{35}/35$	$-\sqrt{10}/5$	
0, −2; $\frac{1}{2}$, 1	$\sqrt{14}/7$	$-3\sqrt{70}/35$	$\sqrt{5}/5$	

$Y = -2 \quad I = 2$

$I_1, Y_1; I_2, Y_2$	64	μ_γ
$\frac{1}{2}$, −1; $\frac{3}{2}$, −1	1	

$Y = -2 \quad I = 1$

$I_1, Y_1; I_2, Y_2$	64	27	μ_γ
$\frac{1}{2}$, −1; $\frac{3}{2}$, −1	$\sqrt{21}/7$	$2\sqrt{7}/7$	
0, −2; 1, 0	$2\sqrt{7}/7$	$-\sqrt{21}/7$	

$Y = -3 \quad I = \frac{3}{2}$

$I_1, Y_1; I_2, Y_2$	64	μ_γ
0, −2; $\frac{3}{2}$, −1	1	

VERIFICATION OF THE TENFOLD ASSIGNMENT OF THE BARYON RESONANCES*

S. Meshkov†
National Bureau of Standards, Washington, D. C.

C. A. Levinson
Department of Nuclear Physics, The Weizmann Institute of Science, Rehovoth, Israel

and

H. J. Lipkin‡
Department of Physics, University of Illinois, Urbana, Illinois
(Received 18 February 1963)

One of the more attractive aspects of the octet model[1] or "eightfold way" has been its ability to accomodate the N^*, Y_1^*, and Ξ^* resonances in one isostrangeness multiplet, the (30).[2] The (30) representation of SU_3 is ten-dimensional as shown in Fig. 1. For this representation, the Okubo unitary mass formula[2,3]

$$m = m_0\{1 + aY + b[T(T+1) - \tfrac{1}{4}Y^2]\} \qquad (1)$$

reduces to

$$m = m_0'(1 + a'Y). \qquad (2)$$

Thus once the masses of the N^*(1238 MeV) and the Y_1^*(1385 MeV) are known, the mass of the Ξ^* is predicted to be 1532 MeV. This predicted mass agrees amazingly well with the experimentally determined value of 1535[4] MeV giving strong support for the validity of the assignment of the baryon resonances to the (30) representation.

The present work proposes experimental tests for the validity of these assignments by examining the processes

$$\text{meson} + \text{baryon} \rightarrow \text{baryon resonance} + \text{vector meson}, \qquad (3a)$$

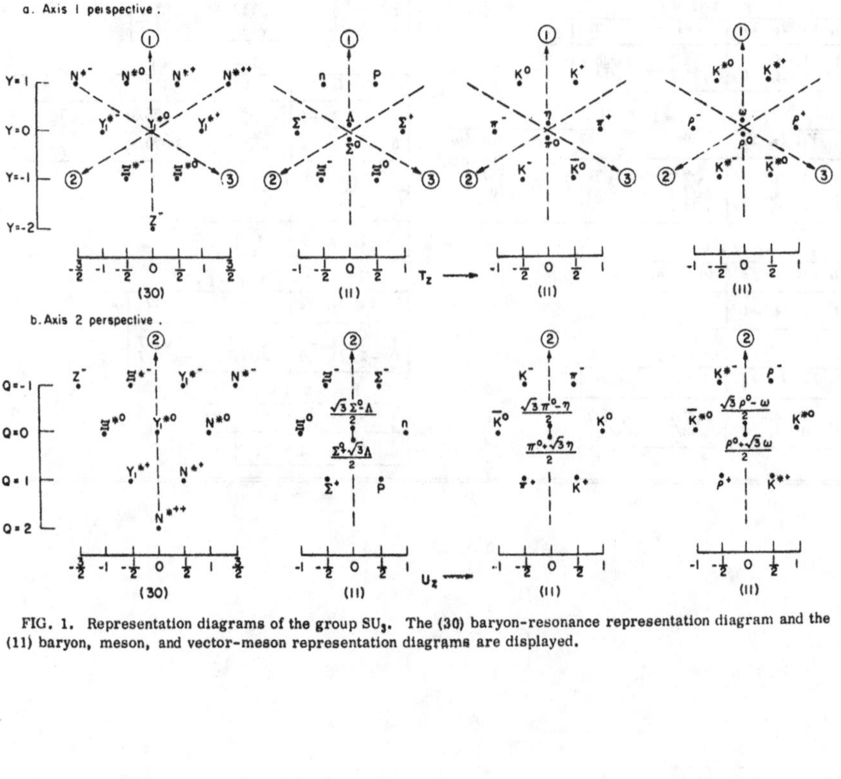

FIG. 1. Representation diagrams of the group SU_3. The (30) baryon-resonance representation diagram and the (11) baryon, meson, and vector-meson representation diagrams are displayed.

meson + baryon → baryon resonance + meson. (3b)

The method used has been described in detail[5,6] for the processes

$$\text{meson} + \text{baryon} \rightarrow \text{meson} + \text{baryon} \qquad (4)$$

according to both the Sakata and octet models.

The meson (1, 1) symmetry and the baryon (1, 1) symmetry can couple together to make product symmetries (22), (11), (11), (30), (03), and (00). Each product symmetry defines a channel through which the reaction can proceed. In a similar fashion the right-hand sides of Eqs. (3a) and (3b) are described by coupling the baryon-resonance (30) symmetry and either the vector-meson or meson (11) symmetries to produce symmetries (41), (30), (22), and (11). One can introduce the energy and angle dependent amplitudes $A^{(\lambda,\mu)}$ which are diagonal elements of the S matrix, and in terms of which one obtains expressions for the amplitude for processes (3a) and (3b). Only those symmetries common to the left and right sides of (3a) and (3b) contribute to the amplitude, i.e., (22), (30), and two (11)'s. The scattering amplitudes have been calculated from the overlaps of the meson-baryon functions (from which the results of reference 6 were obtained) with the functions corresponding to the product (30) × (11). The latter functions have been obtained by combining the tables of Rashid[7] with the appropriate Clebsch-Gordan coefficients.

The following relations and equalities are then seen to hold among the scattering amplitudes:

$$(\pi^- p \,|\, Y_1{}^{*-}K^{*+}) = (K^- p \,|\, \Xi^{*-}K^{*+}) = -(K^- p \,|\, Y_1{}^{*-}\rho^+)$$
$$= -(1/\sqrt{3})(\pi^- p \,|\, N^{*-}\rho^+). \qquad (5a)$$

Alternatively K^{*+} and ρ^+ may be replaced by their meson analogs K^+ and π^+ to yield

$$(\pi^- p \,|\, Y_1{}^{*-}K^+) = (K^- p \,|\, \Xi^{*-}K^+) = -(K^- p \,|\, Y_1{}^{*-}\pi^+)$$
$$= -(1/\sqrt{3})(\pi^- p \,|\, N^{*-}\pi^+). \qquad (5b)$$

As a result of (5a) and (5b), definite predictions about the equalities and ratios of cross sections and polarizations are made, subject to the provisos[5,6] of large enough incident energies and momentum transfer. The experimental angular distributions for the reactions $(\pi^- p \,|\, Y_1{}^{*-}K^+)$[8] and $(K^- p \,|\, Y_1{}^{*-}\pi^+)$[9] show similar behavior, i.e., peaking of the $Y_1{}^{*-}$ in the backward direction. This similarity of the angular distributions is in accord with the relations (5b). Backward peaking for the Ξ^{*-} and the N^{*-} in the remaining processes of (5b) is then predicted.

The relations (5a) and (5b) can be derived in a much more transparent fashion by consideration of a U_2 subgroup of U_3 different from isotopic spin. The usual weight-diagram representations of the U_3 multiplets are given in terms of isospin and hypercharge and are shown in Fig. 1(a). In these diagrams U_2 transformations perpendicular to axis one correspond to the transformations of isospin. However, U_2 transformations perpendicular to axes two or three are also symmetry operations of U_3. We are particularly interested in the transformations perpendicular to axis two and have redrawn the weight diagrams with this axis vertical in Fig. 1(b). Members of the same U_2 multiplets in this scheme all have the same charge and are on the same horizontal line in Fig. 1(b). We shall call the U_2 transformations within these multiplets U-spin transformations, and shall therefore have multiplets of $U = \frac{3}{2}, 1, \frac{1}{2}, 0$ for the (30) representation and $U = \frac{1}{2}, 1, 0, \frac{1}{2}$ for the (11) representation. The components of the U multiplets are labeled by U_z as indicated in Fig. 1(b).

In the (11) diagram there is a charge zero U-spin triplet and U-spin singlet. For the vector-meson octet the U-spin triplet with $U_z = 0$ is

$$M_1 = \tfrac{1}{2}(\rho^0 + \sqrt{3}\omega). \qquad (6a)$$

The U-spin singlet is

$$M_0 = \tfrac{1}{2}(\sqrt{3}\rho^0 - \omega). \qquad (6b)$$

The physical ρ^0 and ω vector mesons are then given by

$$\rho^0 = \tfrac{1}{2}(M_1 + \sqrt{3}M_0), \qquad (7a)$$

$$\omega = \tfrac{1}{2}(\sqrt{3}M_1 - M_0). \qquad (7b)$$

Identical transformations hold for the π^0 and η mesons and for the Σ^0 and Λ particles.

Let us now consider the $\pi^- p$ amplitudes of Eqs. (5). As may be seen from Fig. 1(b), the $\pi^- p$ system has $U_z = 1$, $U = 1$. The processes

$$\pi^- p \rightarrow N^{*-} + \rho^+$$

and

$$\pi^- p \rightarrow Y_1{}^{*-} + K^{*+} \qquad (8)$$

are then determined only by a single $U = 1$ amplitude, T. The branching ratio of these two processes is given solely by the Clebsch-Gordan coefficients, $(U_1 U_2 M_1 M_2 \,|\, UM)$ for the allowed couplings of the members of the $U = \frac{3}{2}$ multiplet (which contains N^{*-} and Y^{*-}) with the $U = \frac{1}{2}$

multiplet (which contains ρ^+ and K^{*+}). It is

$$\frac{(\pi^-p|N^{*-}\rho^+)}{(\pi^-p|Y_1^{*-}K^{*+})}=\frac{(3/2\ 1/2\ 3/2\ -1/2|1\,1)T}{(3/2\ 1/2\ 1/2\ \ 1/2|1\,1)T}$$

$$=\frac{\frac{1}{2}\sqrt{3}T}{-\frac{1}{2}T}=-\sqrt{3}. \tag{9}$$

The processes

$$K^-p\to Y_1^{*-}+\rho^+$$

and

$$K^-p\to\Xi^{*-}+K^{*+} \tag{10}$$

are similarly analyzed in terms of a single $U=1$ amplitude. K^-p can couple to $U=0$ and 1, whereas $Y_1^{*-}\rho^+$ and $\Xi^{*-}K^{*+}$ each may couple to $U=1$ and 2. Therefore, once again only the same $U=1$ amplitude, T, occurs.

The scattering amplitudes for the processes (10) then have the ratio

$$\frac{(K^-p|Y_1^{*-}\rho^+)}{(K^-p|\Xi^{*-}K^{*+})}=\frac{(3/2\ 1/2\ \ 1/2\ -1/2|1\,0)T/\sqrt{2}}{(3/2\ 1/2\ -1/2\ \ 1/2|1\,0)T/\sqrt{2}}$$

$$=\frac{T/2}{-T/2}=-1. \tag{11}$$

Thus the relations (5) follow directly from the U-spin transformation.

In addition, other relations not as simple as those of (5) may be easily obtained using U-spin. Consider the amplitudes for the three processes initiated by π^-p, a pure $U=1$ channel:

$$(\pi^-p|N^{*0}+\rho^0), \tag{12a}$$

$$(\pi^-p|N^{*0}+\omega), \tag{12b}$$

$$(\pi^-p|Y_1^{*0}+K^{*0}). \tag{12c}$$

The amplitudes in (12) are determined by two independent $U=1$ amplitudes. To define them let

$$X=(\Xi^{*0},Y_1^{*0},N^{*0}),$$

$$Y=(\bar{K}^{*0},\tfrac{1}{2}(\rho^0+\sqrt{3}\omega),K^{*0}),$$

and

$$M_0=\tfrac{1}{2}(\sqrt{3}\rho^0-\omega) \tag{13}$$

be the three U-spin multiplets of interest. Let A and B be the U-spin triplet amplitudes leading to (XY) and (XM_0) triplet combinations. We will also need the U-spin singlet amplitude C leading to (XY) in the singlet state.

The processes (12) yield the three relations in two parameters

$$(\pi^-p|N^{*0}\rho^0)=A/2\sqrt{2}+(\sqrt{3}/2)B, \tag{14a}$$

$$(\pi^-p|N^{*0}\omega)=(\sqrt{3}/2\sqrt{2})A-\tfrac{1}{2}B, \tag{14b}$$

$$(\pi^-p|Y_1^{*0}K^{*0})=-A/\sqrt{2}. \tag{14c}$$

A similar analysis of the four K^-p amplitudes yields

$$(K^-p|N^{*0}\bar{K}^{*0})=\tfrac{1}{2}A-C/\sqrt{6}, \tag{15a}$$

$$(K^-p|\Xi^{*0}K^{*0})=-\tfrac{1}{2}A-C/\sqrt{6}, \tag{15b}$$

$$(K^-p|Y_1^{*0}\rho^0)=C/2\sqrt{6}+3B/2\sqrt{6}, \tag{15c}$$

$$(K^-p|Y_1^{*0}\omega)=C/2\sqrt{2}-B/2\sqrt{2}. \tag{15d}$$

From the amplitudes given above, the following relations between differential cross sections hold:

$$\sigma(15c)+\sigma(15d)+\sigma(14c)$$

$$=\tfrac{1}{2}[\sigma(15a)+\sigma(15b)+\sigma(14a)+\sigma(14b)]. \tag{16}$$

The seven processes (14) and (15) are seen to depend on only three independent amplitudes.

As pointed out before, any results obtained involving the vector-meson octet also hold when the vector mesons are replaced by their meson analogs.

We are indebted to George Snow for his encouragement and suggestions and to Gideon Alexander, Manoj Banerjee, and Gaurang Yodh for very useful discussions.

*The research reported in this document has been sponsored in part by the Office of Scientific Research, Office of Air Research, through the European Office, Aerospace Research, U. S. Air Force.

†Work done in part while a visitor at the Weizmann Institute of Science, Rehovoth, Israel.

‡On leave from the Weizmann Institute of Science, Rehovoth, Israel.

[1]Y. Ne'eman, Nucl. Phys. 26, 222 (1961); M. Gell-Mann, California Institute of Technology Report CTSL-20 (unpublished).

[2]S. L. Glashow and J. J. Sakurai, Nuovo Cimento 25, 337 (1962); S. L. Glashow and J. J. Sakurai, Nuovo Cimento 26, 622 (1962); M. Gell-Mann, Proceedings of the International Conference on High-Energy Nuclear Physics, Geneva, 1962 (CERN Scientific Information Service, Geneva, Switzerland, 1962).

[3]S. Okubo, Progr. Theoret Phys. (Kyoto) 27, 949 (1962).

[4]G. M. Pjerrou, D. J. Prowse, P. Schlein, W. E. Slater, D. H. Stork, and H. K. Ticho, Phys. Rev. Letters 9, 114 (1962); L. Bertanza, V. Brisson, P. L. Connolly, E. L. Hart, I. S. Mittra, G. C. Moneti, R. R. Rau, N. P. Samios, I. O. Skillicorn, S. S. Yamamoto, M. Goldberg, L. Gray, J. Leitner, S. Lichtman, and J. Westgard, Phys. Rev. Letters 9, 180 (1962).

[5]C. A. Levinson, H. J. Lipkin, and S. Meshkov, Nuovo Cimento 23, 236 (1962).

[6]C. A. Levinson, H. J. Lipkin, and S. Meshkov, Phys. Letters 1, 44 (1962).

[7]M. A. Rashid, Nuovo Cimento 26, 118 (1962).
[8]A. R. Erwin, R. H. March, and W. D. Walker, Nuovo Cimento 24, 237 (1962).
[9]R. P. Ely, Sun-Yiu Fung, George Gidal, Yu-Li Pan, Wilson M. Powell, and Howard S. White, Phys. Rev. Letters 7, 461 (1961).

COMPARISON OF A NEW SU$_3$ PREDICTION WITH EXPERIMENT

S. Meshkov*
National Bureau of Standards, Washington, D. C.

and

G. A. Snow and G. B. Yodh†
Department of Physics and Astronomy, University of Maryland, College Park, Maryland
(Received 12 December 1963)

The "eightfold way"[1] has been quite successful in describing the masses[2] of elementary particles and resonances, and in explaining the decay widths of various resonances.[3] Various other tests of this symmetry scheme have been proposed which concern themselves with reaction cross sections.[4] Because of the large number of S-matrix elements which usually occur, the simplest proposed tests have taken the form of equalities. Unfortunately, these predictions of equalities have been difficult to verify. This paper proposes a rule for reaction processes, which seems capable of experimental test.

Consider processes of the two types:

meson + proton → baryon resonance + meson, (1)

meson + proton → baryon resonance + vector meson. (2)

For the processes of type (1) the reactions treated are

$$K^+ + p \to N^{*++} + K^0, \tag{1a}$$

$$\pi^+ + p \to N^{*++} + \pi^0, \tag{1b}$$

$$\pi^+ + p \to N^{*++} + \eta, \tag{1c}$$

$$\pi^+ + p \to Y_1^{*+} + K^+. \tag{1d}$$

For the processes of type (2) the reactions treated are

$$K^+ + p \to N^{*++} + K^{*0}, \tag{2a}$$

$$\pi^+ + p \to N^{*++} + \rho^0, \tag{2b}$$

$$\pi^+ + p \to N^{*++} + \varphi^0, \tag{2c}$$

$$\pi^+ + p \to Y_1^{*+} + K^{*+}, \tag{2d}$$

where φ^0 is the $I=0$, $Y=0$ member of the vector-meson octet. Because of ω-φ mixing,[5] it is also necessary to consider the following reactions, in addition to the reactions (2*a*)-(2*d*):

$$\pi^+ + p \to N^{*++} + \omega^0, \tag{2e}$$

$$\pi^+ + p \to N^{*++} + \omega, \tag{2f}$$

$$\pi^+ + p \to N^{*++} + \varphi. \tag{2g}$$

ω^0 is the vector-meson singlet, and ω and φ denote the physically observed particles of masses 782 MeV and 1020 MeV, respectively.

In general, the meson (11) symmetry and the baryon (11) symmetry can couple together to make product symmetries (22), (11), (11), (30), (03), and (00). Each product symmetry defines a channel through which the reaction can proceed. In a similar fashion the right-hand sides of Eqs. (1) and (2) are described by coupling the baryon-resonance (30) symmetry and either the vector-meson or meson (11) symmetries to produce symsymmetries (41), (30), (22), and (11). One can introduce the energy- and angle-dependent amplitudes $A^{(\lambda\mu)}$ which are diagonal elements of the S matrix and in terms of which one obtains expressions for the amplitudes for processes (1) and (2). Only those symmetries common to the left and right sides of (1) and (2) contribute to the amplitude, i.e., (22), (30), and two (11)'s. An important simplification occurs, however, if we restrict ourselves to incident π^+ and K^+ mesons. Then only the two SU$_3$ channels (22) and (30) contribute, and only the N^* and Y_1^* baryon resonances occur. The coefficients of the $A^{(\lambda\mu)}$ for the processes which we consider are given in Table I. The squares of the amplitudes, $|M_a|^2$, $|M_b|^2$, $|M_c|^2$, $|M_d|^2$, for the four reactions of types (1) and (2) depend on two independent complex amplitudes, $A^{(22)}$ and $A^{(30)}$. Therefore, in each case, one obtains the following relation:

$$|M_a|^2 = |M_b|^2 + 3|M_c|^2 - 3|M_d|^2. \tag{3}$$

Table I. Scattering amplitudes $A^{(\lambda\mu)}$ for production of baryon resonances in meson-baryon reactions. The coefficients result from taking scalar products of SU$_3$ functions for the case $\langle(11)\otimes(11)|(30)\otimes(11)\rangle$.

Process \ $(\lambda\mu)$	(22)	(30)
(*a*) $(K^+p\,|N^{*++}K^0)(K^{*0})$	$-\sqrt{3}/2$	
(*b*) $(\pi^+p\,|N^{*++}\pi^0)(\rho^0)$	$-\sqrt{3}/4\sqrt{2}$	$-\sqrt{3}/4$
(*c*) $(\pi^+p\,|N^{*++}\eta)(\varphi^0)$	$-3/4\sqrt{2}$	$1/4$
(*d*) $(\pi^+p\,|Y_1^{*+}K^+)(K^{*+})$	$1/4$	$-1/2\sqrt{2}$

The determination of $|M_{2c}|^2$ is complicated by ω-φ mixing.

Equation (3) is a general result, encompassing the predictions of specific models like the peripheral exchange model[6] which, indeed, do satisfy Eq. (3). The fact that Eq. (3) may or may not be satisfied by the experimental cross sections is a test only of SU_3, not of the peripheral exchange model.

We are now faced with the problem of comparing Eq. (3) with experiment. Since SU_3 symmetry is broken (the masses of particles in an SU_3 multiplet are different), there is at the moment no rigorous way of making this comparison. In the following analysis, we ignore the dynamical effects of the mass-splitting interaction on the reaction amplitudes which lead to Eq. (3). We assume that the best way to test Eq. (3) is to compare experimental cross sections for each channel at the same Q value, where $Q = E^* - M_3 - M_4$, E^* = total energy in the c.m. system, and M_3, M_4 are the masses of the two outgoing particles. This choice has the virtue that the thresholds for each reaction are superimposed at $Q = 0$ and that the opening up of new channels which, through unitarity, will affect the reactions of types (1) or (2) is also superimposed. The kinematic factors that relate cross sections to matrix elements squared are, of course, different for each channel. We assume that the amplitudes of Eq. (3) are relativistically invariant so that the relation between $|M|^2$ and σ for two-body reactions is[7]

$$|M|^2 = (E^{*2} p_{in}/p_{out})\sigma \equiv F\sigma, \tag{4}$$

where p_{in} and p_{out} are the momenta of incident and outgoing particles in the c.m. system. In Fig. 1, the factor $F = (E^{*2} p_{in}/p_{out})$ is plotted vs Q for each of the reactions of type (1) and type (2). The shapes of these curves are nearly the same, again suggesting the usefulness of comparing cross sections for different reactions at similar Q values. With the above assumptions, Eq. (3) predicts the following relation between the experimental cross sections:

$$F_a\sigma_a = F_b\sigma_b + 3F_c\sigma_c - 3F_d\sigma_d. \tag{5}$$

Let us consider the reactions of type (1). Table II lists the pertinent, available experimental cross sections, together with associated values of E^*, Q, and F. In Fig. 2 the four terms of Eq. (5) are plotted individually as functions of Q. The lines are drawn roughly through the

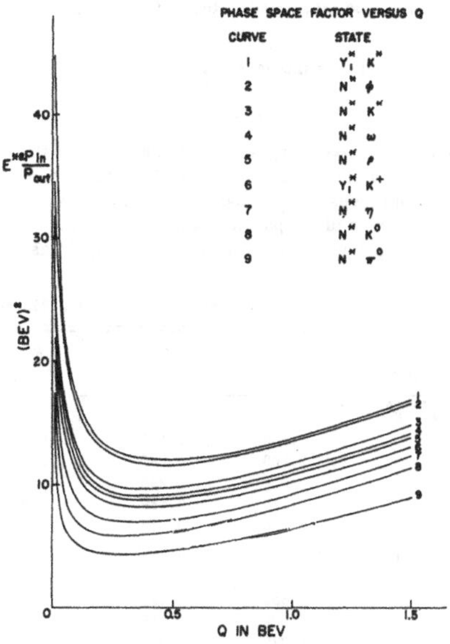

FIG. 1. The factor $F = E^{*2} p_{in}/p_{out}$ vs Q for reactions of type (1a)-(1d) and (2a)-(2f). E^*, p_{in}, p_{out}, and Q are the total energy, the incident momentum, the outgoing momentum, and the total outgoing kinetic energy in the c.m. system.

data points to serve as a guide. At present, a complete comparison of the left-hand side (LHS) and right-hand side (RHS) of Eq. (5) can be made only at $Q \approx 300$ MeV and 500 MeV. As can be seen from Fig. 2, at $Q \approx 300$ MeV (expressing σF in units of BeV^2-mb), LHS = 25 ± 6, and RHS = 28 ± 6. At $Q \approx 500$ MeV, LHS = 21 ± 5 and RHS = 19 ± 5. At Q values below ~120 MeV, the data obviously do not satisfy Eq. (5). In this threshold region, the broad N^{*++} is only partially excited, so that Eq. (5) may not be reliable. Above 500 MeV a detailed test of Eq. (5) requires more and better data than are available, but there is no indication of a large violation of Eq. (5). In general, for reactions of type (1), the SU_3 prediction of Eq. (5) is quite well satisfied.

The analysis of reactions of type (2), involving vector mesons, is complicated by ω-φ mixing.[5] Since the cross section for φ^0 production, σ_{2c}, is not directly measurable, we try to calculate it from $\sigma_{2f}(\omega)$ and $\sigma_{2g}(\varphi)$ (defined previously) which can be measured. Recall that we may write

$$|\omega_0\rangle = \sin\lambda|\varphi\rangle + \cos\lambda|\omega\rangle, \quad (6)$$

$$|\varphi_0\rangle = \cos\lambda|\varphi\rangle - \sin\lambda|\omega\rangle, \quad (7)$$

where the mixing angle λ equals 38°; $\cos^2\lambda = 0.6$, $\sin^2\lambda = 0.4$, and $2\sin\lambda\cos\lambda \cong 1$. The squares of the amplitudes $|M_{2c}|^2$ and $|M_{2e}|^2$ may be written in terms of the physical amplitudes $|M_{2f}|^2$ and $|M_{2g}|^2$ and a cross term $\mathrm{Re}(M_{2f}M_{2g}^*)$, as follows:

$$|M_{2c}|^2 = 0.6|M_{2g}|^2 + 0.4|M_{2f}|^2 - \mathrm{Re}(M_{2g}M_{2f}^*), \quad (6)$$

$$|M_{2e}|^2 = 0.4|M_{2g}|^2 + 0.6|M_{2f}|^2 + \mathrm{Re}(M_{2g}M_{2f}^*). \quad (7)$$

Inasmuch as we have two equations with three unknown quantities, $|M_{2c}|^2$, $|M_{2e}|^2$, and $\mathrm{Re}(M_{2g}M_{2f}^*)$, only lower and upper bounds can be obtained for $|M_{2c}|^2$. However, since $\sigma_{2g} \ll \sigma_{2f}$, the uncertainty introduced by this situa-

Table II. Data for processes of type (1). Meson + proton → baryon resonance + pseudoscalar meson.

Reaction	E^* (BeV)	Q (BeV)	F (BeV^2)	σ (mb)	σF	Reference
$N^{*++}K^0$	1.745	0.014	15.5	1.90 ± 0.20	29.50 ± 5.0	a
	1.859	0.122	6.5	3.60 ± 0.50	23.50 ± 4.0	b
	2.000	0.280	5.9	2.80 ± 1.00	16.52 ± 6.0	c
	2.010	0.290	5.9	4.90 ± 1.00	28.90 ± 6.0	d
	2.225	0.489	6.3	3.10 ± 0.80	21.10 ± 5.0	e
	2.615	0.878	7.9	0.80 ± 0.20	6.32 ± 2.0	f
$N^{*++}\pi^0$	1.387	0.014	9.7	$0.40^{+0.40}_{-0.20}$	$3.90^{+3.9}_{-2.0}$	g
	1.495	0.122	4.9	2.40 ± 0.80	11.80 ± 3.9	h
	1.686	0.313	4.4	5.30 ± 0.40	23.30 ± 1.8	i
	1.692	0.319	4.4	7.40 ± 0.40	32.60 ± 1.8	j
	1.875	0.502	4.7	4.00 ± 1.00	18.80 ± 4.7	j
	2.290	0.917	6.2	0.47 ± 0.06	2.90 ± 0.1	k
	2.500	1.127	7.1	0.36 ± 0.06	2.60 ± 0.1	k
$N^{*++}\eta$	1.790	0.004	37.2	0.04 ± 0.02	1.50 ± 0.8	l
	1.875	0.099	9.2	0.06 ± 0.02	0.55 ± 0.2	l
	2.410	0.624	7.6	0.10 ± 0.03	0.76 ± 0.3	k
	2.715	0.929	8.9	<0.13 ± 0.03	<1.20 ± 0.3	m
$Y_1^{*+}K^+$	2.185	0.306	8.3	0.08 ± 0.03	0.66 ± 0.2	n
	2.410	0.531	8.4	0.11 ± 0.04	0.92 ± 0.3	k
	2.715	0.836	9.5	<0.02 ± 0.01	<0.29 ± 0.1	m

[a] B. Kehoe, Phys. Rev. Letters 11, 93 (1963).
[b] J. Duboc et al., Phys. Letters 6, 233 (1963).
[c] G. B. Chadwick et al., Phys. Letters 6, 309 (1963); D. J. Crennel (private communication).
[d] D. Berley et al., Compt. Rend. 255, 890 (1962).
[e] S. Goldhaber, Proceedings of The Athens Topical Conference on Recently Discovered Resonant Particles, 26-27 April 1963, Ohio University, Athens, Ohio (unpublished), p. 92; G. Goldhaber, ibid., p. 80; G. Goldhaber, W. Chinowsky, S. Goldhaber, W. Lee, and T. O'Halloran, Phys. Letters 6, 62 (1963).
[f] M. Ferro-Luzzi et al., Proceedings of the Siena Conference on Elementary Particles, Siena, Italy, 1963 (unpublished).
[g] Based on interpolation of the $\pi^+ + p \to \pi^+ + \pi^0 + p$ data using the model of M. Olsson and G. B. Yodh, Phys. Rev. Letters 10, 353 (1962).
[h] Peter C. A. Newcomb, Phys. Rev. 132, 1283 (1963).
[i] C. Gensollen, P. Granet, R. Barloutaud, A. Leveque, and J. Meyer, Proceedings of the Siena Conference on Elementary Particles, Siena, Italy, 1963 (unpublished).
[j] D. Stonehill, Yale University dissertation, 1962 (unpublished).
[k] C. Alff et al., Phys. Rev. Letters 9, 322 (1962); N. Gelfand and D. Berley (private communication).
[l] H. J. Foelsche and H. Kraybill (to be published).
[m] M. Abolins et al., Phys. Rev. Letters 11, 381 (1963); N. Xuong (private communication).
[n] F. E. James and H. L. Kraybill (to be published); H. L. Kraybill (private communication).

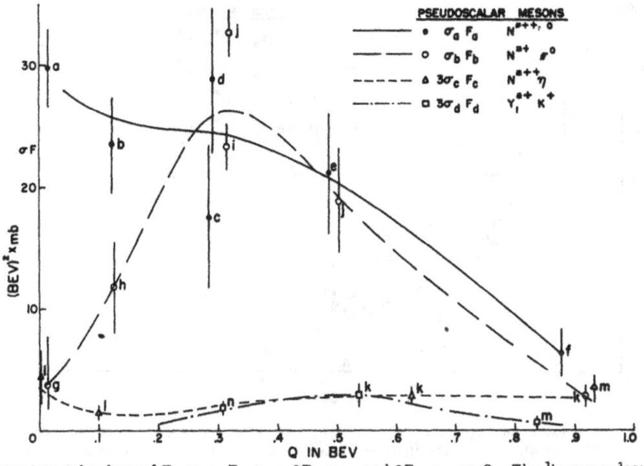

FIG. 2. Experimental values of $F_{1a}\sigma_{1a}$, $F_{1b}\sigma_{1b}$, $3F_{1c}\sigma_{1d}$, and $3F_{1d}\sigma_{1d}$ vs Q. The lines are drawn roughly through the data points simply as a guide. The number next to each data point is the reference number of the data source (Table II).

tion is small. In this discussion of ω-φ mixing, we have ignored the dynamical effects of this mixing on the production amplitudes, analogous to our neglect of the SU_3 mass splitting interaction.

Table III lists the available data on reactions of type (2), together with the associated values of E^*, Q, and F. Note that σ_{2g} is very small,

Table III. Data for processes of type (2). Meson + proton → baryon resonance + vector meson.

Reaction	E^* (BeV)	Q (BeV)	F (BeV²)	σ (mb)	σF	Reference
$N^{*++}K^{*0}$	2.225	0.099	13.4	1.65 ± 0.20	22.10 ± 2.7	a
	2.340	0.214	10.2	1.55 ± 0.20	15.80 ± 2.0	b
	2.615	0.489	9.8	0.65 ± 0.30	6.37 ± 2.9	c
$N^{*++}\rho^0$	2.185	0.214	9.4	1.20 ± 0.20	11.28 ± 1.9	d
	2.290	0.302	8.9	0.61 ± 0.18	5.40 ± 1.6	e
	2.410	0.422	8.8	0.56 ± 0.16	5.00 ± 1.5	e
	2.500	0.512	9.0	0.39 ± 0.12	3.50 ± 1.1	e
	2.715	0.727	9.6	0.85 ± 0.25	8.20 ± 2.1	f
$N^{*++}\omega$	2.185	0.165	10.4	1.10 ± 0.20	11.40 ± 2.1	d
	2.290	0.270	9.4	0.83 ± 0.16	7.80 ± 1.6	e
	2.410	0.390	9.2	0.85 ± 0.17	7.80 ± 1.6	e
	2.500	0.480	9.2	0.50 ± 0.15	4.60 ± 1.0	e
	2.715	0.695	9.9	0.53 ± 0.06	5.20 ± 0.6	f
$N^{*++}\varphi$	2.715	0.457	11.6	<0.02	<0.20	f
$Y_1^{*+}K^{*+}$	2.410	0.160	14.5	0.09 ± 0.03	1.31 ± 0.4	e
	2.715	0.422	12.1	0.02 ± 0.01	0.24 ± 0.1	f

[a]See reference e, Table II.
[b]R. Kraemer et al. (to be published).
[c]See reference f, Table II.
[d]See reference n, Table II.
[e]See reference k, Table II.
[f]See reference m, Table II.

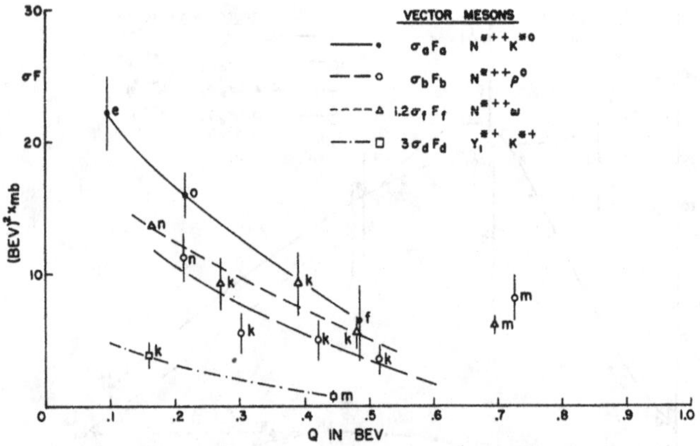

FIG. 3. Experimental values of $F_{2a}\sigma_{2a}$, $F_{2b}\sigma_{2b}$, $3F_{2c}\sigma_{2c}$, and $3F_{2d}\sigma_{2d}$ vs Q. The unmeasured quantity $3F_{2c} \times\sigma_{2c}$ is approximated by $3(0.4)F_{2f}\sigma_{2f}$. The lines are drawn roughly through the data points simply as a guide. The number next to each data point is the reference number of the data source (Table II).

so it seems that it is a good approximation to set

$$F_{2c}\sigma_{2c} = |M_{2c}|^2 \approx 0.4|M_{2f}|^2 = 0.4F_{2f}\sigma_{2f}. \quad (8)$$

Figure 3 displays each of the four terms of Eq. (5) versus Q. Again the lines are drawn through the data points simply as a guide. Figure 3 shows some similarity of the Q dependences of $F_{2a}\sigma_{2a}$, $F_{2b}\sigma_{2b}$, and $F_{2c}\sigma_{2c}$. The $N^*\rho$ and $N^*\omega$ cross sections at $Q\sim 700$ MeV do not follow the downward trend of the other data. At $Q = 210$ MeV the LHS of Eq. (5) is 16 ± 2 and the RHS is 20 ± 3. At $Q = 450$ MeV the LHS is 7.5 ± 2.9, while the RHS is 9.8 ± 3.5. The agreement between the LHS and RHS of Eq. (5) for vector-meson production is quite good, over the whole region for which complete sets of data exist. It should be noted that the extraction of the partial cross sections for quasi two-body reactions from the data is difficult because of the possible presence of complicated interference and background effects.

In conclusion, we find that the reaction predictions of SU_3 embodied in Eq. (5) are, on the whole, in reasonable agreement with experiment, when comparisons are made at equal Q values. Although individual cross sections at a given Q may differ by factors of 10, Eq. (5) is satisfied to better than 50%. The scatter of experimental cross sections points up the difficulty in evaluating the reactions of types (1) and (2) and indicates the need for the accumulation of more precise data over a wide range of Q values. This will allow a more rigorous test of our SU_3 predictions. It would also appear fruitful to test the equalities predicted by SU_3 symmetry,[4] using the same method of comparison described above.

We are indebted to Dr. J. Coyne, Dr. L. Maximon, Dr. B. Kehoe, and Mr. M. Olsson for extensive discussions on the interpretation of the experimental data and to Dr. S. Goldhaber, Dr. G. Goldhaber, N. Xuong, H. Kraybill, N. Gelfand, and D. Berley for private communications of unpublished experimental results.

*Work supported in part by the U. S. Office of Naval Research.

†Work supported by the U. S. Atomic Energy Commission.

[1]Y. Ne'eman, Nucl. Phys. 26, 222 (1961); M. Gell-Mann, California Institute of Technology Report No. CTSL-20, 1961 (unpublished); Phys. Rev. 125, 1067 (1962).

[2]S. Okubo, Progr. Theoret. Phys. (Kyoto) 27, 949 (1962).

[3]S. L. Glashow and A. H. Rosenfeld, Phys. Rev. Letters 10, 192 (1963).

[4]C. A. Levinson, H. J. Lipkin, and S. Meshkov, Phys. Letters 1, 44 (1962); S. Meshkov, C. A. Levin-

son, and H. J. Lipkin, Phys. Rev. Letters 10, 100 (1963); H. J. Lipkin, C. A. Levinson, and S. Meshkov, Phys. Letters 7, 159 (1963); E. C. G. Sudarshan, Proceedings of The Athens Topical Conference on Recently Discovered Resonant Particles, 26-27 April 1963, Ohio University, Athens, Ohio (unpublished), p. 197.

[5]J. J. Sakurai, Phys. Rev. Letters 9, 472 (1962); J. Kalckar, Phys. Rev. 131, 2242 (1963).

[6]L. Stodolsky and J. Sakurai, Phys. Rev. Letters 11, 90 (1963).

[7]R. P. Feynam, Theory of Fundamental Processes (W. A. Benjamin, Inc., New York, 1962), p. 73.

VI

ELECTROMAGNETIC INTERACTIONS OF THE HADRONS

The fact that the electromagnetic current has definite and nontrivial transformation properties under SU(3) has made it possible to derive a number of useful consequences of the symmetry for electromagnetic interactions of hadrons.

In the simplest form of the eightfold way (with no triplets) the electric charge Q in units of e is given by the relation

$$Q = I_3 + \frac{Y}{2} = F_3 + \frac{F_8}{\sqrt{3}}$$

so that charge is just proportional to a component of the F-spin. The current transforms like the same component of an octet.

Let us ignore strong violations of SU(3) for the time being. Then a first-order electromagnetic transition has an independent matrix element for each appearance of **8** in the reduction of the direct product of the initial state and final state representations. For example, an octet of particles has two independent charge form factors, two independent magnetic moment values with their form factors, and so on; one goes with the F-pattern ($F_3 + [F_8/3^{1/2}]$) and the other with the D-pattern ($D_3 + [D_8/3^{1/2}]$) of couplings. We note, however, that a meson octet that goes into itself under charge conjugation lacks the D pattern.

Coleman and Glashow in their paper derive in this way values of all the Λ, Σ, and Ξ magnetic moments, including the $\Lambda \longleftrightarrow \Sigma^0$ transition moment that determines the rate of Σ^0 decay into $\Lambda + \gamma$, from the neutron and proton magnetic moments. The experimental verification of some of these results may soon be forthcoming, when better values of the Λ moment and a value of the Σ^+ moment become available.

In second order in electromagnetism, they study the mass differences within isotopic multiplets, using what is essentially the SU(3) behavior of the square of the current. Besides an invariant piece transforming like **1** (which gives no mass difference), there is a component of an octet and a component of a 27-plet. For the $J = \frac{1}{2}^+$ baryon octet, with its four independent electromagnetic mass differences, there are three independent terms (the octet occurring twice and the 27-plet once) and one relation

$$m(\Xi^-) - m(\Xi^0) = m(\Sigma^-) - m(\Sigma^+) + m(p) - m(n)$$

which has turned out to hold within the present experimental error, around 1 MeV.

Similar results for other supermultiplets and for other electromagnetic processes have been obtained by Cabibbo and Gatto [Nuovo Cimento, **21**, 872 (1961)], Levinson et al. [Phys. Rev. Letters, **10**, 100 (1963)], Oakes [Phys. Rev., **132**, 2349 (1963)], Rosen [Phys. Rev., **132**, 1234 (1963)], Dullemond et al. [Phys. Rev. Letters, **10**, 423 (1963)], and Okubo in his paper below. For example, there are mass difference formulas for the decimet

$$m(\Delta^{++}) + 3m(\Delta^0) = 3m(\Delta^+) + m(\Delta^-)$$

$$m(\Delta^+) - m(\Delta^0) = m(\Sigma^+) - m(\Sigma^0)$$

$$m(\Delta^0) - m(\Delta^-) = m(\Sigma^0) - m(\Sigma^-) = m(\Xi^0) - m(\Xi^-)$$

where the symbols Σ and Ξ refer, of course, to excited states belonging to the decimet. Another example concerns photoproduction amplitudes:

$$[\Delta^0\pi^+ \,|S|\, p\gamma] = -\sqrt{2}\,[\Sigma^0(1385)K^+ \,|S|\, p\gamma]$$

Okubo has also investigated corrections to be applied to some of the relations discussed above when the first-order effect of SU(3) violation is included.

If we consider a version of the eightfold way with real triplets, then the formula for the electric charge has an added term, proportional to the number of triplets minus the number of antitriplets, unless the triplets are quarks (see next part). Some explicit consequences of such an extra term are listed by Nauenberg [Phys Rev. (in press)].

ELECTRODYNAMIC PROPERTIES OF BARYONS IN THE UNITARY SYMMETRY SCHEME

Sidney Coleman* and Sheldon Lee Glashow
California Institute of Technology, Pasadena, California
(Received March 20, 1961)

Gell-Mann[1] recently introduced a theory of strong interactions involving a new symmetry, called "unitary symmetry." The principal purpose of this note is to use this symmetry to express the magnetic moments of all the baryons in terms of those of the neutron and proton. We also derive a relation among baryon electromagnetic mass splittings.

The unitary symmetry scheme proposes that the elementary particles may be represented as tensors in a three-dimensional (generalized isospin) space, and that the strong interactions are invariant under unitary transformations in this space. In particular, the eight baryons form the components of a traceless matrix ψ,

$$\psi = \begin{pmatrix} -(\frac{2}{3})^{1/2}\Lambda & p & n \\ \Xi^- & (\frac{1}{6})^{1/2}\Lambda + (\frac{1}{2})^{1/2}\Sigma^0 & \Sigma^- \\ \Xi^0 & \Sigma^+ & (\frac{1}{6})^{1/2}\Lambda - (\frac{1}{2})^{1/2}\Sigma^0 \end{pmatrix}, \tag{1}$$

while the seven known pseudoscalar mesons (plus a predicted new pseudoscalar meson, χ^0) form the components of a traceless Hermitian matrix ϕ,

$$\phi = \begin{pmatrix} -(\frac{2}{3})^{1/2}\chi^0 & K^+ & K^0 \\ K^- & (\frac{1}{6})^{1/2}\chi^0 + (\frac{1}{2})^{1/2}\pi^0 & \pi^- \\ \bar{K}^0 & \pi^+ & (\frac{1}{6})^{1/2}\chi^0 - (\frac{1}{2})^{1/2}\pi^0 \end{pmatrix}. \tag{2}$$

The scheme also proposes the existence of eight vector mesons which transform in the same way as ϕ. Although of great importance in Gell-Mann's theory, they will not be described here, for their symmetric interactions do not affect our conclusions.

If the strong interactions are to be invariant under unitary symmetry, the possible forms of the Lagrangian density are (assuming *ps-ps* meson-nucleon coupling)

$$L = \mathrm{tr}\bar{\psi}(i\partial_\mu\gamma^\mu - m_0)\psi + \mathrm{tr}(\tfrac{1}{2}\partial_\mu\phi\partial^\mu\phi - \tfrac{1}{2}M_0^2\phi^2)$$
$$+ g\,\mathrm{tr}(\bar{\psi}\gamma_5\psi\phi) + g'\,\mathrm{tr}(\bar{\psi}\gamma_5\phi\psi) + L', \tag{3}$$

where $\bar{\psi}$ is $\psi^\dagger\gamma_0$, tr is the ordinary matrix trace over the generalized isospin space, m_0 is a common baryon bare mass, and M_0 is a common meson bare mass.

There are two terms in L':

(1) A mysterious unknown interaction, weaker than the strong interactions, that breaks the unitary symmetry and causes us to observe the baryons as a singlet, two doublets, and a triplet, rather than as a completely degenerate octet. We know nothing about this interaction—it may be something as simple as a difference in baryon bare masses—and can predict nothing about its effects. Our conclusions are approximations which are exact only in the absence of this mass-splitting interaction. We hope that the mass-splitting interaction is sufficiently weak so that our approximations are good ones; this is the same hope that is fundamental to any quantitative prediction made on the basis of any "higher symmetry" scheme, like global symmetry.

(2) Electromagnetism. In contrast to the mass-splitting interaction, we know the form of the electromagnetic interaction exactly: In our notation it is

$$L_{\text{em}}' = eA^\mu \operatorname{tr}(\bar{\psi}\gamma_\mu[\psi, Q] + i\partial_\mu\phi[\phi, Q]) - \tfrac{1}{2}e^2 A^\mu A_\mu \operatorname{tr}([\phi, Q][\phi, Q]) , \tag{4}$$

where Q is the real traceless diagonal matrix,

$$Q = \tfrac{1}{3}\begin{pmatrix} -1 & 0 & 0 \\ 0 & 2 & 0 \\ 0 & 0 & -1 \end{pmatrix}. \tag{5}$$

Without the mass-splitting interactions, the only departure from unitary symmetry is through the appearance of Q in the electromagnetic interaction. To first order in electromagnetism, but to all orders in the symmetric interactions, the electromagnetic vertex function must be a linear homogeneous invariant function of Q; thus,

$$\begin{aligned}\langle\bar{\psi}|A_\lambda|\psi\rangle = {}& \mu_1(q^2)q_\nu \operatorname{tr}(\bar{u}\sigma_{\lambda\nu}vQ) \\ &+ \mu_2(q^2)q_\nu \operatorname{tr}(\bar{u}\sigma_{\lambda\nu}Qv) \\ &+ e_1(q^2)\operatorname{tr}(\bar{u}\gamma_\lambda vQ) \\ &+ e_2(q^2)\operatorname{tr}(\bar{u}\gamma_\lambda Qv),\end{aligned} \tag{6}$$

where u and v are the incoming and outgoing baryon spinors (they are also traceless 3×3 matrices in generalized isospin space), and q_μ is the 4-momentum transfer. Terms in tr$Q$ are absent from (6) because we have chosen Q to be traceless.

The values of e_1 and e_2 at $q^2 = 0$ are determined by the baryon charges; their higher derivatives yield hyperon form factors which are not likely to be measured soon. Therefore, we fix our attention on the μ_i at $q^2 = 0$. These give, in principle, nine quantities, the eight baryon magnetic moments and the matrix element for $\Sigma^0 \to \Lambda + \gamma$.

Expanding (6) in terms of (1), we find for the magnetic moments,

$$\mu(\Sigma^+) = \mu(p), \tag{7}$$

$$\mu(\Lambda) = \tfrac{1}{2}\mu(n), \tag{8}$$

$$\mu(\Xi^0) = \mu(n), \tag{9}$$

$$\mu(\Xi^-) = \mu(\Sigma^-) = -[\mu(p) + \mu(n)], \tag{10}$$

$$\mu(\Sigma^0) = -\tfrac{1}{2}\mu(n), \tag{11}$$

and for the mixed moment (the multiple of $\bar{\Sigma}^0\sigma_{\mu\nu}\Lambda F_{\mu\nu}$ + H. c.) responsible for the decay $\Sigma^0 \to \Lambda + \gamma$,

$$\mu_m = \tfrac{1}{2}\sqrt{3}\,\mu(n). \tag{12}$$

These equations contain one relation that is independent of the unitary symmetry scheme,

$$\mu(\Sigma^0) = \tfrac{1}{2}[\mu(\Sigma^+) + \mu(\Sigma^-)].$$

This is known to be a consequence of isospin invariance alone.[2]

In the same manner we may consider the electromagnetic corrections to the baryon masses. In the absence of the principal mass-splitting interactions, but in the presence of all symmetric strong interactions, the induced electromagnetic mass difference must be an invariant function of Q. The most general such expression is

$$\langle\bar{\psi}|\delta m|\psi\rangle = \delta_1 \operatorname{tr}(\bar{u}Qu) + \delta_2 \operatorname{tr}(\bar{u}uQ) + \delta_3 \operatorname{tr}(\bar{u}QuQ) + \delta_4 \operatorname{tr}(\bar{u}u). \tag{13}$$

Terms involving Q^2 do not explicitly appear because they are eliminated by the relation $9Q^2 = 2 + 3Q$. The last term in (13) is an addition to the common baryon mass and may be ignored. So, although there are four independent electromagnetic baryon mass splittings, there are only three significant independent parameters in (13). Thus we may obtain a condition on the baryon masses,[3]

$$m(\Xi^-) - m(\Xi^0) = m(\Sigma^-) - m(\Sigma^+) + m(p) - m(n). \tag{14}$$

Gell-Mann[1] has suggested that, in addition to invariance under unitary symmetry, the strong interactions may be invariant under a discrete symmetry, called R. In our notation, R corresponds to replacing every matrix by its transpose. This does not affect the free Lagrangian, but interchanges the two kinds of ps-ps couplings. Thus R invariance demands that $g = g'$.

The electric current is odd under R. Thus we can make the electromagnetic interactions R invariant by extending the definition of R so that A_μ is also odd under R. Should the strong interactions be R invariant as well, it follows in our approximation that $\mu_1 = -\mu_2$ and $\delta_1 = \delta_2$. This yields the manifestly unacceptable results that both the neutron magnetic moment and the charged Σ mass difference vanish.[4] We consider this sufficient grounds for abandoning the R invariance.

*National Science Foundation Predoctoral Fellow.

[1]M. Gell-Mann, Phys. Rev. (to be published). Although our notation differs from that of Gell-Mann, we study the same symmetries as he does. Gell-Mann emphasizes that his symmetry scheme may be seriously broken; experiment indicates that $K\bar{N}\Lambda$ and $K\bar{N}\Sigma$ interactions are much weaker than $\pi\bar{N}N$ interactions, in conflict with the requirements of unitary symmetry.

[2]R. Marshak, S. Okubo, and G. Sudarshan, Phys. Rev. 106, 599 (1957).

[3]Equation (14) gives a predicted value for $m(\Xi^-) - m(\Xi^0)$ of 5.3 ±0.2 Mev. The observed value is 7.4 ±8.0 Mev. We use the data tabulated by W. H. Barkas and A. H. Rosenfeld, Proceedings of the Tenth Annual International Rochester Conference on High-Energy Physics, 1960 (Interscience Publishers, Inc., New York, 1960), p. 877.

[4]It is easy to see this is not true if we abandon R invariance. In this case, we can, for example, set $g' = 0$. Then the only interaction of the neutron with charged mesons is through $g(\bar{n}\gamma_5 p\pi^- + \bar{p}\gamma_5 n\pi^+)$. This gives a nonvanishing moment in order eg^2.

SOME CONSEQUENCES OF UNITARY SYMMETRY MODEL *

S. OKUBO
Department of Physics and Astronomy,
University of Rochester, Rochester, New York

Received 4 February 1963

The purpose of this note is to investigate further consequences of the unitary symmetry model [1,2], based upon invariance of the theory under the three dimensional unitary-unimodular group SU(3) or under the three dimensional unitary group U(3). In this connection, it may be worthwhile to mention the equivalence ** of U(3) and SU(3), if we restrict ourselves to certain irreducible representations as we indeed do in this paper. Hence, we shall use U(3) rather than SU(3) throughout this paper, since the former is more simple to manage.

In a previous paper [2], we have derived the following mass formula for a given irreducible representation of U(3)

$$M = a + b\,Y + c\,[\tfrac{1}{4}\,Y^2 - I(I+1)]\,. \qquad (1)$$

This formula has been proven to the lowest order of perturbation with respect to any U_3-violating mass splitting interaction H_I with a transformation property T_3^3 (of a tensor T_ν^μ), but to all orders with respect to U_3-invariant interactions. If we take into account the second order perturbation correction with respect to the U_3-violating interaction H_I ($= T_3^3$), then we must compute matrix elements of the second order interaction, having the transformation property T_{33}^{33} (of a tensor $T_{\lambda\tau}^{\mu\nu}$). We can evaluate this matrix element in a way analogous † to the derivation of the mass formula eq. (1) to get

$$M = a + b\,Y + c\,[\tfrac{1}{4}\,Y^2 - I(I+1)] + d\,Y^2$$
$$+ e\,[\tfrac{1}{4}\,Y^2 - I(I+1)]\,Y + f\,[\tfrac{1}{4}\,Y^2 - I(I+1)]^2\,. \qquad (2)$$

This second order mass formula contains six para-

* This work has been supported in part by the U.S. Atomic Energy Commission.

** If we specify irreducible representations of U(3) as $U_3(f_1, f_2, f_3)$ ($f_1 \geq f_2 \geq f_3$), then the unimodular restriction of U(3) gives the condition that irreducible representations $U_3(f_1+e, f_2+e, f_3+e)$ for any integers e are exactly the same as $U_3(f_1, f_2, f_3)$ [3]. In our paper, we restrict ourselves to representations satisfying the condition $f_1 + f_2 + f_3 = 0$ and hence this redundancy of representations is immaterial for our case. This can be seen also as follows. In tensor analysis, the difference between U(3) and SU(3) consists of the fact that the completely antisymmetric constant tensors $\epsilon^{\lambda\mu\nu}$ and $\epsilon_{\lambda\mu\nu}$ are invariant under SU(3) but not so under U(3). However, in all problems treated in this paper, $\epsilon_{\lambda\mu\nu}$, if appearing, should always be in the following combination:

$$\epsilon^{\alpha\beta\gamma}\,\epsilon_{\lambda\mu\nu} = \delta_\lambda^\alpha\,[\delta_\mu^\beta\,\delta_\nu^\gamma - \delta_\nu^\beta\,\delta_\mu^\gamma] - \delta_\mu^\alpha\,[\delta_\lambda^\beta\,\delta_\nu^\gamma - \delta_\nu^\beta\,\delta_\lambda^\gamma]$$
$$+ \delta_\nu^\alpha\,[\delta_\lambda^\beta\,\delta_\mu^\gamma - \delta_\mu^\beta\,\delta_\lambda^\gamma]\,.$$

Therefore, the presences of $\epsilon_{\lambda\mu\nu}$ and $\epsilon^{\alpha\beta\gamma}$ are irrelevant.

† Similar to the derivation given in ref. [2], we can show that a tensor T_{33}^{33} in a given irreducible representation of U(3) can be expressed as a linear combination of the following terms: 1, A_3^3, $[A_3^3\,A_3^3]$, $(A\,A)_3^3$, $[A_3^3\,(A\,A)_3^3]$, $[(A\,A)_3^3\,(A\,A)_3^3]$. The rest follows exactly in the same way as in ref. [2].

meters, and does not give any relations among particles belonging to the eight-dimensional representation $U_3(1,0,-1)$ in contrast to the first order mass formula eq. (1). Now, let us consider the ten-dimensional representation [2,4] $U_3(2,-1,-1)$, which may consist of N* ($I=\frac{3}{2}$, $Y=1$), Y_1^* ($I=1$, $Y=0$), Ξ^* ($I=\frac{1}{2}$, $Y=-1$), and Ω ($I=0$, $Y=-2$). In this representation, we have a special relation [5]

$$I=\tfrac{1}{2}\,Y+1\,. \tag{3}$$

Thanks to this constraint, eq. (2) reduces to

$$M=a'+b'\,Y+c'\,Y^2\,, \tag{4a}$$

which is a generalisation of the corresponding first order mass formula eq. (1):

$$M=a'+b'\,Y\,. \tag{4b}$$

As it has been pointed out by several authors [5,6]*, eq. (4b) gives the equal spacing rule for the masses among N*, Y_1^*, Ξ^*, and Ω. In the case of the second order mass formula eq. (4a) we have one relation of the following form **

$$M(\Omega)=M(\mathrm{N}^*)+3\,[M(\Xi^*)-M(Y_1^*)]\,, \tag{5}$$

which predicts $M(\Omega)\approx 1685$ MeV. This value of $M(\Omega)$ differs little from that [5,6] computed by means of eq. (4b), since the equal spacing rule is fairly well satisfied experimentally among $M(\mathrm{N}^*)$, $M(Y_1^*)$ and $M(\Xi^*)$. Therefore, up to the second order perturbation with respect to H_I, we expect an existence of Ω with the mass around 1685 MeV. Experimental verification of its existence will be a crucial test of our theory.

We may give another application of our formula eq. (2) in the 27-dimensional representation $U_3(2,0,-2)$, although we have no particular evidence of any particles belonging to this representation at the moment. If we use the notation $M(I,Y)$ for masses of particles with the isotopic spin I and hypercharge Y, then we shall have the following three relations in $U_3(2,0,-2)$

$$M(\tfrac{3}{2},1)-M(\tfrac{3}{2},-1)=3\,[M(1,2)-M(1,-2)] \\ -5\,[M(\tfrac{1}{2},1)-M(\tfrac{1}{2},-1)]\,, \tag{6a}$$

$$M(2,0)+24\,[M(\tfrac{1}{2},1)+M(\tfrac{1}{2},-1)]=28M(0,0) \\ +9M(1,0)+6\,[M(1,2)+M(1,-2)]\,, \tag{6b}$$

$$[M(\tfrac{3}{2},1)+M(\tfrac{3}{2},-1)]+11\,[M(\tfrac{1}{2},1)+M(\tfrac{1}{2},-1)] \\ =12M(0,0)+6M(1,0)+3\,[M(1,2)+M(1,-2)]\,. \tag{6c}$$

Next, we shall derive a similar formula for the magnetic moment of baryons. In the zeroth order in $H_\mathrm{I}=T_3^3$, we have obtained relations [2,7] like $\mu(\Sigma^+)=\mu(\mathrm{p})$, $\mu(\Xi^0)=\mu(\mathrm{n})$ etc., assuming that the electro-magnetic current j_μ has a transformation property S_1^1 (of a tensor S_β^α) with respect to U(3) (for simplicity, we omit the vector suffix μ of the Lorentz space). If we take account of the mass-splitting interaction H_I up to the first order perturbation, then we have to compute matrix elements of a tensor $S_{13}^{13}=S_1^1\,T_3^3$. Although we can proceed in the same way as in the derivation of the mass formula, we shall use a more direct method. From invariance, we can set

$$\langle\!\langle S_1^1+S_{13}^{13}\rangle\!\rangle=a\,M_\nu^\mu N_\mu^\nu+b\,M_\nu^1 N_1^\nu+c\,M_1^\mu N_\mu^1 \\ +d\,M_1^1 N_3^3+e\,M_3^3 N_1^1+f\,M_3^1 N_1^3+g\,M_1^3 N_3^1 \\ +h\,M_\nu^3 N_3^\nu+j\,M_3^\mu N_\mu^3\,, \tag{7}$$

where we use the same notations as in the previous paper [2], and repeated indices mean summations over 1, 2 and 3. In this case, we have nine parameters † but we get the following two relations

$$\mu(\Sigma^0)=\tfrac{1}{2}\,[\mu(\Sigma^+)+\mu(\Sigma^-)]\,, \tag{8a}$$

$$\mu_\mathrm{T}(\Lambda-\Sigma^0)=\frac{1}{2\sqrt{3}}\,[\mu(\Sigma^0)+3\mu(\Lambda)-2\mu(\Xi^0)-2\mu(\mathrm{n})]\,, \tag{8b}$$

where $\mu_\mathrm{T}(\Lambda-\Sigma^0)$ is the transition magnetic moment between Λ and Σ^0. In this derivation, it is unnecessary to assume the traceless condition $S_\alpha^\alpha=0$ for S_β^α. Eq. (8a) is known to be a consequence [8] of the charge independence alone. Note that we have no longer simple relations like $\mu(\mathrm{p})=\mu(\Sigma^+)$ etc., which are valid in the zeroth order in H_I. Similarly, we may remark that the branching ratio [2,4] of about 15% for $Y_1^*\to\Sigma+\pi$ against $Y_1^*\to\Lambda+\pi$, which is valid in the zeroth order of H_I, does not hold any more in the first order of H_I. Nor is the relation of electro-magnetic mass differences [7]

$$M(\Xi^-)-M(\Xi^0)=M(\Sigma^-)-M(\Sigma^+)+M(\mathrm{p})-M(\mathrm{n})$$

valid if we take account of H_I. In this case, we shall have the following relation up to the first order in H_I

* In refs. [4,5] the symbol Z^- is used instead of Ω.

** This relation can also be derived by a direct tensor calculation as follows: The expectation value of $T_3^3+T_{33}^{33}$ in the $U_3(2,-1,-1)$ representation will be expressed as

$$\langle\!\langle T_3^3+T_{33}^{33}\rangle\!\rangle=a\,\bar{F}_{\mu\nu}^{\alpha\beta}F_{\alpha\beta}^{\mu\nu}+b\,\bar{F}_{\mu\nu}^{\alpha3}F_{\alpha3}^{\mu\nu} \\ +c\,\bar{F}_{\mu3}^{\alpha\beta}F_{\alpha\beta}^{\mu3}+d\,\bar{F}_{\mu\nu}^{33}F_{33}^{\mu\nu}+e\,\bar{F}_{\mu3}^{\alpha3}F_{\alpha3}^{\mu3}$$

where $F_{\mu\nu}^{\alpha\beta}$ satisfies the conditions $F_{\alpha\beta}^{\mu\nu}=F_{\beta\alpha}^{\mu\nu}=-F_{\alpha\beta}^{\nu\mu}$, $F_{\mu\beta}^{\mu\nu}=0$ and $\bar{F}_{\alpha\beta}^{\mu\nu}=[F_{\mu\nu}^{\alpha\beta}]^*$. If we express $F_{\alpha\beta}^{\mu\nu}$ in terms of N*, Y_1^*, Ξ^* and Ω as in ref. [2], the above equation results in eq. (5) after some calculations.

† Actually the time-reversal invariance requires $d=e$ in eq. (7), thus reducing the number of parameters from nine to eight. This can be seen if we compare both terms after expressing them in terms of Λ and Σ^0.

$$M_T(\Lambda - \Sigma^0) = \frac{1}{2\sqrt{3}}[M(\Sigma^0) + 3M(\Lambda) - 2M(n) - 2M(\Xi^0)] , \quad (9)$$

where $M_T(\Lambda - \Sigma^0)$ is the transition mass between Λ and Σ^0 due to the electromagnetic interaction. The proof of eq. (9) can be carried out almost exactly in the same way as that of eq. (8b). Note that the right hand side of eq. (9) will vanish due to the first order mass formula eq. (1) when we switch off the electromagnetic interactions.

Finally, let us investigate electromagnetic decay modes of bosons. In the zeroth order in H_I, we can similarly show the validity of the following relations among transition matrix elements of corresponding processes

$$M(\rho_+ \to \pi_+ + \gamma) = M(\rho_0 \to \pi_0 + \gamma) = M(K_+^* \to K_+ + \gamma)$$
$$= - M(\omega \to \eta + \gamma) = \frac{1}{\sqrt{3}} M(\omega \to \pi_0 + \gamma) \quad (10)$$
$$= \frac{1}{\sqrt{3}} M(\rho_0 \to \eta + \gamma) = - \tfrac{1}{2} M(K_0^* \to K_0 + \gamma) ,$$

$$M(\eta \to \gamma + \gamma) = -\frac{1}{\sqrt{3}} M(\pi_0 \to \gamma + \gamma) . \quad (11)$$

In the derivation of eq. (10), we have utilised the charge-conjugation invariance of the theory together with the traceless condition $S^\mu_\mu = 0$. Again, if we take into account the mass-splitting interaction $H_I = T^3_3$ to the first order of perturbation, then eqs. (10) and (11) do not hold any longer; instead we have the following weaker relation

$$\frac{1}{\sqrt{3}}[M(\omega \to \pi_0 + \gamma) + M(\rho_0 \to \eta + \gamma)] = M(\omega \to \eta + \gamma) - \tfrac{4}{3} M(K_0^* \to K_0 + \gamma) + \tfrac{1}{3} M(\rho_0 \to \pi_0 + \gamma) . \quad (12)$$

Some of these relations may be tested experimentally.

The author would like to express his thanks to Prof. R. E. Marshak for many stimulating discussions. He is also grateful to Professor E. C. G. Sudarshan for reading this manuscript and also for various suggestions.

After having completed this work, it has come to the author's attention that eq. (11) has also been derived by R. Gatto in his lecture given at Trieste 1962.

References

1) Y. Yamaguchi, unpublished (1960).
Y. Ne'eman, Nuclear Phys. 26 (1961) 222.
M. Gell-Mann, Phys. Rev. 125 (1962) 1067.
2) S. Okubo, Progr. Theoret. Phys. (Kyoto) 27 (1962) 949; 28 (1962) 24.
3) H. Weyl, Classical groups (Princeton University Press, 1939), pp. 132 and 208.
4) S. L. Glashow and J. J. Sakurai, Nuovo Cimento 25 (1962) 337.
5) S. L. Glashow and J. J. Sakurai, Nuovo Cimento 26 (1962) 622.
6) M. Gell-Mann, Proc. of the 1962 High-Energy Conf. at CERN, ed. J. Prentki, p. 805.
7) S. Coleman and S. L. Glashow, Phys. Rev. Letters 6 (1961) 423.
8) R. E. Marshak, S. Okubo and E. C. G. Sudarshan, Phys. Rev. 106 (1957) 599.

* * * * *

VII

TRIPLETS AND TRIALITY

We now return to the discussion, begun in the first part, of whether or not triplets [or, in general, spinor representations of SU(3)] will show up among hadrons.

Let us first be more precise about the notion of spinor representations. All representations of SU(3) can be constructed by reducing direct products of the fundamental triplet representation **3** multiplied by itself any number of times. Thus we have

$$\mathbf{3} \times \mathbf{3} = \mathbf{3}^* + \mathbf{6}$$

where $\mathbf{3}^*$ is the representation to which an antitriplet belongs. (Our definition of which is triplet and which is antitriplet is given below.) We also have

$$\mathbf{3} \times \mathbf{3} \times \mathbf{3} = \mathbf{1} + \mathbf{8} + \mathbf{8} + \mathbf{10}$$

$$\mathbf{3} \times \mathbf{3} \times \mathbf{3} \times \mathbf{3} = \mathbf{3} + \mathbf{3} + \mathbf{3} + \mathbf{6}^* + \mathbf{6}^* + \mathbf{15} + \mathbf{15} + \mathbf{15} + \mathbf{15}'$$

(where **15** and **15**′ are two different 15-dimensional representations) and so forth.

If we continue this process indefinitely, we will discover that the representations **3**, $\mathbf{6}^*$, **15**, **15**′, and so forth can occur only in the reduction of a product of 3n + 1 triplets, with n integral, while $\mathbf{3}^*$, **6**, $\mathbf{15}^*$, $\mathbf{15}'^*$, and so forth only come from 3n − 1 triplets and **1**, **8**, **10**, and so forth only come from 3n triplets. Following Baird and Biedenharn (Proceedings of the 1964 Coral Gables Conference on Symmetry Principles at High Energy, W. H. Freeman Publishers, p. 58), we can define the triality τ to be 1, -1, and 0, respectively, in these three cases. Every representation then has a value of τ, and if representation **c** occurs in the reduction of $\mathbf{a} \times \mathbf{b}$, we have

$$\tau_c = \tau_a + \tau_b \pmod 3$$

All the representations that have been identified among the hadrons have triality 0. (These are the ones we have referred to as tensor rather than spinor representations.) The first possibility we may consider is that the other representations, with $\tau = \pm 1$, do not occur. This situation is perfectly compatible with the bootstrap hypothesis and it may very well be the case in reality.

In a Lagrangian field theory, a fairly complicated picture is necessary, as remarked in the first part, to arrive at SU(3) with $\tau = 0$ only; for example, there is the original hypothesis of eight baryons and eight Yang-Mills type vector mesons. A Lagrangian field theory with a fundamental triplet of fields could give a world without real particles of $\tau = \pm 1$ only if something very peculiar happens, which amounts to giving the triplets infinite mass but also infinite binding energy when they combine to form $\tau = 0$ particles!

The remainder of our discussion will be devoted to the hypothetical case that real hadrons exist with $\tau = \pm 1$. Such a situation is clearly compatible with a theory, say a Lagrangian field theory, based on a fundamental triplet (with perhaps some other basic entities), and it may also be compatible with the bootstrap hypothesis.

We shall make two simplifying assumptions that have characterized all work on triplets so far. One is that in the limit of SU(3) symmetry not only is τ conserved modulo 3 but also a quantum number with eigenvalues $\cdots -4, -3, -2, -1, 0, +1, +2, +3, +4 \cdots$ is conserved that is equal to τ modulo 3. We call this quantum number $n_t - n_{\bar{t}}$ (which stands for number of triplets minus number of antitriplets). The other assumption is that there is no other quantum number involved, besides familiar ones.

All known hadron assignments correspond to $n_t - n_{\bar{t}} = 0$. Let us now consider the case $n_t - n_{\bar{t}} = 1$. All representations with $\tau = +1$ are possible for $n_t - n_{\bar{t}} = 1$ if we include enough particles. The representation **3** will occur, containing an isotopic doublet and a singlet, with the electric charge of the singlet in units of e equal to some number, which we call z. The electric charges of the doublet members, in units of e, are z and z + 1. The fact that we take z + 1 instead of z − 1 defines what we mean by a triplet t ($\tau = +1$) rather than an antitriplet $\bar{t}$ ($\tau = -1$); the definition is, of course, arbitrary.

For $n_t - n_{\bar{t}} = -1$, the antitriplet representation $\bar{\mathbf{3}}$ will occur, containing an isotopic singlet with charge $-z$ and an isotopic doublet with charges $-z$ and $-z - 1$. In general, for any supermultiplet, the electric charge Q in units of e will be given by the formula

$$Q = F_3 + \frac{F_8}{\sqrt{3}} + (z + \tfrac{1}{3})\left(n_t - n_{\bar{t}}\right)$$

We may now enumerate four physically distinct situations:

1. The quantum number $n_t - n_{\bar{t}}$ is not a new one, but it is proportional to baryon number $n_B - n_{\bar{B}}$ and thus absolutely conserved. Since the known baryons have triality zero, all baryons must have triality zero in this case, and thus $(n_B - n_{\bar{B}})/(n_t - n_{\bar{t}})$ is an integral multiple of 3. The simplest choice is the one for which the ratio is +3; the triplets are then "quarks," as discussed by Gell-Mann in the paper below. In the formula for the electric charge, we know there is not a term proportional to $n_B - n_{\bar{B}}$, so in this model the coefficient of $n_t - n_{\bar{t}}$ must be zero, giving $z = -\frac{1}{3}$. The quarks thus consist of an isotopic singlet with $Q = -\frac{1}{3}$ and a doublet with $Q = -\frac{1}{3}$ and $Q = \frac{2}{3}$, all with baryon number $\frac{1}{3}$. The known baryons transform under SU(3) like tri-quarks.

2. The quantum number $n_t - n_{\bar{t}}$ is absolutely conserved but is distinct from baryon number. In this case, z can have any value whatever, including simple integral values like -1 and 0. Here, the known mesons transform like combinations $t\bar{t}$, while the known baryons transform like $bt\bar{t}$, $btt\bar{t}\bar{t}$, and so on, where b is a conventional neutral singlet baryon. States of any baryon number and any triality occur in this picture. Some $\tau \neq 0$ particles are stable, as in (1).

3. The quantum number $n_t - n_{\bar{t}}$ is not absolutely conserved, but the violation occurs only through a weak interaction, with triality changing by one unit. Since particles with $\tau = \pm 1$ can now turn into conventional particles with $\tau = 0$, we must have integral z, for example $z = -1$ as in a theory discussed below by Gell-Mann. Again, the known baryons transform like btt, btttt, and so on, but the baryons with $\tau \neq 0$ can decay into them by weak interactions.

4. The conservation of the quantum number $n_t - n_{\bar{t}}$ is violated by the part of the strong interaction that breaks SU(3) symmetry. The symmetry-violating term thus has triality ± 1; only in second order does it give the usual symmetry violation with $\tau = 0$. We want the latter to transform like $\mathbf{8}$, so the $\tau = \pm 1$ symmetry violation must go like $\mathbf{3}$ and $\bar{\mathbf{3}}$, as if it transformed b into t and vice versa. Since the strong violation conserves isotopic spin, an isotopic singlet must go into an isotopic singlet; thus the isotopic singlet in t must have zero charge and hence $z = 0$.

In the foregoing treatment, triplets as such have been discussed mainly for pedagogical reasons; the essential points are the nature of the quantum number $n_t - n_{\bar{t}}$, its relation to triality τ, and the degree of its conservation. These symmetry questions determine the readily observable properties of the hadrons with $\tau \neq 0$, if such exist. The questions of which particles are "fundamental," if any, and how a detailed Lagrangian field theory model could be identified in practice are subtle ones and the whole investigation may not lead anywhere. Certainly, any attempt to describe the known hadrons dynamically as simple pairs or triads of very heavy "fundamental objects" is doomed to failure, because higher-order corrections in the field theory sense will be of the greatest importance and all these dynamical effects would be largely subsumed, in the sense of dispersion theory, in the lower threshold energy channels, such as are included in approximate bootstrap calculations.

Nevertheless, a number of Lagrangian field theory models have been discussed by various authors. Gell-Mann in the accompanying paper treats two, a quark model of type (1) and a model of type (3) with $z = -1$. His main purpose in introducing the models is to abstract from them mathematical relations (involving the weak currents) and symmetry principles like the ones treated above; he is then content to discard the Lagrangians.

Lee, Gürsey, and Nauenberg [Phys. Rev. (in press)] consider a field theory model of type (2) with integral but otherwise unspecified charge. For some reason, they employ as basic units not t and b, but a fermion t and a boson, which transforms like a combination of t and b. These fermion and boson triplets are just like the triplets ℓ and L of the original "The Eightfold Way" paper.

Schwinger [Phys. Rev. Letters, **12**, 237 (1964)] has put forward a rather complicated field theory model which seems to be of type (2).

A model of type (4) is favored by Z. Maki [Prog. Theor. Phys., **31**, 331 (1964)]. The (4) situation is also discussed in the framework of the bootstrap idea by Tarjanne and Teplitz [Phys. Rev. Letters, **11**, 447 (1963)].

Tarjanne and Teplitz emphasize a point that is applicable to any scheme of types (2), (3), and (4), whether in the bootstrap theory or in a theory with a fundamental triplet and singlet. The point is that there may be approximate symmetry under SU(4) when particles with $\tau \neq 0$ are included. SU(4) allows another commuting quantum number along with I_z and Y, which can be just $n_t - n_{\bar{t}}$.

The exciting questions are, of course, experimental. Will hadrons of $\tau \neq 0$ be found and, if so, of what type? If they are of type (4), they may be quite inconspicuous, and some of them may already have been discovered. [See, for example, one of the interpretations of κ(725) in an earlier paper.] If they belong to type (3), then there exist metastable particles with $\tau \neq 0$, decaying by weak interactions into conventional hadrons; we would be dealing with strange particles of a new kind. The most fascinating possibility is that hadrons of type (1) or (2) will be found, for then there must be particles with $\tau \neq 0$ that are absolutely stable.

Such new stable particles must be present on the earth if they exist. Presumably, the cosmic process that is reponsible for the presence of ordinary matter has also produced some admixture of $\tau \neq 0$ matter, but even if it has not, the cosmic radiation, through pair production processes in the atmosphere, must be depositing some $\tau \neq 0$ matter near the surface of the earth. Assuming that these particles exist but are too heavy to be made in present accelerators, they can be sought in analyses of terrestrial material, in cosmic ray experiments, or in accelerator experiments with machines of higher energy.

A SCHEMATIC MODEL OF BARYONS AND MESONS *

M. GELL-MANN
California Institute of Technology, Pasadena, California

Received 4 January 1964

If we assume that the strong interactions of baryons and mesons are correctly described in terms of the broken "eightfold way" [1-3], we are tempted to look for some fundamental explanation of the situation. A highly promised approach is the purely dynamical "bootstrap" model for all the strongly interacting particles within which one may try to derive isotopic spin and strangeness conservation and broken eightfold symmetry from self-consistency alone [4]. Of course, with only strong interactions, the orientation of the asymmetry in the unitary space cannot be specified; one hopes that in some way the selection of specific components of the F-spin by electromagnetism and the weak interactions determines the choice of isotopic spin and hypercharge directions.

Even if we consider the scattering amplitudes of strongly interacting particles on the mass shell only and treat the matrix elements of the weak, electromagnetic, and gravitational interactions by means of dispersion theory, there are still meaningful and important questions regarding the algebraic properties of these interactions that have so far been discussed only by abstracting the properties from a formal field theory model based on fundamental entities [3] from which the baryons and mesons are built up.

If these entities were octets, we might expect the underlying symmetry group to be SU(8) instead of SU(3); it is therefore tempting to try to use unitary triplets as fundamental objects. A unitary triplet t consists of an isotopic singlet s of electric charge z (in units of e) and an isotopic doublet (u, d) with charges z+1 and z respectively. The anti-triplet $\bar{t}$ has, of course, the opposite signs of the charges. Complete symmetry among the members of the triplet gives the exact eightfold way, while a mass difference, for example, between the isotopic doublet and singlet gives the first-order violation.

For any value of z and of triplet spin, we can construct baryon octets from a basic neutral baryon singlet b by taking combinations $(b t \bar{t})$, $(b t t \bar{t} \bar{t})$, etc. **. From $(b t \bar{t})$, we get the representations **1** and **8**, while from $(b t t \bar{t} \bar{t})$ we get **1**, **8**, **10**, $\overline{\mathbf{10}}$, and **27**. In a similar way, meson singlets and octets can be made out of $(t \bar{t})$, $(t t \bar{t} \bar{t})$, etc. The quantum number $n_t - n_{\bar{t}}$ would be zero for all known baryons and mesons. The most interesting example of such a model is one in which the triplet has spin $\frac{1}{2}$ and $z = -1$, so that the four particles d^-, s^-, u^0 and b^0 exhibit a parallel with the leptons.

A simpler and more elegant scheme can be constructed if we allow non-integral values for the charges. We can dispense entirely with the basic baryon b if we assign to the triplet t the following properties: spin $\frac{1}{2}$, $z = -\frac{1}{3}$, and baryon number $\frac{1}{3}$. We then refer to the members $u^{\frac{2}{3}}$, $d^{-\frac{1}{3}}$, and $s^{-\frac{1}{3}}$ of the triplet as "quarks" [6] q and the members of the anti-triplet as anti-quarks $\bar{q}$. Baryons can now be constructed from quarks by using the combinations $(q q q)$, $(q q q q \bar{q})$, etc., while mesons are made out of $(q \bar{q})$, $(q q \bar{q} \bar{q})$, etc. It is assuming that the lowest baryon configuration $(q q q)$ gives just the representations **1**, **8**, and **10** that have been observed, while the lowest meson configuration $(q \bar{q})$ similarly gives just **1** and **8**.

A formal mathematical model based on field theory can be built up for the quarks exactly as for p, n, Λ in the old Sakata model, for example [3] with all strong interactions ascribed to a neutral vector meson field interacting symmetrically with the three particles. Within such a framework, the electromagnetic current (in units of e) is just

$$i\{\tfrac{2}{3}\,\bar{u}\,\gamma_\alpha\,u - \tfrac{1}{3}\,\bar{d}\,\gamma_\alpha\,d - \tfrac{1}{3}\,\bar{s}\,\gamma_\alpha\,s\}$$

or $\mathscr{F}_{3\alpha} + \mathscr{F}_{8\alpha}/\sqrt{3}$ in the notation of ref. [3]. For the weak current, we can take over from the Sakata model the form suggested by Gell-Mann and Lévy [7], namely $i\,\bar{p}\,\gamma_\alpha(1+\gamma_5)(n\cos\theta + \Lambda\sin\theta)$, which gives in the quark scheme the expression ***

$$i\,\bar{u}\,\gamma_\alpha(1+\gamma_5)(d\cos\theta + s\sin\theta)$$

* Work supported in part by the U.S. Atomic Energy Commission.

** This is similar to the treatment in ref. [1]. See also ref. [5].

*** The parallel with $i\,\bar{\nu}_e\,\gamma_\alpha(1+\gamma_5)\,e$ and $i\,\bar{\nu}_\mu\,\gamma_\alpha(1+\gamma_5)\mu$ is obvious. Likewise, in the model with d^-, s^-, u^0, and b^0 discussed above, we would take the weak current to be $i(\bar{b}^0\cos\theta + \bar{u}^0\sin\theta)\,\gamma_\alpha(1+\gamma_5)\,s^- + i(\bar{u}^0\cos\theta - \bar{b}^0\sin\theta)\,\gamma_\alpha(1+\gamma_5)\,d^-$. The part with $\Delta(n_t - n_{\bar{t}}) = 0$ is just $i\,\bar{u}^0\,\gamma_\alpha(1+\gamma_5)(d^-\cos\theta + s^-\sin\theta)$.

or, in the notation of ref. 3),

$$[\mathscr{F}_{1\alpha} + \mathscr{F}^5_{1\alpha} + \mathrm{i}(\mathscr{F}_{2\alpha} + \mathscr{F}^5_{2\alpha})] \cos\theta$$
$$+ [\mathscr{F}_{4\alpha} + \mathscr{F}^5_{4\alpha} + \mathrm{i}(\mathscr{F}_{5\alpha} + \mathscr{F}^5_{5\alpha})] \sin\theta \,.$$

We thus obtain all the features of Cabibbo's picture 8) of the weak current, namely the rules $|\Delta I| = 1$, $\Delta Y = 0$ and $|\Delta I| = \frac{1}{2}$, $\Delta Y/\Delta Q = +1$, the conserved $\Delta Y = 0$ current with coefficient $\cos\theta$, the vector current in general as a component of the current of the F-spin, and the axial vector current transforming under SU(3) as the same component of another octet. Furthermore, we have 3) the equal-time commutation rules for the fourth components of the currents:

$$[\mathscr{F}_{j4}(x) \pm \mathscr{F}^5_{j4}(x),\ \mathscr{F}_{k4}(x') \pm \mathscr{F}^5_{k4}(x')] =$$
$$- 2 f_{jkl}\, [\mathscr{F}_{l4}(x) \pm \mathscr{F}^5_{l4}(x)]\, \delta(x-x') \,,$$

$$[\mathscr{F}_{j4(x)} \pm \mathscr{F}^5_{j4}(x),\ \mathscr{F}_{k4}(x') \mp \mathscr{F}^5_{k4}(x')] = 0 \,,$$

$i = 1, \ldots 8$, yielding the group SU(3) × SU(3). We can also look at the behaviour of the energy density $\theta_{44}(x)$ (in the gravitational interaction) under equal-time commutation with the operators $\mathscr{F}_{j4}(x') \pm \mathscr{F}_{j4}{}^5(x')$. That part which is non-invariant under the group will transform like particular representations of SU(3) × SU(3), for example like $(3, \bar{3})$ and $(\bar{3}, 3)$ if it comes just from the masses of the quarks.

All these relations can now be abstracted from the field theory model and used in a dispersion theory treatment. The scattering amplitudes for strongly interacting particles on the mass shell are assumed known; there is then a system of linear dispersion relations for the matrix elements of the weak currents (and also the electromagnetic and gravitational interactions) to lowest order in these interactions. These dispersion relations, unsubtracted and supplemented by the non-linear commutation rules abstracted from the field theory, may be powerful enough to determine all the matrix elements of the weak currents, including the effective strengths of the axial vector current matrix elements compared with those of the vector current.

It is fun to speculate about the way quarks would behave if they were physical particles of finite mass (instead of purely mathematical entities as they would be in the limit of infinite mass). Since charge and baryon number are exactly conserved, one of the quarks (presumably $u^{\frac{2}{3}}$ or $d^{-\frac{1}{3}}$) would be absolutely stable *, while the other member of the doublet would go into the first member very slowly by β-decay or K-capture. The isotopic singlet quark would presumably decay into the doublet by weak interactions, much as Λ goes into N. Ordinary matter near the earth's surface would be contaminated by stable quarks as a result of high energy cosmic ray events throughout the earth's history, but the contamination is estimated to be so small that it would never have been detected. A search for stable quarks of charge $-\frac{1}{3}$ or $+\frac{2}{3}$ and/or stable di-quarks of charge $-\frac{2}{3}$ or $+\frac{1}{3}$ or $+\frac{4}{3}$ at the highest energy accelerators would help to reassure us of the non-existence of real quarks.

These ideas were developed during a visit to Columbia University in March 1963; the author would like to thank Professor Robert Serber for stimulating them.

References

1) M. Gell-Mann, California Institute of Technology Synchrotron Laboratory Report CTSL-20 (1961).
2) Y. Ne'eman, Nuclear Phys. 26 (1961) 222.
3) M. Gell-Mann, Phys. Rev. 125 (1962) 1067.
4) E.g.: R. H. Capps, Phys. Rev. Letters 10 (1963) 312;
R. E. Cutkosky, J. Kalckar and P. Tarjanne, Physics Letters 1 (1962) 93;
E. Abers, F. Zachariasen and A. C. Zemach, Phys. Rev. 132 (1963) 1831;
S. Glashow, Phys. Rev. 130 (1963) 2132;
R. E. Cutkosky and P. Tarjanne, Phys. Rev. 132 (1963) 1354.
5) P. Tarjanne and V. L. Teplitz, Phys. Rev. Letters 11 (1963) 447.
6) James Joyce, Finnegan's Wake (Viking Press, New York, 1939) p. 383.
7) M. Gell-Mann and M. Lévy, Nuovo Cimento 16 (1960) 705.
8) N. Cabibbo, Phys. Rev. Letters 10 (1963) 531.

* There is the alternative possibility that the quarks are unstable under decay into baryon plus anti-di-quark or anti-baryon plus quadri-quark. In any case, some particle of fractional charge would have to be absolutely stable.

* * * * *

VIII

THE WEAK CURRENT OF THE HADRONS

The first article in this part serves as a kind of review, albeit a lengthy one. In it, the earlier articles by Gell-Mann are criticized and brought up to date and related to the paper by Cabibbo, which is an excellent brief account of the firmer part of the eightfold way theory of the hadronic weak current (or the leptonic weak interactions of hadrons).

The remainder of the theory is more speculative and concerns the hypothesis that the vector and axial vector weak currents, broken up according to the quantum numbers of the strong interactions, generate the algebra of $SU(3) \times SU(3)$.

In the last paper, Gell-Mann and Ne'eman compare this algebra with other possible candidates and conclude that it is favored, because it alone allows a simple definition of the universality of weak interactions, when the hadronic weak current transforms like an octet.

In a theory based on fundamental triplets, of course, $SU(3) \times SU(3)$ is natural, as remarked in the quark paper of the previous part and in earlier work on the Sakata model [for example, Salam and Ward, Nuovo Cimento, **20**, 419 (1961)].

The Symmetry Group of Vector and Axial Vector Currents*

MURRAY GELL-MANN

California Institute of Technology, Pasadena, California

(Received May , 1964)

*Work supported in part by the U. S. Atomic Energy Commission.

ABSTRACT

We review, modify slightly, generalize, and attempt to apply a theory proposed earlier of a higher broken symmetry than the eightfold way. The integrals of the time components of the vector and axial vector current octets are assumed to generate, under equal time commutation, the algebra of SU(3) × SU(3). The energy density of the strong interactions is assumed to consist of a piece invariant under the algebra, a piece that violates conservation of the axial vector currents only and belongs to the representation $(\underset{\sim}{3}, \underset{\sim}{3}^*)$ and $(\underset{\sim}{3}^*, \underset{\sim}{3})$, and a piece that violates the eightfold way and probably belongs to $(\underset{\sim}{1}, \underset{\sim}{8})$ and $(\underset{\sim}{8}, \underset{\sim}{1})$. Assuming the algebraic structure is exactly correct, there is still the question of whether one can assign particles approximately to super-supermultiplets. The pseudoscalar meson octet, together with a pseudoscalar singlet, a scalar octet, and a scalar singlet, may belong to $(\underset{\sim}{3}, \underset{\sim}{3}^*)$ and $(\underset{\sim}{3}^*, \underset{\sim}{3})$. The vector meson octet, together with an axial vector octet, may belong to $(\underset{\sim}{1}, \underset{\sim}{8})$ and $(\underset{\sim}{8}, \underset{\sim}{1})$. The baryon octet with $J = 1/2^+$, together with a singlet with $J = 1/2^-$, may belong to $(\underset{\sim}{3}, \underset{\sim}{3}^*)$ and $(\underset{\sim}{3}^*, \underset{\sim}{3})$, as suggested before. Several crude coupling patterns and mass rules emerge, to zeroth or first order in the symmetry violations. Some are roughly in agreement with experiment, but certain predictions, like that of the existence of a scalar octet, have not been verified. Whether or not they are useful as an approximate symmetry, the equal time commutation rules fix the scale of the weak interaction matrix elements. Further rules of this kind are found to hold in certain Lagrangian field theory models and may be true in reality. In particular, we encounter an algebraic system based on SU(6) that relates quantities with different kinds of behavior under Lorentz transformations.

I. INTRODUCTION

The "eightfold way" theory of a broken higher symmetry for strong interactions was proposed[1,2] at a time when the value of a badly violated symmetry was unclear for two reasons:

1) It was not obvious what real significance could be assigned to the algebraic properties of the higher symmetry.

2) It was not known whether the particle spectrum would show unmistakable evidence of the higher symmetry.

A solution was offered to the first problem when we pointed out[3] that the weak vector currents with $|\Delta \underset{\sim}{I}| = 1/2$, $\Delta Y/\Delta Q = +1$, and $|\Delta \underset{\sim}{I}| = 1$, $\Delta Y = 0$ generate an algebraic system through the equal-time commutation relations of their time-components and that this algebra is preserved even though the conservation of the strangeness-changing currents is violated. We assumed that the algebra in question is that of SU(3); no matter how badly the eightfold way is broken, the vector current octet is then the current of the F-spin. (This result was a simple generalization of the conserved vector current hypothesis, that the $\Delta Y = 0$ vector current is the current of the I-spin.) We went on to consider the equal-time commutation relations of the vector currents with the energy density, which interacts with gravitation. The assumption of the broken eightfold way is that the energy density is the sum of two pieces, one of which is invariant under the F-spin, and the other of which transforms like one component (by definition, the eighth) of an octet:

$$\mathcal{H} = \mathcal{H}_{inv} - c\, \phi_8 \quad . \tag{1}$$

The second problem arose when we attempted to arrange the isotopic multiplets of strongly interacting particles in supermultiplets that correspond, in order c^o, to irreducible representations of SU(3). We then derived, to zeroth and first order in c, coupling patterns and mass rules that were to be compared with experiment. We had no guarantee, however, even if our algebraic system was the right one, that these rules would be sufficiently well obeyed to show traces of the higher symmetry and the manner in which it is violated. Fortunately, ample evidence is now available to support the eightfold way symmetry, together with the pattern of violation in Eq. (1). In fact, the mass rules derived to first order in c are surprisingly accurate.

The success of the broken eightfold way, despite the large violations of symmetry involved, suggests that it may be worthwhile to study in detail the still higher symmetry associated with the axial vector currents. We proposed[3] that the axial vector currents with $|\Delta \underset{\sim}{I}| = 1/2$, $\Delta Y/\Delta Q = +1$, and $|\Delta \underset{\sim}{I}| = 1$, $\Delta Y = 0$ belong to an octet with respect to F-spin and that the time-components of the vector and axial vector octets together generate, under equal time commutation, the algebra $SU(3) \times SU(3)$. Moreover, we suggested that the term $\mathcal{H}_{inv}$ in the energy density in Eq. (1) consists of two parts

$$\mathcal{H}_{inv} = \bar{\mathcal{H}} - \lambda u_o \quad , \tag{2}$$

where $\bar{\mathcal{H}}$ is invariant under the full algebra and leaves both vector and axial vector currents conserved, while the term $-\lambda u_o$, transforming like a particular pair of representations of $SU(3) \times SU(3)$, violates conservation of the axial vector currents, while still commuting with the F-spin. These algebraic statements do not, of course, depend for their proposed validity on the smallness of the parameter λ.

We did not take altogether seriously, in Ref. 3, the idea that these algebraic relations might be applied, like those of the eightfold way, by trying to assign particles to irreducible representations and finding rules to zeroth and first order in λ, to be compared with experiment. The success of the broken eightfold way, however, now makes it less ridiculous to see whether we can find traces of this even more badly broken symmetry. We would try to group the strongly interacting particle supermultiplets into super-supermultiplets, usually including particles of both parities, and to check by experiment very crude relations derived to low order in λ. In Section IV we describe the most plausible scheme of this kind, and note that an octet of scalar mesons is required for its success. Since no such octet has been clearly established at this time, we must reserve judgment on whether the approximation of small λ is of any use in describing the strong interactions.

It was mentioned in Ref. 3 that an algebra is generated by the time-components of the vector and axial vector currents together with the symmetry-breaking term u_o in the energy density. In Section VII we follow up this idea and discuss the possibility that the "extended algebra" may be rather small; we show that in a special model it is the algebra of SU(6) and suggest that such may be the case in reality. It is interesting that the extended algebra ties together quantities with different Lorentz transformation properties, such as the scalar u_o and the four-vector currents.

The algebra (presumably SU(3) x SU(3)) of the vector and axial vector currents and the extended algebra (possibly SU(6)) can be used, in the form of equal time commutation relations, to supplement dispersion relations in the calculation of weak current matrix elements. It may also be true that more and more information about the strongly interacting particles can be expressed in algebraic language by repeated use of the notion of equal-time commutation relations.

II. REVIEW OF THE THEORY

We treat the strong interactions exactly, and the electromagnetic, weak, and gravitational interactions in lowest order. Even though we discuss the scattering amplitudes for strongly interacting particles on the mass shell only (by the method of dispersion relations or "S-matrix theory") we must still acknowledge that in lowest order the matrix elements of the weak and electromagnetic interactions can be determined for arbitrary momentum transfer by measurement and by analytic continuation of the measurable amplitudes. Presumably the same is true in principle of gravitation. Thus we may deal with electromagnetic and weak current operators and a stress-energy-momentum tensor operator $\theta_{\alpha\beta}$, all functions of a space-time variable x, with the matrix elements for any momentum transfer k given by Fourier transform and with time derivatives given by commutation with the space integral of the energy density $\mathcal{H} = -\theta_{44}$.

Incidentally, we notice in this way that the "S-matrix theory" of strong interactions, with electromagnetism, weak interactions, and gravitation treated as small perturbations, is just a branch of abstract field theory, since the current operators and $\theta_{\alpha\beta}$ are all field operators.

The weak current may be broken up, according to quantum numbers conserved by the strong interactions, first into a vector and an axial vector part, and then into pieces characterized by different values of $|\Delta \underset{\sim}{I}|$ and $\Delta Y/\Delta Q$. We restrict our attention here to the familiar terms with $|\Delta \underset{\sim}{I}| = 1$, $\Delta Y = 0$, and $|\Delta \underset{\sim}{I}| = 1/2$, $\Delta Y/\Delta Q = +1$. If there are others, they may lead to bigger algebras than we have here, but need not invalidate our conclusions.

The integrals of the time components of all these currents generate some minimal algebraic system under equal time commutation. For those currents that are conserved, the corresponding integrals (like the electric charge) are constant operators; the others vary with time. But the structure constants of the algebra remain unchanged under all conditions and correspond to a law of nature that specifies the minimal algebra of the vector and axial vector currents that we are studying.

In Ref. 3, we made three assumptions that determine the algebra:

a) The vector weak current, like the electromagnetic current, is a component of the F-spin current $\mathcal{F}_{i\alpha}(x)$, where $i = 1 \ldots 8$ and α is a Lorentz index. We have, then,

$$F_i(t) = -i \int \mathcal{F}_{i4} \, d^3x \quad , \tag{3}$$

$$\left[F_i(t), F_j(t)\right] = i f_{ijk} F_k(t) \quad . \tag{4}$$

Of course, F_1, F_2, and F_3 are conserved by the strong interactions and are just the components of the isotopic spin; thus the conserved vector current hypothesis is included here. The components F_4, F_5, F_6, and F_7 actually vary with time.

b) The axial vector weak current is the same component of another current $\mathcal{F}_{i\alpha}^{5}(x)$ that transforms like an octet with respect to F-spin. We have

$$F_i^{5}(t) = -i \int \mathcal{F}_{i4}^{5} \, d^3x \quad , \tag{5}$$

$$\left[F_i(t), F_j^{5}(t)\right] = i f_{ijk} F_k^{5}(t) \quad . \tag{6}$$

c) The commutation rules of the operators $F_i^{5}(t)$ close the algebraic system by giving

$$\left[F_i^{5}(t), F_j^{5}(t)\right] = i f_{ijk} F_k(t) \quad . \tag{7}$$

We now define

$$2 F_i^{\pm}(t) \equiv F_i(t) \pm F_i^{5}(t) \tag{8}$$

and notice that F_i^{+} and F_i^{-} are two commuting F-spins, so that we are really dealing with the algebra of $SU(3) \times SU(3)$. The two sets of operators, which we may think of as "left-handed" and "right-handed" F-spins respectively, are connected by parity:

$$P F_i^{\pm} P^{-1} = F_i^{\mp} \quad . \tag{9}$$

The total F-spin itself is, according to Eq. (8), just the sum of the left- and right-handed parts:

$$F_i = F_i^{+} + F_i^{-} \quad . \tag{10}$$

We will be concerned with irreducible representations of the system consisting of F_i^{+}, F_i^{-}, and P. We indicate the behavior[3] with respect to (F_i^{+}, F_j^{-}) by a pair of representations, such as $(\underline{3}, \underline{3}^*)$, $(\underline{8}, \underline{8})$, etc.

Since parity interchanges F_i^+ and F_i^-, an irreducible representation with respect to parity and the two F-spins will have such forms as $(\underline{8}, \underline{8})$ or $(\underline{3}, \underline{3}^*)$ and $(\underline{3}^*, \underline{3})$ or $(\underline{1}, \underline{8})$ and $(\underline{8}, \underline{1})$.

If a representation of F_i^+, F_i^-, and P is to contain a component invariant under F_i, it must have the form $(\underline{1}, \underline{1})$ or $(\underline{3}, \underline{3}^*)$ and $(\underline{3}^*, \underline{3})$ or $(\underline{8}, \underline{8})$, etc., so that the product of the two indicated representations will contain $\underline{1}$.

The simplest choice, then, for the term λu_0 in Eq. (2), which violates the conservation of F_i^+ and F_i^- separately while conserving F_i, is to have it belong to $(\underline{3}, \underline{3}^*)$ and $(\underline{3}^*, \underline{3})$, as proposed in Ref. 3. The operator u_0 thus belongs to a set of nine scalar and nine pseudoscalar quantities, in each case forming an octet and a singlet with respect to total F-spin. The scalar octet is labeled $u_1 \ldots u_8$ and the singlet u_0, while the pseudoscalar octet and singlet are labeled $v_1 \ldots v_8$ and v_0 respectively.

To specify the transformation properties of the u's and v's, we introduce[3] a generalization of the symbols f_{ijk} and d_{ijk} to the case $i = 0, 1, \ldots 8$ instead of $i = 1 \ldots 8$. To the 3×3 matrices λ_i $(i = 1 \ldots 8)$ we adjoin the matrix $\lambda_0 = (2/3)^{1/2}\, 1$ and obtain the rules

$$\left[\lambda_i, \lambda_j\right] = 2\, i\, f_{ijk}\, \lambda_k \quad , \tag{11}$$

$$\left\{\lambda_i, \lambda_j\right\} = 2\, d_{ijk}\, \lambda_k \quad , \tag{12}$$

$$\mathrm{Tr}\, \lambda_i\, \lambda_j = 2\, \delta_{ij} \quad , \tag{13}$$

where i, j, and k run from 0 to 8. Here, f_{ijk} vanishes when any index is zero, and d_{ijk} equals $(2/3)^{1/2}\ \delta_{ij}$ when k is zero, etc.

We then obtain, for the transformation properties of the u_i and v_i, the results[3)]

$$\begin{aligned} \left[F_i, u_j\right] &= i\, f_{ijk}\, u_k \quad , \\ \left[F_i, v_j\right] &= i\, f_{ijk}\, v_k \quad , \\ \left[F_i^5, u_j\right] &= -\, i\, d_{ijk}\, v_k \quad , \\ \left[F_i^5, v_j\right] &= i\, d_{ijk}\, u_k \quad . \end{aligned} \tag{14}$$

We note that Eqs. (4), (6), and (7) indicate the representation to which the currents $\mathcal{F}_{i\alpha}(x)$ and $\mathcal{F}_{i\alpha}^5(x)$ belong, namely $(\underset{\sim}{1}, \underset{\sim}{8})$ and $(\underset{\sim}{8}, \underset{\sim}{1})$, with $i = 1 \ldots 8$. At equal times, we have

$$\begin{aligned} \left[F_i, \mathcal{F}_{j\alpha}\right] &= i\, f_{ijk}\, \mathcal{F}_{k\alpha} \quad , \\ \left[F_i, \mathcal{F}_{j\alpha}^5\right] &= i\, f_{ijk}\, \mathcal{F}_{k\alpha}^5 \quad , \\ \left[F_i^5, \mathcal{F}_{j\alpha}\right] &= i\, f_{ijk}\, \mathcal{F}_{k\alpha}^5 \quad , \\ \left[F_i^5, \mathcal{F}_{k\alpha}^5\right] &= i\, f_{ijk}\, \mathcal{F}_{k\alpha} \quad . \end{aligned} \tag{15}$$

The essential physics of the theory is contained in the equations written so far and is taken over directly from Ref. 3. Two more points need to be added, however, which are modifications of the corresponding points in the earlier article. One of these concerns the component of $\mathcal{F}_{i\alpha}^{+}$ utilized for the weak current, and is discussed in the next section. The other point is connected with the transformation properties under

SU(3) × SU(3) of the term in the energy density that violates the eightfold way, namely ϕ_8 in Eq. (1). The simplest possibilities for a unitary octet are, of course, $\left[(\underset{\sim}{3}, \underset{\sim}{3}^*) \text{ and } (\underset{\sim}{3}^*, \underset{\sim}{3})\right]$ and $\left[(\underset{\sim}{1}, \underset{\sim}{8}) \text{ and } (\underset{\sim}{8}, \underset{\sim}{1})\right]$. It now appears that the latter may be more nearly satisfactory than the former, as we shall see in the next section. We had previously assumed not only that ϕ_8 transformed as $\left[(\underset{\sim}{3}, \underset{\sim}{3}^*) \text{ and } (\underset{\sim}{3}^*, \underset{\sim}{3})\right]$, i.e., like u_8, but also that ϕ_8 was equal to u_8.

III. UNIVERSALITY OF THE WEAK INTERACTIONS

We know that the electromagnetic current of the strongly interacting particles (or "hadrons" to use Okun's expression) is given by the formula

$$j_\alpha = e\,(\mathcal{F}_{3\alpha} + \frac{1}{\sqrt{3}}\,\mathcal{F}_{8\alpha}) \quad . \tag{16}$$

(A constant term may have to be added if there exist hadrons corresponding to certain kinds of spinor representations of SU(3), as discussed in Section VI.)

What about the hadron weak current coupled to leptons? It must be a linear combination of $\mathcal{F}^+_{1\alpha} + i\,\mathcal{F}^+_{2\alpha}$ with $\Delta Y = 0$ and $\mathcal{F}^+_{4\alpha} + i\,\mathcal{F}^+_{5\alpha}$ with $\Delta Y/\Delta Q = +1$, if we stick to the assumptions (a) and (b) of Section II. The choice of the linear combination is motivated in part by the requirement of universality of the weak interactions: the algebraic properties of the total weak current should be the same for leptons and for hadrons.[4,5]

We write the effective weak interaction in the local approximation (or at least the part coming from a product of charged currents) as

$$\frac{G}{\sqrt{2}} J_\alpha^+ J_\alpha \quad , \tag{17}$$

where $J_\alpha = J_\alpha(\text{leptons}) + J_\alpha(\text{hadrons})$. The situation is then the following, as described in Ref. 4.

For the now obsolete case of one neutrino for electron and muon, $J_\alpha(\text{leptons})$ has the form

$$\bar{\nu}\,\gamma_\alpha\,(1+\gamma_5)\,e + \bar{\nu}\,\gamma_\alpha\,(1+\gamma_5)\,\mu$$

$$= 2\sqrt{2}\;\bar{\nu}\,\gamma_\alpha\,\frac{(1+\gamma_5)}{2}\,\frac{(e+\mu)}{\sqrt{2}} \quad .$$

The "weak charge" $-i\int d^3x\, J_4(\text{leptons})$ evidently may be written in the form $2\sqrt{2}\,(K_1 + i\,K_2)$, where K_1, K_2, and $-i\left[K_1, K_2\right]$ have the commutation rules of an angular momentum or an isotopic spin. The weak charge and its hermitian conjugate, for leptons, generate the algebra of SU(2), with $\frac{1+\gamma_5}{2}\,\frac{e+\mu}{\sqrt{2}}$ and $\frac{1+\gamma_5}{2}\,\nu$ appearing as the lower and upper components of a spinor.

With distinct neutrinos for electron and muon, as in the real situation, $J_\alpha(\text{leptons})$ becomes

$$\bar{\nu}_e\,\gamma_\alpha\,(1+\gamma_5)\,e + \bar{\nu}_\mu\,\gamma_\alpha\,(1+\gamma_5)\,\mu$$

$$= 2\,\bar{\nu}_e\,\gamma_\alpha\,\frac{1+\gamma_5}{2}\,e + 2\,\bar{\nu}_\mu\,\gamma_\alpha\,\frac{1+\gamma_5}{2}\,\mu \quad . \tag{18}$$

This time the leptonic weak charge has the form $2(K_1 + i\,K_2)$, where again K_1 and K_2 are the first two components of an angular momentum and the algebra of SU(2) is generated. Now $\frac{1+\gamma_5}{2}\,e$ and $\frac{1+\gamma_5}{2}\,\nu_e$

form a spinor and so do $\frac{1+\gamma_5}{2}\ \mu$ and $\frac{1+\gamma_5}{2}\ \nu_\mu$.

Let us now demand universality for the weak interactions. In the one-neutrino case, we would require that the weak charge for hadrons have the form $2\sqrt{2}\,(K_1 + i\,K_2)$ and in the two-neutrino case that it have the form $2(K_1 + i\,K_2)$, where K_1 and K_2 are the first two components of an angular momentum. Writing the weak charge for hadrons in the general form

$$A\,(F_1^+ + i\,F_2^+)\cos\theta + A\,(F_4^+ + i\,F_5^+)\sin\theta \qquad ,$$

we may verify that we have $A\,(K_1 + i\,K_2)$, where

$$K_1 = F_1^+ \cos\theta + F_4^+ \sin\theta \qquad ,$$

$$K_2 = F_2^+ \cos\theta + F_5^+ \sin\theta \qquad ,$$

$$K_3 = F_3^+ \cos^2\theta + \left(\frac{\sqrt{3}}{2}\ F_8^+ + \frac{1}{2}\ F_3^+\right)\sin^2\theta - F_6 \sin\theta\cos\theta \quad .$$

Now from the approximate equality of vector coupling constants in the decay of the muon and the decay of the nucleus O^{14}, we know that $A\cos\theta$ is around 2.

Clearly, then, universality in the one-neutrino case gives us $A = 2\sqrt{2}$, $\theta = 45^\circ$ or

$$J_\alpha = 2\,(\mathcal{F}_{1\alpha}^+ + i\,\mathcal{F}_{2\alpha}^+ + \mathcal{F}_{4\alpha}^+ + i\,\mathcal{F}_{5\alpha}^+)$$

as in Ref. 3. However, for the actual case of two neutrinos, we must take $A = 2$ with θ small and have

$$J_\alpha = 2\cos\theta\,(\mathcal{F}_{1\alpha}^+ + i\,\mathcal{F}_{2\alpha}^+) + 2\sin\theta\,(\mathcal{F}_{4\alpha}^+ + i\,\mathcal{F}_{5\alpha}^+) \quad . \qquad (19)$$

In a recent paper, Cabibbo[6] has combined our assumptions (a) and (b) quoted in Section II with the choice (18) of current components suitable for universality in the two-neutrino case and has shown that such a theory is in reasonable agreement with present information on leptonic decays of hadrons, with $\theta \approx 0.26$. We may therefore adopt Eq. (18) with some confidence, provided experimental leptonic and hadronic weak interactions exhibit no further complications.

The weak charge in general thus has the form $2(K_1 + i K_2)$. Moreover, the electric charge in units of e has the general form $K_3 + K_0$, where K_0 commutes with K_1, K_2, and K_3. The weak and electric charge operators, for both leptons and hadrons, thus generate the algebra of $U(1) \times SU(2)$. It is only when we take these charges for hadrons and break them up according to the quantum numbers conserved by the strong interactions that we get the group $SU(3) \times SU(3)$.

IV. CRUDE RESULTS FOR SMALL λ

We now attempt to make use of the broken symmetry model for rough predictions about the strongly interacting particles. For the most part we shall put $c = 0$ and forget about violations of the eightfold way, concentrating on axial vector current conservation and its violation. In the limit $\lambda \to 0$, where all the axial vector currents are conserved, if the axial vector β-decay coupling constant is not to vanish, we must have either vanishing baryon masses or vanishing pseudoscalar meson masses. We choose vanishing baryon masses, and thus the point of view expressed here differs from that of many authors.[7] Under these conditions, the "renormalization constant" for the axial vector current becomes unity in the limit $\lambda \to 0$. The β-decay interaction in the zero-momentum transfer case is thus completely fixed, in the limit $\lambda \to 0$, by the symmetry pattern.

Since we are going to try assigning dominant representations to the particles, let us begin with the eight baryons having $J = 1/2^+$. For convenience we describe them, in the limit $\lambda \to 0$, by "fields" $\psi_1 \ldots \psi_8$. If they belonged to $(\underset{\sim}{1}, \underset{\sim}{8})$ and $(\underset{\sim}{8}, \underset{\sim}{1})$, then ψ_j and $\gamma_5 \psi_j$ would transform under F_i and F_i^5 in the same way that $\mathcal{F}_{j\alpha}$ and $\mathcal{F}_{j\alpha}^5$ respectively transform in Eq. (15). In the zero momentum transfer case, then, in the limit $\lambda \to 0$, we would have the following pattern for the weak current $\mathcal{F}_{1\alpha}^+$:

$$- i f_{ijk} \bar{\psi}_j \gamma_\alpha (1 + \gamma_5) \psi_k \quad .$$

Both the vector and axial vector currents would be coupled through F rather than D. This seems to be far from the truth.[6]

Instead, we try the other baryon representation suggested in Ref. 3, namely $(\underset{\sim}{3}, \underset{\sim}{3}^*)$ and $(\underset{\sim}{3}^*, \underset{\sim}{3})$. We then have to add a ninth particle, described by ψ_o. Under F_i and F_i^5, ψ_i and $\gamma_5 \psi_i$ transform like u_i and $i v_i$ in Eq. (14). The coupling pattern as $\lambda \to 0$, for any momentum transfer, is then

$$- i f_{ijk} \bar{\psi}_j \gamma_\alpha \psi_k + d_{ijk} \bar{\psi}_j \gamma_\alpha \gamma_5 \psi_k \tag{20}$$

with $i,j,k = 0, 1, \ldots 8$. For the baryon octet, then, the vector current is coupled through F and the axial vector current through D, with equal coefficients. This situation resembles the experimental one,[6] the admixture of F in the axial vector pattern being of the order of 30%. There is a single form factor for the whole expression (20) in the limit $\lambda \to 0$, and the anomalous magnetic and induced pseudoscalar form factors vanish. Since the vector coupling is through F, the neutron has no electrical interaction in the limit.

The interpretation of the ninth baryon depends on whether the mass associated with ψ_o is positive or negative. A negative mass would lead us to treat $\gamma_5 \psi_o$ as the appropriate operator, so that the new particle with positive mass would have $J = 1/2^-$. Now in first order in λ we can compute the ratio of octet and singlet masses, using the transformation property $(\underset{\sim}{3}, \underset{\sim}{3}^*)$ and $(\underset{\sim}{3}^*, \underset{\sim}{3})$ of the mass term u_o. The result is that the mass associated with the singlet is minus twice that of the octet. Thus $\gamma_5 \psi_o$ should describe, to first order in λ, a baryon with $J = 1/2^-$ and twice the average mass of the baryon octet with $J = 1/2^+$. If the extra baryon is identified with $\Lambda(1405)$, the spin and parity assignments may well be right but 1405 MeV is a far cry from twice the average mass of the baryon octet. We should perhaps not expect better agreement, however, with an approximation that treats baryon masses in first order.

The pseudoscalar octet should be assigned to the same representation as the divergence of the axial vector current, so that the Goldberger-Treiman relation can have some validity even to lowest order in λ. As explained in Ref. 3, we have

$$\int d^3x \ \partial_\alpha \mathcal{F}_{i\alpha}^5 = \dot{F}_i^5 = i \left[\int \mathcal{H} \, d^3x, F_i^5\right] = \lambda \, i \left[F_i^5, \int u_o \, d^3x\right]$$

$$= \lambda \, d_{iok} \int v_k \, d^3x = \sqrt{\frac{2}{3}} \, \lambda \int v_i \, d^3x \qquad (21)$$

neglecting c. So the divergence of the current, in lowest order, belongs to $(\underset{\sim}{3}, \underset{\sim}{3}^*)$ and $(\underset{\sim}{3}^*, \underset{\sim}{3})$ and we make the same assignment for the pseudoscalar octet, along with a pseudoscalar singlet, a scalar octet, and a scalar singlet. For convenience, we describe these by "fields" π_i and σ_i, $i = 0, \ldots 8$, transforming like v_i and u_i respectively.

To order λ^0, these eighteen mesons all have a common mass. In order λ, they split according to the following pattern: the scalar singlet has a squared mass equal to $\overline{\mu^2} - 2\Delta$, the pseudoscalar octet has $\overline{\mu^2} - \Delta$, the scalar octet has $\overline{\mu^2} + \Delta$, and the pseudoscalar singlet has $\overline{\mu^2} + 2\Delta$. It is possible that the scalar singlet may be identified with a low-lying $J = 0^+$, $I = 0$ mesonic state decaying quickly into 2π; such states have been reported at various masses and at least one of them may exist. We then expect a scalar octet lying higher than the pseudoscalar one and a pseudoscalar singlet lying still higher.

The matrix elements of the weak current between one of these mesons and another follow the same pattern, in the limit $\lambda \to 0$, as for the baryons in Eq. (20), since these mesons and the nine baryons belong to the same representation:

$$- i f_{ijk} (\pi_j \partial_\alpha \pi_k + \sigma_j \partial_\alpha \sigma_k) + d_{ijk} (\sigma_j \partial_\alpha \pi_k - \pi_j \partial_\alpha \sigma_k) \quad . \qquad (22)$$

In the limit, there is a common form factor for the whole of (22).

If the scalar singlet meson represented by σ_0 (let us call it σ) really has a lower mass than K^+, then the K_{e4} decay $K^+ \to \pi^+ + \pi^- + e^+ + \nu$ should be dominated by the chain $K^+ \to \sigma + e^+ + \nu$, $\sigma \to \pi^+ + \pi^-$. Treating σ as stable and using the zeroth order pattern (22), we may compute the ratio of the rates of $K^+ \to \sigma + e^+ + \nu$ and $K^+ \to \pi^0 + e^+ + \nu$:

$$\frac{K^+ \to \sigma + e^+ + \nu}{K^+ \to \pi^0 + e^+ + \nu} = \frac{8}{3} \; \frac{f\,(m_\sigma/m_K)}{f\,(m_\pi/m_K)} \quad , \qquad (23)$$

where

$$f(\beta) = 1 - 8\,\beta^2 + 24\,\beta^4 \,\ell n\, \beta^{-1} + 8\,\beta^6 - \beta^8 \quad . \qquad (24)$$

Roughly, then, the relative rate of $K^+ \to \pi^+ + \pi^- + e^+ + \nu$ should be given by 2/3 of the expression (23) and the mass of the $\pi^+ \pi^-$ system should be clustered around m_σ. The experimental data[8)] still do not permit any firm conclusions, except that the ratio calculated in (23) must be a few times 10^{-3} if the decay through σ takes place. By contrast, if the σ mass is as low as 310 MeV, our rough formula gives about 0.21, in complete disagreement with observation. With a mass near 400 MeV, agreement is possible.

The symmetry, in the limit $\lambda \to 0$, permits trilinear couplings of the scalar and pseudoscalar mesons to one another and to the nine baryons. In each case the allowed coupling pattern is formed with the symbol d'_{ijk}, which equals d_{ijk} except when one index is zero and the other two equal but not zero; the value of d' is then $-\sqrt{1/6}$. The effective couplings are:

$$2\, g\, d'_{ijk}\ \bar{\psi}_i\, (\sigma_j + i\, \pi_j\, \gamma_5)\, \psi_k \qquad , \tag{25}$$

$$(h/6)\, d'_{ijk}\, (\sigma_i + i\, \pi_i)(\sigma_j + i\, \pi_j)(\sigma_k + i\, \pi_k) + \text{hermitian conjugate.} \tag{26}$$

The coupling of the pseudoscalar octet to the baryon octet through D is, of course, in reasonable agreement with experiment, as we might have expected from the Goldberger-Treiman relation and the axial vector current coupling through D.

Other predictions, however, are not in good agreement with the present experimental situation. No scalar octet has been found. The nuclear forces arising from the scalar particles should be gigantic, according to (25), and evidence for such forces is not convincing at the present time. Also the scalar coupling of the ninth baryon $\gamma_5\, \psi_o$ to the pseudoscalar octet and the baryon octet should also be very strong, according to (25), and lead to a width of several BeV (!) for the ninth baryon if we take the prediction literally. An estimate in the next section of h in Eq. (26) indicates large widths also for the scalar mesons if they are significantly above the thresholds for decay into two pseudoscalar mesons.

We may thus adopt several different attitudes:

(1) The $SU(3) \times SU(3)$ algebraic system, assuming it is correct, does not provide a useful approximate symmetry.

(2) Higher order effects in λ, for example as indicated by the Goldberger-Treiman relation (see next section), may reduce some of the coupling constants.

(3) The scalar mesons will turn up with large couplings, and the ninth baryon will turn up as a very vague bump with a huge width.

(4) Something is wrong with our choices of representations.

At present, it is not easy to choose among these possibilities. They are discussed further in Section V. Meanwhile, we return to the assignment of representations.

The vector meson situation is complicated by the ϕ-ω puzzle and will not be fully treated here, but we should expect a vector meson octet that dominates the vector form factors to transform like the currents, i.e., according to $(\underset{\sim}{1}, \underset{\sim}{8})$ and $(\underset{\sim}{8}, \underset{\sim}{1})$. Thus there should be an axial vector octet nearby. In fact, the splitting between the two octets is of second order in λ, since we cannot make $(\underset{\sim}{3}, \underset{\sim}{3}^*)$ and $(\underset{\sim}{3}^*, \underset{\sim}{3})$ out of $\left[(\underset{\sim}{1}, \underset{\sim}{8}) \text{ and } (\underset{\sim}{8}, \underset{\sim}{1})\right]$ times itself.

In conclusion, we mention the violation of the eightfold way by the term in ϕ_8. If ϕ_8 were to belong to $(\underset{\sim}{3}, \underset{\sim}{3}^*)$ and $(\underset{\sim}{3}^*, \underset{\sim}{3})$, then the splitting of the baryon octet with $J = 1/2^+$ would go mainly with D, since to order $\lambda^o c^1$ the only allowed coupling is analogous to (25). In fact, the splitting is mostly F. We may therefore consider the possibility that ϕ_8 belongs to $(\underset{\sim}{1}, \underset{\sim}{8})$ and $(\underset{\sim}{8}, \underset{\sim}{1})$. The baryon octet is then split only in order $\lambda^1 c^1$ and both F and D come in.

The scalar and pseudoscalar octets are split open in order $\lambda^0 c^1$ by a term ϕ_8 transforming like $(\underset{\sim}{1}, \underset{\sim}{8})$ and $(\underset{\sim}{8}, \underset{\sim}{1})$. Moreover, in that order, the spacing is the same for both octets.

V. THE GOLDBERGER-TREIMAN RELATION

The Goldberger-Treiman relation states in essence that certain matrix elements of the divergence $\partial_\alpha \mathcal{F}^5_{i\alpha}$ of the axial vector current components obey unsubtracted dispersion relations in the invariant momentum squared carried by the current and that these dispersion relations are dominated by the intermediate state with one pseudoscalar meson.[3,9]

The pion decay amplitude is described by the quantity f_π, which is defined, as in Ref. 3, by the formula

$$\langle 0 \,|\partial_\alpha \mathcal{F}^5_{i\alpha}|\, \pi\rangle = \frac{m_\pi^2}{2f_\pi} \Phi_i \quad , \quad (i = 1, 2, 3) \tag{27}$$

where Φ_i is the wave function of the pion. The width for the decay $\pi^+ \to \mu^+ + \nu_\mu$ is then

$$\Gamma_\pi = G^2 \cos^2\theta \, m_\pi \, m_\mu^2 \, (1 - m_\mu^2/m_\pi^2)^2 \left(\frac{f_\pi^2}{4\pi}\right)^{-1} (64\,\pi^2)^{-1} \quad , \tag{28}$$

which differs by the factor $\cos^2\theta$ from the corresponding expression in Ref. 3. Likewise, we have

$$\langle 0 \,|\partial_\alpha \mathcal{F}^5_{i\alpha}|\, K\rangle = \frac{m_K^2}{2f_K} \Psi_i \quad , \quad (i = 4, 5, 6, 7) \tag{29}$$

where ψ_1 is the wave function of the kaon. The width for the decay $K^+ \to \mu^+ + \nu_\mu$ is

$$\Gamma_K = G^2 \sin^2\theta \, m_K \, m_\mu^{\,2} \, (1 - m_\mu^{\,2}/m_K^{\,2})^2 \left(\frac{f_K^{\,2}}{4\pi}\right)^{-1} (64\,\pi^2)^{-1} \quad , \tag{30}$$

where in Ref. 3 we would not have had the factor $\sin^2\theta$. Cabibbo[6] has pointed out that with $\theta \approx 0.26$ we get $f_\pi \approx f_K$. (The attempt[3] to make $f_\pi/f_K \approx m_\pi/m_K$ is thus unnecessary.)

The Goldberger-Treiman relation for neutron β decay now states that

$$2m_N \left(\frac{-\,G_A}{G\cos\theta}\right) \approx g_{NN\pi} \; f_\pi^{\,-1} \quad , \tag{31}$$

where $(-\,G_A/G\cos\theta)$ is the axial vector "renormalization factor" for the nucleon. The matrix element between neutron and proton of the divergence of the axial vector current has been approximately expressed as the product of the pion-nucleon coupling constant $g_{NN\pi}$ and the pion decay constant $f_\pi^{\,-1}$. Experimentally, Eq. (31) is satisfied with an error of around 10%.

If we now generalize to the baryon octet and the pseudoscalar meson octet, in the approximation of the eightfold way (which includes $f_\pi = f_K = f$), we have in general a part of g that goes with the D coupling and a part with the F coupling; the same is true of $(-\,G_A/G\cos\theta)$. Not only should the relation (31) hold, then, but also the F/D ratios should be the same for the meson-baryon coupling and for the axial vector current. Cabibbo's value of 0.30/0.95 for the F/D ratio for the current agrees well with all estimates of F/D for the meson coupling.

If scalar mesons exist, the Goldberger-Treiman relation should apply to the β-decay matrix elements between scalar and pseudoscalar meson states, in relation to the strong coupling constants for the scalar-pseudoscalar-pseudoscalar ($\sigma\ \pi\ \pi$) vertices.

Now let us examine what happens to the Goldberger-Treiman relation for the $NN\pi$ and $\sigma\pi\pi$ cases when we have approximate conservation of both vector and axial vector currents, i.e., $c = 0$ and $\lambda \to 0$, assuming our assignments of both baryons and mesons to ($\underset{\sim}{3}$, $\underset{\sim}{3}^*$) and ($\underset{\sim}{3}^*$, $\underset{\sim}{3}$) are correct. In the limit $\lambda \to 0$, the matrix elements $\langle 0\ |u_i|\ \sigma_i\rangle$ and $\langle 0\ |v_i|\ \pi_i\rangle$ are all non-zero and equal. The quantity f^{-1}, proportional to $\langle 0\ |\partial_\alpha\ \mathcal{F}^5_{i\alpha}|\ \pi_i\rangle$, is evidently of order λ. Likewise, the mass of the nucleon is of order λ. The renormalization ($- G_A/G \cos\Theta$) approaches unity, with the octet pattern becoming pure D. The coupling constant $g_{NN\pi}$ remains finite and the octet coupling pattern here too becomes pure D. Evidently, then, the Goldberger-Treiman relation (27) can hold approximately in the limit $\lambda \to 0$ with both sides of order λ. We have, then, in the double approximation of $\lambda \to 0$ and exact validity of the Goldberger-Treiman relation,

$$2m_N \approx g\,f^{-1} \quad , \tag{32}$$

where g multiplies the whole baryon-baryon-meson coupling pattern, as in (25). The analogous equation for the $\sigma\pi\pi$ vertices is

$$2\,\Delta \approx h\,f^{-1} \quad , \tag{33}$$

where $2\,\Delta$ is the difference in mass squared between the scalar and pseudoscalar octets in order λ and h multiplies the whole trilinear meson coupling pattern, as in (26). This double approximation is equivalent to saying that the violation of SU(3) × SU(3) mass degeneracy is accomplished formally by the displacement

$$\sigma_0 \to \sigma_0 - \sqrt{\frac{3}{8}}\ f^{-1} \qquad (34)$$

in effective couplings such as (25) and (26).

Now if we consider large violations of symmetry, so that higher order effects in λ are important (and even effects involving c, which violate the eightfold way), we may suppose that the Goldberger-Treiman approximation is still good. For example, as we mentioned above, the actual value of $(-G_A/G \cos \theta)$ is around 1.25, with about 0.95 going with the D pattern and about 0.30 with the F pattern, in contrast with the value 1 and the pure D coupling that we would have in the limit $\lambda \to 0$. Likewise, the baryon-baryon-meson coupling departs from the pure D coupling that is acquired in the limit $\lambda \to 0$, but these two departures seem to follow the Goldberger-Treiman relation in that the F/D ratio is similar in the two cases.

We may look, for example, at the decay of the ninth baryon into $\Sigma + \pi$. In order λ, the mass difference between the two baryons is the same as m_N and in order 1 the coupling constant is the same as $g_{NN\pi}$, with the appropriate ratio of d'_{ijk} coefficients. If, now, in higher order in λ and c the mass difference becomes quite different, we might expect the coupling constant to change in proportion, keeping the Goldberger-Treiman equation approximately valid. Instead of having the effective coupling constant g for the decay equal to $2m_N f_\pi$, as in (32), we would have approximately $2(m - m_\Sigma) f_\pi r$, where m is the real mass of the ninth baryon and r is the renormalization factor for the axial vector current matrix element between Σ and the ninth baryon. If m is 1405 MeV, for instance, then the coupling constant g in question is reduced by the factor $r(m - m_\Sigma)\, m_N^{-1} \approx 0.23\, r$ and the width of the ninth baryon is then

$$\Gamma \approx 4\ \frac{\left[2\, f_\pi\, (m - m_\Sigma)\, r\right]^2}{4\pi}\ k\ \frac{m_\Sigma + E_\Sigma}{2m} ,$$

where k is the decay momentum. The width comes out about 200 MeV (r^2) instead of about 3 BeV (r^2); the actual width of $\Lambda(1405)$ is around 60 MeV.

Similar corrections should be applied to the various coupling constants h for the various $(\sigma\,\pi\,\pi)$ vertices; instead of the completely symmetrical formula (33) we can use similar expressions in which the actual differences of mass squared are inserted in place of the first order pattern based on the single quantity Δ. For a scalar K particle of mass μ, the decay into $K + \pi$ would be regulated by a value of h approximately equal to $f_\pi\,(\mu^2 - m_K^2)\, r$, where r is the renormalization factor for the axial vector current matrix element between the scalar K particle and K itself. The decay width for this case is then

$$\Gamma \approx \frac{3}{2}\ \frac{\left[f_\pi\,(\mu^2 - m_K^2)\, r\right]^2}{4\pi}\ \frac{k}{\mu^2} ,$$

where k is the decay momentum. For $\mu \approx 725$ MeV, for example, Γ comes out around 80 MeV (r^2).

None of this is of much use, of course, if the scalar octet does not exist. If it is not found, we will have to abandon the idea of using the group SU(3) × SU(3) of the vector and axial vector currents as an approximate symmetry of the strong interactions.

We should mention one intermediate possibility, which involves a different assignment of representation to the pseudoscalar octet, namely ($\underset{\sim}{1}$, $\underset{\sim}{8}$) and ($\underset{\sim}{8}$, $\underset{\sim}{1}$). No scalar or pseudoscalar unitary singlet would be predicted. The Goldberger-Treiman relation could not in this case be approximately valid for small λ, since the two sides would be of different order in λ, with the pseudoscalar octet not transforming like v_1 under SU(3) × SU(3). Conceivably, however, for a particular value of λ, not particularly small, we could have the relation. The coupling of baryons to the pseudoscalar and scalar mesons would be forbidden as $\lambda \to 0$ and the pseudoscalar octet would acquire its coupling through the violation of the symmetry. The scalar octet would have the opposite properties under charge conjugation to the scalar octet hitherto discussed. Thus in the limit $c \to 0$ of the eightfold way, this scalar octet could have no Yukawa coupling to the baryon octet and no coupling to two mesons of the pseudoscalar octet. When the violation of the eightfold way is turned on, the non-strange members of the scalar octet would still lack these couplings, but the strange members could have them by violating the eightfold way. Such an "abnormal" scalar octet would have very different experimental properties from a normal one, particularly for the $Y = 0$ members, and would be readily identifiable as such.

VI. TRIPLETS, REAL AND MATHEMATICAL

So far, we have concerned ourselves with the assumption that the group of the vector and axial vector octets of currents is SU(3) × SU(3), with the transformation properties under the group of the terms λu_0 and $c \phi_8$ in the energy density, and with the possibility that in a crude approximation λ as well as c might be treated as small. We may, however, go further and ask whether there are additional algebraic relations among the quantities we have introduced. In order to obtain such relations that we may conjecture to be true, we use the method of abstraction from a Lagrangian field theory model. In other words, we construct a mathematical theory of the strongly interacting particles, which may or may not have anything to do with reality, find suitable algebraic relations that hold in the model, postulate their validity, and then throw away the model. We may compare this process to a method sometimes employed in French cuisine: a piece of pheasant meat is cooked between two slices of veal, which are then discarded.[10)]

In Ref. 3, the Sakata model was employed in this way. However, certain adjustments had to be made to get to the eightfold way. Instead, we may employ the quark model,[11)] which gives the eightfold way directly; there are also other models,[11)] based on a fundamental triplet and a fundamental singlet, that are perfectly compatible with the eightfold way.

Any such field-theoretic model must contain some basic set of entities, with non-zero baryon number, out of which the strongly interacting particles can be made. If this set is a unitary octet, the theory is very clumsy; it is hard to arrange any coupling that will reduce the symmetry from SU(8) to SU(3) without introducing[1,2] in addition a Yang-Mills octet of fundamental vector mesons. Thus the only reasonably attractive models are based on unitary triplets and perhaps singlets.

If we adopt such a viewpoint, we should say that the correct dynamical description of the strongly interacting particles requires either the bootstrap theory or else a theory based on a fundamental triplet. In neither case do the familiar neutron and proton play any basic role.

It is, of course, a striking fact that no unitary triplets have so far been identified among the strongly interacting particles; however, they may turn up. Their appearance may, of course, be consistent with either the bootstrap theory or a theory with a fundamental triplet. Their non-appearance could certainly be consistent with the bootstrap idea, and also possibly with a theory containing a fundamental triplet which is hidden, i.e., has effectively infinite mass.

Thus, without prejudice to the independent questions of whether the bootstrap idea is right and whether real triplets will be discovered, we may use a mathematical field theory model containing a triplet in order to abstract algebraic relations.

If we want to use just a triplet and no singlet, we must have quarks, with baryon number 1/3 and electric charges - 1/3, - 1/3, and + 2/3. Such particles presumably are not real but we may use then in our field theory model anyway. Since the quark model is mathematically the simplest, we shall in fact employ it in the next section, as in Ref. 11, for our process of abstraction.

If we consider a model with a basic triplet t and a singlet b, then we are free to take for these particles integral electric charges and baryon number equal to one; say we do so. The singlet must be neutral, and we can then form the known baryon supermultiplets from (b), ($b\, t\, \bar{t}$), ($b\, \bar{b}\, b\, \bar{t}\, t\, \bar{t}\, t$), etc. The triplet can have electric charges q, q, q+1 or -q, -q, -q -1, where q is any integer. In the former case, the electric charge Q in units of e is given by the relation

$Q = (q + 1/3)(n_t - n_{\bar{t}}) + F_3 + \frac{F_8}{\sqrt{3}}$, where $n_t - n_{\bar{t}}$ is the number of triplets minus the number of antitriplets. In the latter case we have the relation

$Q = -(q + 1/3)(n_t - n_{\bar{t}}) + F_3 + \frac{F_8}{\sqrt{3}}$.

With integral charges and baryon number one, there is no reason to expect any member of a real triplet to be absolutely stable; we may permit them all to decay into ordinary baryons. However, the question arises whether such decays take place by weak interactions or by moderately strong interactions that violate SU(3) but not isotopic spin conservation. If the latter, we want the violation of SU(3) to transform like $\underset{\sim}{3}$ and $\underset{\sim}{3}^*$, so that in second order it gives the familiar octet behavior of the violation. But $I = 0$, $Q = 0$ occurs in $\underset{\sim}{3}$ and $\underset{\sim}{3}^*$ only when $q = 0$. Thus if the decay of triplets into baryons is to be attributed to moderately strong interactions that give rise, in second order, to the octet violation of SU(3), we must have $q = 0$. This is the model used by Maki[12] and by Tarjanne and Teplitz.[13]

With other values of q, we presumably have the triplet decaying into ordinary baryons by the weak interaction. As pointed out previously,[11,14] the most interesting case of this kind is the one with $q = -1$, where the four members of the basic triplet and singlet present a perfect analogy with the known leptons.

Absolute stability of one member of a triplet is of course, a possibility for any value of q.

These kinds of triplets with integral charges are, of course, more likely to correspond to real particles than the quarks, and we may also use them in field theory models to abstract algebraic relations, obtaining essentially the same ones as for the quarks in the next section.

VII. FURTHER ALGEBRAIC RELATIONS

We start with the simple Lagrangian model of quarks discussed in Ref. 11. There is a triplet t of fermion fields corresponding to three spin 1/2 quarks: the isotopic doublet u and d, with charges 2/3 and - 1/3 respectively, and the isotopic singlet s, with charge - 1/3. A neutral vector meson field B_α is introduced, too. The Lagrangian is simply

$$-\bar{t}\,\gamma_\alpha\,\partial_\alpha\,t - \mathcal{L}_B - i\,F\,B_\alpha\,\bar{t}\,\gamma_\alpha\,t$$

as $\lambda \to 0$ and $c \to 0$, where $\mathcal{L}_B$ is the free Lagrangian for the field B and

$$\bar{t}\,\gamma_\alpha\,t = \bar{u}\,\gamma_\alpha\,u + \bar{d}\,\gamma_\alpha\,d + \bar{s}\,\gamma_\alpha\,s \quad .$$

Now we may add to the Lagrangian a quark mass term

$$\lambda\,u_o = m_o\,(\bar{u}\,u + \bar{d}\,d + \bar{s}\,s) = m_o\,\bar{t}\,t \quad .$$

The energy density acquires a term that is just the negative of this. In the model we may put

$$u_i = \bar{t}\,\frac{\lambda_i}{2}\,t \quad , \quad v_i = -\,i\,\bar{t}\,\gamma_5\,\frac{\lambda_i}{2}\,t \quad , \quad i = 0, 1, \ldots 8 \quad ,$$

and $\lambda = \sqrt{6}\, m_o$. Likewise, we have, in the model,

$$\mathcal{F}_{i\alpha} = i\, \bar{t}\, \frac{\lambda_i}{2}\, \gamma_\alpha\, t \quad , \quad \mathcal{F}^5_{i\alpha} = i\, \bar{t}\, \frac{\lambda_i}{2}\, \gamma_\alpha \gamma_5\, t \quad .$$

For the moment, we forget the term $c\, \phi_8$ that breaks the eightfold way.

The non-singularity of the model enables us to generalize[3,11] the commutation relations (14) and (15) and Eq. (20) to the local relations

$$\left[\mathcal{F}_{i4}(\underset{\sim}{x}, t), u_j(\underset{\sim}{x}', t)\right] = - f_{ijk}\, u_k(\underset{\sim}{x}, t)\, \delta(\underset{\sim}{x} - \underset{\sim}{x}'), \text{ etc.,} \tag{35}$$

$$\left[\mathcal{F}_{i4}(\underset{\sim}{x}, t), \mathcal{F}_{j4}(\underset{\sim}{x}', t)\right] = - f_{ijk}\, \mathcal{F}_{k4}(\underset{\sim}{x}, t)\, \delta(\underset{\sim}{x} - \underset{\sim}{x}'), \text{ etc.,} \tag{36}$$

$$\partial_\alpha\, \mathcal{F}^5_{i\alpha} = \sqrt{\tfrac{2}{3}}\, \lambda\, v_i + \mathcal{O}(c) \quad . \tag{37}$$

We have proposed that these relations be abstracted from the model and postulated as true. In an exact calculation of the matrix elements of the $\mathcal{F}_{i\alpha}$, u_i, and v_i by means of linear homogeneous dispersion relations without subtractions, the nonlinear relations (35) and (36) supply the scale factors that determine such things as the axial vector current renormalization.

In the quark model, the term $c\, \phi_8$ in the Lagrangian could be put in as a mass difference between singlet and doublet quarks, but ϕ_8 would then be the same as u_8 and would transform like $(\underset{\sim}{3}, \underset{\sim}{3}^*)$ and $(\underset{\sim}{3}^*, \underset{\sim}{3})$. If we want ϕ_8 to belong to $(\underset{\sim}{1}, \underset{\sim}{8})$ and $(\underset{\sim}{8}, \underset{\sim}{1})$, we could put it into the model as a coupling of the meson B_α to the current $\mathcal{F}_{8\alpha}$. Such a term is reminiscent of Ne'eman's "Fifth Interaction"[15] or of Sakurai's use[16] of ϕ-ω mixing as a dynamical mechanism for violating the eightfold way.

We mentioned in Ref. 3 that in the model there are further commutation relations, besides (35) and (36), which we might or might not take seriously, namely the commutation relations of the u's and v's. It is interesting that when these operators are commuted, in the model, they bring back the operators $\mathcal{F}_{i4}$ and $\mathcal{F}_{i4}^5$, along with a new operator, the helicity charge density, which we may call $\mathcal{F}_{o4}^5$. The algebra of the u_i, v_i, $\mathcal{F}_{i4}$, and $\mathcal{F}_{i4}^5$ then closes; we have 18 u's and v's, 8 $\mathcal{F}_{i4}$'s, and 9 $\mathcal{F}_{i4}^5$'s, corresponding to the 35 generators of the algebra SU(6). In the model, the new current $\mathcal{F}_{o\alpha}^5$ is just $(1/2)\,\bar{t}\,\lambda_o\,\gamma_\alpha\,t$.

Evidently, we can look upon all 35 operators as generating infinitesimal unitary transformations among the three left-handed quarks and the three right-handed quarks. This algebraic system connects scalars and pseudoscalars with four-vectors and four-pseudovectors and thus represents a new stage in the generalization of symmetry. Whereas F-spin connects only systems of the same parity and behavior under proper Lorentz transformations, the group SU(3) × SU(3) of the vector and axial vector currents connects systems which may have different parity but must still have the same behavior under the proper Lorentz group, and SU(6) now connects systems with different parity and/or different space-time behavior.

Of course, it is not clear, even in the model, that SU(6) is of any use as an approximate symmetry. If it were, it would arrange particles of various spins and parities in super-super-supermultiplets. However, it does appear to be true that a huge number of special algebraic properties can be abstracted from a field theory model. The situation is reminiscent of the growth of dispersion relations from an obscure equation for forward scattering of light to a huge set of relations among all scattering amplitudes, nearly sufficient to determine the whole S-matrix. Conceivably, the study of algebraic relations will undergo a comparable transformation.

REFERENCES

1. M. Gell-Mann, California Institute of Technology Synchrotron Laboratory Report CTSL-20 (1961) (unpublished).
2. Y. Ne'eman, Nuclear Phys. 26, 222 (1961).
3. M. Gell-Mann, Phys. Rev. 125, 1067 (1962).
4. M. Gell-Mann, Proceedings of the 1960 Annual International Conference on High Energy Physics at Rochester, pp. 508-513 (Interscience Publishers, Inc., New York, 1960).
5. M. Gell-Mann and M. Lévy, Nuovo Cimento 16, 705 (1960).
6. N. Cabibbo, Phys. Rev. Letters 10, 531 (1963).
7. See, for example, Y. Nambu and J. J. Sakurai, Phys. Rev. Letters 11, 42 (1963).
8. R. W. Birge, et al., Phys. Rev. Letters 11, 35 (1963).
9. J. Bernstein, S. Fubini, M. Gell-Mann, and W. Thirring, Nuovo Cimento 17, 757 (1960).
10. I am indebted to Professor V. L. Telegdi for a discussion of this point.
11. M. Gell-Mann, Physics Letters 8, 214 (1964). We note that the mathematical quark model, which illustrates our theory, is not at all symmetrical under the operation R that exchanges N and Ξ, $\underset{\sim}{10}$ and $\underset{\sim}{10}^*$, etc. Thus our selection of D-type couplings in the limit $\lambda \to 0$ is accomplished without necessarily having R invariance in that limit.
12. Z. Maki, Progress of Theoretical Physics 31, 331 (1964).
13. P. Tarjanne and V. L. Teplitz, Phys. Rev. Letters 11, 447 (1963).

14. Y. Hara, Phys. Rev. 134, B701 (1964).

15. Y. Ne'eman, Phys. Rev. (to be published).

16. J. J. Sakurai, Phys. Rev. Letters 9, 472 (1962).

UNITARY SYMMETRY AND LEPTONIC DECAYS

Nicola Cabibbo
CERN, Geneva, Switzerland
(Received 29 April 1963)

We present here an analysis of leptonic decays based on the unitary symmetry for strong interactions, in the version known as "eightfold way,"[1] and the V-A theory for weak interactions.[2,3] Our basic assumptions on J_μ, the weak current of strong interacting particles, are as follows:

(1) J_μ transforms according to the eightfold representation of SU_3. This means that we neglect currents with $\Delta S = -\Delta Q$, or $\Delta I = 3/2$, which should belong to other representations. This limits the scope of the analysis, and we are not able to treat the complex of K^0 leptonic decays, or $\Sigma^+ \to n + e^+ + \nu$ in which $\Delta S = -\Delta Q$ currents play a role. For the other processes we make the hypothesis that the main contributions come from that part of J_μ which is in the eightfold representation.

(2) The vector part of J_μ is in the same octet as the electromagnetic current. The vector contribution can then be deduced from the electromagnetic properties of strong interacting particles. For $\Delta S = 0$, this assumption is equivalent to vector-

current conservation.[3]

Together with the octet of vector currents, j_μ, we assume an octet of axial currents, g_μ. In each of these octets we have a current with $\Delta S=0$, $\Delta Q=1$, $j_\mu^{(0)}$, and $g_\mu^{(0)}$, and a current with $\Delta S=\Delta Q=1$, $j_\mu^{(1)}$, and $g_\mu^{(1)}$. Their isospin selection rules are, respectively, $\Delta I=1$ and $\Delta I=1/2$.

From our first assumption we then get

$$J_\mu=a(j_\mu^{(0)}+g_\mu^{(0)})+b(j_\mu^{(1)}+g_\mu^{(1)}). \tag{1}$$

A restriction $a=b=1$ would not ensure universality in the usual sense (equal coupling for all currents), because if J_μ [as given in Eq. (1)] is coupled, we can build a current, $b(j_\mu^{(0)}+g_\mu^{(0)})-a(j_\mu^{(1)}+g_\mu^{(1)})$, which is not coupled. We want, however, to keep a weaker form of universality, by requiring the following:

(3) J_μ has "unit length," i.e., $a^2+b^2=1$.

We then rewrite J_μ as[4]

$$J_\mu=\cos\theta(j_\mu^{(0)}+g_\mu^{(0)})+\sin\theta(j_\mu^{(1)}+g_\mu^{(1)}), \tag{2}$$

where $\tan\theta=b/a$. Since J_μ, as well as the baryons and the pseudoscalar mesons, belongs to the octet representation of SU_3, we have relations (in which θ enters as a parameter) between processes with $\Delta S=0$ and processes with $\Delta S=1$.

To determine θ, let us compare the rates for $K^+\to\mu^++\nu$ and $\pi^+\to\mu^++\nu$; we find

$$\Gamma(K^+\to\mu\nu)/\Gamma(\pi^+\to\mu\nu)$$
$$=\tan^2\theta M_K(1-M_\mu^2/M_K^2)^2/M_\pi(1-M_\mu^2/M_\pi^2)^2. \tag{3}$$

From the experimental data, we then get[5,6]

$$\theta=0.257. \tag{4}$$

For an independent determination of θ, let us consider $K^+\to\pi^0+e^++\nu$. The matrix element for this process can be connected to that for $\pi^+\to\pi^0+e^++\nu$, known from the conserved vector-current hypothesis (2nd assumption). From the rate[6] for $K^+\to\pi^0+e^++\nu$, we get

$$\theta=0.26. \tag{5}$$

The two determinations coincide within experimental errors; in the following we use $\theta=0.26$.

We go now to the leptonic decays of the baryons, of the type $A\to B+e+\nu$. The matrix element of any member of an octet of currents among two baryon states (also members of octets) can be expressed in terms of two reduced matrix elements[7]

$$\langle A|j_\mu^{(i)}+g_\mu^{(i)}|B\rangle=if_{ABi}O_\mu+d_{ABi}E_\mu; \tag{6}$$

the f's and d's are coefficients defined in Gell-Mann's paper.[1,7] It is sufficient to consider only allowed contributions and write

$$O_\mu, E_\mu=F^{O,E}\gamma_\mu+H^{O,E}\gamma_\mu\gamma_5. \tag{7}$$

From the connection with the electromagnetic current we get the vector coefficients: $F^O=1$, $F^E=0$; from neutron decay we get

$$H^O+H^E=1.25. \tag{8}$$

We remain with one parameter which can be determined from the rate for $\Sigma^-\to\Lambda+e^-+\bar\nu$. The relevant matrix element for this is

$$\cos\theta\langle\Sigma^-|j_\mu^{(0)}+g_\mu^{(0)}|\Lambda\rangle$$
$$=\cos\theta(\tfrac{2}{3})^{1/2}E_\mu=(\tfrac{2}{3})^{1/2}\cos\theta H^E\gamma_\mu\gamma_5. \tag{9}$$

Taking the branching ratio for this mode to be 0.9×10^{-4},[8] we get

$$H^E=\pm0.95. \tag{10}$$

The negative solution can be discarded because it produces a large branching ratio for $\Sigma^-\to n+e^-+\bar\nu$, of the order of 1%. The positive solution ($H^E=0.95$, $H^O=0.30$) is good, because it produces a cancellation of the axial contribution to this process. This explains the experimental result that this mode is more depressed than the $\Lambda\to p+e^-+\bar\nu$ in respect to the predictions of Feynman and Gell-Mann.[2] In Table I we give a summary of our predictions for the electron modes with $\Delta S=1$. The branching ratios for $\Lambda\to p+e^-+\bar\nu$ and $\Sigma^-\to n+e^-+\bar\nu$ are in good agreement with experimental data.[9]

As a final remark, the vector-coupling constant for β decay is not G, but $G\cos\theta$. This gives a correction of 6.6% to the ft value of Fermi transitions, in the right direction to eliminate the discrepancy between O^{14} and muon lifetimes.

Table I. Predictions for the leptonic decays of hyperons.

Decay	Branching ratio From reference 2	Branching ratio Present work	Type of interaction
$\Lambda\to p+e^-+\bar\nu$	1.4 %	0.75×10^{-3}	$V-0.72A$
$\Sigma^-\to n+e^-+\bar\nu$	5.1 %	1.9×10^{-3}	$V+0.65A$
$\Xi^-\to\Lambda+e^-+\bar\nu$	1.4 %	0.35×10^{-3}	$V+0.02A$
$\Xi^-\to\Sigma^0+e^-+\bar\nu$	0.14%	0.07×10^{-3}	$V-1.25A$
$\Xi^0\to\Sigma^++e^-+\bar\nu$	0.28%	0.26×10^{-3}	$V-1.25A$

The correction is, however, too large, leaving about 2% to be explained.[10]

---

[1]M. Gell-Mann, California Institute of Technology Report CTSL-20, 1961 (unpublished); Y. Ne'eman, Nucl. Phys. 26, 222 (1961).

[2]R. P. Feynman and M. Gell-Mann, Phys. Rev. 109, 193 (1958).

[3]R. E. Marshak and E. C. G. Sudarshan, Proceedings of the Padua-Venice Conference on Mesons and Recently Discovered Particles, September, 1957 (Società Italiana di Fisica, Padua-Venice, 1958); Phys. Rev. 109, 1860 (1958).

[4]Similar considerations are forwarded in M. Gell-Mann and M. Lévy, Nuovo Cimento 16, 705 (1958).

[5]The lifetimes from W. H. Barkas and A. H. Rosenfeld, Proceedings of the Tenth Annual International Rochester Conference on High-Energy Physics, 1960 (Interscience Publishers, Inc., New York, 1960), p. 878. The branching ratio for $K^+ \to \mu^+ + \nu$ is taken as 57.4%. W. Becker, M. Goldberg, E. Hart, J. Leitner, and S. Lichtman (to be published).

[6]B. P. Roe, D. Sinclair, J. L. Brown, D. A. Glaser, J. A. Kadyk, and G. H. Trilling, Phys. Rev. Letters 7, 346 (1961). These authors give the branching ratio for $K^+ \to \mu^+ + \nu$ as 64%, from which $\theta = 0.269$. Also this value agrees with that from $K^+ \to \pi^0 + e^+ + \nu$ within experimental errors.

[7]N. Cabibbo and R. Gatto, Nuovo Cimento 21, 872 (1961). Our notation for the currents is different from the one used in this reference and by Gell-Mann; the connection is $j_\mu^{(0)} = j_\mu^1 + ij_\mu^2$, $j_\mu^{(1)} = j_\mu^4 + ij_\mu^5$.

[8]W. Willis et al. reported at the Washington meeting of the American Physical Society, 1963 [W. Willis et al., Bull. Am. Phys. Soc. 8, 349 (1963)] this branching ratio as $(0.9^{+0.5}_{-0.4}) \times 10^{-4}$. If it is allowed to vary between these limits, our predictions for the $\Sigma^- \to ne^-\bar{\nu}$ varies between 0.8×10^{-3} and 4×10^{-3}, and that for $\Lambda^0 \to pe^-\bar{\nu}$ between 1.05×10^{-3} and 0.56×10^{-3}. I am grateful to the members of this group for prepublication communication of their results.

[9]R. P. Ely, G. Gidal, L. Oswald, W. Singleton, W. M. Powell, F. W. Bullock, G. E. Kalmus, C. Henderson, and R. F. Stannard [Proceedings of the International Conference on High-Energy Nuclear Physics, Geneva, 1962 (CERN Scientific Information Service, Geneva, Switzerland, 1962), p. 445] give the branching ratio for $\Lambda \to p + e^- + \bar{\nu}$ as $(0.85 \pm 0.3) \times 10^{-3}$, while that for $\Sigma^- \to n + e^- + \bar{\nu}$ is given (see preceding reference) as $(1.9 \pm 0.9) \times 10^{-3}$.

[10]R. P. Feynman, Proceedings of the Tenth Annual International Rochester Conference on High-Energy Physics, 1960 (Interscience Publishers, Inc., New York, 1960), p. 501. Recent measurements of the muon lifetime have slightly increased the discrepancy. We think that more information will be needed to decide whether our 3rd assumption can be maintained.

From: *Proc Intern Conf High Energy Phys* (Rochester, 1960), pp. 508–513

CONSERVED AND PARTIALLY CONSERVED CURRENTS IN THE THEORY OF WEAK INTERACTIONS

M. Gell-Mann

California Institute of Technology, Pasadena, California

Feynman has said that the experimental situation is not very clear in connection with the conserved vector current. He said also that if the conserved vector current hypothesis crumbles, nothing will crumble with it. It is true that there is not a vast structure of theory that will fall down, but there is some structure, and I would like to talk about that sort of structure which will crumble when the conserved vector current hypothesis is exploded.

The people who have engaged in speculation about partially conserved currents include Schwinger, Bludman, Glashow, Gürsey, Nambu, and Salam. Some remarks relevant to the subject have been made by Polkinghorne and by J. C. Taylor. I have been thinking about such matters during the last year in collaboration with Lévy, Michel, Bernstein, Fubini and Thirring.

If the hypothesis of the conserved $\Delta S = 0$ vector current is correct (and let us assume that it is), does it not suggest something about the behavior of the other currents? Is it not a little peculiar for that one current to be equal to something as important as the isotopic spin current, while there is no corresponding principle for the others? Such asymmetry is unattractive and one reason at least for the kind of research I am going to describe is to remove that assymetry.

One way to do so, which Feynman has discussed, is to adopt the idea that in the world of baryons and mesons there are no other fundamental particles except n, p, and Λ (and perhaps a neutral vector glue). Then $\bar{n}\gamma_\alpha p$ would be the isotopic spin current and if the $\Delta S = 0$ axial current were just $\bar{n}\gamma_\alpha\gamma_5 p$, the two could be quite symmetrical and no asymmetry would exist.

If we think, however, that mesons really exist and that they are present in the Lagrangian, then we find a striking asymmetry between a simple axial current of the form $\bar{n}\gamma_\alpha\gamma_5 p$ and a vector current that has additional terms pertaining to π mesons, etc.:

$$\bar{n}\gamma_\alpha p + \sqrt{2}\left(\pi^0\frac{\partial\pi^+}{\partial x_\alpha} - \pi^+\frac{\partial\pi^0}{\partial x_\alpha}\right) + \ldots$$

There is another question that comes up at the same time. We like to talk about the universality of the weak interactions but what does universality mean? Presumably it means that the same $\sqrt{G}$ occurs everywhere, in the lepton currents, baryon currents, and meson currents. But do we allow a $\sqrt{2}$ sometimes? What if we re-define the fields, for example by choosing $\frac{\Lambda+\Sigma^0}{\sqrt{2}}$ and $\frac{\Lambda-\Sigma^0}{\sqrt{2}}$ instead of Λ and Λ^0? By re-expressing the currents in terms of new fields we can make factors like $\sqrt{2}$ appear and disappear at will. The concept of universality becomes very slippery unless we can define it in a representation-independent way.

If there are no meson terms in the current, then we might try to get a reasonable definition independent of the framework by making a matrix of the coefficients of the bilinear form in the fermion fields that constitutes the current (say with $\gamma_\alpha(1+\gamma_5)$ everywhere) and looking at the eigenvalues of the matrix. The definition of universality would refer to the eigenvalues.

But if there are boson terms as well as fermion terms in the weak current, then the two classes of terms are not easily comparable and the definition of universality is again a problem.

Some of us have been wondering about the possibility of a new kind of principle to determine the weak current, a principle which might help to overcome the difficulties of *V-A* asymmetry and of the vagueness of universality. It was first stated clearly in the literature by Bludman, I believe, and it is based on an analogy with the principle of minimal electromagnetic interaction, which is also a speculation, but a fairly well-founded one.

Consider the electromagnetic interaction to lowest order in e, just as later we will take up the weak interactions to lowest order. We have the electric charge operator Q, and universality simply requires its eigenvalues to be integral multiples of e. Using Q, we can perform an infinitesimal gauge transformation in which every field undergoes the change

$$\psi(x) \to \psi(x) - i\lambda(x)[Q, \psi(x)]$$

where $\lambda(x)$ is a gauge function depending on space and time.

Call the Lagrangian of the leptons or of the strong interactions of baryons and mesons $\mathcal{L}$. Then we have

$$\mathcal{L} \to \mathcal{L} + \frac{\delta \mathcal{L}}{\delta \lambda}\lambda + \frac{\delta \mathcal{L}}{\delta \partial_\alpha \lambda}\partial_\alpha \lambda$$

Even though λ is a gauge variable and not a field variable, you can show that Lagrange's equation holds:

$$\partial_\alpha \frac{\delta \mathcal{L}}{\delta \partial_\alpha \lambda} = \frac{\delta \mathcal{L}}{\delta \lambda}$$

We now define the minimal electric current j_α to be $i\frac{\delta \mathcal{L}}{\delta \partial_\alpha \lambda}$. If the Lagrangian is made invariant under gauge transformations with constant λ (transformations of the first kind), then $\frac{\delta \mathcal{L}}{\delta \lambda} = 0$ and we have the conservation of charge $\partial_\alpha j_\alpha = 0$.

Now, the corresponding proposal for the weak interactions. I shall discuss charge exchange currents only, without prejudice to the question of whether there are also neutral or charge retention currents. Let the charge exchange current (apart from a constant factor) be W_α^+ and its hermitian conjugate W_α^-. From the sum and difference of these we can define $W_{\alpha x}$ and $W_{\alpha y}$; I shall also adjoin for convenience a fictitious charge retention current $W_{\alpha z}$. We suppose that there is a weak charge operator $\mathbf{W}$ analogous to Q and that under $\mathbf{W}$ the fields undergo infinitesimal gauge transformations

$$\psi(x) \to \psi(x) - i\mathbf{t}(x).[\mathbf{W}, \psi(x)]$$

while the Lagrangian changes correspondingly:

$$\mathcal{L} \to \mathcal{L} + \frac{\delta \mathcal{L}}{\delta \mathbf{t}}\cdot \mathbf{t} + \frac{\delta \mathcal{L}}{\delta \partial_\alpha \mathbf{t}}\cdot \partial_\alpha \mathbf{t}.$$

Again we have the Lagrange equation

$$\partial_\alpha \frac{\delta \mathcal{L}}{\delta \partial_\alpha \mathbf{t}} = \frac{\delta \mathcal{L}}{\delta \mathbf{t}}$$

and the weak current is defined as $\mathbf{W}_\alpha \equiv i\frac{\delta \mathcal{L}}{\delta \partial_\alpha \mathbf{t}}$. Then we have

$$\partial_\alpha \mathbf{W}_\alpha = i\frac{\delta \mathcal{L}}{\delta \mathbf{t}}.$$

To the extent that the Lagrangian is invariant under gauge transformations with constant $\mathbf{t}$, the weak current is conserved. The non-invariant terms in $\mathcal{L}$ generate the non-zero divergence of the weak current.

For baryons and mesons we may divide $\mathbf{W}$ into four terms that generate the four pieces of the current:

$$\mathbf{W} = \tfrac{1}{2}(\underset{\substack{\Delta s=0 \\ V}}{\mathbf{C}} + \underset{\substack{\Delta s=0 \\ A}}{\mathbf{D}} + \underset{\substack{\Delta s=1 \\ V}}{\mathbf{F}} + \underset{\substack{\Delta s=1 \\ A}}{\mathbf{G}}).$$

I suppose that $\mathbf{C}$ and $\mathbf{D}$ transform like $|\Delta \mathbf{I}| = 1$ and $\mathbf{F}$ and $\mathbf{G}$ like $|\Delta \mathbf{I}| = 1/2$.

The conserved vector $\Delta S = 0$ current hypothesis is simply the notion that $\mathbf{C} = \mathbf{I}$, the isotopic spin. Since the entire strong-coupling Lagrangian is invariant under $\mathbf{I}$, the divergence of this part of the current is then zero. For the other three operators, there must be terms in the Lagrangian that violate their conservation, but there can still be a strong symmetry among the operators $\mathbf{C}$, $\mathbf{D}$, $\mathbf{F}$, and $\mathbf{G}$, and the various currents are generated in similar ways. The similarity is imperfect only in that $\mathbf{C}$ is exactly conserved and the others not.

The definition of universality no longer presents any difficulties. If $\mathbf{W}$ always has the same commutation rules, it will have only a limited set of possible eigenvalues and universality would pertain to the eigenvalues.

Glashow has examined the commutation rules of $\mathbf{W}$ for the leptons, for which only the free Lagrangians must be included, since there are no strong interactions. Suppose, for the sake of definiteness, that there are two neutrinos, one for the electron and one for the muon. We have weak currents for the pairs $(e\nu)$ and $(\mu\nu')$. Let e_L be the left-handed part of the electron field $e_L \equiv \frac{1+\gamma_5}{2}e$ and $e_R \equiv \frac{1-\gamma_5}{2}e$. For the neutrino, there is only the left-handed part ν_L. We deal then with ν_L, e_L, e_R; ν'_L, μ_L, μ_R. The anti-particles come in automatically in the usual way.

The operator **W** for the leptons can now be written simply as a matrix connecting ν_L, e_L and e_R (or the other three). W_+ takes e_L into ν_L and W_- does the reverse. We have

$$W_x = \begin{array}{c} \\ \nu_L \\ e_L \\ e_R \end{array}\!\!\begin{array}{c} \begin{array}{ccc} \nu_L & e_L & e_R \end{array} \\ \begin{pmatrix} 0 & \frac{1}{2} & 0 \\ \frac{1}{2} & 0 & 0 \\ 0 & 0 & 0 \end{pmatrix} \end{array}$$

$$W_y = \begin{pmatrix} 0 & -i/2 & 0 \\ i/2 & 0 & 0 \\ 0 & 0 & 0 \end{pmatrix}$$

If we put

$$W_z = \begin{pmatrix} \frac{1}{2} & 0 & 0 \\ 0 & -\frac{1}{2} & 0 \\ 0 & 0 & 0 \end{pmatrix}$$

then we have the result that **W** is an angular momentum.

Glashow has pointed out another interesting feature. If we write a fourth matrix

$$W_4 = \begin{pmatrix} -\frac{1}{2} & 0 & 0 \\ 0 & -\frac{1}{2} & 0 \\ 0 & 0 & -1 \end{pmatrix} \quad \text{then } [W_4, \mathbf{W}] = 0$$

and the electric charge $Q = e(W_4 + W_z)$. Also $W_4^2 + W_x^2 + W_y^2 + W_z^2$ is unity—I do not know how important that is.

The symmetry of the lepton Lagrangian under **W** is broken only by the electron and muon mechanical mass terms, which thus give rise to the non-conservation of $\mathbf{W}_\alpha$.

If we really believe in universality, and if all our talk about gauge transformations is not (as it may easily be) total nonsense, then we should expect a **W** operator with similar properties to generate the baryon-meson weak current from the strong-interaction Lagrangian L. At least the kinetic (gradient containing) part of L should be invariant for constant **t** and give rise to the weak current when **t** depends on x.

Various systems can be constructed along these lines. For examples, there is the scheme mentioned by Feynman and favored by Okun, Marshak, and others, based on just n, p, and Λ. Of course, if that is right, we do not need the elaborate machinery I have just described. We simply draw an analogy. If there is only one neutrino we use the correspondence

$$\nu_L \sim p_L, \quad \left(\frac{e+\mu}{\sqrt{2}}\right)_L \sim \left(\frac{n+1}{\sqrt{2}}\right)_L$$

If $\nu \neq \nu'$ then we have

$$\nu_L \sim \nu'_L \sim p_L, \quad e_L \sim \mu_L \sim \left(\frac{n+\varepsilon\Lambda}{\sqrt{1+\varepsilon^2}}\right)_L, \quad \text{with } \varepsilon^2 \ll 1 .$$

In the first case, the extra weakness of the (Λp) current has to be attributed to renormalization. In the second case, it would come from the smallness of ε.

What happens, though, if mesons are really present in the Lagrangian? One possibility is that the mesons we know are all fake and that the real mesons are vector and axial vector mesons that glue nucleons and anti-nucleons together and give rise to the known mesons as bound states. There again a simple model can be constructed.

But I should like to discuss what happens if the known mesons, π and K, are present in the Lagrangian and we have a theory in which the kinetic part of $\mathcal{L}$ conserves $\mathbf{W} = \frac{1}{2}(\mathbf{I}+\mathbf{D}+\mathbf{F}+\mathbf{G})$ and **W** is an angular momentum. (We would also like to have $Q = e(W_Z + W_4)$, where $[\mathbf{W}, W_4] = 0$.)

It cannot be done with π and K alone. We need other fields representing two new kinds of mesons. First there is the σ', proposed by Schwinger, with $I = 0$, $J = 0^+$ (in other words, the properties of the vacuum.) It is a particle capable of virtual disintegration into nothing whatever; of course that does not violate any physical principle. Then there is the K' which is like the K meson ($I = \frac{1}{2}$, $J = 0$) but with opposite parity.

These scalar fields are needed so that together with the pseudoscalar ones they can make axial currents as well as vector currents. Only then can we make $\mathbf{D} \neq 0$ and $\mathbf{G} \neq 0$ for the mesons and have **W** an angular momentum.

The parity doublet (KK') differs from the one suggested some years ago in that the terms that break the conservation of **W** also break the degeneracy of K and K', π and σ'. They all have different masses.

If σ' and K' do exist they must be heavier than their counterparts π and K and we may suppose they are heavy enough to be completely unstable under the strong interactions, decaying in $\sim 10^{-23}$ seconds:

$$\sigma' \to 2\pi \qquad (s\text{-state})$$

$$K' \to K + \pi \qquad (s\text{-state}).$$

These particles would be more or less indistinguishable in simple experiments from dynamical resonances. In order to look for them you would try to find peaks in the distribution of $m^2_{2\pi}$ when two pions are produced together or in the distribution of $m^2_{K+\pi}$ when K and π are produced together.

If the K' exists, it can have a coupling of the form $KK'\pi$ as suggested several years ago in connection with parity doublets. In the $K+N \to K+\pi+N$ reaction, there would be an important dispersion diagram of the form illustrated in Fig. 1 and in associated production of strange particles a strong peaking in angle could be produced by the dispersion diagram illustrated in Fig. 1. Of course, both of these processes would happen without any K', if K' is simply replaced by $K+\pi$. But if the K' has any utility, if it is a reasonably sharp resonance in the $K+\pi$ system, then these diagrams may have an added significance; they may show up more than they would if there were no special interactions between π and K.

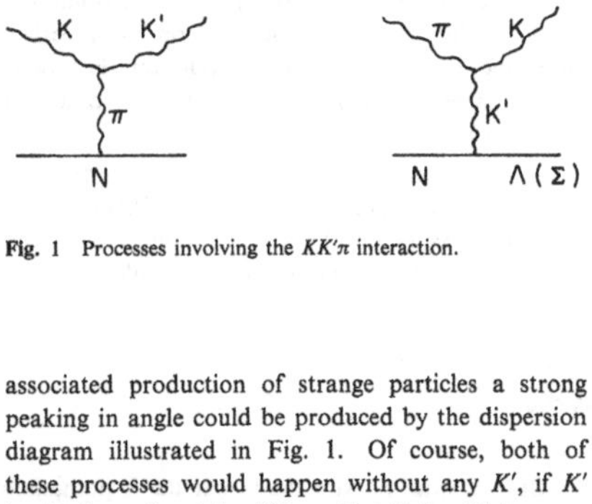

Fig. 1 Processes involving the $KK'\pi$ interaction.

Let me re-emphasize that all of our discussion is completely speculative and that no one should take σ' and K' seriously unless they are actually found.

Now suppose that the new mesons are discovered and our type of speculation encouraged. Then we must ask what terms in the strong coupling Lagrangian violate the conservation of **W**, that is, of **D**, **F**, and **G**.

In the case of the leptons, it was the mechanical mass terms that violated **W** conservation. For the baryons and mesons, we may suppose that in a similar way it is the mechanical mass terms bilinear in the fields that break the symmetry.

Actually, there is an even simpler possibility, since we have a σ' meson that can appear or disappear virtually, namely a linear term in the Lagrangian proportional to σ'. In the case of **C**, in fact, we may even imagine that only the linear term breaks the conservation law.

If we try the same hypothesis for **F** and **G**, however, we find that there is too much symmetry in the system and unwanted degeneracies appear. So probably mechanical mass terms must be invoked to break the symmetries.

It is reasonable, in any case, that only linear and bilinear terms occur in $\delta\mathcal{L}/\delta t$ and thus the divergences of the various weak currents are not very singular operators, as they would be if $\delta\mathcal{L}/\delta t$ had trilinear or quadrilinear terms arising from interaction terms in the Lagrangian.

In saying that the divergence of a current, for instance the $\Delta S = 0$ axial vector current P_α, is nonsingular, I mean particularly that the matrix elements of the operator emphasize low frequencies.

Now if it is really true that $\partial_\alpha P_\alpha$ is an operator that emphasizes low frequencies, if when it acts on the vacuum it emphasizes the creation of low-mass states, then we can find a simple explanation of the remarkable formula of Goldberger and Treiman (G-T).

The G-T relation connects the renormalized axial-vector coupling constant $-G_A$ in β-decay to the amplitude for pion decay into leptons. The latter can be represented by the quantity a_1, where

$$\langle 0|\partial_\alpha P_\alpha|\pi^-\rangle = \frac{ia_1}{\sqrt{2}}\langle 0|\pi_r^-|\pi^-\rangle$$

(π_r^- is the renormalized pion field.)

The rate of $\pi \rightarrow \mu + \nu'$ is

$$\Gamma_\pi = \frac{G^2}{16\pi m_\pi} \cdot \frac{m_\mu{}^2}{m_\pi{}^2}\left(1 - \frac{m_\mu{}^2}{m_\pi{}^2}\right)^2 a_1{}^2$$

in terms of a_1.

The nucleon β-decay amplitude is

$$\langle p|P_\alpha|n\rangle = \frac{-G_A}{G}\alpha(K^2)\tau_+\gamma_\alpha\gamma_5 + iK_\alpha\beta(K^2)\tau_+\gamma_5$$

where $\alpha(0) \equiv 1$. The matrix element of the divergence is

$$2m\left(\frac{-G_A}{G}\right)\alpha + K^2\beta = \frac{-a_1 g_1}{K^2 + m_\pi{}^2} + \text{higher terms},$$

where the higher terms come from the creation of three pions, five pions, $N + \bar{N}$, etc., by the operator $\partial_\alpha P_\alpha$ (Here g_1 is the renormalized π-N coupling constant.)

Now if $\partial_\alpha P_\alpha$ really emphasizes low mass states, the one-pion-pole can dominate at $K^2 = 0$ and we have

$$2m\left(\frac{-G_A}{G}\right) \sim \frac{-a_1 g_1}{m_\pi{}^2},$$

which is the G-T relation.

Another way (emphasized by Nambu) to approach the same situation is to look at the limit as $m_\pi^2 \rightarrow 0$ and $\partial_\alpha P_\alpha \rightarrow 0$, with $\frac{\partial_\alpha P_\alpha}{m_\pi^2}$ remaining fixed. We know (as pointed out by G-T and others) that β has a one-pion pole

$$\frac{a_1 g_1}{m_\pi{}^2}\frac{1}{K^2 + m_\pi{}^2}.$$

In the limit, $2m\left(\frac{-G_A}{G}\right)\alpha + K^2\beta = 0$; so β has a pole

$$\frac{-2m\left(\frac{-G_A}{G}\right)}{K^2}.$$

Evidently, we can identify these poles with each other, and so

$$-2m\left(\frac{-G_A}{G}\right) = g_1 \lim_{m_\pi{}^2 \rightarrow 0} \frac{a_1}{m_\pi{}^2}$$

If the limit is a " gentle " one, then we have once more the G-T relation.

$$2m\left(\frac{-G_A}{G}\right) \approx -\frac{a_1 g_1}{m_\pi{}^2}.$$

It is interesting to ask whether a similar approach can be applied to the other currents, with $\Delta S = 1$. There, too, there are terms analogous to β, an induced pseudoscalar term in the matrix element of the axial vector current and an induced scalar term in that of the vector current. In the limit in which the divergences of the currents approach zero, these terms acquire poles at $K^2 = 0$, like β. It is tempting to suppose that here, too, the poles are simply ones corresponding to virtual mesons, and that the masses of these mesons are approaching zero. In the case of the induced pseudoscalar term, it is presumably the K meson that is involved. If the same situation is obtained for the induced scalar term, then perhaps it is the K' meson at work! (On this last point, I am indebted to J. Bernstein for a stimulating question.)

It will be interesting to see whether there is any substance to the speculations about partially conserved currents. As far as the G-T relation is concerned, I think some real progress has been made.

DISCUSSION

SAKURAI: It is perhaps of interest to examine the effect of the σ' meson on nuclear forces. The exchange of a σ' meson between the two nucleons would lead to a short-range attraction in all angular momentum and parity states. The hypothetical meson does give a spin-orbit force of the correct sign as pointed out by Breit, Duerr and Gupta.

GELL-MANN: So you think that the hypothesis of a σ' meson is not in disagreement with experiment as far as nuclear forces are concerned?

SAKURAI: I cannot be positive because that attractive short range interaction may give some trouble. On the other hand, the gluon or a neutral vector meson can explain both the phenomenological repulsive core and the spin-orbit force.

TIOMNO: With respect to the question of the K' meson I should like to mention that I have also been making speculations about this possibility and its implications. The starting point was, however, the analysis of the angular distribution in ΛK^0 production from π^- on protons. Following Pais we used a $KK'\pi$ vertex but allowing for K'_+ (or K'_0) different from the ordinary K^+ (or K^0). Zagury, Videira and myself tried to find the best value of the mass of this K' particle both in the case of scalar and pseudoscalar coupling for the $K'p\Lambda$ vertex. We found that for the scalar coupling the best fit was for a mass of K' around the sum of the masses of K and π.

GELL-MANN: You would have a resonance decaying fairly slowly?

TIOMNO: Yes, at the energies for which reasonably good angular distributions are available the values of the K' mass ranged from 0.45 to 0.75 BeV being consistent with a unique value. Now, for the pseudoscalar coupling the mass values of K' were of the order of the π-mass. This last possibility is thus ruled out by the fact that the K mesons should then decay into π and K'. In the first case the K' mass could be higher than K plus π and thus would not lead to contradiction with experiment.

GELL-MANN: So you like the K being pseudoscalar and the K' scalar?

TIOMNO: Yes, this is the one which would lead to a stable K particle relatively to strong interactions.

WOLFENSTEIN: Am I correct that in this approach the sign of a_1 and therefore the sign of the induced pseudoscalar terms is not determined?

GELL-MANN: Not the absolute sign of a_1 but the sign of the induced pseudoscalar term relative to the axial vector coupling constant is determined in this approach. In the absorption of μ^- in carbon for example, people have always given two possible signs —one corresponding to the Goldberger effect and the other to a negative Goldberger effect. Which of these alternatives is the correct one? In this approach the second alternative is excluded. The sign is the same sign as proposed by Goldberger.

APPLICATION OF DISPERSION RELATIONS TO WEAK INTERACTIONS

N. N. Khuri

Institute for Advanced Study, Princeton, New Jersey

1. INTRODUCTION

With the exception of μ meson decay, all the weak reactions that we know involve at least one strongly interacting particle. It is well known that about the only tool we have for computing anything about the strong interactions is the dispersion relations, regardless of the many difficulties and disadvantages it presents.

In the past three years, starting with the work of Goldberger and Treiman, the dispersion relations have been used to study the role of strong interactions in weak processes. With varying degrees of success, almost all the known weak decays, both leptonic and non-leptonic, have been considered. The degree of the success depends, as always with the dispersion

Symmetries of Baryons and Mesons*

MURRAY GELL-MANN
California Institute of Technology, Pasadena, California
(Received March 27, 1961; revised manuscript received September 20, 1961)

The system of strongly interacting particles is discussed, with electromagnetism, weak interactions, and gravitation considered as perturbations. The electric current j_α, the weak current J_α, and the gravitational tensor $\theta_{\alpha\beta}$ are all well-defined operators, with finite matrix elements obeying dispersion relations. To the extent that the dispersion relations for matrix elements of these operators between the vacuum and other states are highly convergent and dominated by contributions from intermediate one-meson states, we have relations like the Goldberger-Treiman formula and universality principles like that of Sakurai according to which the ρ meson is coupled approximately to the isotopic spin. Homogeneous linear dispersion relations, even without subtractions, do not suffice to fix the scale of these matrix elements; in particular, for the nonconserved currents, the renormalization factors cannot be calculated, and the universality of strength of the weak interactions is undefined. More information than just the dispersion relations must be supplied, for example, by field-theoretic models; we consider, in fact, the equal-time commutation relations of the various parts of j_4 and J_4. These nonlinear relations define an algebraic system (or a group) that underlies the structure of baryons and mesons. It is suggested that the group is in fact $U(3)\times U(3)$, exemplified by the symmetrical Sakata model. The Hamiltonian density θ_{44} is not completely invariant under the group; the noninvariant part transforms according to a particular representation of the group; it is possible that this information also is given correctly by the symmetrical Sakata model. Various exact relations among form factors follow from the algebraic structure. In addition, it may be worthwhile to consider the approximate situation in which the strangeness-changing vector currents are conserved and the Hamiltonian is invariant under $U(3)$; we refer to this limiting case as "unitary symmetry." In the limit, the baryons and mesons form degenerate supermultiplets, which break up into isotopic multiplets when the symmetry-breaking term in the Hamiltonian is "turned on." The mesons are expected to form unitary singlets and octets; each octet breaks up into a triplet, a singlet, and a pair of strange doublets. The known pseudoscalar and vector mesons fit this pattern if there exists also an isotopic singlet pseudoscalar meson χ^0. If we consider unitary symmetry in the abstract rather than in connection with a field theory, then we find, as an attractive alternative to the Sakata model, the scheme of Ne'eman and Gell-Mann, which we call the "eightfold way"; the baryons N, Λ, Σ, and Ξ form an octet, like the vector and pseudoscalar meson octets, in the limit of unitary symmetry. Although the violations of unitary symmetry must be quite large, there is some hope of relating certain violations to others. As an example of the methods advocated, we present a rough calculation of the rate of $K^+\to\mu^++\nu$ in terms of that of $\pi^+\to\mu^++\nu$.

I. INTRODUCTION

IN connection with the system of strongly interacting particles, there has been a great deal of discussion of possible approximate symmetries,[1] which would be violated by large effects but still have some physical consequences, such as approximate universality of meson couplings, approximate degeneracy of baryon or meson supermultiplets, and "partial conservation" of currents for the weak interactions.

In this article we shall try to clarify the meaning of such possible symmetries, for both strong and weak interactions. We shall show that a broken symmetry, even though it is badly violated, may give rise to certain exact relations among measurable quantities. Furthermore, we shall suggest a particular symmetry group as the one most likely to underlie the structure of the system of baryons and mesons.

We shall treat the strong interactions without approximation, but consider the electromagnetic, weak, and gravitational interactions only in first order.

The electromagnetic coupling is described by the matrix elements of the electromagnetic current operator $ej_\alpha(x)$. Likewise, the gravitational coupling is specified by the matrix elements of the stress-energy-momentum tensor $\theta_{\alpha\beta}(x)$, particularly the component $\theta_{44}=H$, the Hamiltonian density.

The weak interactions of baryons and mesons with leptons are assumed to be given (ignoring possible nonlocality) by the interaction term[2]

$$GJ_\alpha^\dagger J_\alpha^{(l)}/\sqrt{2}+\text{H.c.},\tag{1.1}$$

where the leptonic weak current $J_\alpha^{(l)}$ has the form

$$J_\alpha^{(l)}=i\bar{\nu}\gamma_\alpha(1+\gamma_5)e+i\bar{\nu}\gamma_\alpha(1+\gamma_5)\mu.\tag{1.2}$$

We shall refer to $J_\alpha(x)$ as the weak current of baryons and mesons. Its matrix elements specify completely the weak interactions with leptons.

It is possible that the full weak interaction may be given simply by the term

$$G(J_\alpha+J_\alpha^{(l)})^\dagger(J_\alpha+J_\alpha^{(l)})/\sqrt{2},\tag{1.3}$$

although this form provides no explanation of the approximate rule $|\Delta I|=\frac{1}{2}$ in the nonleptonic decays of strange particles. If we can find no *dynamical* explanation of the predominance of the $|\Delta I|=\frac{1}{2}$ amplitude in these decays, we may be forced to assume that in addition to (1.3) there is a weak interaction involving the product

$$GL_\alpha^\dagger L_\alpha/\sqrt{2},\tag{1.4}$$

of charge-retention currents (presumably not involving leptons); or else we may be compelled to abandon (1.3)

* Research supported in part by U. S. Atomic Energy Commission and Alfred P. Sloan Foundation. A report of this work was presented at the La Jolla Conference on Strong and Weak Interactions, June, 1961.

[1] For example, see the "global symmetry" scheme of M. Gell-Mann, Phys. Rev. 106, 1296 (1957) and J. Schwinger, Ann. Phys. 2, 407 (1957).

[2] We use $\hbar=c=1$. The Lorentz index α takes on the values 1, 2, 3, 4. For each value of α, the Dirac matrix γ_α is Hermitian; so is the matrix γ_5.

altogether. In any case, we shall define the weak current J_α by the coupling to leptons.

We shall assume microcausality and hence the validity of dispersion relations for the matrix elements of the various currents and densities. In addition, we shall sometimes require the special assumption of highly convergent dispersion relations.

Our description of the symmetry group for baryons and mesons is most conveniently given in the framework of standard field theory, where the Lagrangian density L of the strong interactions is expressed as a simple function of a certain number of local fields $\psi(x)$, which are supposed to correspond to the "elementary" baryons and mesons. Recently this type of formalism has come under criticism[3]; it is argued that perhaps none of the strongly interacting particles is specially distinguished as "elementary," that the strong interactions can be adequately described by the analyticity properties of the S matrix, and that the apparatus of field theory may be a misleading encumbrance.

Even if the criticism is justified, the field operators $j_\alpha(x)$, $\theta_{\alpha\beta}(x)$, and $J_\alpha(x)$ may still be well defined (by all their matrix elements, including analytic continuations thereof) and measurable in principle by interactions with external electromagnetic or gravitational fields or with lepton pairs. Since the Hamiltonian density H is a component of $\theta_{\alpha\beta}$, it can be a physically sensible quantity.

In order to make our description of the symmetry group independent of the possibly doubtful details of field theory, we shall phrase it ultimately in terms of the properties of the operators H, j_α, and J_α. In introducing the description, however, we shall make use of field-theoretic models. Moreover, in describing the behavior of a particular group, we shall refer extensively to a special example, the symmetrical Sakata model of Ohnuki *et al.*,[4] Yamaguchi,[5] and Wess.[6]

The order of presentation is as follows: We treat first the hypothesis of highly convergent dispersion relations for the matrix elements of currents; and we show that the notion of a meson being coupled "universally" or coupled to a particular current or density means simply that the meson state dominates the dispersion relations for that current or density at low momenta. Next we discuss the universality of strength of the currents themselves; evidently it cannot be derived from homogeneous linear dispersion relations for the matrix elements of the currents. We show that equal-time commutation relations for the currents fulfill this need (or most of it), and that, in a wide class of model field theories, these commutation rules are simple and reflect the existence of a symmetry group, which underlies the structure of the baryon-meson system even though some of the symmetries are badly violated. We present the group properties in an abstract way that does not involve the details of field theory.

Next, it is asked what group is actually involved. The simplest one consistent with known phenomena is the one suggested. It is introduced, for clarity, in connection with a particular field theory, the symmetrical Sakata model, in which baryons and mesons are built up of fundamental objects with the properties of n, p, and Λ. For still greater simplicity, we discuss first the case in which Λ is absent.

We then return to the question of broken symmetry in the strong interactions and show how some of the symmetries in the group, if they are not too badly violated, would reveal themselves in approximately degenerate supermultiplets. In particular, there should be "octets" of mesons, each consisting of an isotopic triplet with $S=0$, a pair of doublets with $S=\pm1$, and a singlet with $S=0$. In the case of pseudoscalar mesons, we know of π, K, and $\bar{K}$; these should be accompanied by a singlet pseudoscalar meson χ^0, which would decay into 2γ, $\pi^++\pi^-+\gamma$, or 4π, depending on its mass.

In Sec. VIII, we propose, as an alternative to the symmetrical Sakata model, another scheme with the same group, which we call the "eightfold way." Here the baryons, as well as mesons, can form octets and singlets, and the baryons N, Λ, Σ, and Ξ are supposed to constitute an approximately degenerate octet.

In Sec. IX, some topics are suggested for further investigation, including the possibility of high energy limits in which non-conserved quantities become conserved, and we give, as an example of methods suggested here, an approximate calculation of the rate of $K^+\rightarrow\mu^++\nu$ decay from that of $\pi^+\rightarrow\mu^++\nu$ decay.

II. MESONS AND CURRENTS

To introduce the connection between meson states and currents or densities, let us review the derivation[7] of the Goldberger-Treiman relation[8] among the charged pion decay amplitude, the strength of the axial vector weak interaction in the β decay of the nucleon, and the pion-nucleon coupling constant.

The axial vector term in J_α with $\Delta S=0$, $|\Delta \mathbf{I}|=1$, $GP=-1$, can be written as $P_{1\alpha}+iP_{2\alpha}$, where $\mathbf{P}_\alpha$ is an axial vector current that transforms like an isotopic vector. We have, for nucleon β decay,

$$\langle N|\mathbf{P}_\alpha|N\rangle=\bar{u}_f[i\gamma_\alpha F_{\text{ax}}(s)+k_\alpha\beta(s)]\gamma_5(\boldsymbol{\tau}/2)u_i, \quad (2.1)$$

where u_i and u_f are the initial and final spinors, k_α is the four-momentum transfer, and $s=-k^2=-k_\alpha k_\alpha$. At

[3] G. F. Chew, Talk at La Jolla Conference on Strong and Weak Interactions, June, 1961 (unpublished).

[4] M. Ikeda, S. Ogawa, and Y. Ohnuki, Progr. Theoret. Phys. (Kyoto) 22, 715 (1959); Y. Ohnuki, *Proceedings of the 1960 Annual International Conference on High-Energy Physics at Rochester* (Interscience Publishers, Inc., New York, 1960).

[5] Y. Yamaguchi, Progr. Theoret. Phys. (Kyoto) Suppl. No. 11, 1 (1959).

[6] J. Wess, Nuovo cimento 10, 15 (1960).

[7] J. Bernstein, S. Fubini, M. Gell-Mann, and W. Thirring, Nuovo cimento 17, 757 (1960). See also Y. Nambu, Phys. Rev. Letters 4, 380 (1960); and Chou Kuang-Chao, Soviet Phys.—JETP 12, 492 (1961).

[8] M. Goldberger and S. Treiman, Phys. Rev. 110, 1478 (1958).

$s=0$ we have just

$$F_{ax}(0)=-G_A/G, \qquad (2.2)$$

the axial vector renormalization constant.

The axial vector current is not conserved; its divergence $\partial_\alpha P_\alpha$ has the same quantum numbers as the pion ($J=0^-$, $I=1$). Between nucleon states we have

$$\langle N|\partial_\alpha P_\alpha|N\rangle=\bar{u}_f i\gamma_5(\tau/2)u_i[2m_N F_{ax}(s)+s\beta(s)]. \qquad (2.3)$$

We may compare this matrix element with that between the vacuum and a one-pion state

$$\langle 0|\partial_\alpha P_\alpha|\pi\rangle=m_\pi^2(2f_\pi)^{-1}\phi, \qquad (2.4)$$

where ϕ is the pion wave function and the constant f_π (or at least its square) may be measured by the rate of $\pi^+\rightarrow\mu^++\nu$:

$$\Gamma_\pi=G^2 m_\pi m_\mu^2(1-m_\mu^2/m_\pi^2)^2(f_\pi^2/4\pi)^{-1}(64\pi^2)^{-1}. \qquad (2.5)$$

It is known that the matrix element (2.3) has a pole at $s=m_\pi^2$ corresponding to the virtual emission of a pion that undergoes leptonic decay. The strength of the pole is given by the product of m_π^2/f_π and the pion-nucleon coupling constant $g_{NN\pi}$. If we assume that the expression in brackets vanishes at large s, we have an unsubtracted dispersion relation for it consisting of the pole term and a branch line beginning at $(3m_\pi)^2$, the next lowest mass that can be virtually emitted:

$$2m_N F_{ax}(s)+s\beta(s)=(g_{NN\pi}/f_\pi)m_\pi^2(m_\pi^2-s)^{-1}$$
$$+\int\sigma_{ax}(M^2)M^2dM^2\,(M^2-s-i\epsilon)^{-1}. \qquad (2.6)$$

At $s=0$, we have, using (2.2), the sum rule

$$2m_N(-G_A/G)=g_{NN\pi}/f_\pi+\int\sigma_{ax}(M^2)dM^2. \qquad (2.7)$$

Now if the dispersion relation (2.6) is not only convergent but dominated at low s by the term with the lowest mass, then we have the approximate Goldberger-Treiman relation

$$2m_N(-G_A/G)\approx g_{NN\pi}/f_\pi, \qquad (2.8)$$

which agrees with experiment to within a few percent.

The success of the relation suggests that other matrix elements of $\partial_\alpha P_\alpha$ may also obey unsubtracted dispersion relations dominated at low s by the one-pion term. For example, if we consider the matrix element between Λ and Σ, we should arrive at the relation

$$(m_\Lambda+m_\Sigma)(-G_A^{\Lambda\Sigma}/G)\approx g_{\Lambda\Sigma\pi}/f_\pi, \qquad (2.9)$$

if Λ and Σ have the same parity, or an analogous relation if they have opposite parity.

If such a situation actually obtains, then it may be said that the pion is, to a good approximation, coupled "universally" to the divergence of the axial vector current. To calculate any g approximately, we multiply the universal constant f_π, the sum of the initial and final masses, and the renormalization factor for the axial vector current.

Now let us turn to the case of a current that is conserved, say the isotopic spin current $\mathfrak{J}_\alpha$ with quantum numbers $J=1^-$, $I=1$. Acting on the vacuum, the operator $\mathfrak{J}_\alpha$ does not lead to any stable one-meson state, but it does lead to the unstable vector meson state ρ at around 750 Mev, which decays into 2π or 4π. For simplicity, let us ignore the rather large width ($\Gamma_\rho\sim 100$ Mev) of the ρ state and treat it as stable. The mathematical complications resulting from the instability are not severe and have been discussed elsewhere.[9,10]

In place of (2.4), then, we have the definition

$$\langle 0|\mathfrak{J}_\alpha|\rho\rangle=m_\rho^2(2\gamma_\rho)^{-1}\phi_\alpha, \qquad (2.10)$$

of the constant γ_ρ, where ϕ_α is the wave function of the ρ meson. In place of (2.1) or (2.3) ,we consider the matrix element between nucleon states of the isotopic spin current:

$$\langle N|\mathfrak{J}_\alpha|N\rangle=\bar{u}_f i\gamma_\alpha(\tau/2)u_i F_1^V(s)+\text{magnetic term}, \qquad (2.11)$$

where $F_1^V(s)$ is the familiar isovector form factor of the electric charge of the nucleon, since the electromagnetic current has the form

$$j_\alpha=\mathfrak{J}_{3\alpha}+\text{isoscalar term}. \qquad (2.12)$$

If we continue to ignore the width of ρ, we get a dispersion relation like (2.6) with a pole term at m_ρ^2:

$$F_1^V(s)=(\gamma_{NN\rho}/\gamma_\rho)m_\rho^2(m_\rho^2-s)^{-1}$$
$$+\int\sigma_1^V(M^2)dM^2\,M^2(M^2-s-i\epsilon)^{-1}. \qquad (2.13)$$

Here $\gamma_{NN\rho}$ is the coupling constant of ρ to $\bar{u}_f i\tau\gamma_\alpha u_i$, just as $g_{NN\pi}$ is the coupling constant of π to $\bar{u}_f i\tau\gamma_5 u_i$. In this case, we have used an unsubtracted dispersion relation just for convenience.

Since the current is conserved, there is no renormalization and we have

$$F_1^V(0)=1, \qquad (2.14)$$

giving, in place of (2.7), the sum rule

$$1=\gamma_{\rho NN}/\gamma_\rho+\int\sigma_1^V(M^2)dM^2. \qquad (2.15)$$

If the dispersion relation is dominated at low s by the ρ term, then we obtain the analog of the Goldberger-Treiman formula:

$$1\approx\gamma_{\rho NN}/\gamma_\rho. \qquad (2.16)$$

[9] G. F. Chew, University of California Radiation Laboratory Report No. UCRL-9289, 1960 (unpublished).

[10] M. Gell-Mann and F. Zachariasen, Phys. Rev. 124, 953 (1961).

Now the same reasoning may be applied to the isovector electric form factor of another particle, for example the pion:

$$\langle\pi|\mathfrak{J}_\alpha|\pi\rangle=[i\phi_f^*\times\partial_\alpha\phi_i-i\partial_\alpha\phi_f^*\times\phi_i]F_\pi(s), \quad (2.17)$$

$$F_\pi(s)=(\gamma_{\rho\pi\pi}/\gamma_\rho)m_\rho^2(m_\rho^2-s)^{-1}$$

$$+\int\sigma_\pi(M^2)dM^2\,M^2(M^2-s-i\epsilon)^{-1}, \quad (2.18)$$

and

$$1=\gamma_{\rho\pi\pi}/\gamma_\rho+\int\sigma_\pi(M^2)dM^2. \quad (2.19)$$

If this dispersion relation, too, is dominated by the ρ pole at low s, then we find

$$1\approx\gamma_{\rho\pi\pi}/\gamma_\rho. \quad (2.20)$$

To the extent that the ρ pole gives most of the sum rule in each case, we have ρ coupled *universally* to the isotopic spins of nucleon, pion, etc., with coupling parameter $2\gamma_\rho$. Such universality was postulated by Sakurai,[11] within the framework of a special theory, in which ρ is treated as an elementary vector meson described by a Yang-Mills field. It can be seen that whether or not such a field description is correct, the *effective* universality ($\gamma_{\rho\pi\pi}\approx\gamma_{\rho NN}\approx\gamma_{\rho KK}$, etc.) is an approximate rule the validity of which depends on the domination of (2.15), (2.19), etc., by the ρ term.

The various coupling parameters $\gamma_{\rho\pi\pi}$, $\gamma_{\rho NN}$, etc., can be determined from the contribution of the ρ "pole" to various scattering processes, for example $\pi+N\rightarrow\pi+N$. But the factors $\gamma_{\rho\pi\pi}/\gamma_\rho$, $\gamma_{\rho NN}/\gamma_\rho$, etc., can also be measured, using electromagnetic interactions.[10]

An approximate determination of $\gamma_{\rho NN}/\gamma_\rho$ was made by Hofstadter and Herman[12] as follows The masses M^2 in the integral in Eq. (2.13) are taken to be effectively vary large, so that (2.13) becomes approximately

$$F_1^V(s)\approx(\gamma_{NN\rho}/\gamma_\rho)m_\rho^2(m_\rho^2-s)^{-1}$$
$$+1-(\gamma_{\rho NN}/\gamma_\rho). \quad (2.21)$$

Fitting the experimental data on $F_1^V(s)$ with such a formula and using $m_\rho\approx750$ Mev, we obtain $\gamma_{\rho NN}/\gamma_\rho\approx1.4$. (Hofstadter and Herman, with a smaller value of m_ρ, found 1.2.)

III. EQUAL-TIME COMMUTATION RELATIONS

The dispersion relations for the matrix elements of weak or electromagnetic currents are linear and homogeneous. For example, Eq. (2.6) may be thought of as an expression for the matrix element of $\mathbf{P}_\alpha$ between the vacuum and a nucleon-antinucleon pair state. On the right-hand side, the pole term contains the product of the matrix element of $\mathbf{P}_\alpha$ between the vacuum and a one-pion state multiplied by the transition amplitude for the transition from π to $N\bar{N}$ by means of the strong interactions. The weight function $\sigma_{n\pi}(M^2)$ is just the sum of such products over many intermediate states (such as 3π, 5π, etc.) with total mass M.

Now such linear, homogeneous equations may determine the dependence of the current matrix elements on variables such as s, but they cannot fix the scale of these matrix elements; constants like $-G_A/G$ cannot be calculated without further information. A field theory of the strong interactions, with explicit expressions for the currents, somehow contains more than these dispersion relations. In what follows, we shall extract some of this additional information in the form of equal-time commutation relations between components of the currents. Since these are nonlinear relations, they can help to fix the scale of each matrix element. Moreover, these relations may be the same for the lepton system and for the baryon-meson system, so that universality of strength of the weak interactions, for example, becomes meaningful.[13]

Let us begin our discussion of equal-time commutation relations with a familiar case—that of the isotopic spin **I**. Its components I_i obey the well-known commutation relations,

$$[I_i,I_j]=ie_{ijk}I_k. \quad (3.1)$$

In terms of the components $\mathfrak{J}_{i\alpha}$ of the isotopic spin current, we have

$$I_i=-i\int\mathfrak{J}_{i4}d^3x, \quad (3.2)$$

and the conservation law

$$\partial_\alpha\mathfrak{J}_{i\alpha}=0 \quad (3.3)$$

tells us that

$$\dot{I}_i=\int\partial_\alpha\mathfrak{J}_{i\alpha}d^3x=0, \quad (3.4)$$

at all times.

Now the commutator of $\mathfrak{J}_{i4}(\mathbf{x},t)$ and $\mathfrak{J}_{j4}(\mathbf{x}',t)$ must vanish for $\mathbf{x}\neq\mathbf{x}'$, in accorance with microcausality. (Note we have taken the times equal.) If the commutator is not more singular than a delta function, then (3.1) and (3.2) give us the relation

$$[\mathfrak{J}_{i4}(\mathbf{x},t),\mathfrak{J}_{j4}(\mathbf{x}',t)]=-ie_{ijk}\mathfrak{J}_{k4}(\mathbf{x},t)\delta(\mathbf{x}-\mathbf{x}'), \quad (3.5)$$

which can also be obtained in any simple field theory by explicit commutation.[14]

In discussing the various parts of the weak current J_α, we shall have to deal with currents like $\mathbf{P}_\alpha$ that are not

[11] J. J. Sakurai, Ann. Phys. **11**, 1 (1960).

[12] R. Hofstadter and R. Herman, Phys. Rev. Letters **6**, 293 (1961). See also S. Bergia, A. Stanghellini, S. Fubini, and C. Villi, Phys. Rev. Letters **6**, 367 (1961).

[13] M. Gell-Mann, *Proceedings of the 1960 Annual International Conference on High-Energy Physics at Rochester* (Interscience Publishers, Inc., New York, 1960).

[14] In some cases explicit commutation may be ambiguous and misleading. For example, a superficial consideration of $[j_i(\mathbf{x},t), j_4(\mathbf{x}',t)]$ for $i=1, 2, 3$ may lead to the conclusion that the expression vanishes. Yet the vacuum expectation value of the commutator can be shown to be a nonzero quantity times $\partial_i\delta(\mathbf{x}-\mathbf{x}')$, and that result is confirmed by more careful calculation. See J. Schwinger, Phys. Rev. Letters **3**, 296 (1959).

conserved.[15] Here, too, we may define a quantity analogous to **I**:

$$D_i = -i\int P_{i4}d^3x, \tag{3.6}$$

but D_i is *not* independent of time:

$$\dot{D}_i = \int \partial_\alpha P_{i\alpha}d^3x \neq 0. \tag{3.7}$$

For the moment, let us restrict our attention to the currents $\mathfrak{J}_\alpha$ and $\mathbf{P}_\alpha$ and the operators **I** and **D**(t). Since **D** is an isovector, we have the relations

$$[I_i,D_j]=[D_i,I_j]=ie_{ijk}D_k, \tag{3.8}$$

but what is the commutator of two components of **D**? Since $\mathbf{P}_\alpha$ is a physical quantity, so is **D** and the question is one with direct physical meaning. We shall give both a general and a specific answer.

In general, we may take the commutators of D's (divided by i), the components of **I** and **D**, the commutators of all of these with one another (divided by i), etc., until we obtain a system of Hermitian operators that is closed under commutation. Any of these operators can be written as a linear combination of N linearly independent Hermitian operators $R_i(t)$, where N might be infinite, and where the commutator of any two R_i is a linear combination of the R_i:

$$[R_i(t),R_j(t)]=ic_{ijk}R_k(t), \tag{3.9}$$

with c_{ijk} real. Such a system is called an algebra by the mathematicians. If we consider the set of infinitesimal unitary operators $1+i\epsilon R_i(t)$ and all possible products of these, we obtain an N-parameter continuous group of unitary transformations. We can refer to (3.9) as the algebra of the group. It is a physically meaningful statement to specify what group or what algebra is generated in this way by the currents $\mathfrak{J}_\alpha$ and $\mathbf{P}_\alpha$. Since a commutation relation like (3.9) is left invariant by a unitary transformation such as $\exp(-it\int Hd^3x)$, the numbers c_{ijk} are independent of time.

A second mathematical statement is also in order, i.e., the specification of the transformation properties of the Hamiltonian density $H(\mathbf{x},t)$ under the group or the algebra. Those R_i for which $[R_i(t),H(\mathbf{x},t)]=0$ are independent of time, but some of them, like D_i, do not commute with H. If all of the R_i commuted with H, then H would belong to the trivial one-dimensional representation of the group. In fact, H behaves in a more complicated way. By commuting all of the $R_i(t)$ with $H(\mathbf{x},t)$, we obtain a linear set of operators, containing H, that form a representation of the group; it may be broken up into the direct sum of irreducible representations. We want to know, then, what group is generated by **I** and **D** and to what irreducible representations of this group H belongs. Suggested are specific answers to both questions.

Let us look at the vector and axial vector weak currents for the leptons. For the time being, we shall consider only ν and e, ignoring the muon. (In the same way, we shall, in this section, ignore strange particles, and consider only baryons and mesons with $S=0$.) The vector weak current $i\bar{\nu}\gamma_\alpha e$ and the axial current $i\bar{\nu}\gamma_\alpha\gamma_5 e$ can be regarded formally as components of two "isotopic vector" currents for the leptons:

$$\mathfrak{J}_\alpha^{(l)}=i\bar{\xi}\boldsymbol{\tau}\gamma_\alpha\xi/2, \quad \mathbf{P}_\alpha^{(l)}=i\bar{\xi}\boldsymbol{\tau}\gamma_\alpha\gamma_5\xi/2, \tag{3.10}$$

where ξ stands for (ν,e). We can also form the mathematical analogs of **I** and **D**:

$$\mathbf{I}^{(l)}=-i\int\mathfrak{J}_4^{(l)}d^3x, \quad \mathbf{D}^{(l)}=-i\int\mathbf{P}_4^{(l)}d^3x. \tag{3.11}$$

Now in this leptonic case we can easily compute the commutation rules of $\mathbf{I}^{(l)}$ and $\mathbf{D}^{(l)}$:

$$\begin{aligned}
&[I_i^{(l)},I_j^{(l)}]=ie_{ijk}I_k^{(l)},\\
&[I_i^{(l)},D_j^{(l)}]=[D_i^{(l)},I_j^{(l)}]=ie_{ijk}D_k^{(l)}, \qquad (3.12)\\
&[D_i^{(l)},D_j^{(l)}]=ie_{ijk}I_k^{(l)}.
\end{aligned}$$

Another way to phrase these commutation rules is to put

$$\begin{aligned}
\mathbf{I}^{(l)}&=\mathbf{L}_+^{(l)}+\mathbf{L}_-^{(l)},\\
\mathbf{D}^{(l)}&=\mathbf{L}_+^{(l)}-\mathbf{L}_-^{(l)},
\end{aligned} \tag{3.13}$$

and to notice that $\mathbf{L}_+^{(l)}$ and $\mathbf{L}_-^{(l)}$ are two commuting angular momenta [essentially $\boldsymbol{\tau}(1+\gamma_5)/4$ and $\boldsymbol{\tau}(1-\gamma_5)/4$]. The weak current $i\bar{\nu}\gamma_\alpha(1+\gamma_5)e$ is just a component of the current of $\mathbf{L}_+^{(l)}$.

We now suggest that the algebraic structure of **I** and **D** is exactly the same in the case of baryons and mesons. To (3.1) and (3.8), we add the rule[16,17]

$$[D_i,D_j]=ie_{ijk}I_k, \tag{3.14}$$

which closes the system and makes $\mathbf{I}^+\equiv(\mathbf{I}+\mathbf{D})/2$ and $\mathbf{I}^-\equiv(\mathbf{I}-\mathbf{D})/2$ two commuting angular momenta. Again, we make the weak current a component of the current of $\mathbf{I}^+$. Evidently the statement that $(\mathbf{I}+\mathbf{D})/2$ is an angular momentum and not some factor times an angular momentum, fixes the scale of the weak current. It makes universality of strength between baryons and leptons meaningful, and it specifies, together with the dispersion relations, the value of such constants as $-G_A/G$.

The simplest way to realize the algebraic structure under discussion in a field-theory model of baryons and mesons is to construct the currents $\mathfrak{J}_\alpha$ and $\mathbf{P}_\alpha$ out of p and n fields just as $\mathfrak{J}_\alpha^{(l)}$ and $\mathbf{P}_\alpha^{(l)}$ are made out of ν and e fields:

$$\mathfrak{J}_\alpha=i\bar{N}\boldsymbol{\tau}\gamma_\alpha N/2, \quad \mathbf{P}_\alpha=i\bar{N}\boldsymbol{\tau}\gamma_\alpha\gamma_5 N/2, \tag{3.15}$$

[15] We assume that the vector weak current with $\Delta S=0$ is just a component of the isotopic spin current $\mathfrak{J}_\alpha$ and thus conserved.

[16] F. Gursey, Nuovo cimento 16, 230 (1960).

[17] M. Gell-Mann and M. Lévy, Nuovo cimento 16, 705 (1960)

where N means (p,n). We then obtain not only the commutation rules (3.1), (3.8), and (3.14), but the stronger rule (3.5) and its analogs:

$$\begin{aligned} [\mathfrak{J}_{i4}(\mathbf{x},t),P_{j4}(\mathbf{x}',t)] &= -ie_{ijk}P_{k4}(\mathbf{x},t)\delta(\mathbf{x}-\mathbf{x}'), \\ [P_{i4}(\mathbf{x},t),P_{j4}(\mathbf{x}',t)] &= -ie_{ijk}\mathfrak{J}_{k4}(\mathbf{x},t)\delta(\mathbf{x}-\mathbf{x}'). \end{aligned} \quad (3.16)$$

Next we want to use a field-theory model to suggest an answer to the second question—how H behaves under the group or, what is the same thing, under the algebra consisting of **I** and **D** or of $\mathbf{I}^+$ and $\mathbf{I}^-$. Since $\mathbf{I}^+$ and $\mathbf{I}^-$ are two commuting angular momenta, any irreducible representation of the algebra is specified by a pair of total angular momentum quantum numbers: i_+ for $\mathbf{I}^+$ and i_- for $\mathbf{I}^-$. The total isotopic spin quantum number I is associated with $\mathbf{I}^+ + \mathbf{I}^- = \mathbf{I}$.

Now we want the vector weak current $\mathfrak{J}_\alpha$ to be the isotopic spin current and to be conserved. Thus H must commute with **I**; it transforms as an isoscalar, with $I=0$. In order to couple to zero, i_+ and i_- must be equal. So H can consist of terms with $(i_+,i_-)=(0,0)$, $(\frac{1}{2},\frac{1}{2})$, $(1,1)$, $(\frac{3}{2},\frac{3}{2})$, etc. Which of these are in fact present?

The simplest model in which the total isotopic current is given by just (3.15) is the Fermi-Yang[18] model, in which the pion is a composite of nucleon and antinucleon. To write an explicit Lagrangian, it must be decided what form the binding interaction takes. Since a direct four-fermion coupling leads to unpleasant singularities, we whall use a massive neutral vector meson field B^0 coupled to the nucleon current, as proposed by Teller[18] and Sakurai[11]; the exchange of a B^0 gives attraction between nucleon and antinucleon, permitting binding, and it also gives repulsion between nucleons, contributing to the "hard core." The model Lagrangian is then[19]

$$\begin{aligned} L = &-\bar{N}\gamma_\alpha\partial_\alpha N - (\partial_\alpha B_\beta - \partial_\beta B_\alpha)^2/4 \\ &-\mu_0^2 B_\alpha B_\alpha/2 - ih_0 B_\alpha \bar{N}\gamma_\alpha N - m_0\bar{N}N. \end{aligned} \quad (3.17)$$

If the mass term for the nucleon were absent, then both $\mathfrak{J}_\alpha$ and $\mathbf{P}_\alpha$ would be conserved; **I** and **D** would both commute with L and with H. Thus,

$$H = H(0,0) - u_0, \quad (3.18)$$

where $H(0,0)$ transforms according to $(i_+,i_-)=(0,0)$ and the noninvariant term u_0 is just $-m_0\bar{N}N$. To what representation does it belong?

It is easy to see that the field B^0 belongs to $(0,0)$, while $N_L \equiv (1+\gamma_5)N/2$ belongs to $(\frac{1}{2},0)$ and $N_R \equiv (1-\gamma_5)N/2$ belongs to $(0,\frac{1}{2})$. One can thus verify that all terms of (3.17) except the last belong to $(0,0)$, since $\bar{N}\gamma_\alpha N$ or $\bar{N}\gamma_\alpha\partial_\alpha N$ couples $\bar{N}_L$ to N_L and $\bar{N}_R$ to N_R. But the Dirac matrix β, unlike $\beta\gamma_\alpha$, anticommutes with γ_5, so that the last term $-m_0\bar{N}N$ couples $\bar{N}_L$ to N_R and $\bar{N}_R$ to N_L. Thus u_0 belongs to $(\frac{1}{2},\frac{1}{2})$. We have $H=H(0,0)+H(\frac{1}{2},\frac{1}{2})$.

There are four components to the representation $(\frac{1}{2},\frac{1}{2})$ to which $u_0 = -H(\frac{1}{2},\frac{1}{2})$ belongs. By commuting **D** with u_0, we generate the other three easily and see that they are proportional to $-i\bar{N}\boldsymbol{\tau}\gamma_5 N$. In fact **D** acts like $\boldsymbol{\tau}\gamma_5/2$, **I** like $\boldsymbol{\tau}/2$, u_0 like β, and the other three components like $-i\beta\gamma_5\boldsymbol{\tau}$. Denoting the three new components by v_i, we have

$$\begin{aligned} [I_i,u_0] &= 0, & [D_i,u_0] &= -iv_i, \\ [I_i,v_j] &= ie_{ijk}v_k, & [D_i,v_j] &= i\delta_{ij}u_0. \end{aligned} \quad (3.19)$$

In the model, there are the even stronger relations for the densities

$$\begin{aligned} [\mathfrak{J}_{i4}(\mathbf{x},t),u_0(\mathbf{x}',t)] = 0, \quad & [P_{i4}(\mathbf{x},t),u_0(\mathbf{x}',t)] \\ &= -iv_i(\mathbf{x},t)\delta(\mathbf{x}-\mathbf{x}'), \text{ etc.} \end{aligned} \quad (3.20)$$

The noninvariant term u_0 is what prevents the axial vector current from being conserved. Thus one can express the divergence $\partial_\alpha\mathbf{P}_\alpha$ of the current in terms of the commutator of **D** with u_0. The conditions for this relation to hold are treated in the appendix and are applicable to all models we discuss. We find simply

$$\partial_\alpha\mathbf{P}_\alpha = -i[\mathbf{D},H] = i[\mathbf{D},u_0] = \mathbf{v}, \quad (3.21)$$

and, of course,

$$\partial_\alpha\mathfrak{J}_\alpha = -i[\mathbf{I},H] = 0. \quad (3.22)$$

It is precisely the operator **v**, then, that we used in a dispersion relation in order to obtain the Goldberger-Treiman relation in Sec. II. Acting on the vacuum, it leads mostly to the one-pion state, so that the pion is effectively coupled universally to the divergence of the axial vector current. Thus **v** is a sort of effective pion field operator for the Fermi-Yang theory, which has no explicit pion field.

If we insist on a model in which there is a field variable $\boldsymbol{\pi}(x,t)$ then we must complicate the discussion. The total isotopic spin current is no longer given by just (3.15); there is a pion isotopic current term as well. In order to preserve the same algebraic structure of **I** and **D**, one must then modify $\mathbf{P}_\alpha$ as well. Such a theory was described by Gell-Mann and Lévy,[17] who called it the "σ-model".[20] Along with the field $\boldsymbol{\pi}$, we must introduce a scalar, isoscalar field σ' in such a way that $\boldsymbol{\pi}$, σ' transform under the group like **v**, u_0. Then, just as $\mathfrak{J}_\alpha$ has an additional term quadratic in $\boldsymbol{\pi}$, $\mathbf{P}_\alpha$ requires an additional term bilinear in $\boldsymbol{\pi}$ and σ'.

As we shall see in the next section, the introduction of

---

[18] E. Fermi and C. N. Yang, Phys. Rev. 76, 1739 (1949); E. Teller, *Proceedings of the Sixth Annual Rochester Conference on High-Energy Nuclear Physics, 1956* (Interscience Publishers, Inc., New York, 1956).

[19] Conceivably a massive B^0 meson can be described by (3.17) even with $\mu_0=0$. [J. Schwinger, lectures at Stanford University, summer, 1961 (unpublished)]. In that case the noninvariant term in (3.17) is just equal to $\theta_{\alpha\alpha}$ and the traceless part of $\theta_{\alpha\beta}$ commutes with the group elements at equal times. In any case, whether μ_0 is zero or not, the off-diagonal terms in $\theta_{\alpha\beta}$ commute with the group.

[20] In the σ model, explicit commutation of u_0 and **v** at equal times gives zero, while in the Fermi-Yang model this is not so; if we take these results seriously, they give us definite physical distinctions among models.

strange particles makes the group much larger. The term u_0 is then a member of a much larger representation, with eighteen components. Thus if a pion field is introduced, fifteen more components are needed as well. Such a theory is too complicated to be attractive; we shall therefore ignore it and concentrate on the simplest generalization of the Fermi-Yang model to strange particles, namely the symmetrical Sakata model.

IV. SYMMETRICAL SAKATA MODEL AND UNITARY SYMMETRY

In the previous section, we proceeded inductively. We showed that starting from physical currents like $\mathfrak{J}_\alpha$ and $\mathbf{P}_\alpha$ we may construct a group and its algebra and that it is physically meaningful to specify the group and also the transformation properties of H under the group. We chose the algebraic structure by analogy with the case of leptons and we saw that the simplest field theory model embodying the structure is just the Fermi-Yang model, in which p and n fields are treated just like the ν and e fields for the leptons, except that they are given a mass and a strong "gluon" coupling. The transformation properties of H were taken from the model; H consists, then, of an invariant part $H_{0,0}$ plus a term $(-u_0)$, where u_0 and a pseudoscalar isovector quantity $\mathbf{v}$ belong to the representation $(\frac{1}{2},\frac{1}{2})$ of the group. We then have the commutation rules (3.1), (3.8), (3.14), and (3.19). Microcausality with the assumption of commutators that are not too singular, or else direct inspection of the model, gives the stronger commutation rules (3.5), (3.16), and (3.20) for the densities. The model also gives specific equal-time commutation rules for u_0 and $\mathbf{v}$, which we did not list. All of these properties can be abstracted from the model and considered on their own merits as proposed relations among the currents and the Hamiltonian density.

Now, to argue deductively, we want to include the strange particles and all parts of the weak current J_α and the electromagnetic current j_α. We generalize the Fermi-Yang description to obtain the symmetrical Sakata model and abstract from it as many physically meaningful relations as possible.

It has long been recognized that the qualitative properties of baryons and mesons could be understood in terms of the Sakata model,[21] in which all strongly interacting particles are made out of N, Λ, $\bar{N}$, and $\bar{\Lambda}$ (or at least out of basic fields with the same quantum numbers as these particles).

We write the Lagrangian density for the Sakata model as a generalization of (3.17):

$$L=-\bar{p}\gamma_\alpha p-\bar{n}\gamma_\alpha\partial_\alpha \bar{n}-\bar{\Lambda}\gamma_\alpha\partial_\alpha\Lambda-\tfrac{1}{4}(\partial_\alpha B_\beta-\partial_\beta B_\alpha)^2 -\tfrac{1}{2}\mu_0^2 B_\alpha B_\alpha-ih_0(\bar{p}\gamma_\alpha p+\bar{n}\gamma_\alpha n+\bar{\Lambda}\gamma_\alpha\Lambda)B_\alpha -m_{0N}(\bar{n}n+\bar{p}p)-m_{0\Lambda}\bar{\Lambda}\Lambda. \quad (4.1)$$

According to this picture, the baryons present a striking parallel with the leptons,[22] for which we write the Lagrangian density

$$L_l=-\bar{\nu}\gamma_\alpha\nu-\bar{e}\gamma_\alpha\partial_\alpha e-\bar{\mu}\gamma_\alpha\partial_\alpha\mu-0\cdot(\bar{\nu}\nu+\bar{e}e)-m_\mu\bar{\mu}\mu, \quad (4.2)$$

if we turn off the electromagnetic and weak couplings, along with the ν-e mass difference. Here it is assumed there is only one kind of neutrino.

The only real difference between baryons and leptons in (4.1) and (4.2), respectively, is that the baryons are coupled, through the baryon current, to the field B. It is tempting to suppose that the weak current of the strongly interacting particles is just the expression.

$$i\bar{p}\gamma_\alpha(1+\gamma_5)n+i\bar{p}\gamma_\alpha(1+\gamma_5)\Lambda, \quad (4.3)$$

analogous to Eq. (1.2) for the leptonic weak current $J_\alpha^{(l)}$. Now (4.3) is certainly a reasonable expression, qualitatively, for weak currents of baryons and mesons. As Okun has emphasized,[23] the following properties of the weak interactions, often introduced as postulates, are derivable from (1.1), (1.2), (4.1), and (4.3):

(a) The conserved vector current.[24] In the model under discussion, as in that of Fermi and Yang, $i\bar{p}\gamma_\alpha n$ is a component of the total isotopic spin current.

(b) The rules $|\Delta S|=1$, $\Delta S/\Delta Q=+1$, and $|\Delta\mathrm{I}|=\frac{1}{2}$ for the leptonic decays of strange particles.[25]

(c) The invariance under GP of the $\Delta S=0$ weak current.[26]

(d) The rules $|\Delta S|=1$, $|\Delta\mathrm{I}|=\frac{1}{2}$ or $\frac{3}{2}$ in the nonleptonic decays of strange particles; along with $|\Delta S|=1$, we have the absence of a large K_1^0-K_2^0 mass difference.

The quantitative facts that the effective coupling constants for $|\Delta S|=1$ leptonic decays are smaller than those for $|\Delta S|=0$ leptonic decays and that in nonleptonic decays of strange particles the $|\Delta\mathrm{I}|=\frac{1}{2}$ amplitude greatly predominates over the $|\Delta\mathrm{I}|=\frac{3}{2}$ amplitude are not explained in any fundamental way.[27]

---

[21] S. Sakata, Progr. Theoret. Phys. (Kyoto) 16, 686 (1956).

[22] A. Gamba, R. E. Marshak, and S. Okubo, Proc. Natl. Acad. Sci. U. S. 45, 881 (1959).

[23] L. Okun, Ann. Rev. Nuclear Sci. 9, 61 (1959).

[24] R. P. Feynman and M. Gell-Mann, Phys. Rev. 109, 193 (1958). See also S. S. Gershtein and J. B. Zeldovich, Soviet Phys.—JETP 2, 576 (1957).

[25] M. Gell-Mann, *Proceedings of the Sixth Annual Rochester Conference on High-Energy Nuclear Physics, 1956* (Interscience Publishers, Inc., New York, 1956). These rules were in fact suggested on the basis of the idea that N and Λ are fundamental. Should the rules prove too restrictive (for example should $\Delta S/\Delta Q=+1$ be violated), then we would try a larger group; in the language of the field-theoretic model, we would assume more fundamental fields. For a discussion of possible larger groups, see M. Gell-Mann and S. Glashow, Ann. Phys. 15, 437 (1961) and S. Coleman and S. Glashow (to be published).

[26] S. Weinberg, Phys. Rev. 112, 1375 (1958).

[27] A possible dynamical explanation of the predominance of $|\Delta\mathrm{I}|=\frac{1}{2}$ is being investigated by Nishijima (private communication). For example, consider the decay $\Lambda\to N+\pi$. A dispersion relation without subtractions is written for the matrix element of $J_\alpha{}^\dagger J_\alpha$ between the vacuum and a state containing $N+\bar{\Lambda}+\pi$. The parity-violating part leads to intermediate pseudoscalar states with $S=+1$ and with $|\Delta\mathrm{I}|=\frac{1}{2}$ or $\frac{3}{2}$. In the case of $|\Delta\mathrm{I}|=\frac{1}{2}$, there is an intermediate K particle, which may give a large contribution, swamping the term with $|\Delta\mathrm{I}|=\frac{3}{2}$, which has no one-meson state. For the same argument to apply to the parity-conserving part, we need the K' meson of Table III.

TABLE I. A set of matrices λ_i.

$$\lambda_1=\begin{pmatrix}0&1&0\\1&0&0\\0&0&0\end{pmatrix}\quad \lambda_2=\begin{pmatrix}0&-i&0\\i&0&0\\0&0&0\end{pmatrix}\quad \lambda_3=\begin{pmatrix}1&0&0\\0&-1&0\\0&0&0\end{pmatrix}$$

$$\lambda_4=\begin{pmatrix}0&0&1\\0&0&0\\1&0&0\end{pmatrix}\quad \lambda_5=\begin{pmatrix}0&0&-i\\0&0&0\\i&0&0\end{pmatrix}\quad \lambda_6=\begin{pmatrix}0&0&0\\0&0&1\\0&1&0\end{pmatrix}$$

$$\lambda_7=\begin{pmatrix}0&0&0\\0&0&-i\\0&i&0\end{pmatrix}\quad \lambda_8=\begin{pmatrix}1/\sqrt{3}&0&0\\0&1/\sqrt{3}&0\\0&0&-2/\sqrt{3}\end{pmatrix}$$

The electromagnetic properties of baryons and leptons are not exactly parallel in the Sakata model. The electric current (divided by e), which are denoted by j_α, is given by

$$i\bar{p}\gamma_\alpha p \tag{4.4}$$

for the baryons and mesons and by

$$-i(\bar{e}\gamma_\alpha e+\bar{\mu}\gamma_\alpha\mu) \tag{4.5}$$

for the leptons.

Now, we return to the Lagrangian (4.1) and separate it into three parts:

$$L=\bar{L}+L'+L'', \tag{4.6}$$

where $\bar{L}$ stands for everything except the baryon mass terms, while L' and L'' are given by the expressions

$$\begin{aligned}L'&=(2m_{0N}+m_{0\Lambda})(\bar{N}N+\bar{\Lambda}\Lambda)/3,\\ L''&=(m_{0N}-m_{0\Lambda})(\bar{N}N-2\bar{\Lambda}\Lambda)/3.\end{aligned} \tag{4.7}$$

If we now consider the Lagrangian with the mass-splitting term L'' omitted, we have a theory that is completely symmetrical in p, n, and Λ. We may perform any unitary linear transformation (with constant coefficients) on these three fields and leave $\bar{L}+L'$ invariant. Thus in the absence of the mass-splitting term L'' the theory is invariant under the three-dimensional unitary group $U(3)$; we shall refer to this situation as "unitary symmetry."

If we now turn on the mass-splitting, the symmetry is reduced. The only allowed unitary transformations are those involving n and p alone or Λ alone. The group becomes $U(2)\times U(1)$, which corresponds, as we shall see, to the conservation if isotopic spin, strangeness, and baryon number.

For simplicity, let us return briefly to the simpler case in which there is no Λ. The symmetry group is then just $U(2)$, the set of unitary transformations on n and p. We can factor each unitary transformation uniquely into one which multiplies both fields by the same phase factor and one (with determinant unity) which leaves invariant the product of the phase factors of p and n. Invariance under the first kind of transformation corresponds to conservation of nucleons n and p; it may be considered separately from invariance under the class of transformations of the second kind [called by mathematicians the unitary unimodular group $SU(2)$ in two dimensions]. In mathematical language, we can factor $U(2)$ into $U(1)\times SU(2)$.

Each transformation of the first kind can be written as a matrix 1 $\exp i\phi$, where 1 is the unit 2×2 matrix. The infinitesimal transformation is $1+i1\delta\phi$, and so the unit matrix is the infinitesimal generator of these transformations. Those of the second kind are generated in the same way by the three independent traceless 2×2 matrices, which may be taken to be the Pauli isotopic spin matrices τ_1, τ_2, and τ_3. We thus have

$$N\to(1+i\sum_{k=1}^{3}\delta\theta_k\tau_k/2)N, \tag{4.8}$$

as the general infinitesimal transformation of the second kind. Symmetry under all the transformations of the second kind is the same as symmetry under isotopic spin rotations. The whole formalism of isotopic spin theory can then be constructed by considering the transformation properties of the doublet or spinor (p,n) and of more complicated objects that transform like combinations of two or more such nucleons (or antinucleons).

The Pauli matrices τ_k are Hermitian and obey the rules

$$\begin{aligned}\mathrm{Tr}\,\tau_i\tau_j&=2\delta_{ij},\\ [\tau_i,\tau_j]&=2ie_{ijk}\tau_k,\\ \{\tau_i,\tau_j\}&=2\delta_{ij}1.\end{aligned} \tag{4.9}$$

The invariance under the group $SU(2)$ of isotopic spin rotations corresponds to conservation of the isotopic spin current

$$\mathfrak{J}_\alpha=i\bar{N}\boldsymbol{\tau}\gamma_\alpha N/2,$$

while the invariance under transformations of the first kind corresponds to conservation of the nucleon current $i\bar{N}\gamma_\alpha N/2=n_\alpha$.

Defining the total isotopic spin $\mathbf{I}$ as in (3.2), we obtain for I_i the commutation rules (3.1), which are the same as those for $\tau_i/2$. Likewise the nucleon number is defined as $-i\int n_4d^3x$ and commutes with $\mathbf{I}$.

We now generalize the idea of isotopic spin by including the third field Λ. Again we factor the unitary transformations on baryons into those which are generated by the 3×3 unit matrix 1 (and which correspond to baryon conservation) and those which are generated by the eight independent traceless 3×3 matrices [and which form the unitary unimodular group $SU(3)$ in three dimensions]. We may construct a typical set of eight such matrices by analogy with the 2×2 matrices of Pauli. We call then $\lambda_1\cdots\lambda_8$ and list them in Table I. They are Hermitian and have the properties

$$\begin{aligned}\mathrm{Tr}\,\lambda_i\lambda_j&=2\delta_{ij},\\ [\lambda_i,\lambda_j]&=2if_{ijk}\lambda_k,\\ \{\lambda_i,\lambda_j\}&=2d_{ijk}\lambda_k+\tfrac{4}{3}\delta_{ij}1,\end{aligned} \tag{4.10}$$

where f_{ijk} is real and totally antisymmetric like the

TABLE II. Nonzero elements of f_{ijk} and d_{ijk}. The f_{ijk} are odd under permutations of any two indices while the d_{ijk} are even.

ijk	f_{ijk}	ijk	d_{ijk}
123	1	118	$1/\sqrt{3}$
147	1/2	146	1/2
156	−1/2	157	1/2
246	1/2	228	$1/\sqrt{3}$
257	1/2	247	−1/2
345	1/2	256	1/2
367	−1/2	338	$1/\sqrt{3}$
458	$\sqrt{3}/2$	344	1/2
678	$\sqrt{3}/2$	355	1/2
···	···	366	−1/2
···	···	377	−1/2
···	···	448	$-1/(2\sqrt{3})$
···	···	558	$-1/(2\sqrt{3})$
···	···	668	$-1/(2\sqrt{3})$
···	···	778	$-1/(2\sqrt{3})$
···	···	888	$-1/\sqrt{3}$

Kronecker symbol e_{ijk} of Eq. (4.9), while d_{ijk} is real and totally symmetric. These properties follow from the equations

$$\mathrm{Tr}\lambda_k[\lambda_i,\lambda_j]=4if_{ijk},$$
$$\mathrm{Tr}\lambda_k\{\lambda_i,\lambda_j\}=4d_{ijk}, \tag{4.11}$$

derived from (4.10).

The nonzero elements of f_{ijk} and d_{ijk} are given in Table II for our choice of λ_i. Even and odd permutations of the listed indices correspond to multiplication of f_{ijk} by ±1, respectively, and of d_{ijk} by $+1$.

The general infinitesimal transformation of the second kind on the three basic baryons b is, of course,

$$b\rightarrow(1+i\sum_{i=1}^{8}\delta\theta_i\lambda_i/2)b, \tag{4.12}$$

by analogy with (4.8). Together with conservation of baryons, invariance under these transformations corresponds to complete "unitary symmetry" of the three baryons. We have factored $U(3)$ into $U(1)\times SU(3)$.

The invariance under transformations of the first kind gives us conservation of the baryon current

$$i\bar{b}\gamma_\alpha b=i\bar{n}\gamma_\alpha n+i\bar{p}\gamma_\alpha p+i\bar{\Lambda}\gamma_\alpha\Lambda, \tag{4.13}$$

while invariance under the second class of transformations would give us conservation of the eight-component "unitary spin" current

$$\mathfrak{F}_{i\alpha}=i\bar{b}\lambda_i\gamma_\alpha b/2\quad(i=1,\cdots,8). \tag{4.14}$$

Now in fact L'' is not zero and so not all the components of $\mathfrak{F}_{i\alpha}$ are actually conserved. This does not prevent us from defining $\mathfrak{F}_{i\alpha}$ as in (4.14), nor does it affect the commutation rules of the unitary spin density. The total unitary spin F_i is defined by the relation

$$F_i=-i\int\mathfrak{F}_{i4}d^3x, \tag{4.15}$$

at any time and at equal times the commutation rules for F_i follow those for $\lambda_i/2$

$$[F_i,F_j]=if_{ijk}F_k. \tag{4.16}$$

The baryon number, of course, commutes with all components F_i.

It will be noticed that λ_1, λ_2, and λ_3 agree with τ_1, τ_2, and τ_3 for p and n and have no matrix elements for Λ. Thus the first three components of the unitary spin are just the components of the isotopic spin. The matrix λ_8 is diagonal in our representation and has one eigenvalue for the nucleon and another for the Λ. Thus F_8 is just a linear combination of strangeness and baryon number. It commutes with the isotopic spin.

The matrices λ_4, λ_5, λ_6, and λ_7 connect the nucleon and Λ. We see that the components F_4, F_5, F_6, and F_7 of the unitary spin change strangeness by one unit and isotopic spin by a half unit. When the mass-splitting term L'' is "turned on," it is these components that are no longer conserved, while the conservation of F_1, F_2, F_3, F_8, and baryon number remains valid.

V. VECTOR AND AXIAL VECTOR CURRENTS

We may unify the mathematical treatment of the baryon current and the unitary spin current if we define a ninth 3×3 matrix

$$\lambda_0=(\tfrac{2}{3})^{\frac{1}{2}}1, \tag{5.1}$$

so that the *nine* matrices λ_i obey the rules

$$[\lambda_i,\lambda_j]=2if_{ijk}\lambda_k\quad(i=0,\cdots,8),$$
$$\{\lambda_i,\lambda_j\}=2d_{ijk}\lambda_k\quad(i=0,\cdots,8), \tag{5.2}$$
$$\mathrm{Tr}\lambda_i\lambda_j=2\delta_{ij}\quad(i=0,\cdots,8).$$

Here, f_{ijk} is defined as before, except that it vanishes when any index is zero; d_{ijk} is also defined as before, except that it has additional nonzero matrix elements equal to $(\frac{2}{3})^{\frac{1}{2}}$ whenever any index is zero and the other two indices are equal. The baryon current is now $(\frac{3}{2})^{\frac{1}{2}}\mathfrak{F}_{0\alpha}$.

The definitions (4.15) and the equal-time commutation relations (4.16) now hold for $i=0,\cdots,8$. Moreover, there are the equal-time commutation relations

$$[\mathfrak{F}_{i4}(\mathbf{x},t),\mathfrak{F}_{j4}(\mathbf{x}',t)]=-if_{ijk}\mathfrak{F}_{k4}(\mathbf{x},t)\delta(\mathbf{x}-\mathbf{x}') \tag{5.3}$$

for the densities.

The electric current j_α is then

$$j_\alpha=(\sqrt{2}\mathfrak{F}_{0\alpha}+\mathfrak{F}_{8\alpha}+\sqrt{3}\mathfrak{F}_{3\alpha})/2\sqrt{3}, \tag{5.4}$$

while the vector weak current is

$$\mathfrak{F}_{1\alpha}+i\mathfrak{F}_{2\alpha}+\mathfrak{F}_{4\alpha}+i\mathfrak{F}_{5\alpha}. \tag{5.5}$$

We now wish to set up the same formalism for the axial vector currents. We recall that the presence of the symmetry-breaking term L'' did not prevent us from defining the $\mathfrak{F}_{i\alpha}$ and obtaining the commutation rules (5.3) characteristic of the unitary symmetry group $U(3)$.

In the same way, we now remark that if both L''

and L' are "turned off," we have invariance under the infinitesimal unitary transformations

$$b \to (1+i\sum_{i=0}^{8} \delta\psi_i\gamma_5\lambda_i/2)b, \quad (5.6)$$

as well as the infinitesimal transformations

$$b \to (1+i\sum_{i=0}^{8} \delta\theta_i\lambda_i/2)b \quad (5.7)$$

we have used before.[28] Thus the axial vector currents

$$\mathfrak{F}_{i\alpha}{}^5 = i\bar{b}\lambda_i\gamma_\alpha\gamma_5 b/2 \quad (5.8)$$

would be conserved if both L' and L'' were absent. Even in the presence of these terms, we have the commutation rules

$$[\mathfrak{F}_{i4}{}^5(\mathbf{x},t),\mathfrak{F}_{j4}(\mathbf{x}',t)] = -if_{ijk}\mathfrak{F}_{k4}{}^5(\mathbf{x},t)\delta(\mathbf{x}-\mathbf{x}') \quad (5.9)$$

and

$$[\mathfrak{F}_{i4}{}^5(\mathbf{x},t),\mathfrak{F}_{j4}{}^5(\mathbf{x}',t)] = -if_{ijk}\mathfrak{F}_{k4}(\mathbf{x},t)\delta(\mathbf{x}-\mathbf{x}') \quad (5.10)$$

at equal times, We may use the definition

$$F_i{}^5(t) \equiv -i\int \mathfrak{F}_{i4}{}^5 d^3x, \quad (5.11)$$

along with (4.15).

Just as we put $\mathbf{I}=\mathbf{L}_+ + \mathbf{L}_-$ and $\mathbf{D}=\mathbf{L}_+ - \mathbf{L}_-$ in the discussion following Eq. (3.16), so we now write

$$F_i(t) = F_i{}^+(t) + F_i{}^-(t),$$
$$F_i{}^5(t) = F_i{}^+(t) - F_i{}^-(t), \quad (5.12)$$

and it is seen that $F_i{}^+$ and $F_i{}^-$ separately obey the commutation rules

$$[F_i{}^\pm, F_j{}^\pm] = if_{ijk}F_k{}^\pm, \quad (5.13)$$

while they commute with each other:

$$[F_i{}^\pm, F_j{}^\pm] = 0. \quad (5.14)$$

Thus we are now dealing with the group $U(3)$ taken twice: $U(3)\times U(3)$. Factoring each $U(3)$ into $U(1)\times SU(3)$, we have[29] $U(1)\times U(1)\times SU(3)\times SU(3)$. Thus we have defined a left- and a right-handed baryon number and a left- and right-handed unitary spin.

The situation is just as in Sec. III, where we defined a left- and a right-handed isotopic spin and we could have defined a left- and a right-handed nucleon number. The left- and right-handed quantities are connected to each other by the parity operation P:

$$PF_i{}^\pm P^{-1} = F_i{}^\mp. \quad (5.15)$$

Now that we have constructed the mathematical apparatus of the group $U(3)\times U(3)$ and its algebra, we may inquire how the Hamiltonian density H behaves under the group, i.e., under commutation with the algebra.

In the model, there is, corresponding to (4.6), the formula

$$H = \bar{H} - L' - L'', \quad (5.16)$$

where $\bar{H}$ is the Hamiltonian density derived from the Lagrangian density $\bar{L}$ and is completely invariant under the group. Instead of defining u_0 as in Sec. III, let us put

$$u_0 = L' \propto \bar{b}\lambda_0 b. \quad (5.17)$$

We can easily see that by commutation of u_0 with F_i and $F_i{}^5$ $(i=0,\cdots,8)$ at equal times we obtain a set of eighteen quantities:

$$u_i \propto \bar{b}\lambda_i b,$$
$$v_i \propto -i\bar{b}\lambda_i\gamma_5 b. \quad (5.18)$$

In fact F_i acts like $\lambda_i/2$, $F_i{}^5$ like $\lambda_i\gamma_5/2$, u_i like $\beta\lambda_i$, and v_i like $-i\beta\gamma_5\lambda_i$. Thus we have at equal times[30]

$$[F_i,u_j] = if_{ijk}u_k,$$
$$[F_i,v_j] = if_{ijk}v_k,$$
$$[F_i{}^5,u_j] = -id_{ijk}v_k,$$
$$[F_i{}^5,v_j] = id_{ijk}u_k, \quad (5.19)$$

and the stronger relations

$$[\mathfrak{F}_{i4}(\mathbf{x},t),u_j(\mathbf{x}',t)] = -f_{ijk}u_k(\mathbf{x},t)\delta(\mathbf{x}-\mathbf{x}'), \text{ etc.} \quad (5.20)$$

for the densities. All indices run from 0 to 8.

Note that we can now express not only L' (which is defined to be u_0) but L'' as well, since by (4.7) it is proportional to u_8. We have, then,

$$H = \bar{H} - u_0 - cu_8, \quad (5.21)$$

where c is of the order $(m_{0N}-m_{0\Lambda})/m_{0N}$ in the model.

We may now make a series of abstractions from the model. First, we suppose that currents $\mathfrak{F}_{i\alpha}$ and $\mathfrak{F}_{i\alpha}{}^5$ are defined, with commutation rules (5.3), (5.9), and (5.10), and with the weak current given by the analog of (5.5)[31]:

$$J_\alpha = \mathfrak{F}_{1\alpha} + \mathfrak{F}_{1\alpha}{}^5 + i\mathfrak{F}_{2\alpha} + i\mathfrak{F}_{2\alpha}{}^5 + \mathfrak{F}_{4\alpha} + \mathfrak{F}_{4\alpha}{}^5 + i\mathfrak{F}_{5\alpha} + i\mathfrak{F}_{5\alpha}{}^5, \quad (5.22)$$

---

[28] Actually the Lagrangian (4.1) without the nucleon mass terms is invariant under a larger continuous group of transformations than the one $[U(3)\times U(3)]$ that we treat here. For example, there are infinitesimal transformations in which the baryon fields b acquire small terms in $\bar{b}$. Invariance under these is associated with the conservation of currents carrying baryon number 2. The author wishes to thank Professor W. Thirring for a discussion of these additional symmetries and of conformal transformations, which give still more symmetry.

[29] The groups $U(1)$, $SU(3)$, and $SU(2)$ cannot be further factored in this fashion. They are called *simple.*

[30] Note that even if we use just F_i and $F_i{}^5$ for $i=1,\cdots,8$, or $SU(3)\times SU(3)$ only, we still generate all eighteen u's and v's. [In the two-dimensional case described in Sec. III the situation is different. Using $SU(2)\times SU(2)$, we generate from u_0 only itself and v_1, v_2, v_3; if we then bring in $F_0{}^5$ as well, we obtain three more u's and one more v.] This remark is interesting because the group that gives currents known to be physically interesting is just $U(1)\times SU(3)\times SU(3)$; there is no known physical coupling to $\mathfrak{F}_{0\alpha}{}^5$, the axial vector baryon current.

[31] Note that the *total* weak current, whether for baryons and mesons or for leptons, is just a component of the current of an angular momentum. See reference 13.

while the electric current is given by (5.4). Next, we may take the Hamiltonian density to be of the form (5.21), with $\bar{H}$ invariant and u_i and v_i transforming as in (5.20). Then, if the theory is of the type described in Appendix A, we can calculate the divergences of the currents in terms of the equal-time commutators

$$\begin{aligned} \partial_\alpha \mathfrak{F}_{i\alpha} &= i[F_i, u_0] + ic[F_i, u_8], \\ \partial_\alpha \mathfrak{F}_{i\alpha}{}^5 &= i[F_i{}^5, u_0] + ic[F_i{}^5, u_8], \end{aligned} \tag{5.23}$$

or, explicitly,

$$\begin{aligned} &\partial_\alpha \mathfrak{F}_{i\alpha} = 0, \quad (i=0,1,2,3,8) \\ &\partial_\alpha \mathfrak{F}_{4\alpha} = (\tfrac{3}{2})^{\frac{1}{2}} u_5, \text{ etc.,} \\ &\partial_\alpha \mathfrak{F}_{0\alpha}{}^5 = (\tfrac{2}{3})^{\frac{1}{2}} v_0 + (\tfrac{2}{3})^{\frac{1}{2}} c v_8, \\ &\partial_\alpha \mathfrak{F}_{1\alpha}{}^5 = [(\tfrac{2}{3})^{\frac{1}{2}} + (\tfrac{1}{3})^{\frac{1}{2}} c] v_1, \text{ etc.,} \\ &\partial_\alpha \mathfrak{F}_{4\alpha}{}^5 = [(\tfrac{2}{3})^{\frac{1}{2}} - (\tfrac{1}{12})^{\frac{1}{2}} c] v_4, \text{ etc.,} \\ &\partial_\alpha \mathfrak{F}_{8\alpha}{}^5 = [(\tfrac{2}{3})^{\frac{1}{2}} - (\tfrac{1}{3})^{\frac{1}{2}} c] v_8 + (\tfrac{2}{3})^{\frac{1}{2}} c v_0. \end{aligned} \tag{5.24}$$

Finally, if we taken the model really seriously, we may abstract the equal-time commutation relations of the u_i and v_i as obtained by explicit commutation in the model.

The relations of Sec. III are all included in those of this section, except that what was called u_0 there is now called $u_0 + cu_8$ and what was called v_i is now called $[(\frac{2}{3})^{\frac{1}{2}} + (\frac{1}{3})^{\frac{1}{2}} c] v_i$ for $i=1, 2, 3$.

All of the relations used here are supposed to be exact and are not affected by the symmetry-breaking character of the non-invariant term in the Hamiltonian. In the next section, we discuss what happens if c can be regarded as small in any sense. We may then expect to see some trace of the symmetry under $U(3)$ that would obtain if c were 0 and L'' disappeared. In this limit, N and Λ are degenerate, and all the components F_i of the unitary spin are conserved. The higher symmetry would show up particularly through the existence of degenerate baryon and meson supermultiplets, which break up into ordinary isotopic multiplets when L'' is turned on. These supermultiplets have been discussed previously for baryons and pseudoscalar mesons[4,6] and then for vector mesons.[32–34]

We shall not discuss the case in which both L' and L'' are turned off; that is the situation, still more remote from reality, in which all the axial vector currents are conserved as well as the vector ones.

VI. BROKEN SYMMETRY—MESON SUPERMULTIPLETS

We know that because of isotopic spin conservation the baryons and mesons form degenerate isotopic multiplets, each corresponding to an irreducible representation of the isotopic spin algebra (3.1). Each multiplet has $2I+1$ components, where the quantum number I distinguishes one representation from another and gives us the eigenvalue $I(I+1)$ of the operator $\sum_{i=1}^3 I_i^2$, which commutes with all the elements of the isotopic spin group. The operators I_i are represented, within the multiplet, by Hermitian $(2I+1)\times(2I+1)$ matrices having the commutation rules (3.1) of the algebra.

If we start from the doublet representation, we can build up all the others by considering combinations of particles that transform like the original doublet. Just as (p,n) form a doublet representation for which the I_i are represented by $\tau_i/2$, the antiparticles $(\bar{n}, -\bar{p})$ also form a doublet representation that is equivalent. (Notice the minus sign on the antiproton state or field.) Now, if we put together a nucleon and an antinucleon, we can form the combination

$$\bar{N}N = \bar{p}p + \bar{n}n,$$

which transforms like an isotopic singlet, or the combinations

$$\bar{N}\tau_i N, \quad (i=1,2,3)$$

which form an isotopic triplet. The direct product of nucleon and antinucleon doublets gives us a singlet and a triplet. Any meson that can dissociate virtually into nucleon and antinucleon must be either a singlet or a triplet. For the singlet state, the components I_i are all zero, while for the three triplet states the three 3×3 matrices, $I_i{}^{jk}$ of the components I_i, are given by

$$I_i{}^{jk} = -ie_{ijk}. \tag{6.1}$$

Now let us generalize these familiar results to the unitary spin and the three basic baryons b (comprising n, p, and Λ). These three fields or particles form a three-dimensional irreducible representation of the unitary spin algebra (4.16) from which all the other representations may be constructed.

For example, consider a meson that can dissociate into b and $\bar{b}$. It must transform either like

$$\bar{b}b = \bar{p}p + \bar{n}n + \bar{\Lambda}\Lambda,$$

a unitary singlet, or else like

$$\bar{b}\lambda_i b, \quad (i=1,\cdots,8)$$

a unitary octet.

The unitary singlet is evidently neutral, with strangeness $S=0$, and forms an isotopic singlet. But how does the unitary octet behave with respect to isotopic spin? We form the combinations

$$\left.\begin{aligned} &\bar{b}(\lambda_1 - i\lambda_2)b/2 = \bar{n}p, \\ &\bar{b}\lambda_3 b/\sqrt{2} = (\bar{p}p - \bar{n}n)/\sqrt{2}, \\ &\bar{b}(\lambda_1 + i\lambda_2)b/2 = \bar{p}n, \end{aligned}\right\} \; I=1, S=0$$
$$\left.\begin{aligned} &\bar{b}(\lambda_4 - i\lambda_5)b/2 = \bar{\Lambda}p, \\ &\bar{b}(\lambda_6 - i\lambda_7)b/2 = \bar{\Lambda}n, \end{aligned}\right\} \; I=\tfrac{1}{2}, S=+1 \tag{6.2}$$
$$\left.\begin{aligned} &\bar{b}(\lambda_4 + i\lambda_5)b/2 = \bar{p}\Lambda, \\ &\bar{b}(\lambda_6 + i\lambda_7)b/2 = \bar{n}\Lambda, \end{aligned}\right\} \; I=\tfrac{1}{2}, S=-1$$
$$\bar{b}\lambda_8 b/\sqrt{2} = (\bar{p}p + \bar{n}n - 2\bar{\Lambda}\Lambda)/\sqrt{6}, \quad I=0, S=0,$$

[32] M. Gell-Mann, California Institute of Technology Synchrotron Laboratory Report No. CTSL-20, 1961 (unpublished).
[33] Y. Ne'eman, Nuclear Phys. 26, 222 (1961).
[34] A. Salam and J. C. Ward, Nuovo cimento 20, 419 (1961).

and we see immediately that the unitary octet comprises an isotopic triplet with $S=0$, a pair of isotopic doublets with $S=\pm 1$, and an isotopic singlet with $S=0$. All these are degenerate only in the limit of unitary symmetry ($L''=0$); when the mass-splitting term is turned on, the singlet, the triplet, and the pair of doublets should have three somewhat different masses.

The known pseudoscalar mesons (π, K, and $\bar{K}$) fit very well into this picture, provided there is an eighth pseudoscalar meson to fill out the octet. Let us call the hypothetical isotopic singlet pseudoscalar meson χ^0. Since it is pseudoscalar, it cannot dissociate (virtually or really) into 2π. It has the value $+1$ for the quantum number G, so that it cannot dissociate into an odd number of pions either. Thus in order to decay by means of the strong interactions, it must have enough energy to yield 4π. It would then appear as a 4π resonance. The decay into 4π is, however, severely hampered by centrifugal barriers.

If the mass of χ^0 is too low to permit it to decay readily into 4π, then it will decay electromagnetically. If there is sufficient energy, the decay mode $\chi^0 \to \pi^+ + \pi^- + \gamma$ is most favorable; otherwise[34a] it will decay into 2γ like π^0.

Let us now turn to the vector mesons. The best known vector meson is the $I=1, J=1^-$ resonance of 2π, which we shall call ρ. It has a mass of about 750 Mev. According to our scheme, it should belong, like the pion, to a unitary octet. Since it occupies the same position as the π ($I=1, S=0$), we denote it by the succeeding letter of the Greek alphabet.

The vector analog of χ^0 we shall call ω^0 (skipping the Greek letter ψ). It must have $I=0$, $J=1^-$, and $G=-1$ and so it is capable of dissociation into $\pi^+ + \pi^- + \pi^0$. Presumably it is the 3π resonance found experimentally[35] at about 790 Mev.

In order to complete the octet, we need a pair of strange doublets analogous to K and $\bar{K}$. In the vector case, we shall call them M and $\bar{M}$ (skipping the letter L). Now there is a known $K\pi$ resonance with $I=\frac{1}{2}$ at about 884 Mev. If it is a p-wave resonance, then it fits the description of M perfectly.

In the limit of unitary symmetry, we can have, besides the unitary octet of vector mesons, a unitary singlet. The hypothetical B^0 that we discussed in Sec. III would have such a character. If B^0 exists, then the turning-on of the mass-splitting term L'' mixes the states B^0 and ω^0, which are both *isotopic* singlets.

Other mesons may exist besides those discussed, for example, scalar and axial vector mesons. All those that can associate into $b+\bar{b}$ should form unitary octets or singlets or both, with each octet splitting into isotopic multiplets because of the symmetry-breaking term L''.

TABLE III. Possible meson octets and singlets.

Unitary spin	Isotopic spin	Strangeness	Pseudoscalar	Vector	Scalar	Axial vector
Octet	1	0	π	ρ	π'	ρ'
	1/2	+1	K	M	K'	M'
	1/2	−1	$\bar{K}$	$\bar{M}$	$\bar{K}'$	$\bar{M}'$
	0	0	χ	ω	χ'	ω'
Singlet	0	0	A	B	A'	B'

A list of some possible meson states is given in Table III, along with suggested names for the mesons.

It is interesting that we can predict not only the degeneracy of an octet in the limit $L'' \to 0$ but also a sum rule[32] that holds in first order in L'':

$$(m_K+m_{\bar{K}})/2=(3m_\chi+m_\pi)/4,$$
$$(m_M+m_{\bar{M}})/2=(3m_\omega+m_\rho)/4. \tag{6.3}$$

If M is at about 884 Mev and ρ at about 750 Mev, then ω should lie at about 930 Mev according to the sum rule; since it is actually at 790 Mev, the sum rule does not seem to give a good description of the splitting. Perhaps an important effect is the repulsion between the ω^0 and B^0 levels, pushing ω^0 down and B^0 up. For what it is worth, (6.3) gives a χ^0 mass of around 610 Mev.

In the limit of unitary symmetry, not only are the supermultiplets degenerate but their effective couplings are symmetrical. For example, the effective coupling of the unitary pseudoscalar octet to N and Λ takes the form

$$ig_1\{\bar{N}\boldsymbol{\tau}\gamma_5 N\cdot\boldsymbol{\pi}+\bar{N}\gamma_5\Lambda K+\bar{\Lambda}\gamma_5 N\bar{K}$$
$$+3^{-\frac{1}{2}}\bar{N}\gamma_5 N\chi-2\times 3^{-\frac{1}{2}}\bar{\Lambda}\gamma_5\Lambda\chi\}, \tag{6.4}$$

in terms of renormalized "fields." Now, as the term L'' is turned on, the various coupling constants become unequal; instead of calling them all g_1, we refer to them as $g_{NN\pi}$, $g_{N\Lambda K}$, $g_{NN\chi}$, and $g_{\Lambda\Lambda\chi}$, respectively, each of these constants being the measurable renormalized coupling parameter at the relevant pole.

We have written the effective coupling (6.4) as if there were renormalized fields for all the particles involved, but that is only a matter of notation; the mesons can perfectly well be composite. We may simplify the notation still further by constructing a traceless 3×3 matrix Π containing the pseudoscalar "fields" in such a way that (6.4) becomes

$$ig_1\bar{b}\Pi\gamma_5 b. \tag{6.5}$$

We may now write, in a trivial way, other effective couplings in the limit of unitary symmetry. We define a traceless 3×3 matrix W_α containing the "fields" for the vector meson octet just as Π^* contains those for the pseudoscalar octet. We then have the invariant

[34a] *Note added in proof.* H. P. Duerr and W. Heisenberg (preprint) have pointed out the importance of the decay mode $\chi^0 \to 3\pi$ induced by electromagnetism. For certain χ masses, it may be a prominent mode.

[35] B. C. Maglić, L. W. Alvarez, A. H. Rosenfeld, and M. L. Stevenson, Phys. Rev. Letters 7, 178 (1961).

effective coupling

$$i\gamma_1 \,\mathrm{Tr} W_\alpha(\Pi\partial_\alpha\Pi - \partial_\alpha\Pi\Pi)/2 \tag{6.6}$$

in the symmetric limit. When the asymmetry is turned on, the single coupling parameter γ_1 is replaced by the set of different parameters $\gamma_{\rho\pi\pi}$, $\gamma_{\rho KK}$, $\gamma_{\omega KK}$, $\gamma_{MK\pi}$, and $\gamma_{MK\chi}$.

In the same way, we have another effective coupling

$$ih_1 \,\mathrm{Tr}\Pi(\partial_\alpha W_\beta - \partial_\beta W_\alpha)(\partial_\gamma W_\delta - \partial_\delta W_\gamma)e_{\alpha\beta\gamma\delta} \tag{6.7}$$

in the symmetric limit; in the actual asymmetric case, we define the distinct constants $h_{\pi\omega\rho}$, $h_{\pi MM}$, $h_{\chi\omega\omega}$, $h_{\chi\rho\rho}$, $h_{\chi MM}$, $h_{KM\rho}$, and $h_{KM\omega}$. All of these constants can be measured, in principle, in "pole" experiments, except that for the broad resonances like ρ the poles are well off the physical sheet.

We have generalized the definitions of constants like $g_{NN\pi}$ and $\gamma_{\rho\pi\pi}$, as used in Sec. II, to other particles. The constants γ_ρ and f_π of Sec. II also have analogs, of course, and we define γ_ω, f_K, etc., in the obvious way. In the limit of unitary symmetry, of course, we would have $f_\pi = f_K = f_\chi$ and $\gamma_\rho = \gamma_\omega = \gamma_M$. Likewise, the constant $-G_A/G$ for nucleon β decay would equal the corresponding quantity $-G_A{}^{\Lambda N}/G$ for the β decay of Λ.

VII. BROKEN SYMMETRY—BARYON SUPERMULTIPLETS

What has been done in the previous section may be described mathematically as follows. We considered a three-dimensional representation of the unitary spin algebra (4.16) or of the group $SU(3)$ that is generated by the algebra. It is the representation to which b belongs (that is, n, p, and Λ) and we may denote it by the symbol **3**.

The antiparticles $\bar{b}$ belong to the conjugate representation **3***, which is inequivalent[36] to **3**. We have then taken the direct product $\mathbf{3}\times\mathbf{3}^*$ and found it to be given by the rule

$$\mathbf{3}\times\mathbf{3}^* = \mathbf{8}+\mathbf{1}, \tag{7.1}$$

where **8** is the octet representation and **1** the singlet representation of unitary spin. Each of these is its own conjugate; that is a situation that occurs only when the dimension is the cube of an integer.

There are, of course, more complicated representations to which mesons might belong that are incapable (in the limit of unitary symmetry) of dissociation into $b+\bar{b}$ but capable of dissociation into $2b+2\bar{b}$ or higher configurations. But we might guess that at least the mesons of lowest mass would correspond to the lowest configurations.

Now we want to examine the simplest configurations for baryons, apart from just b. Evidently the next simplest is $2b+\bar{b}$, which poses the problem of reducing the direct product $\mathbf{3}\times\mathbf{3}\times\mathbf{3}^*$; the result is the following:

$$\mathbf{3}\times\mathbf{3}\times\mathbf{3}^* = \mathbf{3}\times\mathbf{1}+\mathbf{3}\times\mathbf{8} = \mathbf{3}+\mathbf{3}+\mathbf{6}+\mathbf{15}. \tag{7.2}$$

The six-dimensional representation **6** is composed of an isotopic triplet with $S=-1$, a doublet with $S=0$, and a singlet with $S=+1$; the fifteen-dimensional representation **15** is composed of a doublet with $S=-2$, a singlet and a triplet with $S=-1$, a doublet and a quartet with $S=0$, and a triplet with $S=+1$.

According to the scheme, then, Ξ should belong to **15**. Where are the other members of the supermultiplet? For $S=-1$ and $S=0$, there are many known resonances, some of which might easily have the same spin and parity as Ξ. For $S=+1$, $I=1$, however, no resonance has been found so far (in K^+-p scattering, for example).

The hyperon Σ should also be placed in a supermultiplet, which may or may not be the same one to which Ξ belongs; we do not know if the spin and parity of Σ and Ξ are the same, with K taken to be pseudoscalar. If Σ belongs to **6** in the limit of unitary symmetry, then there should be a KN resonance in the $I=0$ state.

It is difficult to say at the present time if the baryon states can be reconciled with the model. Further knowledge of the baryon resonances is required.

One curious possibility is that the fundamental objects b are hidden and that the physical N and Λ, instead of belonging to **3**, belong, along with Σ and Ξ, to the representation **15** in the limit of unitary symmetry. That would require the spins and parities of N, Λ, Σ, and Ξ to be equal, and it would require a πN resonance in the $p_{\frac{1}{2}}$, $I=\frac{3}{2}$ state as well as a KN resonance in the $p_{\frac{1}{2}}$, $I=1$ state to fill out the supermultiplet.

VIII. THE "EIGHTFOLD WAY"

Unitary symmetry may be applied to the baryons in a more appealing way if we abandon the connection with the symmetrical Sakata model and treat unitary symmetry in the abstract. (An abstract approach is, of course, required if there are no "elementary" baryons and mesons.) Of all the groups that could be generated by the vector weak currents, $SU(3)$ is still the smallest and the one that most naturally gives rise to the rules $|\Delta I| = \frac{1}{2}$ and $\Delta S/\Delta Q = 0, +1$.

There is no longer any reason for the baryons to belong to the **3** representation or the other spinor representations of the group $SU(3)$; the various irreducible spinor representations are those obtained by reducing direct products like $\mathbf{3}\times\mathbf{3}\times\mathbf{3}^*$, $\mathbf{3}\times\mathbf{3}\times\mathbf{3}\times\mathbf{3}^*\times\mathbf{3}^*$, etc.

Instead, the baryons may belong, like the mesons, to representations such as **8** or **1** obtained by reducing the direct products of equal numbers of **3**'s and **3***'s. It is then natural to assign the stable and metastable baryons N, Λ, Σ, and Ξ to an octet, degenerate in the limit of unitary symmetry. We thus obtain the scheme

[36] In other words, no unitary transform can convert the representations 3 and 3* into each other. That is easy to see, since the eigenvalues of λ_8 are opposite in sign for the two representations, and changing the signs changes the set of eigenvalues. In the case of the group $SU(2)$ of isotopic spin transformations, the basic spinor representation $I=\frac{1}{2}$ is equivalent to the corresponding antiparticle representation.

of Gell-Mann[32] and Ne'eman[33] that we call the "eightfold way." The component F_8 of the unitary spin is now $(\sqrt{3}/2)Y$, where Y is the hypercharge (equal to strangeness plus baryon number).

The baryons of the octet must have the same spin and parity (treating K as pseudoscalar). To first order in the violation of unitary symmetry, the masses should obey the sum rule analogous to (6.3):

$$(m_N+m_\Xi)/2=(3m_\Lambda+m_\Sigma)/4, \qquad (8.1)$$

which agrees surprisingly well with observations, the two sides differing by less than 20 Mev.

To form mesons that transform like combinations of these baryons and their antiparticles, we reduce the direct product 8×8 (remembering that $8=8^*$) and obtain

$$8\times 8=1+8+8+10+10^*+27, \qquad (8.2)$$

where **1** and **8** are the singlet and octet representations already discussed; **10** consists of an isotopic triplet with $Y=0$, a doublet with $Y=-1$, a quartet with $Y=+1$, and a singlet with $Y=-2$; **10*** has the opposite behavior with respect to Y; and **27** consists of an isotopic singlet, triplet, and quintet with $Y=0$, a pair of doublets with $Y=\pm 1$, a pair of quartets with $Y=\pm 1$, and a pair of triplets with $Y=\pm 2$. Evidently the known mesons are to be assigned to octets and perhaps singlets, as in Sec. VI. The meson-nucleon scattering resonances must then also be assigned representations among those in (8.2); the absence so far of any observed structure in K-N scattering makes it difficult to place the $I=3/2$, $J=3/2$, π-N resonance in a supermultiplet.

The fact that **8** occurs twice in Eq. (8.2) means that there are two possible forms of symmetrical Yukawa coupling of a meson octet to the baryon octet in the limit of unitary symmetry. As in Sec. VI for the mesons, we form a 3×3 traceless matrix out of the formal "fields" of the baryon octet; call it $\mathcal{B}$. The effective symmetrical coupling of pseudoscalar mesons may then be written as

$$ig_1\alpha \operatorname{Tr}(\bar{\mathcal{B}}\pi\gamma_5\mathcal{B}+\pi\bar{\mathcal{B}}\gamma_5\mathcal{B})/2 + ig_1(1-\alpha)\operatorname{Tr}(\bar{\mathcal{B}}\pi\gamma_5\mathcal{B}-\pi\bar{\mathcal{B}}\gamma_5\mathcal{B})/2. \qquad (8.3)$$

The two types of coupling differ in their behavior under the operation R that exchanges N and Ξ, K and $\bar{K}$, M and $\bar{M}$, etc.; the first term is symmetric while the second is antisymmetric under R. The parameter α just specifies how much of each effective coupling is presented in the limit of unitary symmetry. When we take into account violations of the symmetry, we must define separate coupling constants $g_{NN\pi}$, $g_{NK\Lambda}$, etc., in a suitable way.

Likewise the vector mesons have the general symmetrical coupling

$$i\gamma_1\beta \operatorname{Tr}(\bar{\mathcal{B}}W_\alpha\gamma_\alpha\mathcal{B}+W_\alpha\bar{\mathcal{B}}\gamma_\alpha\mathcal{B}) + i\gamma_1(1-\beta)\operatorname{Tr}(\bar{\mathcal{B}}W_\alpha\gamma_\alpha\mathcal{B}-W_\alpha\bar{\mathcal{B}}\gamma_\alpha\mathcal{B}), \qquad (8.4)$$

where we ignore Pauli moment terms for simplicity. To the extent that the vector meson octet W_α dominates the dispersion relations for the unitary spin current $\mathfrak{F}_{i\alpha}$, then the mesons of W_α are coupled effectively to the components of $\mathfrak{F}_{i\alpha}$, and we have $\beta=0$ in (8.5). Then ρ is effectively coupled to the isotopic spin current, ω to the hypercharge current, and M to the strangeness-changing vector current. The first two of these currents are conserved, and so we have the approximate universality of ρ and ω couplings proposed by Sakurai[11] and discussed in Sec. II. In the limit of unitary symmetry, under the assumptions just mentioned, ρ is effectively coupled to the current of $2\gamma_1\mathbf{I}$ and ω to the current of $2\gamma_1 F_8=\sqrt{3}\gamma_1 Y$.

The electromagnetic current is now given by the formula

$$j_\alpha=\mathfrak{F}_{3\alpha}+3^{-\frac{1}{2}}\mathfrak{F}_{8\alpha} \qquad (8.5)$$

instead of (5.4), while the weak vector current is still described by Eq. (5.5). If we are to treat the vector and axial vector currents by means of $SU(3)\times SU(3)$, as we did earlier, then the entire weak current is given by (5.22) and we have the commutation rules (5.3), (5.9), and (5.10) for the various components of the currents. The question of the behavior of H under the group $SU(3)\times SU(3)$ should, however, be re-examined for the eightfold way; we shall not go into that question here. But let us consider how the baryon octet transforms in the limit of conserved vector *and* axial vector currents [invariance under $SU(3)\times SU(3)$]. In the Sakata model, the left-handed baryons transformed under (F_i^+,F_j^-) like (3, 1), while the right-handed baryons transformed according to (1, 3). For the eightfold way, there are two simple possibilities for these transformation properties. Either we have (8, 1) and (1, 8) or else we adjoin a ninth neutral baryon (which need not be degenerate with the other eight in the limit of conserved *vector* currents and which need not have the same parity) and use the transformation properties (3, 3*) and (3*, 3). In the first case, the baryons transform like the quantities $\mathfrak{F}_{i\alpha}$ and $\mathfrak{F}_{i\alpha}{}^5$ $(i=1,\cdots,8)$ and in the second case they transform like u_i and v_i $(i=0,\cdots,8)$ of Sec. V.

IX. REMARKS AND SUGGESTIONS

Our approach to the problem of baryon and meson couplings leads to a number of suggestions for new investigations, both theoretical and experimental.

First, the equal-time commutation relations for currents and densities lead to exact sum rules for the weak and electromagnetic matrix elements. As an example, take the commutation rules (3.5) for the isotopic spin current. These do not, of course, depend on any higher symmetry, but they can be used to illustrate the results that can be obtained from the more general relations like (5.3).

Consider the electromagnetic form factor $F_\pi(s)$ of the charged pion, which is just the form factor of the isotopic spin current between one-pion states. Let p and p' be the initial and final pion four-momenta, with $s=-(p-p')^2$. Let K be any four-momentum with $K^2=-m_\pi^2$. Then, taking the matrix element of (3.5) between one-pion states, we obtain the result

$$2(p_0+p_0')K_0F_\pi(-(p-p')^2)$$
$$=(p_0+K_0)(p_0'+K_0)F_\pi(-(p-K)^2)F_\pi(-(p'-K)^2)$$
$$-(p_0-K_0)(p_0'-K_0)F_\pi(-(p+K)^2)F_\pi(-(p'+K)^2)$$
$$+\text{inelastic terms}, \quad (9.1)$$

where the inelastic terms come from summing over bilinear forms in the inelastic matrix elements of the current. We see that if there is no inelasticity the form factor is unity. Thus the departure from unity of $F_\pi(s)$ is related to the amount of inelasticity.

A similar relation is familiar in nonrelativistic quantum mechanics:

$$\langle e^{i(\mathbf{p}-\mathbf{p}')\cdot\mathbf{x}}\rangle_{00}=\sum_n\langle e^{i(\mathbf{p}-\mathbf{k})\cdot\mathbf{x}}\rangle_{0n}\langle e^{i(\mathbf{k}-\mathbf{p}')\cdot\mathbf{x}}\rangle_{n0}. \quad (9.2)$$

If we apply relations like (9.1) to the matrix elements of non-conserved currents like $\mathbf{P}_\alpha$, along with the linear homogeneous dispersion relations for these matrix elements, we can in principle determine constants like $-G_A/G$.

A second line of theoretical investigation is suggested by the vanishing at high momentum transfer of matrix elements of divergences of non-conserved currents, like $\partial_\alpha\mathbf{P}_\alpha$. We should try to find limits involving high energies and high momentum transfers in which we can show that the conservation of helicity, unitary spin, etc., becomes valid. A preliminary effort in this direction has been made by Gell-Mann and Zachariasen.[37]

A third topic of study is the testing of broken symmetry at low energy. Do the mesons fall into unitary octets and singlets? An experimental search for χ^0 is required and also a determination of the spin of K^* at 884 Mev to see if it really is our M meson.

Let us discuss briefly the properties of χ^0. An $I=0$ state of 4π can have two types of symmetry: either totally symmetric (partition [4]) in both space and isotopic spin or else the symmetry of the partition [2+2] in space and in isotopic spin. For a pseudoscalar state, the first type of wave function in momentum space is very complicated. If $\mathbf{p}$, $\mathbf{q}$, and $\mathbf{r}$ are the three momentum differences, it must look like

$$\mathbf{p}\cdot\mathbf{q}\times\mathbf{r}(E_1-E_2)(E_2-E_3)(E_3-E_4)$$
$$\times(E_1-E_3)(E_1-E_4)(E_2-E_4),$$

times a symmetric function of the energies E_1, E_2, E_3, E_4 of the four pions. On the basis of any reasonable dynamical picture of χ^0, such a wave function should have a very small amplitude. In particular, dispersion theory suggests that the wave function of χ^0 should have large contributions from virtual dissociation into 2ρ, which gives a wave function with [2+2] symmetry.

[37] M. Gell-Mann and F. Zachariasen, Phys. Rev. 123, 1065 (1961).

If [2+2] predominates, then the charge ratio in decay is 2:1 in favor of $2\pi^0+\pi^++\pi^-$ over $2\pi^++2\pi^-$, with $4\pi^0$ absent. If virtual dissociation into 2ρ actually predominates, then the matrix element of the 4π configuration is easily written down and the spectrum of the decay $\chi^0\rightarrow4\pi$ can be calculated.

If χ^0 is lighter than 4π, it will, of course, decay electromagnetically. Even if it is above threshold for 4π, however, the matrix element for decay contains so many powers of pion momenta that electromagnetic decay should be appreciable over a large range of masses. The branching ratio $(\pi^++\pi^-+\gamma)/(4\pi)$ is approximately calculable by dispersion methods. In both cases χ^0 first dissociates into 2ρ. Then either both virtual ρ mesons decay into 2π, or else (in the case where both are neutral) one may decay into $\pi^++\pi^-$, while the other turns directly into γ. If we draw a diagram for such a process, then the constant $\gamma_{\rho\pi\pi}$ is inserted whenever we have a $\rho\pi\pi$ vertex and the constant $em_\rho^2/2\gamma_\rho$ at a ρ-γ junction.[10]

If the meson spectrum is consistent with broken unitary symmetry, we should examine the baryons, and see whether the various baryon states fit into the representations **3**, **6**, and **15** (or the representations **1**, **8**, **10**, **10***, and **27** that arise in the alternative form of unitary symmetry).

If some states are lacking in a given supermultiplet, it does not necessarily prove that the broken symmetry is wrong, but only that it is badly violated. We assume that baryon isobars like the $\pi N\ \frac{3}{2},\frac{3}{2}$ resonance are dynamical in nature; there may be some attractive and some repulsive forces in this channel, and the attractive ones have won out, producing the resonance. In the KN channel with $I=1$, for example, it is conceivable that the repulsive ones are stronger (because of symmetry violation), and the analogous $p_{\frac{3}{2}}$ resonance disappears. In such a situation, the concept of broken symmetry at low energies is evidently of little value.

Suppose, however, that the idea of broken unitary symmetry is confirmed for both mesons and baryons, say according to the Sakata picture, in which N and Λ belong to the representation **3** in the limit of unitary symmetry. There are, nevertheless, gross violations of unitary symmetry, and the elucidation of these, both theoretical and experimental, is a fourth interesting subject.

If unitary symmetry were exact, then not only would m_K/m_π equal unity, instead of about 3.5, but f_K^2/f_π^2 would be 1 instead of about 6, and $3G_A^2+G_V^2$ for the β decay of Λ would be equal to $3G_A^2+G_V^2$ for the nucleon instead of being 1/15 as large. All these huge departures from unity represent very serious violations of unitary symmetry.

Yet the relatively small mass difference of N and Λ compared to their masses would seem to indicate, if our model is right, that the constant c in Eq. (5.21) is considerably smaller than unity. It is conceivable that the large mass ratio of K to π comes about because the total mass of the system is so small. It is possible that even with a fairly small c (say $\sim -\frac{1}{10}$) we might explain the gross violations of unitary symmetry. We might try to interpret the large values of $g_{NN\pi}^2/g_{NK\Lambda}^2$, f_K^2/f_π^2, etc., in terms of the large value of m_K^2/m_π^2.

An example of such a calculation, and one that illustrates the various methods suggested in this article, is the following. We try to calculate f_K^2/f_π^2 in terms of m_K^2/m_π^2.

Consider the following vacuum expectation value, written in parametric representation:

$$\langle[\mathfrak{F}_{1\alpha}{}^5(x),\partial_\beta\mathfrak{F}_{1\beta}{}^5(x')]\rangle_0=i/(2\pi)^3\int d^4K\, e^{iK\cdot(x-x')}$$
$$\times K_\alpha\epsilon(K)\int dM^2/M^2\delta(K^2+M^2)\rho(M^2). \quad (9.3)$$

Here x and x' are arbitrary space-time points. In terms of (9.3), we have

$$\langle[\partial_\alpha\mathfrak{F}_{1\alpha}{}^5(x),\partial_\beta\mathfrak{F}_{1\beta}{}^5(x')]\rangle=1/(2\pi)^3\int d^4K\, e^{iK\cdot(x-x')}$$
$$\times\epsilon(K)\int dM^2\,\delta(K^2+M^2)\rho(M^2). \quad (9.4)$$

Now the contribution of the one-pion intermediate state is easily obtained in terms of the constant f_π^2:

$$\rho(M^2)=\delta(M^2-m_\pi^2)m_\pi^4/4f_\pi^2+\text{higher terms}. \quad (9.5)$$

If $\int\rho(M^2)dM^2/M^2$ converges and if the one-pion term dominates, we have

$$\int\rho(M^2)dM^2/M^2\approx m_\pi^2/4f_\pi^2. \quad (9.6)$$

But from (9.3) we can extract the expectation value of the equal-time commutator of the fourth component of $\mathfrak{F}_{1\alpha}{}^5$ with $\partial_\beta\mathfrak{F}_{1\beta}{}^5$; making use of (5.20) and (5.24), we can express the result in terms of $\langle u_0\rangle$ and $\langle u_8\rangle$. Thus we find

$$\int\rho(M^2)dM^2/M^2=[(2/3)^{\frac{1}{2}}+(1/3)^{\frac{1}{2}}c]$$
$$\times[(2/3)^{\frac{1}{2}}\langle u_0\rangle_0+(1/3)^{\frac{1}{2}}\langle u_8\rangle_0], \quad (9.7)$$

assuming convergence.

Now we can do exactly the same thing for $\mathfrak{F}_{4\alpha}{}^5$ and the K meson, obtaining, in place of the formula

$$m_\pi^2/4f_\pi^2\approx[(\tfrac{2}{3})^{\frac{1}{2}}+(\tfrac{1}{3})^{\frac{1}{2}}c][(\tfrac{2}{3})^{\frac{1}{2}}\langle u_0\rangle_0+(\tfrac{1}{3})^{\frac{1}{2}}\langle u_8\rangle_0], \quad (9.8)$$

the analogous result

$$m_K^2/4f_K^2\approx[(\tfrac{2}{3})^{\frac{1}{2}}-(\tfrac{1}{12})^{\frac{1}{2}}c][(\tfrac{2}{3})^{\frac{1}{2}}\langle u_0\rangle_0-(\tfrac{1}{12})^{\frac{1}{2}}\langle u_8\rangle_0]. \quad (9.9)$$

If c is really small, presumably $\langle u_8\rangle_0$ is also small compared to $\langle u_0\rangle_0$. Then we can, roughly, set (9.8) equal to (9.9), obtaining

$$f_K^2/f_\pi^2\approx m_K^2/m_\pi^2. \quad (9.10)$$

The left-hand side is about 6 and the right-hand side about 10. Thus we can, in a crude approximation, calculate the rate of $K^+\to\mu^++\nu$ in terms of that for $\pi^+\to\mu^++\nu$ and explain one large violation of symmetry in terms of another.

The Goldberger-Treiman formula relating f_π, $g_{NN\pi}$, and $(-G_A/G)$ can also be used for the K particle to give a relation among f_K, $g_{N\Lambda K}$, and $(-G_A/G)$ for the β decay of Λ. Of course, the K-particle pole is much closer to the branch line beginning at $(m_K+2m_\pi)^2$ than the pion pole is to the branch line beginning at $9m_\pi^2$; thus the Goldberger-Treiman formula may be quite bad for the K meson. Still, we may try to use it to discuss the coupling of N and Λ to K and to leptons. We have

$$(m_N+m_\Lambda)(-G_A{}^{N\Lambda}/G)\approx g_{N\Lambda K}/f_K, \quad (9.11)$$

by analogy with (2.8). Comparing the two formulas, we have

$$(-G_A{}^{N\Lambda}/G)^2(-G_A/G)^{-2}$$
$$\approx g_{N\Lambda K}{}^2g_{NN\pi}{}^{-2}(2m_Nf_\pi)^2[(m_\Lambda+m_N)f_K]^{-2}. \quad (9.12)$$

The ratio of g^2 factors is thought to be $\sim$0.1 from photoproduction of K, while the remaining factor on the right is also $\sim$0.1, so that the Goldberger-Treiman relation leads us to expect a very small axial vector β-decay rate for the Λ, much smaller than the observed one. The observed β decay would be nearly all vector; this prediction of the Goldberger-Treiman formula can easily be checked by observing the electron-neutrino angular correlation in the β decay of Λ, using bubble chambers.

We should, of course, try to predict the value of $g_{N\Lambda K}{}^2g_{NN\pi}{}^{-2}$ in terms of m_K^2/m_π^2 just as we did above for f_K^2/f_π^2; however, it is a much harder problem.

When we know more about the coupling constants of the vector mesons (strong coupling constants such as $\gamma_{\omega NN}$, $h_{\omega\pi\rho}$, etc., and coupling strengths of currents such as γ_ω, γ_M, etc.) we will be able to make a survey of the pattern of coupling constants as well as the pattern of masses and see whether the higher symmetry has any relevance. Also it should become clear how well the approximation of dominant low-mass states works, in terms of universality of meson couplings and Goldberger-Treiman relations.[38]

In summary, then, we suggest the use of the equal-time commutators to predict sum rules, attempts to derive high-energy conservation laws and to check them

[38] An interesting relation of the Goldberger-Treiman type is one that holds if the trace $\theta_{\alpha\alpha}$ of the stress-energy-momentum tensor has matrix elements obeying highly convergent dispersion relations. Because of the vanishing of the self-stress, the expectation value of $\theta_{\alpha\alpha}$ in the state of a particle at rest gives the mass of the particle. Rewriting the matrix element as one between the vacuum and a one-pair state, we see that the dispersion relation involves intermediate states with $I=0$, $J=0^+$, $G=+1$. If there is a resonance or quasi-resonance in this channel (like the k' meson of Table III) and if that resonance dominates the dispersion relation at low momentum transfers, then the coupling of the resonant state to different particles is roughly proportional to their masses. That is just the situation discussed by Schwinger in reference 1 and by Gell-Mann and Lévy in reference 17 for the "σ meson."

experimentally, the search for broken symmetry at low energies, attempts to calculate some violations in terms of others, and efforts to check the highly convergent dispersion relations dominated by low-mass states.

Nowhere does our work conflict with the program of Chew *et al.* of dynamical calculation of the S matrix for strong interactions, using dispersion relations. If something like the Sakata model is correct, then most of the mesons are dynamical bound states or resonances, and their properties are calculable according to the program. Those particles for which there are fundamental fields (like n, p, Λ, and B^0 in the specific field-theoretic model) would presumably occur as CDD poles or resonances in the dispersion relations.[39]

If there are no fundamental fields and no CDD poles, all baryons and mesons being bound or resonant states of one another, models like that of Sakata will fail; the symmetry properties that we have abstracted can still be correct, however. This situation would presumably differ in two ways[10] from the one mentioned above. First, all the masses and coupling constants could be calculated from coupled dispersion relations. Second, certain scattering amplitudes at high energies would show different behavior, corresponding to different kinds of subtractions in the dispersion relations. The second point should be investigated further, as it could lead to experimental tests of the "fundamental" character of various particles.[10,40]

ACKNOWLEDGMENTS

It is a pleasure to thank R. P. Feynman, S. L. Glashow, and R. Block for many stimulating discussions of symmetry, and to acknowledge the great value of conversations with G. F. Chew, S. Frautschi, R. Haag, R. Schroer, and F. Zachariasen about the explanation of approximate universality in terms of highly convergent dispersion relations.

APPENDIX

The field theories of the Fermi-Yang and Sakata models, given by Eqs. (3.17) and (4.1), respectively, belong to a general class of theories, which we now describe.

The Lagrangian density L is given as a function of a number of fields ψ_A and their gradients. The kinetic part of the Lagrangian (consisting of those terms containing gradients) is invariant under a set of infinitesimal unitary transformations generated by N independent Hermitian operators R_i, which may depend on the time. Under the transformations, the various fields ψ_A undergo linear recombinations:

$$\psi_A(\mathbf{x},t)\rightarrow\psi_A(\mathbf{x},t)-i\Lambda_i[R_i(t),\psi_A(\mathbf{x},t)]$$
$$=\psi_A(\mathbf{x},t)+i\Lambda_i\sum_B M_i^{AB}\psi_B(\mathbf{x},t), \quad \text{(A1)}$$

where Λ_i is the infinitesimal gauge constant associated with the ith transformation. The equal-time commutation rules of the R_i are the same as those of the matrices M_i. Moreover, the set of R_i and linear combinations of R_i is algebraically complete under commutation; in other words, we have an algebra. The matrices M_i are the basis of a representation of the algebra (in general, a reducible representation). It is convenient to take the matrices of the basis to be orthonormal,

$$\mathrm{Tr}M_iM_j=(\mathrm{const})\delta_{ij}, \quad \text{(A2)}$$

redefining the R_i accordingly. The structure constants c_{ijk} in the commutation rules

$$[M_i,M_j]=ic_{ijk}M_k,$$
$$[R_i(t),R_j(t)]=ic_{ijk}R_k(t), \quad \text{(A3)}$$

are now real and totally antisymmetric in i, j, and k. We may still perform real rotations in the N-dimensional space of the R_i or the M_i. Suppose, after performing such a rotation, that we can split the R_i into two sets that commute with each other. Then our algebra is the direct sum of two commuting algebras. We continue this process until no further splitting is possible, even after performing rotations. The algebra has then been expressed as the direct sum of *simple* algebras. All the simple algebras have been listed by Cartan.[40] Besides the trivial one-dimensional algebra of $U(1)$ (which is not included by the mathematicians), there are the three-dimensional algebra of $SU(2)$, the eight-dimensional algebra of $SU(3)$, and so forth.

Now let us construct the currents of the operators R_i. We consider the gauge transformation of the second kind

$$\psi_A\rightarrow\psi_A(\mathbf{x},t)-i\Lambda_i(\mathbf{x},t)[R_i(t),\psi_A(\mathbf{x},t)], \quad \text{(A4)}$$

and ask what change it induces in the Lagrange density L. There will be a term in Λ_i and a term in $\partial_\alpha\Lambda_i$, so adjusted[17] that the total change is just the divergence of a four-vector:

$$L\rightarrow L(\mathbf{x},t)-\partial_\alpha\Lambda_i(\mathbf{x},t)R_{i\alpha}(\mathbf{x},t)-\Lambda_i(\mathbf{x},t)\partial_\alpha R_{i\alpha}(\mathbf{x},t). \quad \text{(A5)}$$

We define $R_{i\alpha}$ to be the current of R_i. It can be shown that R_i is in fact given by the relation

$$R_i=-i\int R_{i4}d^3x. \quad \text{(A6)}$$

Now if, for constant Λ_i, the whole Lagrangian is invariant under R_i, then the term in Λ_i in (A4) must vanish; we have $\partial_\alpha R_{i\alpha}=0$. In other words, if there is exact symmetry under R_i, the current $R_{i\alpha}$ is conserved.

If there is a noninvariant part of L with respect to the symmetry operation R_i, then the current will not be conserved. By hypothesis, the noninvariant term (call it u) contains no gradients. Therefore, the effect

[39] L. Castillejo, R. H. Dalitz, and F. J. Dyson, Phys. Rev. 101, 453 (1956).

[40] S. C. Frautschi, M. Gell-Mann, and F. Zachariasen (to be published).

[40] E. Cartan, *Sur la Structure des groupes de transformations finis et continus*, thèse (Paris, 1894; 2nd ed., 1933).

of the transformation (A3) for *constant* Λ_i will be simply to add a term $-i\Lambda_i[R_i,u]$ to the Lagrangian density. We have, then, using (A4), the result

$$\partial_\alpha R_{i\alpha}(\mathbf{x},t)=i[R_i(t),u(\mathbf{x},t)]. \quad \text{(A7)}$$

Since u contains no gradients, it is not only the noninvariant term in the Lagrangian density, but also the negative of the noninvariant term in the Hamiltonian density. The invariant part of H evidently commutes with R_i. Thus we have

$$\partial_\alpha R_{i\alpha}(\mathbf{x},t)=-i[R_i(t),H(\mathbf{x},t)]. \quad \text{(A8)}$$

By considering the transformation properties of H under commutation with the algebra, we generate the divergences of all the currents. The formula obtained by integrating (A6) over space is, of course, very familiar:

$$\dot{R}_i=\int\partial_\alpha R_{i\alpha}d^3x=-i\left[R_i,\int Hd^3x\right]. \quad \text{(A9)}$$

Current-Generated Algebras*

MURRAY GELL-MANN and YUVAL NE'EMAN**

California Institute of Technology, Pasadena, California

(Received June , 1964)

* Work supported in part by the U. S. Atomic Energy Commission.

**On leave of absence from Tel Aviv University and the Israel Atomic Energy Commission.

ABSTRACT

We assume that a component of the F-spin current is utilized as part or all of the vector weak current for strongly interacting particles. Likewise we assume that the same component of an axial vector current octet is part or all of the axial vector weak current. The space integrals of the time components of these two octets then generate, by repeated equal time commutation relations, some algebra. We describe the classification of these possible algebras and give some examples. The algebra $SU(3) \times \dot{S}U(3)$ proposed earlier is not only the smallest of these possible candidates. It is also uniquely selected if the two octets exhaust the whole weak current of strongly interacting particles and if that current has the same algebraic properties as the weak current of the known leptons, in accord with the principle of universality of weak interactions.

The weak current of hadrons (strongly interacting particles) can be broken up according to the quantum numbers conserved by the strong interactions. There is a vector and a pseudovector part, and each of these contains a piece with $|\Delta \underset{\sim}{I}| = 1$, $GP = -1$, $\Delta Y = 0$, and a piece with $|\Delta \underset{\sim}{I}| = 1/2$, $\Delta Y/\Delta Q = +1$. It is still possible that there are other pieces, too, but we shall concentrate our attention on the ones mentioned, whether or not others exist.

Specific proposals have been made regarding the relation of these pieces of the weak current to the eightfold way.[1,2] It has been suggested[3-6] that the relation is the following:

(a) The vector currents belong to the octet consisting of the F-spin current. This idea is a simple generalization of the conserved vector current hypothesis. It means that the space integrals of the time-components of the vector current octet constitute the F-spin octet, obeying at equal times the commutation relations

$$\left[F_i, F_j\right] = i\, f_{ijk}\, F_k \quad . \tag{1}$$

(b) The axial vector currents belong to an octet, too. The space integrals of the time-components of the axial vector octet are called F_i^5 and obey the equal-time commutation relations

$$\left[F_i, F_j^5\right] = i\, f_{ijk}\, F_k^5 \quad . \tag{2}$$

(c) The operators F_i^5 obey the equal-time commutation relations

$$\left[F_i^5, F_j^5\right] = i\, f_{ijk}\, F_k \quad , \tag{3}$$

so that the algebraic system of F_i and F_i^5 closes to form the sixteen parameter Lie algebra of $SU(3) \times SU(3)$.

The relations (1), (2), and (3) are supposed to hold to all orders in the strong interactions, no matter how strong the violation of the conservation of F_i^5 or of F_4, F_5, F_6, and F_7.

The hypotheses (a) and (b) have been compared with experiment by Cabibbo,[7] with fairly good success. We shall assume them to be true in everything that follows.

We pose ourselves the problem of seeing whether there are any reasonable theoretical alternatives to the third hypothesis (c), which is more difficult to check by experiment. Are there other simple commutation rules besides (3) that we might assume? Evidently any other rules will lead to a larger algebra than $SU(3) \times SU(3)$; in each case we must investigate what algebra would be generated. We look for the minimal algebra, i.e., the one obtained by commuting $\underset{\sim}{F}$ and $\underset{\sim}{F}^5$ and then their commutators, and so forth, until the system closes under commutation.[8]

In Sections II and III we show how to classify the possible sets of commutation rules that might replace (3), according to the minimal algebra that would result in each case. In most cases, a minimal algebra corresponds to a family of sets of commutation rules, with a number of continuous parameters.

In Section IV we prove that the originally suggested algebra $SU(3) \times SU(3)$, which is also the smallest possible minimal algebra of $\underset{\sim}{F}$ and $\underset{\sim}{F}^5$, is uniquely selected under the following conditions:

(1) There are no other terms in the hadronic weak current besides the vector and axial vector octets.

(2) The total weak current operator for leptons is proportional to $\bar{\nu}_e \gamma_\alpha (1+\gamma_5) e + \bar{\nu}_\mu \gamma_\alpha (1+\gamma_5) \mu$, which generates the algebra of SU(2).

(3) The total weak current operator for hadrons has the property of universality, i.e., it generates the same algebra as in the lepton case. (Also the strength of the two currents in the complete weak interaction is the same. This statement has meaning only if the leptonic and hadronic weak currents share a common nonlinear relation, like the commutation rules of the algebra.)

We need look at other candidates only if one of these assumptions is to be dropped. We do not consider that a likely situation, but we discuss alternatives anyway, for the sake of completeness.[9]

II. CLASSIFICATION OF MINIMAL ALGEBRAS

We want to classify the possible candidates for the minimal algebra generated from the octets $\underset{\sim}{F}$ and $\underset{\sim}{F}^5$ by repeated commutation. We may factor the group G of the minimal algebra into simple Lie groups: $G = G^{(1)} \times G^{(2)} \times \ldots . G^{(r)}$. Each of the octets $\underset{\sim}{F}^5$ and $\underset{\sim}{F}$ is the sum of octets of generators from each of the groups $G^{(1)} \ldots . G^{(r)}$. (If $\underset{\sim}{F}$ lacked a term from one of the $G^{(i)}$, then it could not commute properly with an $\underset{\sim}{F}_5$ that includes such a term. If $\underset{\sim}{F}_5$ lacked a term from one of the $G^{(i)}$, we could leave all the commutation rules unchanged by dropping that factor group; the group G would not then be minimal.) Thus each group $G^{(i)}$ contains SU(3) and includes among its generators in at least one way a set $\underset{\sim}{F}^{(i)}$ with the commutation rules of an $\underset{\sim}{F}$-spin. In order for F to obey the same rules, we must have

$$\underset{\sim}{F} = \sum_{i=1}^{r} \underset{\sim}{F}^{(i)} \quad ; \tag{4}$$

in other words, $\underset{\sim}{F}$ is the total F-spin. Similarly, $\underset{\sim}{F}_5$ can be written in the form

$$\underset{\sim}{F}^5 = \sum_{i=1}^{r} c_i \, \underset{\sim}{H}^{(i)} \quad , \tag{5}$$

where $\underset{\sim}{H}^{(i)}$ is an octet of generators of $G^{(i)}$ transforming like an octet under commutation with $\underset{\sim}{F}^{(i)}$; under certain conditions $\underset{\sim}{H}^{(i)}$ may be equal to $\underset{\sim}{F}^{(i)}$, but in other cases it is a distinct octet.

Now we must impose the condition that $\underset{\sim}{F}^5$ be a pseudoscalar; $\underset{\sim}{F}$ is, of course, scalar. We can have first of all a situation in which a single $\underset{\sim}{H}^{(i)}$ obeys the rule $P \, \underset{\sim}{H}^{(i)} \, P^{-1} = - \underset{\sim}{H}^{(i)}$, where P is parity. In that case we say $G^{(i)}$ is a simple factor. The next case is one in which two terms $\underset{\sim}{H}^{(i)}$ and $\underset{\sim}{H}^{(j)}$ obey the rule $P \, \underset{\sim}{H}^{(i)} \, P^{-1} = \underset{\sim}{H}^{(j)}$, $P \, \underset{\sim}{H}^{(j)} \, P^{-1} = \underset{\sim}{H}^{(i)}$. To make $\underset{\sim}{F}^5$ a pseudoscalar, we must have $c_i = - c_j$. In this case we say that $G^{(i)} \times G^{(j)}$ is a double factor. The word "factor" will be used for a simple or a double factor.*

---

*As pointed out in Reference 9, the parity operation can take a $G^{(i)}$ into itself or interchange a $G^{(i)}$ and a $G^{(j)}$. Since these are the only alternatives, no triple or higher factors can exist. We are indebted to Professor S. Coleman for this remark.

We note that if G is minimal, then for each i, $\underset{\sim}{F}^{(i)}$ and $\underset{\sim}{H}^{(i)}$ give rise to the entire algebra of $G^{(i)}$ by repeated commutation. It is thus clear that any factor of a minimal group is itself a minimal group; it can be referred to as a minimal factor. Evidently all minimal groups can be obtained by stringing together products of minimal factors. Moreover, one can see that any such product is, in general, minimal, except for special values of the parameters c_i. Thus we need classify only minimal factors.

Let us begin with simple factors. A simple factor $G^{(i)}$ must have more than 16 generators, with eight constituting the scalar F-spin octet $\underset{\sim}{F}^{(i)}$ and eight more making the pseudoscalar set $\underset{\sim}{H}^{(i)}$ that behaves like an octet under commutation with $\underset{\sim}{F}^{(i)}$. Apart from the exceptional groups,[10)] $G^{(i)}$ can be a unitary group SU(n), a rotation group R(n), or a symplectic group Sp(2k), where $2k \equiv n$. For each case, we can display the algebra of the group explicitly in terms of its basic n-dimensional representation. The n basis vectors of this representation we call the substrate. To exhibit the subgroup of $G^{(i)}$ generated by the F-spin $\underset{\sim}{F}^{(i)}$, we must be able to arrange the substrate so that it consists of various irreducible representations of SU(3). The octet $\underset{\sim}{F}^{(i)}$ then constitutes the F-spin for these representations; it does not connect one representation with another.

In the Appendix we demonstrate that the representations of SU(3) given by the various pieces of the substrate of a simple minimal factor must all have the same "triality"; if the factor is a rotation group or a symplectic group, this common triality must be zero.

In the next section we consider a number of candidates for simple minimal factors, arranged according to the SU(3) representations occurring in the substrate, for example SU(8) with $\underset{\sim}{8}$, SU(6) with $\underset{\sim}{3} + \underset{\sim}{3}$, SU(9) with $\underset{\sim}{3} + \underset{\sim}{6}^*$, SU(9) with $\underset{\sim}{8} + \underset{\sim}{1}$, and R(9) with $\underset{\sim}{8} + \underset{\sim}{1}$, and we determine which of these are actually minimal factors.

One property that must be guaranteed is that the generators of the simple group can be divided into two classes, scalar and pseudoscalar, such that the commutator of two scalars or two pseudoscalars is a scalar, while the commutator of a scalar and a pseudoscalar is a pseudoscalar. Thus the scalar operators (associated with vector currents) generate a subgroup; the pseudoscalar operators, which are outside the algebra of the subgroup, must have all their commutators lying inside it. We know of only two ways to arrange that this condition be fulfilled. One way is to have the substrate divided into more than one piece and to assign parity + to some pieces and parity - to the rest; evidently the scalar operators connect the pieces of the substrate with the same parity and the pseudoscalars connect those with opposite parity. The other method applies to a unitary group SU(n). The matrices of the algebra of the subgroup R(n) are antisymmetric in the $n \times n$ representation, while the other matrices of the algebra of SU(n) are symmetric; the antisymmetric matrices can correspond to the scalar operators and the symmetric matrices to the pseudoscalar operators. We have not proved that these two are the only ways to satisfy the parity condition, but they suffice for our examples.**

Now let us turn to double minimal factors $G^{(i)} \times G^{(j)}$. The group $G^{(i)} \times G^{(j)}$ must be generated by $\underset{\sim}{F}^{(i)} + \underset{\sim}{F}^{(j)}$ and $\underset{\sim}{H}^{(i)} - \underset{\sim}{H}^{(j)}$, where $P\, \underset{\sim}{H}^{(i)} P^{-1} = \underset{\sim}{H}^{(j)}$ and $\underset{\sim}{F}^{(i)}$ and $\underset{\sim}{F}^{(j)}$ are scalar. Evidently parity

---

**Professor S. Coleman informs us that a rigorous statement on this point can be derived from the work of Reference 9.

merely interchanges the two groups $G^{(i)}$ and $G^{(j)}$, which must therefore be identical. Again, we deal with R(n), SU(n), or Sp(n) and display each in terms of its basic n-dimensional representation. The substrate for each of the groups $G^{(i)}$ and $G^{(j)}$ must be broken into irreducible representations of SU(3), with $\underset{\sim}{F}^{(i)}$ and $\underset{\sim}{F}^{(j)}$ equal respectively to the F-spins for these sets of representations. The list of irreducible representations of SU(3) must be the same for the substrate of $G^{(i)}$ and that of $G^{(j)}$. The parity operation then can be thought of as merely interchanging the two substrates arranged in this way. For example, we might have SU(9) x SU(9) with each substrate broken into $\underset{\sim}{8}$ + $\underset{\sim}{1}$; parity would just interchange the octets and interchange the singlets.

As in the case of simple minimal factors, the pieces of the substrate must have equal triality, and this triality must be zero if we are dealing with rotation groups.

In the next section, after studying some candidates for simple minimal factors, we examine a list of candidates for double minimal factors and determine which ones are really minimal.

III. CANDIDATES FOR SIMPLE AND DOUBLE MINIMAL FACTORS

We start with simple factors, arranged according to the SU(3) representations occurring in the substrate.

$\underset{\sim}{8}$: SU(8). Minimal. The scalar sub-algebra is that[11] of R(8), and is spanned by $\underset{\sim}{F}$ and the $\underset{\sim}{10}$ and $\overline{\underset{\sim}{10}}$ operators; the pseudoscalars are $\underset{\sim}{D}$ and the $\underset{\sim}{27}$. Evidently we take $\underset{\sim}{H} = \underset{\sim}{D}$.

R(8) contains only one octet and Sp(8) has none, so they are unsuitable.

$\underset{\sim}{3} + \underset{\sim}{3}$: SU(6). Not minimal. We must assign opposite parities to the two $\underset{\sim}{3}$'s and let $\underset{\sim}{H}$ connect one with the other. The commutator of $\underset{\sim}{H}$ with $\underset{\sim}{H}$ is then proportional to $\underset{\sim}{F}$, so that $\underset{\sim}{F}$ and $\underset{\sim}{H}$ generate the algebra of SU(3) × SU(3), which is a double minimal factor. We say SU(6) "reduces" to SU(3) × SU(3).

R(6) and Sp(6) are obviously unsuitable.

$\underset{\sim}{3} + \underset{\sim}{6}^*$: SU(9). Minimal. We assign opposite parities to $\underset{\sim}{3}$ and $\underset{\sim}{6}^*$ and let $\underset{\sim}{H}$ connect them. The commutator of $\underset{\sim}{H}$ with $\underset{\sim}{H}$ is proportional to $\underset{\sim}{F}$ within the $\underset{\sim}{3}$ representation and within the $\underset{\sim}{6}$ representation, but the coefficients are different; thus $\left[\underset{\sim}{H}, \underset{\sim}{H}\right]$ generates a new octet not equal to $\underset{\sim}{F}$ and repeated commutation gives the whole of SU(9). The scalar subalgebra is that of SU(3) × SU(6) × U(1).

R(9) is unsuitable because the triality is not zero.

$\underset{\sim}{1} + \underset{\sim}{8}$: R(9). Minimal. We assign opposite parities to $\underset{\sim}{1}$ and $\underset{\sim}{8}$ and let $\underset{\sim}{H}$ connect them. The scalar subalgebra is that of R(8); all parts of it are obtained by commutation of $\underset{\sim}{H}$ with itself.

SU(9). Also minimal, but not if we assign opposite parities to $\underset{\sim}{1}$ and $\underset{\sim}{8}$. (In that case, SU(9) reduces to R(9).) Instead, we let the scalar subalgebra consist of the anti-symmetric matrices in the 9×9 representation of SU(9), in other words the algebra of R(9). The pseudoscalar operators correspond to the symmetric matrices, which include the $\underset{\sim}{D}$ octet within $\underset{\sim}{8}$ and one of the octets (call it $\underset{\sim}{L}$) connecting $\underset{\sim}{1}$ and $\underset{\sim}{8}$. We take $\underset{\sim}{H} = \cos \phi \, \underset{\sim}{D} + \sin \phi \, \underset{\sim}{L}$. If $\phi = 0$, SU(9) reduces to SU(8) and if

$\phi = \pi/2$ it reduces to R(9), but for intermediate values of ϕ, SU(9) is minimal.

$\underline{3} + \underline{3} + \underline{3}$: SU(9). Not minimal, reduces to SU(3) × SU(3).

We now list some candidates for double minimal factors. In each case, we give the representations of SU(3) occurring in one of the two substrates.

$\underline{3}$: SU(3) × SU(3). Minimal.

$\underline{6}$: SU(6) × SU(6). Not minimal, reduces to SU(3) × SU(3).

$\underline{8}$: R(8) × R(8). Not minimal, reduces to SU(3) × SU(3).

SU(8) × SU(8). Minimal, with a one-parameter family of possibilities for $\underline{H}$.

$\underline{8} + \underline{1}$: R(9) × R(9). Not minimal, reduces to R(9).

SU(9) × SU(9). Minimal with a two-parameter family of possibilities for $\underline{H}$.

Of these double factors, SU(3) × SU(3) has already been mentioned and will be further discussed in the next section. It is very likely the correct choice. In the notation $G^{(1)} \times G^{(2)}$, we note that $\underline{F} = \underline{F}^{(1)} + \underline{F}^{(2)}$ and $\underline{F}^5 \propto \underline{F}^{(1)} - \underline{F}^{(2)}$; in other words, $\underline{H}^{(1)} = \underline{F}^{(1)}$ and $\underline{H}^{(2)} = \underline{F}^{(2)}$. The commutation rules (1), (2), and (3) are then evident, if the scale of $\underline{F}^5$ is fixed so that $\underline{F}^5 = \underline{F}^{(1)} - \underline{F}^{(2)}$.

Another example, SU(8) × SU(8), may also be described with the notation $G^{(1)} \times G^{(2)}$. We have $\underline{F} = \underline{F}^{(1)} + \underline{F}^{(2)}$ and $\underline{H}^{(1)} = \underline{F}^{(1)} \cos\phi + \underline{D}^{(1)} \sin\phi$, $\underline{H}^{(2)} = \underline{F}^{(2)} \cos\phi + \underline{D}^{(2)} \sin\phi$, with $\underline{F}^5 \propto \underline{H}^{(1)} - \underline{H}^{(2)}$. If $\phi = 0$, our algebra reduces to that of SU(3) × SU(3), and for $\phi = \pi/2$ it reduces to that of SU(8), but for intermediate values of ϕ we generate the whole algebra of SU(8) × SU(8). There are 63 vector currents, transforming under SU(3) like 8, 8, 10, 10^*, and 27, and also 63 axial vector currents, transforming in the same way. (Each scalar operator corresponds to a vector current, and each pseudoscalar operator to an axial vector current.)

The reader may continue for himself the process we have begun, and construct more and more complicated simple and double minimal factors. Then he can use these minimal factors and their products, ad libitum, to construct minimal algebras for F and F^5.

Of all the compound minimal groups, we mention one that has recently been discussed by Freund and Nambu,[12] namely $\left[SU(3) \times SU(3)\right] \times \left[SU(3) \times SU(3)\right]$. Here $F = F^{(1)} + F^{(2)} + F^{(3)} + F^{(4)}$ and
$F^5 \propto \cos \phi \, (F^{(1)} - F^{(2)}) + \sin \phi \, (F^{(3)} - F^{(4)})$. For $\phi = 0$, $\pi/4$, or $\pi/2$, our group reduces to SU(3) × SU(3), but for intermediate values of ϕ the entire algebra of $\left[SU(3) \times SU(3)\right] \times \left[SU(3) \times SU(3)\right]$ is generated by F and F^5. There are sixteen vector currents, in two octets, and sixteen axial vector currents, likewise in two octets.

Many of the groups we have discussed in this section were mentioned in the list of "chiral symmetries" given by Coleman and Glashow.[9] The assumptions underlying their classification are different, however, from the ones used in this article.***

***Nevertheless, our problem can be considered within the framework of chiral symmetries; in fact, the work can be simplified and certain proofs supplied if the analysis of Coleman and Glashow is used.

IV. SU(3) × SU(3) FOR UNIVERSALITY

Let us now demonstrate that SU(3) × SU(3) is the only minimal algebra of $\underset{\sim}{F}$ and $\underset{\sim}{F}^5$ that will give universality of weak interactions along with the leptonic weak current as now understood, provided the hadronic weak current includes only the vector and axial vector octets.

The total leptonic weak current and its hermitian conjugate generate the group SU(2). We must add strangeness-preserving and strangeness-changing parts of $\underset{\sim}{F}$ and $\underset{\sim}{F}^5$ and construct an operator that, together with its hermitian conjugate, also generates SU(2). The operator has the form

$$W \equiv A \left\{\cos\theta\,(F_1 + i\,F_2 + F_1{}^5 + i\,F_2{}^5) + \sin\theta\,(F_4 + i\,F_5 + F_4{}^5 + i\,F_5{}^5)\right\} \quad . \tag{6}$$

Here the scale of $\underset{\sim}{F}^5$ is arbitrarily chosen so that $\underset{\sim}{F}$ and $\underset{\sim}{F}^5$ appear with equal coefficients. The proportion of strangeness-changing and strangeness-preserving terms is arbitrary and equal to $\tan\Theta$; likewise the overall scale is arbitrary and is given by A. The vector and axial vector currents are assumed to have the same value of $\tan\Theta$; actually nothing new results if we allow this quantity to be different for the two cases.

We now require, for universality,[5,6,13] that W and W^+ be components $2(K_1 + i\,K_2)$ and $2(K_1 - i\,K_2)$ of an angular momentum operator (K_1, K_2, K_3). In other words, we require that W and W^+ generate the algebra of SU(2).

A unitary transformation belonging to SU(3) will transform $F_1\cos\theta + F_4\sin\theta$ into F_1, $F_2\cos\theta + F_5\sin\theta$ into F_2, and likewise

$F_1^5 \cos\theta + F_4^5 \sin\theta$ into F_1^5, and $F_2^5 \cos\theta + F_5^5 \sin\theta$ into F_2^5. After the unitary transformation is performed, we obtain

$$W' = A\left\{F_1 + i\,F_2 + F_1^5 + i\,F_2^5\right\} \quad . \qquad (7)$$

We are now demanding that

$$\left[A(F_1 + F_1^5),\ A(F_2 + F_2^5)\right] = 2i\ A(F_3 + F_3^5) \quad , \qquad (8)$$

and so forth for cyclic permutations of 1, 2, 3. In other words, we must have

$$\left[A(F_i + F_i^5),\ A(F_j + F_j^5)\right] = 2i\ e_{ijk}\ A(F_k + F_k^5) \qquad (9)$$

for $i, j, k = 1, 2, 3$. Equating pseudoscalar terms on the two sides, and using the known equation of commutation of $\underset{\sim}{F}$ and $\underset{\sim}{F}^5$, we obtain $A = 1$. We then equate scalar terms on both sides, using the commutation rules of the F-spin, and we find the condition

$$\left[F_i^5,\ F_j^5\right] = i\ e_{ijk}\ F_k \qquad (10)$$

for $i, j, k = 1, 2, 3$.

Now, in general for $i, j, k = 1, \ldots 8$, we have

$$\left[F_i^5,\ F_j^5\right] = i\ f_{ijk}\ E_k + i\ g_{ijn}\ T_n + i\ g^*_{ijn}\ U_n \quad , \qquad (11)$$

where E_k is an octet ($k = 1, \ldots 8$), T_n is a decimet ($n = 1, \ldots 10$), and U_n is an anti-decimet ($n = 1, \ldots 10$). These are the only representations of SU(3) that can be made by an antisymmetric combination of two identical octets. The symbol g_{ijn} is essentially a "Clebsch-Gordan" coefficient for two octets forming a decimet. Now the octet, decimet, and anti-decimet each contain one isotopic triplet with $Y = 0$. Let that correspond to $n = 1, 2, 3$ as well as $k = 1, 2, 3$. Within the range $i, j, k, n = 1, 2, 3$, not only do we have $f_{ijk} = e_{ijk}$ but also (with suitable normalization) $g_{ijn} = g^*_{ijn} = e_{ijn}$. Thus for $i, j, k = 1, 2, 3$ we have

$$\left[F_i^5, F_j^5\right] = i\, e_{ijk}\,(E_k + T_k + U_k) \qquad . \tag{12}$$

Comparing with (10), we see that the octet E_k equals F_k and that the decimet and anti-decimet are absent ($T_k = U_k = 0$). Thus (11) reads

$$\left[F_i^5, F_j^5\right] = i\, f_{ijk}\, F_k \tag{13}$$

for $i, j, k = 1, \ldots 8$ and our minimal algebra is just $SU(3) \times SU(3)$.

APPENDIX

Exceptional Groups

We ignore the exceptional algebras because one of them, G_2, is unsuitable and the others, F_4, E_6, E_7, and E_8, have only representations of very large dimension, starting at 26, 27, 56, and 248, respectively.[14]

G_2 is unsuitable because it contains SU(3) only in the sense that the fourteen generators of G_2 can be broken up into an F-spin, a unitary triplet of operators, and a unitary anti-triplet of operators. However, triplets and anti-triplets of operators can never appear in the minimal algebra of $\underset{\sim}{F}$ and $\underset{\sim}{F}^5$ (see below).

Triality Content of Substrates

For the classical groups R(n), SU(n), and Sp(2k) with $2k = n$, we introduce the n-component substrate and the $n \times n$ matrix representation for simplicity. The substrate is then broken up into pieces that form representations of SU(3).

The triality t of an SU(3) representation as introduced by Baird and Biedenharn[15] corresponds in the quark model[5] to the number of quarks modulo 3, with $-1 \leq t \leq 1$. In the direct product of representations, the triality is always the sum of the triality of the factors (modulo 3). The octet representation has $t = 0$ and so have all representations obtained by decomposing any product of octets. The triplet representation $\underset{\sim}{3}$ has $t = 1$ while $\underset{\sim}{3}^*$ has $t = -1$; if we write $\underset{\sim}{3} \times \underset{\sim}{3} = \underset{\sim}{6} + \underset{\sim}{3}^*$, we see that $\underset{\sim}{6}^*$ has $t = 1$ and $\underset{\sim}{6}$ has $t = -1$.

It is clear that by repeated commutation of $\underset{\sim}{F}$ and $\underset{\sim}{F}^5$, we always generate sets of operators that belong to SU(3) representations with $t = 0$, like $\underset{\sim}{8}$, $\underset{\sim}{10}$, $\underset{\sim}{10}^*$, and $\underset{\sim}{27}$. A minimal factor must never have any generators that belong to a representation of SU(3) with $t \neq 0$.

Now consider the n-dimensional representation of R(n), Sp(n), or SU(n). The matrices connecting a piece of the substrate of triality t with a piece having triality t' will belong to representations of SU(3) with triality $t - t'$. Since this operator triality $t - t'$ must always be zero (modulo 3) for a minimal factor, we see that the substrate trialities must all be equal.

When the group is R(n) or Sp(2k), we can prove the stronger result that the common triality of the pieces of the substrate is zero. The $n \times n$ matrices representing the algebra of R(n) are just the antisymmetric $n \times n$ matrices. The $2k \times 2k$ matrices representing the algebra of Sp(2k) are just the direct products of linear combinations of σ_x, σ_y, and σ_z with the symmetric $k \times k$ matrices and the direct products of the two-dimensional unit matrix with the antisymmetric $k \times k$ matrices. Consider a basis of hermitian $n \times n$ matrices in either of these two cases. We see that the negative complex conjugates of the hermitian matrices form a basis for an equivalent representation of the algebra of R(n) or Sp(2k) respectively. Thus if we complex-conjugate the substrate, we obtain an equivalent representation of the group R(n) or Sp(2k). The SU(3) content of the substrate must therefore correspond to self-conjugate SU(3) representations and conjugate pairs of SU(3) representations. This is consistent with a common triality t only if $t = 0$.

REFERENCES

1. Y. Ne'eman, Nuclear Phys. 26, 222 (1961).

2. M. Gell-Mann, California Institute of Technology Synchrotron Laboratory Report CTSL-20 (1961).

3. M. Gell-Mann, Phys. Rev. 125, 1067 (1962). See also, "Symmetries, Currents, and Resonances", paper presented at La Jolla Conference on Strong and Weak Interactions, 1961, and "Symmetry Properties of Fields", Proceedings of the 1961 Solvay Conference, Interscience Publishers, Inc., New York.

4. For a discussion of SU(3) × SU(3) in connection with the Sakata model, see A. Salam and J. C. Ward, Nuovo Cimento 20, 419 (1961).

5. M. Gell-Mann, Physics Letters 8, 214 (1964).

6. M. Gell-Mann, Physics (in press).

7. N. Cabibbo, Phys. Rev. Letters 10, 531 (1963).

8. We use bold-faced type for octets of operators like $\mathbf{F}$ and $\mathbf{F}^5$.

9. Many "chiral symmetries" are considered as possible alternatives to SU(3) × SU(3) by S. Coleman and S. Glashow, Ann. Phys. 17, 41 (1962), and by S. Coleman, Ph.D. dissertation, California Institute of Technology (1961). See also S. Coleman, Ann. Phys. 24 37 (1963).

10. See Appendix.

11. Y. Ne'eman, Physics Letters 4, 81 (1962) and 5, 312 (1963).

12. P. Freund and Y. Nambu, to be published.

13. M. Gell-Mann, Proceedings of the 1960 Annual International Conference on High Energy Physics at Rochester, pp. 508-513 (Interscience Publishers, Inc., New York, 1960).

14. E. Dynkin, "Semi-Simple Subgroups of the Classical Groups", (Supplement), Amer. Math. Soc. Translations, Series 2, 6 (1957).

15. G. E. Baird and L. C. Biedenharn, Proceedings of the Miami Conference on Symmetry Principles at High Energy (1964), to be published.

IX

THE NONLEPTONIC WEAK INTERACTION

In this part, as in the previous one, the first paper serves as a lengthy introduction. Although a special point of view is adopted, various ideas about the interpretation of the nonleptonic interaction according to the eightfold way are discussed, including those expressed in the succeeding articles by d'Espagnat and Gell-Mann.

The first paper also treats some of the subject matter of the next part—namely, the problem of how the strong violation of SU(3) symmetry is to be explained in the framework of the bootstrap hypothesis.

Octet Enhancement*

ROGER F. DASHEN, STEVEN C. FRAUTSCHI, MURRAY GELL-MANN, and YASUO HARA

California Institute of Technology, Pasadena, California

ABSTRACT

A simple universal current-current theory of weak interactions now accounts fairly well for all weak phenomena except the nonleptonic $|\Delta \underline{I}| = 1/2$ rule, which presumably should be generalized to a unitary octet rule, covering not only the familiar $|\Delta Y| = 1$ nonleptonic interactions but also the $|\Delta Y| = 0$ nonleptonic interactions (for which experimental evidence in heavy nuclei has been presented by Boehm and Kankeleit).

One may account for the octet nonleptonic rule either (a) by a theory that adds extra current-current products for strongly interacting particles alone, or else (b) by a dynamical mechanism that enhances octets by means of strong interactions. It is interesting that we can distinguish any reasonable theory of type (a) from any theory of type (b) by the amount of $|\Delta \underline{I}| = 1$ in the $\Delta Y = 0$ nonleptonic interaction; the $|\Delta \underline{I}| = 1$ component is large in the former case and small in the latter case. Difficult experiments involving light nuclei may be able to resolve the two possibilities.

*Work supported in part by the U. S. Atomic Energy Commission.

If octet enhancement is indeed the explanation, then the same mechanism can operate (as remarked by Coleman and Glashow) to enhance the octet portion of electromagnetic mass splittings in various supermultiplets, and can also build up the octet violation of SU(3) symmetry of strong interactions.

In the bootstrap theory of strong interactions, a particularly simple description of enhancement is available, which we illustrate for the electromagnetic mass differences. The electromagnetic interaction is introduced as a driving term, which affects the masses of external particles and exchanged particles in the diagrams that account for the structure of particles. Crudely, one has a matrix relation like

$$\Delta m = D + A\,(\Delta m) \quad ,$$

where D is the direct effect of the driving force and $A\,(\Delta m)$ is the effect on external and exchanged particles. Strong octet enhancement corresponds to having octet eigenvalues ${}_8a$ of A close to unity, while 27-plet eigenvalues ${}_{27}a$ are far from unity.

In the case of the breaking of SU(3) symmetry of the strong interactions (for which a related approach was originally suggested by Cutkosky) one might consider spontaneous violation, which corresponds in the linear approximation to $D = 0$ and ${}_8a = 1$. However, the direction in SU(3) space of the octet violation seems very difficult to explain in terms of the directions defined by the electromagnetic and weak perturbations. Therefore, even in an otherwise pure bootstrap theory, a driving term something like Ne'eman's "fifth interaction" may be needed. Weak nonleptonic effects, electromagnetic mass differences, and strong violations of SU(3) symmetry would then really be analogous to one another.

I. THE WEAK INTERACTION

Let us start from a simple picture of the weak interaction, which explains nearly all of the experimental evidence. The weak coupling is represented, in the local limit, as the product of a single current and its hermitian conjugate:

$$\frac{G}{\sqrt{2}} J_\alpha^+ J_\alpha \quad . \tag{1}$$

The interaction may, of course, be nonlocal -- that is, mediated by an intermediate boson of finite mass rather than infinite mass -- for our purposes we need not emphasize this distinction.

The current J_α may be written in the form

$$J_\alpha = J_\alpha^\ell + J_\alpha^h \quad , \tag{2}$$

where J_α^ℓ is the leptonic part and has, as far as we know, the simple structure

$$J_\alpha^\ell = \bar{\nu}_e \gamma_\alpha (1+\gamma_5) e + \bar{\nu}_\mu \gamma_\alpha (1+\gamma_5) \mu \quad , \tag{3}$$

where J_α^h is the part that concerns hadrons (strongly interacting particles) and is assumed to be given by the formula

$$J_\alpha^h = (\mathcal{F}_{1\alpha} + i\,\mathcal{F}_{2\alpha} + \mathcal{F}_{1\alpha}^5 + i\,\mathcal{F}_{2\alpha}^5) \cos\theta + (\mathcal{F}_{4\alpha} + i\,\mathcal{F}_{5\alpha} + \mathcal{F}_{4\alpha}^5 + i\,\mathcal{F}_{5\alpha}^5) \sin\theta, \tag{4}$$

where $\mathcal{F}_{i\alpha}$ $(i = 1 \ldots 8)$ is the current of the F-spin and $\mathcal{F}_{i\alpha}^5$ $(i = 1 \ldots 8)$ is an octet of axial vector currents.[1-4] As Cabibbo[2] has shown, Eq. (4) gives a good description of leptonic weak interactions with θ of the order of 15°.

It is useful to introduce[5] a unitary transformation in the space of SU(3), which transforms the familiar isotopic spin and hypercharge operators as follows:

$$I_1 = F_1 \to I_1' = F_1' = F_1 \cos\theta + F_4 \sin\theta \qquad ,$$

$$I_2 = F_2 \to I_2' = F_2' = F_2 \cos\theta + F_5 \sin\theta \qquad ,$$

$$I_3 = F_3 \to I_3' = F_3' = F_3 \cos^2\theta + (\tfrac{\sqrt{3}}{2} F_8 + \tfrac{1}{2} F_3) \sin^2\theta - F_6 \sin\theta\cos\theta \qquad ,$$

$$\frac{\sqrt{3}}{2} Y = F_8 \to \frac{\sqrt{3}}{2} Y' = F_8' = F_8 (1 - \tfrac{3}{2}\sin^2\theta) + F_3 (\tfrac{\sqrt{3}}{2}\sin^2\theta)$$

$$+ \sqrt{3}\, F_6 \sin\theta\cos\theta \qquad . \tag{5}$$

The nonleptonic part of the weak coupling (1) can then be written

$$\frac{G}{\sqrt{2}} \left\{(\mathcal{F}_{1\alpha}' + \mathcal{F}_{1\alpha}^{5'})(\mathcal{F}_{1\alpha}' + \mathcal{F}_{1\alpha}^{5'}) + (\mathcal{F}_{2\alpha}' + \mathcal{F}_{2\alpha}^{5'})(\mathcal{F}_{2\alpha}' + \mathcal{F}_{2\alpha}^{5'})\right\}$$

$$\equiv Z_{1\alpha}' \; Z_{1\alpha}' + Z_{2\alpha}' \; Z_{2\alpha}' \tag{6}$$

The main experimental fact that is not accounted for by the simple theory we have sketched is the $|\Delta I| = 1/2$ rule for the $|\Delta Y| = 1$ part of the nonleptonic interaction. The coupling (6) contains, of course, both $|\Delta I| = 1/2$ and $|\Delta I| = 3/2$. From the point of view of the eightfold way, the expression (6) consists of three parts, transforming like 1, 8, and 27 respectively. The $|\Delta I| = 3/2$ contribution comes entirely from the 27 part, and it is natural to suppose that whatever is responsible for the $|\Delta I| = 1/2$ nonleptonic rule actually makes the entire octet contribution much more important than the entire 27 contribution.[5-8] We shall henceforth assume that the $|\Delta I| = 1/2$ rule is actually a rule of predominance of 8 (and perhaps 1) over 27.

If we display the interaction (6) explicitly as a sum of $\underset{\sim}{1}$, $\underset{\sim}{8}$, and $\underset{\sim}{27}$ terms, it becomes:

$$Z'_{1\alpha} Z'_{1\alpha} + Z'_{2\alpha} Z'_{2\alpha} = (\tfrac{1}{4})\, S + (\tfrac{2}{5})\, O_8 + (\tfrac{1}{20})\, T \tag{7}$$

where

$$S = Z'_{1\alpha} Z'_{1\alpha} + Z'_{2\alpha} Z'_{2\alpha} + Z'_{3\alpha} Z'_{3\alpha} + Z'_{4\alpha} Z'_{4\alpha} + Z'_{5\alpha} Z'_{5\alpha} + Z'_{6\alpha} Z'_{6\alpha} + Z'_{7\alpha} Z'_{7\alpha} + Z'_{8\alpha} Z'_{8\alpha} \quad , \tag{8}$$

$$O_8 = Z'_{1\alpha} Z'_{1\alpha} + Z'_{2\alpha} Z'_{2\alpha} + Z'_{3\alpha} Z'_{3\alpha} - \tfrac{1}{2}\, Z'_{4\alpha} Z'_{4\alpha} - \tfrac{1}{2}\, Z'_{5\alpha} Z'_{5\alpha} - \tfrac{1}{2}\, Z'_{6\alpha} Z'_{6\alpha} - \tfrac{1}{2}\, Z'_{7\alpha} Z'_{7\alpha} - Z'_{8\alpha} Z'_{8\alpha} \quad , \tag{9}$$

$$T = 7\, Z'_{1\alpha} Z'_{1\alpha} + 7\, Z'_{2\alpha} Z'_{2\alpha} - 13\, Z'_{3\alpha} Z'_{3\alpha} - Z'_{4\alpha} Z'_{4\alpha} - Z'_{5\alpha} Z'_{5\alpha} - Z'_{6\alpha} Z'_{6\alpha} - Z'_{7\alpha} Z'_{7\alpha} + 3\, Z'_{8\alpha} Z'_{8\alpha} \quad . \tag{10}$$

(Note that O_8 transforms like the eighth component of a primed octet.) There are, of course, two different ways to assure predominance of $\underset{\sim}{8}$ over $\underset{\sim}{27}$. One is to add to our basic interaction (1) some new products of currents (presumably involving hadrons alone) so as to cancel the $\underset{\sim}{27}$ portion. The other is to find a mechanism that enhances the octet (and perhaps singlet) contributions to nonleptonic processes.

Let us begin with the first mechanism, which involves adding extra current products to the expression $Z'_{1\alpha} Z'_{1\alpha} + Z'_{2\alpha} Z'_{2\alpha}$. We assume that these further products also have positive coefficients, so as to correspond to positive probabilities for exchanged intermediate bosons. Then, if we are restricted to linear combinations of the unitary singlet S and the eighth component O_8 of a primed unitary octet, the only expressions we can obtain are sums of either seven or eight current products. No theory of this type seems to have any distinctive features to recommend it. We may, however, also make use of the third component of a primed octet, proportional to

$$O_3 = \frac{\sqrt{3}}{2}(Z'_{4\alpha} Z'_{4\alpha} + Z'_{5\alpha} Z'_{5\alpha} - Z'_{6\alpha} Z'_{6\alpha} - Z'_{7\alpha} Z'_{7\alpha}) + 2 Z'_{3\alpha} Z'_{8\alpha} \quad . \qquad (11)$$

In that case we can obtain the more elegant interaction of d'Espagnat:[9]

$$\frac{2}{3} S + \frac{1}{3} O_8 - \frac{1}{\sqrt{3}} O_3 = Z'_{1\alpha} Z'_{1\alpha} + Z'_{2\alpha} Z'_{2\alpha}$$

$$+ (Z'_{3\alpha} - \frac{1}{\sqrt{3}} Z'_{8\alpha})(Z'_{3\alpha} - \frac{1}{\sqrt{3}} Z'_{8\alpha}) + Z'_{6\alpha} Z'_{6\alpha} + Z'_{7\alpha} Z'_{7\alpha} \quad . \qquad (12)$$

Here, there are only five current products; moreover, if the interaction is expressed in terms of intermediate bosons, they belong to a triplet and anti-triplet and the coupling transforms under SU(3) like $\underset{\sim}{3}$ and $\overline{\underset{\sim}{3}}$.

We have seen, then, that an attractive theory involving extra current products involves not just S and O_8 (Eqs. (8) and (9)) but also an admixture, with coefficient of order unity, of O_3 (Eq. (11)). Now O_3 has the property $|\Delta \underset{\sim}{I}'| = 1$, while S and O_8 have $|\Delta \underset{\sim}{I}'| = 0$. Translating into the language of ordinary isotopic spin and hypercharge, we may say that the $\cos^2\theta$ term in O_3 has $\Delta Y = 0$, $|\Delta \underset{\sim}{I}| = 1$, while the $\cos^2\theta$ term in O_8 (like all of S) has $\Delta Y = 0$, $|\Delta \underset{\sim}{I}| = 0$. Thus in an elegant theory of added current products, the $\cos^2\theta$ term in the $\Delta Y = 0$ nonleptonic interaction should have an admixture of $|\Delta \underset{\sim}{I}| = 1$ of order unity.

Now let us contrast such a situation with the case in which the $\underset{\sim}{27}$ part of the nonleptonic interaction is unimportant because of octet (and perhaps singlet) enhancement. Such enhancement by strong interactions cannot introduce any $|\Delta I| = 1$ into the $\cos^2\theta$ term in the original interaction (6), which contains[10] only $|\Delta I| = 0$ and $|\Delta I| = 2$. The $|\Delta I| = 0$ part of the $\cos^2\theta$ term will be enhanced and the $|\Delta I| = 2$ part will not.

Since $\cot^2\theta \approx 15$, the amount of $|\Delta I| = 1$, $\Delta Y = 0$ in the extra current model is around 15 times larger than in the octet enhancement model. If the isotopic spin dependence can be tested experimentally, it would help to distinguish the two possibilities. At present, our evidence comes entirely from the $|\Delta Y| = 1$ nonleptonic decays, and consists mostly in the observation that the $|\Delta I| = 1/2$ octet amplitudes seem to be rather larger than one might expect from crude estimates based on the current-current product without enhancement, while the $|\Delta I| = 3/2$, 27-plet amplitudes are roughly of the expected order of magnitude.[11] Such evidence, of course, supports the octet enhancement model, but is far from conclusive.[12]

II. NUCLEAR PHYSICS TESTS OF THE STRUCTURE OF NON-LEPTONIC WEAK INTERACTIONS

The effects of nonleptonic $\Delta Y = 0$ weak interactions are hard to study experimentally because they are generally masked by the stronger interactions. Nevertheless, many attempts have been made, and recently Boehm and Kankeleit[13] have claimed detection of circular polarization in 482-keV γ-rays emitted by unpolarized Ta^{181}, thus providing evidence for parity violation in nuclear forces. Although information about the isotopic spin of the weak interaction cannot be gained from such a heavy nucleus as Ta^{181}, the work of Boehm and Kankeleit encourages the hope that similar experiments can eventually be done on light nuclei, where weak $|\Delta I| = 0$, 1, and 2 effects can be distinguished from one another.

In evaluating possible experiments on light nuclei, we need to estimate the magnitude of parity-mixing to be expected in nuclear states on the basis of the octet enhancement and extra-current models respectively. The usual steps in such an estimate are to give a weak two-nucleon potential, deduce an effective weak potential for one nucleon in nuclear matter, and then study the parity-mixing that results in nuclear wave functions. The last step requires detailed analysis of specific nuclear states, but the first two steps can be treated in a slightly more general fashion.

One finds the two-nucleon potential as a function of internucleon distance by summing the various possible exchanges such as π, η, ρ, ω, ϕ, W, W+π, and so forth. (Here, W is the hypothetical intermediate boson.) This procedure closely parallels the usual treatment of strong interactions except that now one vertex is weak (or, as in W exchange, each vertex is half-weak). Exchanges of more massive systems can be neglected with better assurance than usual because they contribute to the weak potential mainly inside the repulsive core of the strong interactions, where the nucleons rarely penetrate.[14)]

Classifying the various long-range, parity-violating terms, one finds that CP forbids π^0 and η exchange and forces a purely $|\Delta \underset{\sim}{I}| = 1$ character upon $\pi^{\pm}$ exchange,[10)] whereas ρ, ω, and ϕ exchange contribute only to $|\Delta \underset{\sim}{I}| = 0$ or 2. The long-range effects of $W^{\pm}$ exchange are already included in $\rho^{\pm}$ exchange, since the NNW vertex acquires a form factor, extending the original range of W exchange, from $W \to \rho \to N\bar{N}$. These contributions from $\rho^{\pm}$ exchange have been estimated by Michel.[15)] The space-spin dependence turns out to have the important property that the interactions of a single unpaired nucleon with the other nucleons, including those in closed shells, add constructively.

For the $|\Delta I| = 1$ potential, we consider $\pi^{\pm}$ exchange and again find that the interactions of a single unpaired nucleon with the other nucleons add. The magnitude of the weak $NN\pi$ vertex does not follow immediately from the current-current formalism, but experimental numbers are available for its $\Delta Y = 1$ counterparts $\Xi \to \Lambda + \pi$, $\Sigma \to N + \pi$, and $\Lambda \to N + \pi$, and each of the models we are comparing relates these values for $\Delta Y = 1$ to the $\Delta Y = 0$, $|\Delta I| = 1$ amplitude. Working out the implications of the d'Espagnat model, one finds that in nuclear matter $\pi^{\pm}$ exchange contributes a $|\Delta I| = 1$, parity-violating, single-nucleon effective potential with just about the same strength as the $|\Delta I| = 0$ parity-violating potential estimated by Michel.[15)] By contrast, in the octet enhancement model with no extra currents, the $|\Delta I| = 1$ potential is weaker by the factor $\cot^2\theta \approx 15$, as described previously. The two models also lead to different expectations for the absolute magnitude of the $|\Delta I| = 0$ amplitude, which would be raised above Michel's estimate by octet enhancement,[16)] and the $|\Delta I| = 2$ amplitude, which may be wiped out completely by extra currents but not by octet enhancement.

Thus the isospin properties of the effective parity-violating potential in nuclear matter closely parallel those of the basic weak interaction discussed in Section I, and experimental information on the magnitudes of $|\Delta I| = 0$, 1, and 2 effects would be of great value for deciding whether extra currents exist. Although we have no immediately feasible experiments to propose, we will mention briefly a couple of examples[17)] of the kind of experiment which may prove successful after further work. The general idea is to find cases where the parity-conserving amplitude is strongly inhibited and dominates the parity-violating amplitude by less than the usual factor of 10^6 or 10^7.

One case which has already been studied intensively[18] is the parity forbidden α decay from the 8.8 MeV 2^-, $I = 0$ level of O^{16} to $C^{12}(0^+, I=0)+\alpha$. Upper limits of order 10^{-12} of normal parity-conserving α decay rates have been placed on its occurrence. If any decays were seen they would provide sure evidence that the initial or final state has an admixture of the opposite parity and same I spin. Thus if the search can be pushed somewhat farther and parity-violating α decays found, it should be possible to deduce the $|\Delta I| = 0$ weak interaction strength which will be particularly large if the octet is enhanced.

Needless to say, developments making it possible to observe a weak amplitude instead of a rate in α decay, or to observe $|\Delta I| = 1$ terms, would also be very useful.

For an example of another line of attack, which can be directed towards either $|\Delta I| = 0$ or $|\Delta I| = 1$ terms, consider the nucleus B^{10}. Gamma rays emitted in transitions from the 5.11-MeV, $I = 0$, 2^- level to lower 1^+, 2^+, or 3^+ levels with $I = 0$ are likely to exhibit an especially large circular polarization because of the following factors: (i) the parity-conserving E1 matrix element is inhibited. If the states were pure $I = 0$, the E1 matrix element would vanish for long wavelength; in practice, it is rescued from this fate by the small Coulomb admixture of $I = 1$ in the states.

(ii) The parity-violating M1 amplitude, which interferes with E1 to produce circular polarization, is enhanced by parity mixing between our 5.11-MeV, $I = 0$, 2^- level and the very nearby 5.16-MeV, $I = 1$, 2^+ level provided the weak interaction has a large $|\Delta I| = 1$ component to connect the two states. As a bonus, the M1 electromagnetic transition is then

$|\Delta I| = 1$ and makes use of the large nucleon isovector magnetic moment.

Thus, a survey of the gamma transition rates from the 5.11-MeV level of B^{10}, which are not very well understood at present, might suggest a favorable case for detection of a large $|\Delta I| = 1$ component in the weak interaction.

III. OCTET ENHANCEMENT IN WEAK AND ELECTROMAGNETIC INTERACTIONS

Coleman and Glashow[19)] have pointed out the analogy between the nonleptonic weak decay amplitudes and the electromagnetic mass differences of strongly interacting particles. In each case, the interaction is a product of an octet current with itself; the product contains $\underset{\sim}{27}$ as well as $\underset{\sim}{8}$. In the case of weak nonleptonic decays, the $\underset{\sim}{8}$ part has been found to be much larger than the $\underset{\sim}{27}$ part in all known processes. In the electromagnetic case, there are two supermultiplets for which good information is available. For the $J = 1/2^+$ baryon octet, the $\underset{\sim}{8}$ part of the electromagnetic mass difference strikingly predominates over the $\underset{\sim}{27}$, as evidenced by the approximately equal spacing of the Σ masses; for the $J = 0^-$ meson octet, the $\underset{\sim}{8}$ part of the difference in mass squared is only slightly bigger than the $\underset{\sim}{27}$ part. Despite the counterexample of the pseudoscalar octet, we may guess that there is a systematic tendency toward octet enhancement for the electromagnetic perturbation as well as the weak nonleptonic one.

The usual mechanism of enhancement, often referred to in terms of "tadpoles", has been discussed for many years in a variety of ways.[19-21)] We may describe it briefly as follows. The nonleptonic weak interaction acts like a scalar and a pseudoscalar "spurion" carrying zero energy and momentum. We may, however, write an unsubtracted dispersion relation in the four-momentum squared t carried by the nonleptonic weak interaction and evaluate the dispersion integral at $t = 0$. We may furthermore suppose that the integral is dominated by low-lying meson states with suitable quantum numbers. Now if scalar and pseudoscalar meson octets of low mass exist, while 27-plets do not, then we have an explanation of octet predominance over the 27. Moreover, this explanation can be generalized, more or less, to the electromagnetic as well as strong mass differences.[19]

An important feature[8] of the "tadpole" mechanism is that the charge-conjugation properties of the parity-violating spurion octet and of the known pseudoscalar meson octet are opposite. For the $|\Delta Y| = 1$ part of the nonleptonic interaction, this means that the "tadpole" mechanism of octet enhancement is forbidden by SU(3) symmetry; of course, the enhancement effect could still be considerable despite the SU(3)-forbiddenness, like the ratio of $K_1^0 \to 2\pi$ and $K^+ \to 2\pi$ rates.[12,22] If we turn to the $\Delta Y = 0$ nonleptonic interaction, however, we see that the disagreement in charge conjugation properties between the parity-violating spurion and the pseudoscalar mesons completely forbids the tadpole mechanism of octet enhancement. For $\Delta Y = 0$, then, tadpoles are powerless to make $|\Delta I| = 0$ more important than $|\Delta I| = 2$ in parity-violating weak forces.

A somewhat different mechanism for octet enhancement has been proposed by Dashen and Frautschi,[23] extending earlier suggestions by Cutkosky and Tarjanne.[24] For purposes of describing the mechanism, we consider the specific case of electromagnetic mass corrections.

We assume the bootstrap theory and describe each strongly interacting particle as a bound state of strongly interacting particles. For simplicity, we consider two-body channels and single-body exchange forces only, treating unstable particles and stable ones on the same footing. Also, again for simplicity, we ignore strong violations of SU(3) symmetry.

Each supermultiplet now appears as a set of degenerate poles in the various two-body scattering amplitudes with appropriate quantum numbers. The electromagnetic interaction causes shifts in the positions of the poles, in each supermultiplet, from their original common value, by amounts that can be expressed by a mass shift matrix[25] δM_{ij}. Here, $i = 1, \ldots \nu$, where ν is the supermultiplicity. The shift in mass of the bound state occurs in response to electromagnetic mass shifts of the external particles and of the exchanged particles, as well as more direct electromagnetic effects involving photon exchange. To order e^2 in the electromagnetic perturbations, we have, for the various supermultiplets such as B ($J = 1/2^+$ baryon octet) and Π ($J = 0^-$ meson octet), a relation of the form

$$\delta M_{ij}^{B} = A_{ij,k\ell}^{BB}\, \delta M_{k\ell}^{B} + A_{ij,k\ell}^{B\Pi}\, \delta M_{k\ell}^{\Pi} + \ldots + D_{ij}^{B} \quad ,$$

$$\delta M_{i}^{\Pi} = A_{ij,k\ell}^{\Pi B}\, \delta M_{k\ell}^{B} + A_{ij,k\ell}^{\Pi\Pi}\, \delta M_{k\ell}^{\Pi} + \ldots + D_{ij}^{\Pi} \quad , \qquad (13)$$

etc.,

where the photon exchange effects have been lumped together in the driving terms D. In these equations, the A coefficients, because of SU(3) symmetry, have the important property that the various representations $\underset{\sim}{1}$, $\underset{\sim}{8}$, $\underset{\sim}{27}$, etc., do not mix. Let us consider one of these representations, say $\underset{\sim}{8}$. We enumerate all the relevant sets of matrices transforming like 8, namely D_{ij}^{n} and F_{ij}^{n} for the baryons B, D_{ij}^{n} for the mesons Π, and so forth, with $n = 1, \ldots 8$. We may now write

$$\text{Octet part of } \delta M_{ij}^{B} = {}_{8}\epsilon_{n}^{(1)} D_{ij}^{n} + {}_{8}\epsilon_{n}^{(2)} F_{ij}^{n} \quad ,$$

$$\text{Octet part of } \delta M_{ij}^{\Pi} = {}_{8}\epsilon_{n}^{(3)} D_{ij}^{n} \quad , \quad \text{etc.} \tag{14}$$

where the index α in $\epsilon_{n}^{(\alpha)}$ runs over all possible independent octets of mass matrices. Likewise, we break up the driving force in the same way:

$$\text{Octet part of } \Delta_{ij}^{B} = {}_{8}d_{n}^{(1)} D_{ij}^{n} + {}_{8}d_{n}^{(2)} F_{ij}^{n} \quad ,$$

$$\text{Octet part of } \Delta_{ij}^{\Pi} = {}_{8}d_{n}^{(3)} D_{ij}^{n} \quad , \quad \text{etc.} \tag{15}$$

We may now rewrite the octet part of the system of equations (13), replacing the coefficients $A_{ij,k\ell}^{BB}$ and so on by new coefficients ${}_{8}a_{\alpha\beta}$:

$${}_{8}\epsilon_{n}^{(\alpha)} = {}_{8}a_{\alpha\beta} \; {}_{8}\epsilon_{n}^{(\beta)} + {}_{8}d_{n}^{(\alpha)} \quad , \quad n = 1, \ldots 8 \quad , \tag{16}$$

where the ${}_{8}\epsilon$'s correspond to octet mass shifts and the ${}_{8}d$'s to octet driving terms.

In the same fashion, we can let an index λ run over all the independent 27-plets of mass matrices, and obtain a set of equations

$${}_{(27)}\epsilon_{m}^{(\lambda)} = {}_{(27)}a_{\lambda\mu} \; {}_{(27)}\epsilon_{m}^{(\mu)} + {}_{(27)}d_{m}^{(\lambda)} \quad , \quad m = 1, \ldots 27 \quad , \tag{17}$$

and so forth. For the electromagnetic perturbation in order e^2, there are driving terms only for 27 and 8, apart from the trivial singlet portion which does not split supermultiplets.

It is easy to see what is a natural enhancement mechanism in this formalism. We invert one of the equations for ϵ, say (16), obtaining

$$ {}_8\epsilon_n^{(\alpha)} = \left(\frac{1}{1 - {}_8\underline{a}} \right)_{\alpha\beta} {}_8d_n^{(\beta)} \quad , \tag{18} $$

with an octet mass difference operator pointing in the electric charge direction in SU(3) space, just like the octet part of the electromagnetic perturbation. If one (or more) of the eigenvalues of ${}_8\underline{a}$ is near unity, then ${}_8\epsilon_n^{(\alpha)}$ will contain a large term multiplying the associated eigenvector (s). If the matrix ${}_{27}\underline{a}$ lacks eigenvalues near unity, then the octet is preferentially enhanced.

Dashen and Frautschi have studied this mechanism in the particular case of the B octet. Using their N/D perturbation theory,[26] they can make rough estimates of the driving terms and the eigenvalues of $\underline{a}$, and they indeed find a strong octet enhancement. In their estimate, the B octet is treated as a bound state of ΠB only, with static model kinematics, strong forces largely from B, decimet, and ρ exchange, and the F/D ratio in BBΠ coupling suggested both by bootstrap calculations and by inference from observations. They investigate one eigenvalue for ${}_{27}\underline{a}$, which comes out negative, and two eigenvalues for ${}_8\underline{a}$, one of which is very close to unity; its associated eigenvector has a large component in the direction of B mass difference, with a well-defined F/D ratio. Thus they predict a strongly enhanced octet effect in baryon electromagnetic mass differences, with an F to D ratio which turns out to agree with experiment.

The formalism we have discussed for mass differences can evidently be generalized to include changes in coupling constants together with mass differences, so that our equations would be replaced by linear equations involving δg's and δM's. This extended formalism can presumably be applied to the weak nonleptonic perturbation and made to yield weak parity-conserving and parity-violating coupling constants. Again, octet enhancement can result from finding octet eigenvalues near unity, with 27-plet eigenvalues far away.

If this mechanism does in fact predict octet enhancement for the nonleptonic weak interaction, then the enhancement, even in the parity violating case, should apply to both $\Delta Y = 0$ and $|\Delta Y| = 1$ amplitudes, and without necessarily violating SU(3) symmetry. This is in contrast to the "tadpole" mechanism utilizing the Π octet, which fails to enhance the $|\Delta \underset{\sim}{I}| = 0$, $\Delta Y = 0$ amplitude above the $|\Delta \underset{\sim}{I}| = 2$, and fails to give the sum rule derived in Ref. 8 for $|\Delta Y| = 1$ parity-violating nonleptonic baryon decays.

IV. OCTET VIOLATION OF SU(3) SYMMETRY

We now turn to the violation of SU(3) symmetry in strong interactions. As is well known, the mass splitting among members of a supermultiplet occurs in an octet pattern, and we wish to emphasize how this property may be related to the octet violations of SU(3) symmetry in electromagnetic and weak interactions.

One possible explanation of SU(3) symmetry breaking, advanced by Cutkosky and Tarjanne[24)] and by Glashow,[25)] is that of spontaneous violation. To first order in such a violation, their theory can be described with the aid of our Eqs. (13) to (18); there are no driving terms "d" and one or more eigenvalues of ${}_8\underline{a}$ must be unity, while eigenvalues corresponding to other representations are not close to one. (When we take into account nonlinear effects, the octet eigenvalue in the linear perturbation theory need not necessarily be exactly one.)

If only one eigenvalue of ${}_8\underline{a}$ lies near one and provides the dominant term, the relative amounts of mass splitting in the π octet, Δ decimet, B octet with D and F terms, and so forth can be read off from the components of the associated eigenvector. Dashen and Frautschi[23] have extended their treatment of B to a reciprocal bootstrap on B and Δ, and find that in this simple model there are actually two eigenvalues of ${}_8\underline{a}$ lying near one. The correspondence between the Δ and B components of the associated eigenvectors and the observed Δ and B mass splittings is promising, although a detailed comparison with experiment must await more precise determination of the eigenvalues. Note that this result, while encouraging for the bootstrap mechanism, does not prove that spontaneous violation has occurred; a non-zero driving term would also be enhanced preferentially along the same eigenvectors.

In an effort to compare spontaneous violation with SU(3) mass splittings induced by a special force, let us turn from the orientation of the octet perturbation eigenvector in the space of Δ, B, π, etc., to the direction of the perturbation in the space of SU(3). We have seen that in the linear theory with a driving term, the resultant mass splitting has the same direction in SU(3) space as the perturbing force. If there is a special force responsible for the strong violations of SU(3), it must transform like the eighth component of an octet. In spontaneous breakdown of symmetry without any electromagnetic or weak perturbation, the linear approximation provides a homogeneous equation for the mass splitting;

with an eigenvalue of ${}_{8}\underline{a}$ equal to unity, the octet splitting can point in any direction in SU(3) space and can have any size.

We must, however, consider the nonlinearity of the problem and also the presence of electromagnetic and weak perturbations, which define directions in SU(3) space that have no simple relation to the direction finally chosen by the large mass splitting. The nonlinearity alone presumably introduces a scale for the splitting, but does not affect its arbitrariness of direction. The perturbations alone would be more and more enhanced the closer the eigenvalue of ${}_{8}\underline{a}$ is to unity, and the net result would be a splitting oriented in accordance with these perturbations. What happens when both are included is not clear, but it is not easy to see how the splitting finally emerges as the eighth component of an octet.

In an attempt to avoid this problem, Ne'eman[27)] has suggested that the bootstrap mechanism does not give rise to SU(3) violation, and must be supplemented by a special force of the current-current type that induces the large mass splittings. He refers to this force as the "fifth interaction", although we might just as well think of it as the "fourth interaction", since in the bootstrap scheme there is no other explicit strong coupling. The current presumably would transform partly like a singlet and partly like the eighth component of an octet, so that the interaction would contain $\underset{\sim}{1}$, $\underset{\sim}{8}$, and $\underset{\sim}{27}$. If we forget[28)] the suggestion that the force be long-range, there is no reason why the singlet part of the current should not be bigger than the octet part, so that the $\underset{\sim}{8}$ part of the interaction is more important than the $\underset{\sim}{27}$. In any case, the $\underset{\sim}{8}$ part would be preferentially enhanced.

We do not know whether the mass splitting can be described as spontaneous in the bootstrap theory, with some explanation of the direction in SU(3) space, or whether a force like Ne'eman's will turn out to cause it, but in the latter case we can say that octet enhancement is a common feature of weak nonleptonic effects, electromagnetic mass differences, and strong violations of SU(3) symmetry.

REFERENCES

1. M. Gell-Mann, Phys. Rev. 125, 1067 (1962).
2. N. Cabibbo, Phys. Rev. Letters 10, 531 (1963).
3. M. Gell-Mann, Physics Letters 8, 214 (1964).
4. The beauty of Eqs. (3) and (4) lies in the principle of universality. We may define, for each part of the current, a charge operator $W = -i \int J_4 \, d^3x$. The leptonic weak charge operator W^ℓ and its hermitian conjugate obey the equal time commutation relations characteristic of an angular momentum; that is, $W^\ell = Z(K_1{}^\ell + i K_2{}^\ell)$, where $\left[K_1{}^\ell, K_2{}^\ell\right] = i K_3{}^\ell$, etc. Now for universality we want to have $W^h = 2(K_1{}^h + i K_2{}^h)$, where $\left[K_1{}^h, K_2{}^h\right] = i K_3{}^h$, etc. Equation (4) assures us of this relation if and only if the operators $F_i = -i \int \mathcal{F}_{i4} \, d^3x$ and $F_i{}^5 = -i \int \mathcal{F}_{i4}^5 \, d^3x$ generate at equal times the algebra of SU(3) × SU(3). See M. Gell-Mann, Physics (in press), and M. Gell-Mann and Y. Ne'eman, Ann. Phys. (in press).
5. N. Cabibbo, Phys. Rev. Letters 12, 62 (1964).
6. B. Lee, Phys. Rev. Letters 12, 83 (1964).
7. H. Sugawara, Progr. Theoret. Phys. (Kyoto) 31, 213 (1964).
8. M. Gell-Mann, Phys. Rev. Letters 12, 155 (1964).
9. B. d'Espagnat, Physics Letters 7, 209 (1963).
10. G. Barton, Nuovo Cimento 19, 512 (1961).
11. For recent versions of these arguments, see A. Salam, Physics Letters 8, 216 (1964), and J. Schwinger, Phys. Rev. Letters 12, 630 (1964).

12. There is also an argument that seems to favor slightly the hypothesis of extra currents.[5,8] With the octet part O_8 of the nonleptonic interaction, the decay $K_1^{\,0} \to 2\pi$ is forbidden by SU(3) symmetry and CP conservation, even though it is allowed by the $|\Delta I| = 1/2$ rule. The ratio of the rate of $K_1^{\,0} \to 2\pi$ to that of $K^+ \to 2\pi$, which is around 500, would be even larger if it were not for the SU(3) forbiddenness. A proponent of the extra current hypothesis could then argue that the $|\Delta I| = 3/2$, 27-plet term in the $K \to 2\pi$ amplitude comes only from electromagnetic corrections, with the 27-plet otherwise cancelled by extra currents as in Eq. (12). However, as emphasized by Salam,[11] if we believe the crude estimates that show the $|\Delta I| = 3/2$, 27-plet amplitude to have a normal order of magnitude, then we must in any case attribute the large $K_1^{\,0} \to 2\pi/K^+ \to 2\pi$ ratio to enhancement of the octet rather than cancellation of the 27-plet.

13. F. Boehm and E. Kankeleit, to be published.

14. R. J. Blin-Stoyle, Phys. Rev. 118, 1605 (1960).

15. C. Michel, Phys. Rev. 133, B329 (1964).

16. An enhancement of this type is consistent with the result of Boehm and Kankeleit, who find a circular polarization substantially larger than the polarization predicted by Wahlborn (to be published) on the basis of Michel's estimate of the weak potential. However, the nuclear theory connecting the two-nucleon weak potential to circular polarization is not sufficiently reliable, for a heavy nucleus like Ta^{181}, to permit us to draw firm conclusions at the present time.

17. We wish to thank Professor C. A. Barnes for discussions of these problems in nuclear physics.

18. W. Kaufmann and H. Waffler, Nuclear Phys. 24, 62 (1961); D. E. Alburger, R. E. Pixley, D. H. Wilkinson, and P. Donovan, Phil. Mag. 6, 171 (19); R. E. Segel, J. W. Olness, and E. L. Sprenkel, Phys. Rev. 123, 1382 (1961).

19. S. Coleman and S. Glashow, Phys. Rev. 134, B671 (1964).

20. A. Salam and J. C. Ward, Phys. Rev. Letters 5, 390 (1960).

21. E. McCliment and K. Nishijima, Phys. Rev. 128, 1970 (1962).

22. If the tadpole mechanism really dominates parity-violating amplitudes through breaking of SU(3), the sum rule derived from SU(3) in Ref. 8 for parity-violating baryon decays should not hold.

23. R. F. Dashen and S. C. Frautschi, to be published.

24. R. E. Cutkosky and P. Tarjanne, Phys. Rev. 132, 1355 (1963).

25. S. L. Glashow, Phys. Rev. 130, 2132 (1963).

26. R. F. Dashen and S. C. Frautschi, Phys. Rev. (in press).

27. Y. Ne'eman, Phys. Rev. 134, B1355 (1964).

28. D. Beder, R. Dashen, and S. Frautschi, (to be published). We may also ignore the suggestion in Ref. 27 that the strong current-current interaction has a connection with the muon mass problem.

POSSIBLE UNITARY SYMMETRY PROPERTIES OF THE INTERMEDIATE BOSONS AND THEIR INTERACTIONS

B. D'ESPAGNAT

Laboratoire de Physique Théorique et Hautes Energies, Orsay (S & O) France and CERN, Geneva, Switzerland

Received 21 October 1963

Two assumptions currently made are that
a) the bulk of the strong interactions is invariant under the group U_3 (unitary symmetry) and
b) the weak interactions are mediated by weakly coupled vector mesons. The question then is raised whether or not the weak interactions and their intermediate bosons also have something to do with the group U_3. Assuming that they do, one of course has to inquire what these connections could be and how they could be observed.

One very natural and mild hypothesis is that the *currents* J_i of strongly interacting particles which appear in the weak Lagrangian are members of a representation of U_3. Because of the fact that some of these currents are conserved this representation must then be "8". As shown by Cabbibo [1)] this hypothesis, as regards the leptonic decays of the strange particles, already allows quite specific predictions which seem moreover to agree with experiment. The hypothesis which we here propose to add to this one is that also the intermediate mesons, w_k, be ascribed to a definite representation of U_3. This new assumption is in a sense complementary to the former since the w's and the J's both appear in the Lagrangian. Its justification however lies not only there but also in the fact that, as shown below, it makes possible to account simultaneously, and in what looks a natural way, for the two main features of the non-leptonic decays, namely the $|\Delta S| \neq 2$ and $|\Delta I| = \frac{1}{2}$ rules.

Of course, ascribing the w's to a definite representation makes sense only if their interactions have some group property. This therefore implies some further assumption. The simplest possible choice that we can make in that respect is to associate *the weak interaction Lagrangian* $\mathcal{L}$ *with the three dimensional representations of the group*. More precisely speaking, let us assume that

$$\mathcal{L} = L + L^{\dagger} \,, \tag{1}$$

L being in "3" and all boson fields being complex (see eq. (2) below for more details). It is, we believe, worth noticing that this simplest and most straightforward assumption is sufficient to account for the two empirical rules stated above. The proof is quite simple. Let H be the Hamiltonian corresponding to $\mathcal{L}$ and let us first imagine that the strong interactions are completely switched off. Then a non-leptonic decay is described by a second-order matrix element which is a sum of products of first order ones and which therefore is necessarily a combination of members of the "1" and "8" representations that constitute the product of "3" by "$\bar{3}$" (the bosons w, of course, must be emitted and absorbed *). This matrix element then obeys $|\Delta S| \leq 1$ and $|\Delta I| \leq 1$ since neither "1" nor "8" contain objects with $|\Delta S|$ or $|\Delta I| > 1$. It can be, if one so chooses, derived from a (somewhat non-local) Fermi type Hamiltonian H_f involving four fields and which itself satisfies the above rules. If we now switch on the strong interactions (and if we remember that the w's themselves have no strong interactions so that all renormalizations and such like effects act the same way on H and H_f) we see by the usual argument that these cannot spoil the rules under discussion, which therefore hold.

If $\mathcal{L}$ is in "3" and the J_i are in "8", then the w_k are either in "3" in "$\bar{6}$" or in "15". This gives then three possible models and no more. The choice between these models is, we feel, primarily an experimental problem. The first one for instance cannot accomodate any $\Delta S = -\Delta Q$ interaction. On the contrary the second and the first one can, because it is then possible to introduce some admixture of currents belonging to the representation "10". This admixture should be small enough so as to preserve an approximate conservation of the overall $\Delta S = 0$ currents: but it *can* be coupled to "$\bar{6}$" or to "15" in order to give "3" for the Lagrangian, as is the case for the main term. As some of these "10" currents allow for $\Delta S = -\Delta Q$ it is clear that a certain amount of $\Delta S = -\Delta Q$ transitions can easily be accounted for. This however is at the price of increasing the number of complex meson fields from three to at least six, and the ex-

* L being in "3", if L contains the emission operator of a boson w_k, the absorption operator of this same boson is necessarily in $L^{\dagger}$ and not in L, as is easily seen.

perimental evidence on these interactions is still too scarce to make this model based on "6" particularly likely.

As a final remark we would like to point out the following. In each of these models it is of course very easy to write down $\mathcal{L}$ explicitly. The "$\bar{3}$" representation to which $\mathcal{L}$ belongs is chosen so as to have two electrically neutral components, so that $\mathcal{L}$ is made of two terms

$$\mathcal{L} = a F_{"3","n"} (J_i, w_k) + b F_{"3","\Lambda"} (J_i, w_k) + \text{h.c.}\,, \tag{2}$$

where a and b are parameters and where the functions $F_{"3","n"}$ and $F_{"3","\Lambda"}$ are the two members of the "3" representation that behave as the "n" and the "Λ" of the Sakata model respectively. a and b may then be regarded as the components of a universal "spurion in "$\bar{3}$"" as shown in fig. 1. Formally $\mathcal{L}$ can be written as

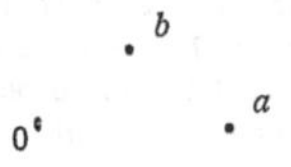

Fig. 1. A spurion in $\bar{3}$.

$$\mathcal{L} = S_{"\bar{3}"} F_{"3"} + \text{h.c.} \tag{3}$$

and is then to be considered as an invariant in U_3. Now it is possible to define new currents $J_i^!$ and new bosons w_k' which are linear combinations of the old ones and which are obtained by applying to the latter one of the charge conserving transformations of the group. Let this transformation * be characterized by an angular parameter θ and let us call it $R(\theta)$. The spurion $S(a, b)$ transforms as a doublet under $R(\theta)$ as is easily seen by inspection of fig. 1 (R acts parallel to the third root E_3 of SU_3 in the notations of ref. [3]). Therefore

$$a' = a \cos\theta + b \sin\theta, \quad b' = -a \sin\theta + b \cos\theta\,,$$

and there is a choice of θ for which a' vanishes. Let us make this choice and let us consider the subgroup "G" of U_3 conjugated to the isospin group "I" by means of $R(\theta)$. $\mathcal{L}$ (eq. (3)), being an invariant in U_3, is also a scalar in "G". On the other hand, it is of course a linear and homogeneous function of a' and b' and, because of our choice $a' = 0$, b' factors out. Expressed in terms of the primed quantities $J_i^!$ and w_k', the fundamental Lagrangian (1) of the theory is therefore a scalar in "G". This result shows in particular that the present theory is a special case - adapted to the intermediate boson picture - of a more general description of possible links between the two $|\Delta I| = \frac{1}{2}$ and $|\Delta S| \neq 2$ rules that has recently been proposed [2].

As regards the leptonic decays these depend of course on the particular model that is selected. It is easily seen however that generally speaking the theory is very similar in that respect to that of ref. [1], with θ playing the same role in both cases, and that it can easily be made to produce also the same results.

We would like to conclude by stressing the generality of the present suggestion and of the accompanying proof, which does not depend on the particular representation chosen for the w's among the three possible ones. The theory of course will have to be particularized and it is somewhat irritating that because of the absence of neutral leptonic currents the leptons themselves cannot be usefully inserted into the group. On the other hand the non leptonic $|\Delta I| = \frac{1}{2}$ rule seems here to emerge rather naturally from the general assumptions and, as in ref. [2], is tightly linked with $|\Delta S| \neq 2$ instead of the necessitating *ad hoc* Lagrangian terms. A kind of symmetry, moreover, concerning the properties of the different types of interactions with respect to the group is noteworthy: all the three types of interactions, strong, electromagnetic and weak are invariant under an SU_2 subgroup of U_3 when the other two can be neglected. This kind of symmetry, which was postulated in ref. [2] is here derived from our postulates.

In the model where the w's are in "3" one combination of the six mesons does not interact, as can easily be seen. In the present approach the minimum number of real distinct bosons is therefore five. Detailed quantitative investigations of the three models are in progress.

It is a pleasure to acknowledge many illuminating discussions with M. Gourdin and J. Prentki. Part of this work was carried out at the 1963 Summer Institute for Theoretical Physics, University of Wisconsin. The author would like to thank Professor R. G. Sachs for the hospitality of the Institute.

References

1) N. Cabbibo, Phys. Rev. Letters 10 (1963) 531.
2) B. d'Espagnat and J. Prentki, Nuovo Cimento 24 (1962) 497.
3) R. E. Behrends et al., Revs. Modern Phys. 34 (1962) 1.

* A more explicit investigation of $R(\theta)$ can be found in ref. [2].

* * * * *

NONLEPTONIC WEAK DECAYS AND THE EIGHTFOLD WAY*

Murray Gell-Mann
California Institute of Technology, Pasadena, California
(Received 2 January 1964)

Assuming that the strongly interacting particles approximately obey the "eightfold way" symmetry,[1,2] we may try to investigate the consequences of the further assumption that the weak interactions of baryons and mesons have simple transformation properties under the same symmetry.

It was suggested[1] that the weak current for interaction with leptons consists of a vector term, which is the current of a component of the F spin, plus an axial-vector term that constitutes the same component of another unitary octet. This model leads to the selection rules $\Delta Y = 0$, $|\Delta \vec{I}| = 1$ and $\Delta Y/\Delta Q = +1$, $|\Delta \vec{I}| = 1/2$ for the current, as well as a conserved vector current for $\Delta Y = 0$ and a nearly conserved vector current for $\Delta Y/\Delta Q = +1$. Let us suppose that the model is correct.

In reference 1, the component of the F spin involved was taken to be $F_1 + iF_2 + F_4 + iF_5$ (equal strength for $\Delta Y = 0$ and $\Delta Y/\Delta Q = +1$). However, an alternative and, in some ways, more attractive form for universality was suggested by Gell-Mann and Lévy,[3] which amounts in the eightfold way to taking a different component, namely, $\cos\theta(F_1 + iF_2) + \sin\theta(F_4 + iF_5)$ with θ small. This form of the eightfold-way model has been studied in an elegant paper of Cabibbo,[4] who finds that several experiments indicate a consistent value of θ around 0.26 radian.

Let us now apply the model to the nonleptonic weak interactions formed from the weak current times its Hermitian conjugate; such a current-current interaction may, of course, be carried by an intermediate boson. The strangeness-changing part of this nonleptonic interaction is then the sixth component of a unitary octet plus a term that transforms like the representation 27. The octet part obeys $|\Delta \vec{I}| = 1/2$ while the 27 part has $|\Delta \vec{I}| = 1/2, 3/2$. Both obey $|\Delta Y| = 1$, of course.

There are two simple possibilities that may explain the predominance[5] of $|\Delta \vec{I}| = 1/2$ in nonleptonic decays: (a) There are other terms, formed from neutral current products, such that $|\Delta \vec{I}| = 3/2$ is canceled out, with $|\Delta Y|$ still equal to 1, or (b) some mechanism, such as the predominance of scalar and pseudoscalar meson intermediate states, selectively enhances the contribution of $|\Delta \vec{I}| = 1/2$ "spurions" compared with that of $|\Delta \vec{I}| = 3/2$ "spurions." Let us assume that in either case the mechanism that gets rid of the $|\Delta \vec{I}| = 3/2$ term gets rid of the whole 27 and leaves just the octet,[6,7] for $|\Delta Y| = 1$.

Now every octet that goes into itself under charge conjugation has a characteristic number $\mathcal{C} = \pm 1$, which is the charge-conjugation quantum number C of its 1, 3, 4, 6, and 8 components; the charge-conjugation quantum number C of the 2, 5, and 7 components is $-\mathcal{C}$. When two octets are combined to make a third octet, the value of $\mathcal{C}$ for the third one is the product of the $\mathcal{C}$ values for the first two octets times η, where η is +1 when the octets are coupled through the symmetric symbol d_{ijk} (which is nonvanishing only when an even number of 2's, 5's, and 7's occur among the indices) and -1 when they are coupled through the antisymmetric symbol f_{ijk} (which is nonvanishing only for an odd number of 2's, 5's, and 7's).

The vector current has $\mathcal{C} = -1$, like the vector-meson octet, while the axial-vector current has $\mathcal{C} = +1$ (with no $\mathcal{C} = -1$ impurity ever proved) like the pseudoscalar meson octet. In forming the nonleptonic interaction, the symmetric coupling through the d_{ijk} symbol is what comes in and thus the octet that results has $\mathcal{C} = +1$ for the parity-conserving (p-c) term and $\mathcal{C} = -1$ for the parity-violating (p-v) term. The sixth component thus has $CP = +1$ in each case.

By contrast, the K_1^0 particle, in the eightfold

way, is the seventh component of a pseudoscalar octet with $\mathcal{C} = +1$. Again $CP = +1$, but the transition between the p-v spurion and $K_1{}^0$ is forbidden by SU(3). This fact tends to make alternative (b) above somewhat less attractive than it was without unitary symmetry.

Next, consider the transition from the p-v spurion to $K_1{}^0 + \pi^+ + \pi^-$ or $K_1{}^0 + \pi^0 + \pi^0$ in a symmetrical state, which is responsible for the decays $K_1{}^0 \to 2\pi$. Since the K and π's all belong to the same octet, the unitary spin coupling of the three octets to form the fourth must be totally symmetric. With this coupling, we can form out of $K_1{}^0 + 2\pi$ the seventh component of a pseudoscalar octet with $\mathcal{C} = +1$ (like $K_1{}^0$) but not the sixth component of a pseudoscalar octet with $\mathcal{C} = -1$ (like the p-v spurion). Thus, $K_1{}^0 \to 2\pi$ is forbidden[8] by SU(3).

Finally, the property $\mathcal{C} = -1$ of the p-v spurion gives one condition on the four independent observable p-v amplitudes A in the weak decay of strange baryons into baryon plus pion:

$$-A(\Lambda \to p + \pi^-) + 2A(\Xi^- \to \Lambda + \pi^-) = \sqrt{3}A(\Sigma^+ \to p + \pi^0).$$

This condition, compatible with experimental evidence, is valid in the limit of unitary symmetry in the model presented here. In references 6 and 7, the same condition is found for both p-v and p-c amplitudes, but on the basis of R invariance, which does not appear to be a good approximation, at any rate for baryons.

---

*Work supported in part by the U. S. Atomic Energy Commission.

[1]M. Gell-Mann, California Institute of Technology Synchrotron Laboratory Report CTSL-20, 1961 (unpublished); Phys. Rev. 125, 1067 (1962).

[2]Y. Ne'eman, Nucl. Phys. 26, 222 (1961).

[3]M. Gell-Mann and M. Lévy, Nuovo Cimento 16, 705 (1960). See also M. Gell-Mann, Proceedings of the Tenth Annual International Rochester Conference on High-Energy Physics, 1960 (Interscience Publishers, Inc., New York, 1960).

[4]N. Cabibbo, Phys. Rev. Letters 10, 531 (1963).

[5]The current evidence for this rule is reviewed by R. H. Dalitz (to be published).

[6]Benjamin Lee (to be published).

[7]Hirotaka Sugawara (to be published).

[8]The author is much indebted to Professor Cabibbo for a discussion of this matter. N. Cabibbo (to be published) points out that with $K_1{}^0 \to 2\pi$ forbidden by SU(3), the comparatively high rate of $K^+ \to 2\pi$ need not be an obstacle to a theory of purely electromagnetic violation of the nonleptonic rule $|\Delta\vec{I}| = 1/2$, in accordance with possibility (a) above. One simple theory of type (a) would involve a single neutral current, transforming like $(-F_3 - 3^{-1/2}F_8)\cos\theta + (F_6 + iF_7)\sin\theta$, to accompany the single charged current transforming like $(F_1 + iF_2) \times \cos\theta + (F_4 + iF_5)\sin\theta$. Each would be multiplied by its Hermitian conjugate.

X

BROKEN SYMMETRY AND THE BOOTSTRAP HYPOTHESIS

The bootstrap idea has already been discussed extensively in several places, including the introduction to Part I and the leading article in Part IX.

The most hopeful attitude to the bootstrap hypothesis leads to the expectation that it will explain the whole system of hadrons with no free parameters except the scale of the masses. Such an attitude may be one of extravagant optimism, considering the primitive state of our present knowledge. Let us see what would be involved in the matter of symmetry: (1) The notion of symmetry under a Lie group would have to emerge as a consequence. (2) The particular symmetry group SU(3) would have to be selected, as well as the correct behavior (whatever it is), with respect to triality. (3) The violation of SU(3), following the octet pattern and reducing the symmetry to SU(2), would have to come out as well. (4) The direction of the violation in SU(3) space is perfectly arbitrary in the bootstrap model as long as electromagnetic and weak perturbations are ignored. When they are turned on, however, the dynamics of these perturbations would have to permit the violation to lie in the observed direction, which is skew with respect to the directions defined by the electromagnetic and the weak current.

The first paper, by Cutkosky, addresses itself to the first point. He looks at the reduced problem of vector mesons bootstrapping themselves and concludes that symmetry under a Lie algebra appears naturally, much as in the Yang-Mills formalism. Unfortunately, he has imposed some restrictions, namely degenerate masses and conservation of electric charge. Indications have since appeared that perhaps the first restriction can be removed [Abers, Zachariasen, and Zemach, Phys. Rev., 132, 1831 (1963)]. Considerations similar to those of Cutkosky were put forward by R. H. Capps [Phys. Rev. Letters, 10, 312 (1963) and elsewhere].

None of our papers attacks the second point. It is conceivable that the choice of the symmetry group requires information external to the pure bootstrap, such as the algebra generated by the weak and electromagnetic currents. On the other side, there has been some speculation by Chew and his collaborators that this whole problem may be avoided if there is in fact a hierarchy of SU(n) symmetries, with larger violations as n gets larger. [It should be noted that the adjoint representation of SU(4) contains SU(3) triplets, so hadrons of triality $\tau \neq 0$ would be required for the success of this idea.]

The third point is treated in the papers by Glashow and by Cutkosky and Tarjanne and in some recent work of Capps. All these authors consider the possibility of spontaneous violation of SU(3) symmetry, with octet violation reinforcing itself and other violations damping themselves out. A discussion of these ideas has been given in the first article of the previous part.

The fourth point is taken up by Ne'eman in the last article. (This matter also has been reviewed in the paper by Dashen, Frautschi, Gell-Mann, and Hara.) Ne'eman suggests that it may be impossible for the pure bootstrap, with electromagnetic and weak perturbations only, to explain the direction of SU(3) violation. He therefore proposes a "fifth interaction," outside the bootstrap, as a mechanism for the violation. Actually, we might call it the "fourth interaction," since there is no explicit strong coupling in the pure bootstrap model, just dynamics resulting from kinematics and consistency. In the article, it is mentioned that the new interaction might be of long range, might be connected with the muon mass, and should be coupled to a particular combination of hypercharge and baryon number; all these features are inessential and should not be considered part of the basic proposal.

A Mechanism for the Induction of Symmetries Among the Strong Interactions

R. E. Cutkosky
Carnegie Institute of Technology, Pittsburgh, Pennsylvania
(Received 4 April 1963)

A model is constructed in which there are N equally massive vector mesons, which are self-consistent bound states of pairs of these same vector mesons. It is shown that N must be equal to the number of parameters of some compact, semisimple Lie group, and that the renormalized coupling constants must be proportional to the structure constants of the group.

THE strong interactions are known to exhibit isotopic spin symmetry, which is based on the group SU_2. There is also evidence that a further symmetry described by SU_3 exists, although this further symmetry is certainly much more approximate. In this paper, we raise the question of whether these symmetries might have a simple dynamical origin.

A phenomenological symmetry related to a Lie group is understood to mean two things. First, the mass spectrum of particles which have the same spin and parity should show a clustering which can be identified with the multiplet structure corresponding to representations of the group. Second, the S-matrix elements, and in particular, the renormalized coupling constants referring to different particles from the same multiplet should be related, approximately, through Clebsch-Gordan coefficients. It is clear that the origin of such a phenomenological symmetry could be established only through the development of a rather complete understanding of strong-interaction dynamics. Nevertheless, there is a possibility that the examination of simple models might help to clarify the problem. We are encouraged in this investigation by the observation that the ratios of the masses and coupling constants of particles which have the same spins and parities are somewhat easier to calculate than are other quantities.

We consider here a very simple model in which it is assumed that there are a number (N) of vector mesons which have the same mass. That is, it is assumed that the mass spectrum does exhibit a clustering effect. Then, with a number of additional assumptions that will be made clear, it will be shown that a Lie group can be associated with these particles. The essential feature of the model is that only these N vector mesons are introduced. They are all supposed to arise as self-consistent bound states of pairs of vector mesons, and the binding force is mediated by the exchange of single vector mesons. In this paper we shall assume that the N states are exactly degenerate.

We assume the spatial invariances usually associated with strong interactions, as well as invariance under charge conjugation. We may then represent the particles by real vector fields A_a, $a=1, \cdots, N$. The three-meson vertex has the form

$$g(F_{abc}\Theta+F_{abc}'\Theta')A_aA_bA_c, \quad (1)$$

where Θ and Θ' are certain functions of the momenta. Symmetry with respect to interchange of all three particles leads to the result that there are only two independent terms, and that F_{abc} and F_{abc}' are imaginary and antisymmetric, one in two, the other in all indices. We shall investigate only the case that $F_{abc}'\equiv F_{abc}$.

Two mesons can interact by exchanging another between them. The Born-approximation scattering amplitude is proportional to

$$V_{ab,cd}=F_{adr}F_{bcr}-F_{acr}F_{bdr} \quad (2)$$

in states which are antisymmetric in the spatial variables. To the extent that the sign of the force can be deduced from the "Coulombic" part of the interaction, it will be attractive between particles which have opposite charges, that is, when the expectation of $V_{ab,cd}$ is positive. In fact, the attraction is very singular at short distances, so we must cut it off at an effective mass value denoted by Λ.

The problem of actually solving the partial-wave dispersion relations following, for example, the method of Zachariasen and Zemach,[1] would be extremely complicated. Fortunately, we do not need to solve them; we only need to show that solutions having the desired properties exist. For this, it is sufficient to observe that if we have an attractive force, we can always adjust Λ so that a bound state of any desired mass—in particular, the vector-meson mass, which we take to be unity—could be obtained. The coupling constants for the bound

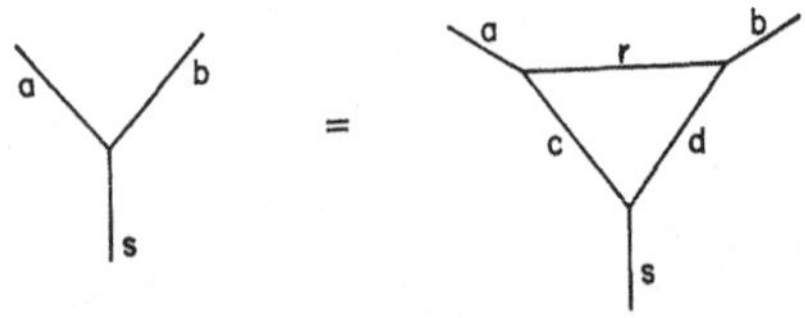

FIG. 1. A graphical representation of Eq (3). Another interpretation of Eq. (3) is that the renormalized coupling constants are considered to be generated by the simplest irreducible vertex part, with the "bare-coupling constants" set equal to zero.

state to the two vector mesons are obtained from the residue of the bound-state pole. We do not examine the different helicity amplitudes of the bound-state pole, because these are sensitive to the way the cutoff is introduced.

Since all the particles which are being bound together, and also all of the exchanged particles, have the same mass, it is clear that we can obtain N degenerate bound states only if V has N degenerate eigenvalues. Moreover, the F_{abs} must themselves be the eigenvectors;

$$\lambda F_{abs}=V_{ab,cd}F_{cds}, \quad (3)$$

where $\lambda>0$. Equation (3) can be pictured as in Figure 1. In addition, since the same residue will be obtained in each of these dynamically equivalent states, we can impose the normalization condition

$$F_{abr}F_{bas}=\delta_{rs}. \quad (4)$$

In terms of elementary wave-mechanical ideas, this can be interpreted as saying that in each vector-meson state, the total probability of all virtual pairs is unity. The actual value of the residue must be g^2, which leads to a second relation between g^2 and Λ.

There are $\frac{1}{2}N(N-3)$ antisymmetric eigenvectors ψ_{ab}^i of $V_{ab,cd}$ which are orthogonal to F_{abs}. We denote their eigenvalues by λ_i and the degeneracy by d_i. There are, therefore, $\frac{1}{2}N(N-3)$ vector states in which the dynamics is identical to that in the N self-consistent states, except that the effective strength of the force is multiplied by a factor λ_i/λ. It seems plausible that, for given values of g^2 and Λ, the mass of a bound state would depend monotonically upon the eigenvalue of V. We must require that no other vector particles which have a lower mass (or even a slightly larger mass) than the N we started with should arise from the potential, because otherwise our model would not really be self-consistent; we would have to start over again and include the extra particles from the beginning. We, therefore, require that

$$\lambda>\lambda_i. \quad (5)$$

We shall see that this inequality is crucial to the determination of the F's.

It is clear that strong interactions in states with other spins and parities might also arise, and if so, these other states ought to be incorporated into the model.

[1] F. Zachariasen and C. Zemach, Phys. Rev. **128**, 849 (1962).

We ignore such states here, with the understanding that the generality of our conclusions depends on the validity of the assumption that the influence of other states upon the vector particles can be adequately represented by the single adjustable parameter Λ.

We have now defined our model, and proceed to the determination of F_{abc} which satisfy the conditions (3), (4), and (5). We first observe that

$$N\lambda = F_{bas}V_{ab,cd}F_{cds} = 2F_{bas}F_{adr}F_{bcr}F_{cds}. \tag{6}$$

Now, we have from the definition of the trace that

$$\mathrm{Tr}\, V^2 = N\lambda^2 + \sum_i d_i\lambda_i^2, \tag{7}$$

but in our case, the explicit form (2) gives, with the help of (4) and (6),

$$\mathrm{Tr}\, V^2 = 2N - N\lambda. \tag{8}$$

Therefore, it follows that

$$\lambda(\lambda+1) = 2 - N^{-1}\sum_i d_i\lambda_i^2, \tag{9}$$

so that

$$\lambda \leq 1. \tag{10}$$

The equality can hold only if all λ_i vanish.

Under the orthogonal transformations $A_a' = A_b O_{ab}$ the equations (3) and (4) are covariant. For infinitesimal transformations $O_{ab} = \delta_{ab} + i\epsilon^\alpha G_{ab}^\alpha$, the F_{abc} transform according to

$$F_{abc}' = F_{abc} + i\epsilon^\alpha f_{abc}^\alpha,$$

$$f_{abc}^\alpha = F_{xbc}G_{xa}^\alpha + F_{axc}G_{xb}^\alpha + F_{abx}G_{xc}^\alpha. \tag{11}$$

We shall now make a further assumption, that the interactions satisfy a nontrivial additive conservation law. For instance, we might assume that some of the particles have electric charges, and that charge is conserved at each vertex. This requirement gives additional information about the F_{abc}. It is not clear whether this additional information is necessary, but it greatly simplifies our analysis. If we assume such an additive conservation law, we can make gauge transformations of the first kind which leave the F_{abc} invariant. In the representation in which real A_a are used, these gauge transformations are just a special case of the orthogonal transformations introduced above. If we assume the existence of r independent additive conservation laws, (for example, charge and hypercharge) the F_{abc} will be invariant under an r parameter Abelian subgroup of O_N.

We denote by G_{ab}^A a generator of the Abelian gauge transformations. Then we have

$$F_{xbc}G_{xa}^A + F_{axc}G_{xb}^A + F_{abx}G_{xc}^A = 0. \tag{12}$$

If we multiply (12) by F_{bad}, and use the fact that $G_{ab}^A = -G_{ba}^A$, we obtain

$$G_{cd}^A = V_{cd,ab}G_{ab}^A. \tag{13}$$

The generators G_{ab}^A are eigenvectors of V with a unit eigenvalue.

There are now two cases to be considered. If $F_{bac}G_{ab}^A = 0$, we have found another eigenvalue: $\lambda_A = 1$. Then, we see from (9) that $\lambda < 1$, so that $\lambda < \lambda_A$, and the condition (5) is violated. If, on the other hand, $F_{bac}G_{ab}^A \neq 0$ for some c, we have $\lambda = 1$, so that all λ_i vanish. The completeness of the eigenvectors of V then allows us to write

$$V_{ab,cd} = F_{abr}F_{dcr}, \tag{14}$$

which can be rewritten as

$$F_{abr}F_{cdr} + F_{bcr}F_{adr} + F_{car}F_{bdr} = 0. \tag{15}$$

Equations (15) and (4) are the necessary and sufficient conditions for the F_{abc} to be (apart from a factor i) the structure constants of a compact, semisimple Lie group. It is evident that the rank of the group must be at least as great as r. If we assume that the N particles cannot be divided into subsets which interact only among themselves, in some representation, the group must be simple.

It may be of interest to observe that examples of F_{abc} which satisfy (3) and (4) but violate (5) are provided by the Clebsch-Gordan coefficients of many group representations. For example, for SU_2, any representation corresponding to an odd-integral value of the isospin T can be used. In particular, one calculates for $T=3$, $\lambda = -1$; for $T=5$, $\lambda = +0.37$. For all T, in accordance with (13), in the state $T'=1$ we have $\lambda_1 = 1$.

Our conclusion is that a set of N degenerate vector mesons, which interact in accordance with the model set forth, can always be associated with a Lie group. Moreover, the association is necessarily with a particular representation of the group, the adjoint representation. Note that this association with the adjoint representation is exactly the same as that implied by a Yang-Mills type of theory, in which invariance under a simple group of transformations is assumed at the very beginning.[2,3]

Our model also leads to certain negative conclusions. We observe that λ, and hence, the cutoff Λ, is independent of the group. This suggests that it might be hard to use an extension of the self-consistent model, in which an attempt was made to calculate an *a priori* value of Λ from dynamical arguments, to eliminate certain groups from consideration. One must keep in mind the possibility that, in addition to certain additive quantum numbers which might be supposed to be exactly conserved, there might be other additive quantum numbers which arose out of the strong interactions themselves. However, it would certainly be very hard to determine the number of such quantum numbers from the self-consistency requirement.

ACKNOWLEDGMENTS

The author wishes to thank Jorgen Kalckar, Professor Alfred Schild, and Dr. Pekka Tarjanne for helpful discussions.

[2] C. N. Yang and R. Mills, Phys. Rev. **96**, 191 (1954).
[3] S. L. Glashow and M. Gell-Mann, Ann. Phys. (N.Y.) **15**, 437 (1961).

Spontaneous Breakdown of Octet Symmetry*

S. L. GLASHOW†
Physics Department, University of California, Berkeley, California
(Received 7 December 1962)

A model of strong interactions with the octet symmetry of Gell-Mann and Ne'eman is considered, and a spontaneous breakdown of this symmetry leading to nondegenerate baryon masses is sought. The Gell-Mann mass formula is deduced for the physically relevant symmetry-breaking solutions.

INTRODUCTION

CONSIDER the possibility that octet symmetry[1] exactly characterizes strong interactions—that there are no medium-strong symmetry-breaking interactions. Observed departures from octet symmetry are attributed to the dynamic instability of fully symmetric solutions of the quantum field theory. Baker and Glashow[2] showed the possibility for such a spontaneous symmetry breakdown in the symmetric Sakata model.[3] They showed that solutions could exist which retain only the reduced symmetries of isospin and hypercharge conservation. Applying related considerations to a model with octet symmetry, we find that there may be solutions with only these reduced symmetries, but that a mass sum rule must be satisfied in the approximation where the mass splittings are small compared to the cutoff. If the solutions are required to violate R symmetry (thus, to break the $N-\Xi$ degeneracy), this sum rule is the Gell-Mann mass formula.

I. MEANING OF THE MASS FORMULA

Gell-Mann's formula relates the masses of the eight baryons,

$$\tfrac{1}{2}m_N+\tfrac{1}{2}m_\Xi=\tfrac{3}{4}m_\Lambda+\tfrac{1}{4}m_\Sigma, \quad (1.1)$$

and the squared masses of the eight pseudoscalar mesons,

$$\mu_K^2=\tfrac{3}{4}\mu_\chi^2+\tfrac{1}{4}\mu_\pi^2. \quad (1.2)$$

Both formulas are well satisfied—to 0.5% for the baryons, and to 2% in mass for the mesons. Okubo[4] generalized Gell-Mann's formula for any irreducible unitary multiplet, obtaining

$$m=a+bY+c[T(T+1)-\tfrac{1}{4}Y^2] \text{ for fermions,} \quad (1.3)$$

$$\mu^2=\alpha+\beta[T(T+1)-\tfrac{1}{4}Y^2] \text{ for bosons.} \quad (1.4)$$

These relations are equivalent to the physical mass Lagrangian (or, the inverse noninteracting renormalized Green's function) having transformation properties under SU_3 of the superposition of a unitary singlet and a $\mathbf{T}=0$, $Y=0$ member of a unitary octet. Because the reduction of a direct product of an irreducible representation of SU_3 with its adjoint always contains a singlet just once and an octet at most twice, it follows that the masses within such a multiplet may be expressed [as in (1.3)] in terms of no more than three parameters. For some multiplets like the 10, the decomposition yields but one octet and the mass formula simplifies to[5]

$$m=a+bY.$$

Gell-Mann got his mass formula by introducing simple symmetry-breaking interactions, e.g.,

$$g\phi\sum_{ij}\bar{\psi}_i(\delta_{ij}+\epsilon O_{ij})\psi_j,$$

where ϕ is a scalar unitary singlet meson, and $\mathbf{O}$ transforms like the $\mathbf{T}=0$, $Y=0$ member of a unitary octet (i.e., like the χ meson) and ϵ is a small parameter. Departures from baryon degeneracy to order ϵ satisfy (1.3), but the mass formula does not persist to order ϵ^2.

Here, we try to obtain split masses satisfying the mass formula without breaking the symmetry of the dynamics.

II. THE MODEL

We discuss eight baryons with four-fermion interactions invariant under SU_3, the baryons behaving like a unitary octet. Our considerations are otherwise independent of the form of the interaction. We regard mass as entirely dynamical in origin; therefore, we take the bare baryon masses as zero. Solutions to the quantum field theory depend parametrically only upon the square of the cutoff momentum Λ^2, and upon the dimensionless couplings $g\Lambda^2$ (g standing for the various symmetric four-fermion coupling strengths). In the approximations of reference 2, the mass operator is a constant, the masses being determined by coupled algebraic equations

$$m_i=\sum_j g_{ij}m_jh(\Lambda^{-2}m_j^2), \quad (2.1)$$

which results from putting the expression for the bare mass equal to zero. The g_{ij} are linear combinations of

* Research supported in part by the U. S. Air Force Office of Scientific Research.
† Alfred P. Sloan Foundation Fellow.
[1] M. Gell-Mann, California Institute of Technology Synchrotron Laboratory Report No. 20, 1961 (unpublished); Y. Ne'eman, Nucl. Phys. **26**, 222 (1961).
[2] M. Baker and S. L. Glashow, Phys. Rev. **128**, 2462 (1962).
[3] M. Ikeda, S. Ogawa, and Y. Ohnuki, Progr. Theoret. Phys. (Kyoto) **22**, 715 (1959); J. Wess, Nuovo Cimento **15**, 52 (1960); Y. Yamaguchi, Progr. Theoret. Phys. (Kyoto) Suppl. **11**, 1 (1959).
[4] S. Okubo, Progr. Theoret. Phys. (Kyoto) **27**, 949 (1962).
[5] M. Gell-Mann, in *Proceedings of the International Conference on High-Energy Physics at Geneva, 1962* (CERN, Geneva, 1962); S. L. Glashow, in International Summer School on Group Theoretical Methods in Elementary Particle Physics at Istanbul, 1962 (Gordon and Breach, London, to be published).

the various $g\Lambda^2$ and the function $h(x)$ depends on how the cutoff is introduced. We do not confine ourselves to this approximate expression, for the *exact* solution to the model is expected to yield an analogous result,

$$m_i = \sum_j \mathfrak{F}_{ij}(g\Lambda^2, (m/\Lambda)^2) m_j, \tag{2.2}$$

in which $\mathfrak{F}_{ij}$ depends upon all the baryon masses and is not a linear function of $g\Lambda^2$. Invariance of the dynamics under SU_3 is bound up in the allowed structure of the functions $\mathfrak{F}_{ij}$.

Only in a special basis of SU_3 are the masses diagonal (i.e., does the choice of fields correspond to particles of definite mass). What is required in order to study the group-theoretic properties of (2.2) is to re-express these equations in a general basis, where they take the form of 8×8 matrix equations,

$$M_{ij} = \sum_{kl} \mathfrak{F}_{ij;kl}(g\Lambda^2, (\mathbf{M}/\Lambda)^2) M_{kl}, \tag{2.3}$$

in terms of a Hermitian mass matrix whose eigenvalues are m_i. Let the 8×8 matrix U_{ij} represent an element of SU_3. Under this transformation,

$$M_{ij} \to M_{ij}' = \sum_{ab} U_{ia} M_{ab} U_{bj}^\dagger, \tag{2.4}$$

and symmetry of the dynamics under SU_3 requires that $\mathbf{M}'$ also satisfy (2.3).

Expand (2.3) in powers of Λ^{-2} about $\mathbf{M}=0$,

$$M_{ij} = \sum_{kl} F_{ij;kl}(g\Lambda^2) M_{kl} + O(\mathbf{MMM}\Lambda^{-2}). \tag{2.5}$$

Invariance of (2.5) under (2.4) requires that $F_{ij;kl}$, and higher matrices in the expansion, are ***invariant*** tensorial operators, i.e.,

$$F_{ij;kl} = \sum_{abcd} U_{ia} U_{ld} F_{ab;cd} U_{bj}^\dagger U_{ck}^\dagger \tag{2.6}$$

for any transformation $\mathbf{U}$ of SU_3. It may be shown that $\mathbf{F}$ has the general form,

$$\mathbf{F} = \lambda^{(1)}\mathbf{P}^{(1)} + \lambda^{(10)}\mathbf{P}^{(10)} + \lambda^{(\overline{10})}\mathbf{P}^{(\overline{10})} + \lambda^{(27)}\mathbf{P}^{(27)} + \lambda^{(D)}\mathbf{P}^{(D)} + \lambda^{(F)}\mathbf{P}^{(F)} + \eta\mathbf{N} + \eta'\mathbf{N}^\dagger, \tag{2.7}$$

where the eight parameters λ and η are real functions of $g\Lambda^2$, and the $\mathbf{P}^{(\alpha)}$ are projection operators for the various irreducible families of 8×8 matrices. $\mathbf{P}^{(D)}$ projects onto the completely symmetric 8_D, while $\mathbf{P}^{(F)}$ projects onto the completely antisymmetric 8_F. Because the two eights are equivalent, the operator $\mathbf{N}$ mapping 8_D onto corresponding members of 8_F (and also its adjoint $\mathbf{N}^\dagger$, mapping 8_F onto 8_D) also appears.

All members of the 27 are, thus, degenerate eigenmatrices of $\mathbf{F}$ with eigenvalue $\lambda^{(27)}$, and analogously for the members of the 10, the $\overline{10}$, and the singlet. Only if R symmetry, as well as unitary symmetry, characterizes the dynamics is the situation for the eights so simple. Then, $\eta=\eta'=0$, and the 8_D (8_F) are eigenmatrices of $\mathbf{F}$ belonging to $\lambda^{(D)}(\lambda^{(F)})$. In general, the supermatrix referring to the two eights may be triangularized, and the existence is assured of at least one family of eight matrices, irreducible under SU_3, whose members are degenerate eigenmatrices of $\mathbf{F}$. By group theory alone, we cannot further determine these eigenmatrices, for the result depends upon the detailed dynamics.

III. THE MASS FORMULA

In the basis where M_{ij} is diagonal, (2.5) becomes

$$m_i = \sum_j f_{ij}(g\Lambda^2) m_j + O(m^3\Lambda^{-2}). \tag{3.1}$$

Define the generators both of hypercharge Y, and of electrical charge $T_3+\frac{1}{2}Y$, so that they are diagonal in this basis. It is only a matter of convention that no breakdown of these conservation laws results from the asymmetry of the masses. This is not so for the total isospin, but we may look for solutions to (3.1) which are at least approximately invariant under the isospin subgroup, thus ignoring the possible existence of other solutions in which isospin is grossly violated. There are just four diagonal 8×8 matrices giving masses compatible with isospin and hypercharge conservation (accommodating the four isotopic submultiplets within the unitary octet): the unit matrix, two matrices with octet transformation properties, and one member of the 27-plet. We denote these diagonal 8×8 matrices by four-dimensional vectors

$$\mathbf{u} = (u_1, u_2, u_3, u_4), \tag{3.2}$$

whose entries refer, respectively, to N, Λ, Σ, and Ξ, and the isotopic multiplicities are implicit so that the norm is defined by

$$|\mathbf{u}|^2 = 2u_1^2 + u_2^2 + 3u_3^2 + 2u_4^2. \tag{3.3}$$

To the unitary singlet corresponds the normalized vector

$$\mathbf{u}^{(1)} = 8^{-1/2}(1,1,1,1), \tag{3.4}$$

and to the $\mathbf{T}=0$, $Y=0$ member of the 27-plet corresponds

$$\mathbf{u}^{(27)} = (3/40)^{1/2}(1, -3, -1/3, 1). \tag{3.5}$$

The remaining two-dimensional subspace normal to $\mathbf{u}^{(1)}$ and $\mathbf{u}^{(27)}$ has octet transformation properties.

From (3.1), with the neglect of nonlinear terms in masses, it follows that m_i must be an eigenvector of f_{ij} belonging to eigenvalue one. This requirement determines the coupling strength g. Barring an accidental degeneracy between $\lambda^{(1)}$ and either $\lambda^{(8)}$ or $\lambda^{(27)}$, the only physically admissible solution in this approximation is $\mathbf{m}\sim\mathbf{u}^{(1)}$, corresponding to complete degeneracy, for the other eigenvectors of f_{ij} have negative entries. Only when nonlinearities are included do we obtain an equation determining the value of the degenerate mass. We conclude that to order $(m/\Lambda)^2$ the only meaningful solutions to (3.1) are completely degenerage.

A more satisfactory result is obtained from the expansion of (2.2) about the mean baryon mass $\bar{m}$. We find

$$\bar{m}+\delta_i = h(g\Lambda^2, (\bar{m}/\Lambda)^2)\bar{m} + \sum_j \bar{f}_{ij}(g\Lambda^2, (\bar{m}/\Lambda)^2)\delta_j + O(\bar{m}\delta^2\Lambda^{-2}), \tag{3.6}$$

where $\bar{m}=m_i-\delta_i=\frac{1}{8}(2m_1+m_2+3m_3+2m_4)$, and $\boldsymbol{\delta}$ is normal to $\mathbf{u}^{(1)}$ with the norm (3.3). The discussion of Sec. II applies equally well to $\mathcal{J}_{ij}$, so that we conclude from the linearized approximation to (3.6) that

$$h(g\Lambda^2,(\bar{m}/\Lambda)^2)=1, \tag{3.7}$$

and that $\boldsymbol{\delta}$ must be an eigenvector of $\mathcal{J}_{ij}$ normal to $\mathbf{u}^{(1)}$ belonging to eigenvalue one. These two requirements determine the mean mass $\bar{m}$ and the coupling strength g. There are three possibilities:

(i) $\boldsymbol{\delta}=0$ This gives the fully symmetric and supposedly unstable solution to (2.2).

(ii) $\boldsymbol{\delta}\sim\mathbf{u}^{(27)}$ For this type of solution, N and Ξ remain degenerate.

(iii) $\boldsymbol{\delta}\cdot\mathbf{u}^{(27)}=0$ In this case, $\boldsymbol{\delta}$ has octet transformation properties. Comparison with (3.3) and (3.5) gives the Gell-Mann mass rule.

This approximation [to all orders in $(\bar{m}/\Lambda)^2$, and linear in $\boldsymbol{\delta}$] does not determine the magnitude of $\boldsymbol{\delta}$, but from our earlier discussion we know that

$$\delta/\bar{m}=O(\bar{m}/\Lambda)^2. \tag{3.8}$$

Deviations from the mass rule are comparable to the neglected nonlinear terms of (3.4), and are of order $\bar{m}(\delta/\Lambda)^2$. Both the size of the observed mass splittings and of the deviations from the mass formula are compatible with a cutoff energy of several BeV. Of course, we have not demonstrated that spontaneous asymmetries exist, but only that when they do they must satisfy an approximate sum rule.

Electromagnetic mass splittings are of similar size to deviations from the mass formula. Perhaps violation of isospin, and even electromagnetism, is *already implicit* in the model we have considered, possibly along the lines discussed by Bjorken.[6]

We also remark on the derivation of the mass rule for the masses of mesons μ, or for the masses of multiplets of resonances m^*. Such derivative phenomena should satisfy *inhomogeneous* equations of the form

$$m_i^*=\sum_j \mathcal{F}_{ij}^* m_j^*+\sum_k \mathcal{F}_{ik}' m_k, \tag{3.9}$$

$$\mu_i^2=\Lambda^2\mathcal{H}+\sum_j \mathcal{H}_{ij}\mu_j^2+\sum_j \mathcal{G}_{ij}m_j^2, \tag{3.10}$$

and an analysis similar to that of (2.2) shows that m_i^* and μ_i^2 satisfy analogous mass formulas.[7]

It must be emphasized that our approach depends hardly at all upon the use of a field-theoretic model. The starting point, Eq. (2.2), could equally well have arisen from reasoning akin to that of Zachariasen and Zemach,[8] wherein the eight nucleon masses (through their symmetric interactions) are required to support themselves self-consistently. In this case, no cutoff appears in the bootstrap equations analogous to Eq. (2.2).

ACKNOWLEDGMENTS

We gratefully acknowledge helpful discussions with Professor N. Burgoyne, Professor H. P. Duerr, Professor C. Schwartz, and Professor C. Zemach. We thank Professor M. Gell-Mann for stressing the probable independence of our result of its field-theoretic beginnings.

[6] J. Bjorken (to be published).

[7] Analogous, but not identical, sum rules characterize the m_i^* and the μ_i. Thus, if $m_i-\bar{m}$ has octet transformation properties, so must $m_i^*-\bar{m}^*$. For the 10-plet of $J=3/2^+$ resonances, this gives an equal-spacing rule. See reference 5 and S. L. Glashow and J. J. Sakurai, Nuovo Cimento 26, 622 (1962).

[8] F. Zachariasen and C. Zemach, Phys. Rev. 128, 849 (1962).

Self-Consistent Deviations from Unitary Symmetry*

R. E. Cutkosky and Pekka Tarjanne†
Carnegie Institute of Technology, Pittsburgh, Pennsylvania
(Received 24 June 1963)

A method of investigating the possible dynamic origin of symmetries among the strong interactions is illustrated by application to a model with vector mesons that are self-consistently bound states of one another. The SU_3 model, with eight vector mesons, is concentrated upon. All possible types of first-order perturbations are treated in the ladder approximation, and some second-order effects are also considered. The results emerging from a qualitative discussion uniquely suggest the possibility (in addition to the degenerate mass solution) of a self-supporting small mass splitting structure of the type leading to the Gell-Mann–Okubo mass formula. Moreover, SU_2 symmetry is necessarily retained, although the differentiation between charge and hypercharge is not possible in a theory which does not include electromagnetism.

I. INTRODUCTION

THERE are many indications that SU_3 symmetry provides a useful way to correlate the properties of the strongly interacting particles.[1–7] Although the departures from perfect symmetry are large, they have themselves a characteristic structure exhibited through the retention of isotopic spin symmetry and through the Gell-Mann–Okubo formula for the mass differences.[8–12] One of the central questions of strong interaction physics is whether these characteristic relations among the masses and coupling constants must be obtained from *ad hoc* postulates, or whether they arise automatically through the interworkings of dynamical effects.[13,14,14a] We consider here a restricted aspect of this question using a simplified treatment. We shall discuss interrelations among the various manifestations of a small dissymmetry, and shall suggest a possible dynamical reason for the particular form which the dissymmetry takes.

Charge and hypercharge conservation, as well as charge conjugation symmetry, are assumed to hold. We do not examine perturbations leading to the possibility that these concepts break down, because they seem to have validity outside the domain of the strong interactions.

In the investigation of this problem, we have been forced, as a result of the limitations in our present understanding of strong interaction dynamics, to confine our attention to a qualitative discussion of simple models having the following general features: (1) We consider only states with two relatively light particles; (2) we use the ladder approximation; (3) we introduce an over-all cutoff Λ, which we adjust to enable the equations for the masses and coupling constants to be self-consistently satisfied, but which we do not attempt to calculate from the particle size implied by our model; (4) we study only those solutions of the nonlinear self-consistency equations in which the dissymmetries are small.

We shall examine in this paper a simple model in which only vector mesons appear. It has been shown, for this model, that if the masses are equal, the mesons must fit into the adjoint representation of some semi-simple Lie group.[9] In our discussion of the possibility of nonequal masses, we consider primarily the case of eight vector mesons, which corresponds to SU_3 being taken as a first approximation.

By considering this simple vector meson model, we will be able to describe the underlying physical concepts as well as the calculational techniques in the simplest situation possible. A remarkable limitation on the possible types of dissymmetry arises from our

* Supported in part by the U. S. Atomic Energy Commission.
† Present address: Department of Technical Physics, Finland Institute of Technology, Otaniemi, Finland.
1 M. Gell-Mann, California Institute of Technology Report CTSL-20, 1961 (unpublished).
2 M. Gell-Mann, Phys. Rev. **125**, 1067 (1962).
3 Y. Ne'eman, Nucl. Phys. **26**, 222 (1961).
4 S. L. Glashow and A. H. Rosenfeld, Phys. Rev. Letters **10**, 192 (1963).
5 R. E. Cutkosky, Ann. Phys. (N. Y.) (to be published).
6 R. E. Behrends, J. Dreitlein, C. Fronsdal, and B. W. Lee, Rev. Mod. Phys. **34**, 1 (1962).
7 Pekka Tarjanne, Ann. Acad. Sci. Fennicae: Ser. A VI 105 (1962).
8 S. L. Glashow, Phys. Rev. **130**, 2132 (1963).
9 R. E. Cutkosky, Phys. Rev. **131**, 1888 (1963).
10 R. E. Cutkosky, J. Kalckar, and P. Tarjanne, Phys. Letters **1**, 93 (1962).
11 R. E. Cutkosky, J. Kalckar, and P. Tarjanne, in *Proceedings of the 1962 Annual International Conference on High-Energy Physics at CERN* (CERN, Geneva, 1962), p. 653.
12 S. Okubo, Progr. Theoret. Phys. (Kyoto) **27**, 949 (1962).
13 M. Gell-Mann, in *Proceedings of the Sixth Annual Rochester Conference on High-Energy Nuclear Physics* (Interscience Publishers, Inc., New York, 1956), Sec. III, p. 30.
14 G. F. Chew, La Jolla Conference on Weak and Strong Interactions, 1961 (unpublished).
14a *Note added in proof.* After we submitted this paper, we learned of a paper by E. Abers, F. Zachariasen, and G. Zemach [Phys. Rev. **132** (1963), (to be published)] in which the possibility of obtaining interaction symmetries from bootstrap mechanisms was analyzed in detail. The general similarity of outlook in this paper and in ours will be clear to the reader, but we feel that it would be helpful if we take this opportunity to point out two differences of emphasis. First, we have assumed that the homogeneity of the self-consistency equations would preclude the existence of any solution to their approximated versions, unless an adjustable parameter were introduced, and that with a free parameter, there would be a large number of solutions (cf. Refs. 9, 11). Second, we have, in consequence, concentrated upon the delimitation of such solutions, in particular, assuming the existence of a symmetrical solution, we have here examined solutions with broken symmetry, whereas Abers, Zachariasen, and Zemach have focused upon establishment of the plausibility of the existence of symmetrical solutions.

TABLE I. Mass dissymmetry matrices $D(r,T)$ for octuplets.

	Diagonal elements								Off-diag.	Norm.
(Y,T,T_3)	$(1,\frac{1}{2},\frac{1}{2})$	$(1,\frac{1}{2},-\frac{1}{2})$	(0,1,1)	(0,1,0)	(0,1,−1)	(0,0,0)	$(-1,\frac{1}{2},\frac{1}{2})$	$(-1,\frac{1}{2},-\frac{1}{2})$	$(0,T,0)$	
$D(8,0)$	1	1	−2	−2	−2	2	1	1	0	$2\sqrt{5}$
$D(8,1)$	$\sqrt{3}$	$-\sqrt{3}$	0	0	0	0	$-\sqrt{3}$	$\sqrt{3}$	2	$2\sqrt{5}$
$D(27,0)$	3	3	−1	−1	−1	−9	3	3	0	$2\sqrt{30}$
$D(27,1)$	−1	1	0	0	0	0	1	−1	$\sqrt{3}$	$\sqrt{10}$
$D(27,2)$	0	0	1	−2	1	0	0	0	0	$\sqrt{6}$
$D(S)$	1	1	1	−3	1	−3	1	1	0	$2\sqrt{6}$
$D(1,0)$	1	1	1	1	1	1	1	1	0	$2\sqrt{2}$

qualitative discussion of this model, which encourages the belief that the problem of strong interaction symmetry can be solved by considering it as a problem of self-consistency. The discussion of the vector-meson model which is given here can be generalized without difficulty to a more realistic model in which there are boson-baryon couplings, leading to similar results which will be reported in a subsequent paper.

In the generalized vector-meson model (generalized to nonequal masses) the masses and coupling constants are determined by self-consistency equations which could be derived (for example) by the method of Zachariasen and Zemach,[15] or from the Bethe-Salpeter equation,[16] and which are similar to those in Ref. 9. If we eliminate the coupling constants from these equations, the self-consistency equation for the mass matrix $(M^2)_{ab}$ can be written in the form[17]:

$$M^2=k(M^2,\Lambda), \tag{1}$$

which just represents the dependence of the calculated mass on the masses of the particles being bound together and of the exchanged particles.

Following Glashow,[8] we represent M^2 in terms of normalized tensor operators $D(rT)$:

$$M^2=\sum_{rT} a(rT)D(rT), \tag{2}$$

where r denotes a representation of SU_3 and T the total isotopic spin. Since charge Q and hypercharge Y are conserved, the D's correspond to $Q=Y=0$. For the vector mesons, charge-conjugation symmetry and Hermiticity limit r to (2,2) (27-fold), $(1,1)_S$ (8-fold), and (0,0) (singlet). The explicit forms of the $D(rT)$ are given in Table I. The coefficient $a(0,0)$, which is related to the average $\mathfrak{M}^2$, is determined by a choice of scale and will be ignored hereafter. A self-consistent solution which is completely symmetric corresponds to $a(rT)=0$ for $r\neq(0,0)$. We assume that such a solution of (1) exists.

If we restrict our attention to small dissymmetries, we may expand (1) to second order as

$$a(rT)=K(r)a(rT)+\sum_{(r_1T_1,r_2T_2)} L(rT,r_1T_1,r_2T_2)a(r_1T_1)a(r_2T_2). \tag{3}$$

We have denoted by $K(r)$ the eigenvalue of $K_{ab,cd}=\partial k_{ab}/\partial(M^2)_{cd}$ in the representation r. Equation (3), which consists of five simultaneous quadratic equations, has a large number of solutions [not all of which necessarily correspond to solutions of (1)].

We shall classify the solutions of (1) or (3) according to their relation to the root diagram of SU_3 (Fig. 1). In expanding the mass dissymmetry matrix, we have labelled the tensor operators by their isotopic spin. Now, SU_2 is contained as a subgroup in SU_3 in three distinct ways, corresponding to the three sets of reciprocal roots; any of these sets could be interpreted as the isotopic spin operators. If we express the mass matrix corresponding to one of the solutions of (1) in terms of tensor operators labeled according to one of the other two sets of reciprocal roots, we obtain new coefficients $a'(rT)$ which will also correspond to a solution of (1). In other words, the solutions of (1) or (3) form a representation of the symmetry group of the root diagram.

The nonlinear terms of (3) are, in part, determined by the Clebsch-Gordan coefficients of SU_3; in particular, they must be consistent with the subgroups of SU_3. We look especially at dissymmetries which are invariant under a subgroup of SU_3; the direct product of one-dimensional representations of a group is one dimensional, so other types of dissymmetry cannot be mixed in through the nonlinearities. The possibility that all the a's vanish, i.e., that complete symmetry under SU_3 is maintained, is, of course, obtained as a trivial

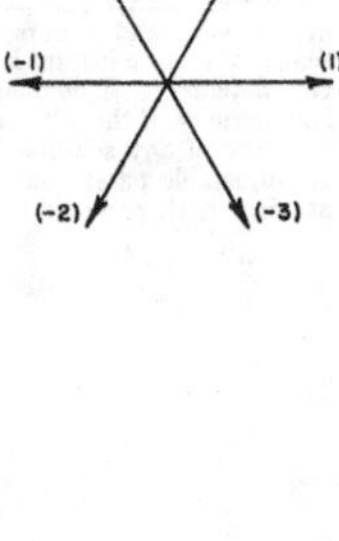

FIG. 1. The root diagram for SU_3.

[15] F. Zachariasen and C. Zemach, Phys. Rev. 128, 849 (1962).
[16] E. E. Salpeter and H. A. Bethe, Phys. Rev. 84, 1232 (1951).
[17] We follow the conventional practice in considering the squares of the boson masses to be the variables.

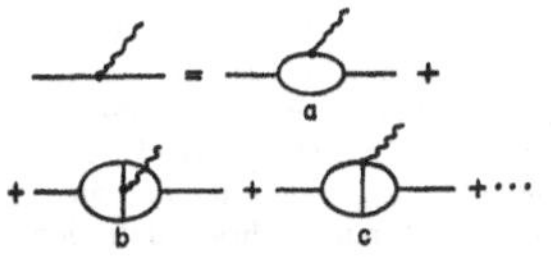

FIG. 2. First-order mass perturbation of vector mesons as an expansion in self-energy graphs. The solid lines represent the vector mesons. The perturbation can act on a propagator or a vertex (see Fig. 3) and it is described by a wriggly line. This suggests its interpretation as an external field with $Q=Y=0$. We wish to emphasize, however, that the "bare coupling constants" of this field need not be considered to be different from zero.

solution of (3). We see that there must also be solutions of (3) [with $a((8),0)$ and $a((27),0)$ being the only non-vanishing coefficients] which maintain isotopic spin symmetry, and further solutions [with $a((27),2)$ also different from zero] which exhibit charge symmetry but not full isotopic spin symmetry. From these solutions, others can be generated by permuting the roots, as described above. The dissymmetry given by

$$D(S)=(\tfrac{1}{3}\sqrt{5})D(27,0)+\tfrac{2}{3}D(27,2) \tag{4}$$

is invariant under permutation of the roots, and therefore, will necessarily arise as one of the solutions.

II. THE SU_2 MODEL

Before examining the possibility of unsymmetrical solutions in the SU_3 model, we shall illustrate the approach by discussing SU_2, which is much simpler. The only possible dissymmetry in the adjoint representation corresponds to $T=2$: $M_{\pm}^2=\mathfrak{M}^2+a/\sqrt{6}$, $M_0^2=\mathfrak{M}^2-2a/\sqrt{6}$. Then upon eliminating Λ by requiring that self-consistency be attained with a given fixed value of $\mathfrak{M}^2$, (1) takes the form

$$\begin{aligned} a&=k(a)\\ &=Ka+La^2+\cdots. \end{aligned} \tag{5}$$

From the truncated expansion one obtains the solutions

$$\begin{aligned} a&=0,\\ a&=(1-K)/L. \end{aligned} \tag{6}$$

If it should turn out that K were close to unity, the neglect of higher-order terms in (5) would be justified, and we would have found a second self-consistent solution which was slightly unsymmetrical. If, on the other hand, $1-K$ were not small, one would need to look at the exact expression in order to see whether additional solutions existed. We suggest that if the dissymmetry predicted by (6) is large ($a\gg\mathfrak{M}^2$), the existence of an unsymmetrical solution is implausible.

One of the contributions to K is described by graph 2a, which represents the change in the mass of a bound state arising from a change in the mass of one constituent. The charged states are made up of one neutral and one charged meson, and the neutral state of two charged mesons. Therefore, we have

$$\begin{aligned} \Delta_{\pm}(2a)&=\alpha(\Delta_{\pm}+\Delta_0)=-\alpha a/\sqrt{6},\\ \Delta_0(2a)&=\alpha(2\Delta_{\pm})=2\alpha a/\sqrt{6}, \end{aligned} \tag{7}$$

where α is a constant, presumably positive, which characterizes the internal structure of the particles. From (7), we obtain $K_{2a}=-\alpha$. We wish to emphasize that the effect represented by Eq. (7) is not to be looked upon as a modification of the ordinary second-order self-energy term, even though we have pictured it by such a graph. In fact, the end vertices in graph (a) in Fig. 2 could be interpreted as standing for ladder graphs with an indefinite number of vertical rungs, in accordance with our view of the vector states as states bound by the ladder-approximation potential.

If the mass of the exchanged quantum is increased by an amount Δ, the potential is presumably decreased in magnitude, leading to a resultant energy increased by $\beta\Delta$. This effect is represented by graph 2(b). Since for SU_2 the exchange quantum has the same charge as the bound state, we obtain $K_{2b}=+\beta$.

We continue to assume, as in Ref. 9, that at every vertex the independent covariants have fixed ratios. Then the question of a change in the ratios of the coupling constants does not arise for SU_2, because there is only one.

The values of α and β cannot be calculated without a detailed dynamical model; in particular, they depend on the way the cutoff is introduced. A simple way is to assume that for each line, the propagator is regularized with auxiliary masses which are taken to be proportional to the physical mass. One then obtains for the ladder approximation, by considering the effect of a change in the unit of mass, the identity

$$2\alpha+\beta=1. \tag{8}$$

One expects, moreover, that $\alpha\approx\beta$, so that each would be about $\tfrac{1}{3}$. In the ladder approximation, therefore, one anticipates that $K\approx 0$.

The coefficient L of the second-order term in (5) represents nonlinear effects on the bound-state energy of a change in the masses of the constituents or of the exchanged quantum. The size, and even the sign of these terms appears to be quite model-dependent. We expect, in any event, that $|L|\lesssim\mathfrak{M}^{-2}$. The SU_2 model could not, then, have a solution with a small dissymmetry, and we should like to suggest the possibility that it does not admit any second unsymmetrical solution. A detailed investigation of this question would be of considerable interest.

III. THE SU_3 MODEL: FIRST-ORDER PERTURBATIONS

We now return to the quantity $K(r)$ appearing in (3), and evaluate it within the ladder approximation.

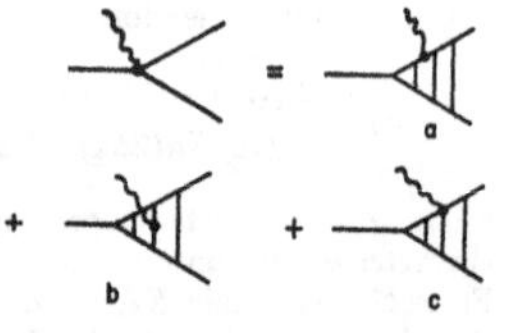

FIG. 3. First-order vertex modifications showing explicitly the use of the ladder approximation. Symmetrization with respect to the three legs is implied.

Graph 2(a) gives the following contribution to $K_{ab,cd}\Delta_{cd}$:

$$\Delta_{ab}(2a)=\alpha F_{cra}F_{sdb}(\delta_{cd}\Delta_{rs}+\delta_{rs}\Delta_{cd}) = -2\alpha F_{acr}F_{bdr}\Delta_{cd}, \qquad (9)$$

where α is the same quantity appearing in (7). The eigenvalues of $F_{acr}F_{bdr}$ are as calculated in Refs. 5, 7, and 10, apart from a different normalization, which we take to be $F_{abr}F_{bas}=\delta_{rs}$ in the present work, as in Ref. 9. One finds that $K_{2a}(8)=\alpha$, $K_{2a}(27)=-\frac{2}{3}\alpha$.

Graph (b) of Fig. 2 leads to:

$$\Delta_{ab}(2b)=\beta F_{rma}(-2F_{rsc}\Delta_{cd}F_{mnd})F_{nsb}. \qquad (10)$$

The factor (-2) is introduced into (10) so that β will have the same meaning as before. This equation can be written as

$$\Delta_{ab}(2b)=\beta[F_{acr}F_{bdr}+2F_{amr}F_{bnr}F_{mcs}F_{nds}]\Delta_{cd},$$

from which we find $K_{2b}(8)=0$ and $K_{2b}(27)=5\beta/9$. Comparing these results with those obtained in the previous section, we see that the contributions to $K(27)$ are similar to the corresponding ones of the SU_2 model. The additional possibility (8) of the SU_3 model, on the other hand, corresponds to a $K(8)$ which, on the basis of these graphs, is positive. This encourages us to go further with the calculations.

In the SU_3 model there is also a contribution to K from graph 2(c), although the calculation of the vertex modifications described by Fig. 3 requires that, at least implicitly, we go beyond the ladder approximation. It is also, unfortunately, much more difficult. The method by which we shall calculate the change in the coupling constants can be described as follows: The self-energy corrections of Fig. 2 can be thought of as the expectation value of the "mass energy" [Fig. 2(a)] and of the potential energy [Figs. 2(b) and 2(c)] in the unperturbed states. At the same time, we can calculate the admixture of components from the representation (20) $\equiv(3,0)\oplus(0,3)$ into the eigenstates. These admixtures give directly the vertex modifications which are to be used in calculating the perturbed potential. We are only interested in the changes in the ratios of the coupling constants, because their average is determined by the equation for self-consistency of the average of the masses.

We know that the vertex must involve the three lines symmetrically. However, our method of calculation treats them in an unsymmetrical way, because one particle is thought of as a bound state of the other two. This would not matter, if we used the exact potential in our calculation and treated the change in the normalization condition properly, because then the symmetry would arise automatically. In our calculation, we use the ladder approximation and an artificial normalization; to overcome these limitations we must symmetrize explicitly at the end.

We denote by $-V$ the coefficient of the momentum-dependent factors in the one-particle exchange potential. It is assumed that these momentum-dependent factors are approximately the same in all elements of the potential, so that they may be adequately represented by a suitable average. Then we may assume that the mass will be given by an equation having the form

$$M^2=S-f(V)+\alpha A, \qquad (11)$$

which is a matrix equation in which M^2-S is an eigenvalue and A represents the change in the masses of the bound particles. Since for exact SU_3 symmetry, V has only one nonvanishing eigenvalue, we may write, to a sufficient approximation, $f(V)=Vf(\lambda)/\lambda$, where λ denotes the larger eigenvalue of V. This may be expanded about the unperturbed potential V_0 as $f(V)=t(V_0+\delta V)+t'\delta V_d$, where δV_d is the part of the perturbation which is diagonal in the unperturbed representation. We define t so that V_0 has a unit eigenvalue.

We next separate V into two parts, corresponding to the two graphs (b) and (c) of Fig. 2 [as well as to Figs. 3(b) and 3(c)]. We write

$$V=v+\beta B/(t+t'),$$

where B is the change in the mass of the exchanged quantum (the coefficient is chosen so that our previous definition of β is retained) and where v depends only on the coupling constants. We, finally, obtain

$$f(V)=tv+[t/(t+t')]\beta B+t'(v_d-V_0)+[t'/(t+t')]\beta B_d. \qquad (12)$$

In calculating for the (1,1) type of dissymmetry, we look at $D(8,0)$: $\Delta_\rho=-2a'$, $\Delta_\varphi=+2a'$, and $\Delta_M=+a'$ [with $a(8,0)=(2\sqrt{5})a'$]. The modified coupling constants are:

$$g(\rho^3)=g_0(\rho^3)(1+f_0),$$
$$g(\bar{M}M\varphi)=g_0(\bar{M}M\varphi)(1+f_1), \qquad (13)$$
$$g(\bar{M}M\rho)=g_0(\bar{M}M\rho)(1+f_2).$$

For a dissymmetry described by $D(8,0)$, it can be shown by general arguments that to first order $f_2=0$ and $f_1=-\frac{2}{3}f_0$. We will demonstrate shortly that these ratios form a self-consistent choice.

For the ρ state, we write $|\rho\rangle=\xi|\rho\rho\rangle+\eta|\bar{M}M\rangle$. As a matrix acting on the column vector (ξ,η), we have

$$A=\begin{pmatrix}-4a' & 0\\ 0 & 2a'\end{pmatrix}. \qquad (14)$$

The matrix v can be calculated from the explicit reduction coefficients for SU_3 (we do not give the details here), and has the form:

$$v=\begin{pmatrix}\frac{2}{3}(1+f_0)^2 & \frac{1}{3}\sqrt{2}(1+f_2)^2\\ \frac{1}{3}\sqrt{2}(1+f_2)^2 & \frac{1}{2}(1+f_1)^2-\frac{1}{6}(1+f_2)^3\end{pmatrix}. \quad (15)$$

In the present section, only the terms in v which are linear in the f_i are relevant, but we have given the complete form so it can be referred to later. The matrix B is gotten by multiplying each term in v by the mass change of the exchanged particle which is responsible for that term. The result is easily seen to be:

$$B=a'\begin{pmatrix}-\frac{4}{3}(1+f_0)^2 & \frac{1}{3}\sqrt{2}(1+f_2)^2\\ \frac{1}{3}\sqrt{2}(1+f_2)^2 & (1+f_1)^2+\frac{1}{3}(1+f_2)^2\end{pmatrix}. \quad (16)$$

As a check, it can be verified that the expectation values of A and B in the unperturbed state agree with the results obtained above.

In v, we introduce the relations between the f_i mentioned following Eq. (13). Then, to first order in f_0 and a', we find the following admixture into the state of the ρ:

$$\langle(20)|\rho\rangle=\sqrt{2}h, \quad (17)$$

where

$$h=2\alpha a'/t+\beta a'/(t+t')+\tfrac{2}{3}f_0. \quad (18)$$

From (17), one easily calculates:

$$\xi=(\tfrac{2}{3})^{1/2}(1+h), \quad \eta=(\tfrac{1}{3})^{1/2}(1-2h). \quad (19)$$

Next, we consider the state $|M\rangle=\xi|\rho M\rangle+\eta|\varphi M\rangle$, for which the matrices are

$$A=\begin{pmatrix}-a' & 0\\ 0 & 3a'\end{pmatrix}, \quad (14')$$

$$v=\begin{pmatrix}\frac{2}{3}(1+f_0)(1+f_2)-\frac{1}{6}(1+f_2)^2 & \frac{1}{2}(1+f_1)(1+f_2)\\ \frac{1}{2}(1+f_1)(1+f_2) & \frac{1}{2}(1+f_1)^2\end{pmatrix}, \quad (15')$$

$$B=a'\begin{pmatrix}-\frac{4}{3}(1+f_0)(1+f_2)-\frac{1}{6}(1+f_2)^2 & \frac{1}{2}(1+f_1)(1+f_2)\\ \frac{1}{2}(1+f_1)(1+f_2) & \frac{1}{2}(1+f_1)^2\end{pmatrix}. \quad (16')$$

Again calculating to first order, we have

$$\langle(20)|M\rangle=h, \quad (17')$$

and

$$\xi=\tfrac{1}{2}\sqrt{2}(1+h), \quad \eta=\tfrac{1}{2}\sqrt{2}(1-h). \quad (19')$$

There is no admixture into the state $|\varphi\rangle=|\bar{M}M\rangle$. In this state, A, v, and B are just numbers:

$$A=2a', \quad (14'')$$

$$v=\tfrac{1}{2}(1+f_1)^2+\tfrac{1}{2}(1+f_2)^2, \quad (15'')$$

$$B=a'[(1+f_1)^2-(1+f_2)^2]. \quad (16'')$$

We now symmetrize the vertices $\langle\rho|\bar{M}M\rangle$ and $\langle M|\rho M\rangle$, as well as $\langle\varphi|\bar{M}M\rangle$ and $\langle M|\varphi M\rangle$, among the three lines. Then we obtain

$$f_0=h, \quad f_1=-\tfrac{2}{3}h, \quad f_2=0. \quad (20)$$

On comparing (20) with (18), we find

$$h=6\alpha a'/t+3\beta a'/(t+t'). \quad (21)$$

When these values are substituted into (15), (15′), or (15″), we obtain from the expectation values

$$K_{2c}(8)=2\alpha'+\beta, \quad (22)$$

where $\alpha'=\alpha(t+t')/t\approx\alpha$.

The calculation of $K_{2c}(27)$ proceeds in the same way as above. In fact, if we considered a perturbation proportional to $D(27,0)$, we could use the same matrices v and merely make appropriate modifications in A and B. However, for later reference, we shall discuss $D(S)$, which is actually simpler. The two particles ($\mathfrak{i}$) with $Q=Y=0$ have $\Delta_{\mathfrak{i}}=-3a'$, and the remaining six ($\mathfrak{o}$) have $\Delta_{\mathfrak{o}}=a'$ (with $a(S)=2a'\sqrt{6}$). There are only two coupling constants:

$$g(\mathfrak{o}^2\mathfrak{i})=g_0(\mathfrak{o}^2\mathfrak{i})(1+f_1),$$
$$g(\mathfrak{o}^3)=g_0(\mathfrak{o}^3)(1+f_2), \quad (23)$$

which are related by $f_2=-3f_1$.

In the state $|\mathfrak{i}\rangle$ there is no admixing: $|\mathfrak{i}\rangle=|\mathfrak{o}\mathfrak{o}\rangle$. The values of A, v, and B are

$$A=2a',$$
$$v=\tfrac{1}{3}(1+f_2)^2+\tfrac{2}{3}(1+f_1)^2, \quad (24)$$
$$B=a'[\tfrac{1}{3}(1+f_2)^2-2(1+f_1)^2].$$

For the $\mathfrak{o}$ we write $|\mathfrak{o}\rangle=\xi|\mathfrak{o}\mathfrak{i}\rangle+\eta|\mathfrak{o}\mathfrak{o}\rangle$. The matrices are

$$A=\begin{pmatrix}-2a' & 0\\ 0 & 2a'\end{pmatrix},$$

$$v=\begin{pmatrix}\frac{2}{3}(1+f_1)^2 & \frac{1}{3}\sqrt{2}(1+f_1)(1+f_2)\\ \frac{1}{3}\sqrt{2}(1+f_1)(1+f_2) & \frac{1}{3}(1+f_1)^2\end{pmatrix}, \quad (25)$$

$$B=a'\begin{pmatrix}\frac{2}{3}(1+f_1)^2 & \frac{1}{3}\sqrt{2}(1+f_1)(1+f_2)\\ \frac{1}{3}\sqrt{2}(1+f_1)(1+f_2) & -(1+f_1)^2\end{pmatrix}.$$

From these we calculate the following admixture:

$$\langle(20)|\mathfrak{o}\rangle=\sqrt{2}h', \quad (26)$$

where

$$h'=4\alpha a'/3t-4\beta a'/9(t+t')+4f_1/9. \quad (27)$$

This leads to

$$\xi=(\tfrac{2}{3})^{1/2}(1+h'), \quad \eta=(\tfrac{1}{3})^{1/2}(1-2h'). \quad (28)$$

Symmetrizing as before, we obtain $f_1=\frac{2}{3}h'$, $f_2=-2h'$, which gives

$$h' = \frac{12a'}{19}\left(\frac{3\alpha}{t} - \frac{\beta}{t+t'}\right). \tag{29}$$

The expectation values of v then lead to

$$K_{2c}(27) = -16\alpha'/3\cdot 19 + 16\beta/9\cdot 19.$$

Adding together our estimates of the various contributions to K, we have

$$\begin{aligned} K(8) &= \alpha + 2\alpha' + \beta, \\ K(27) &= -\tfrac{2}{3}\alpha - (16/57)\alpha' + (111/171)\beta. \end{aligned} \tag{30}$$

These formulas should be considered in the light of our estimate that α, α', and β should be near to $\frac{1}{3}$. It is certainly consistent with (30) to have a second type of solution of the self-consistency equation (1) in which the predominate dissymmetry is of the (1,1) type; on the other hand, a (2,2) type of dissymmetry is not favored. We suggest, as the origin of the Gell-Mann–Okubo rule, that

$$|1-K(8)| \ll |1-K(27)|. \tag{31}$$

While our calculations indicate the plausibility of obtaining (31) from a more complete theory, we cannot claim to have established it.

IV. SECOND-ORDER TERMS

Before we examine the relative sizes of the different second order terms in (3), we shall point out an important result which follows directly from the assumption that it is the nonlinear terms which determine the magnitudes of small dissymmetries. This result also depends on the dominance of the (1,1) type of dissymmetry. If the suggestion (31) is correct, we are justified in neglecting the terms in (3) referring to the (2,2) dissymmetry when we calculate $a(8,0)$ and $a(8,1)$. Then (3) is reduced to

$$\begin{aligned} (1-K)a(0) &= L[a(0)^2 - a(1)^2], \\ (1-K)a(1) &= -2La(0)a(1), \end{aligned} \tag{32}$$

where we have simplified the notation by omitting reference to the representation, and where the ratios of the terms on the right are obtained from Clebsch-Gordan coefficients. The solutions to (32) are

$$a(0) = a(1) = 0, \tag{33a}$$

$$a(1) = 0, \quad a(0) = a \equiv (1-K)/L, \tag{33b}$$

$$a(0) = -\tfrac{1}{2}a, \quad a(1) = \pm\tfrac{1}{2}\sqrt{3}a. \tag{33c}$$

Solution (33b) corresponds to retention of isotopic spin symmetry. The two solutions (33c) also correspond to SU_2 symmetry, but with different pairs of the roots shown in Fig. 1 being interpreted as the isotopic-spin displacement operators. This feature of the solutions to (32) is actually a direct consequence of the general discussion given in the introduction, and is accordingly more general than Eq. (32); we only need to assume that the nonlinear terms involving the $a(8,T)$ are more important than those involving the $a(27,T)$.

We now ask, what *a priori* criteria might be used to distinguish among the solutions (33)? We may note that the value of the cutoff parameter Λ which is fixed by the self-consistency requirement will be different for solution (33a) and for the other three; this may ultimately lead to a way to discriminate against (33a). However, the three solutions (33b,c) have completely identical properties as far as the strong interactions are concerned. In other words, there is no way, as long as only the strong interactions are considered, to decide which of the conserved quantities should be called the "charge" and which the "hypercharge." It is the electromagnetic interactions which distinguish between these quantities. The ambiguity is, therefore, a necessary feature of any attempt to derive isotopic spin within the strong interactions; the explanation of the relation between isotopic spin and electromagnetic interactions must lie in the nature of electromagnetism.

In looking more closely at the second-order terms, we are chiefly interested in the following points: (1) We want to know something about the magnitude of the coefficient L in (32); in particular, whether there might be reason to suspect it of being anomalously small. (2) A small (1,1) dissymmetry will induce, in second order, a still smaller (2,2) dissymmetry; choosing $T=0$, we write

$$[1-K(27)]a(27,0) = L'a(8,0)^2, \tag{34}$$

under the assumption that second-order terms involving $a(27,0)$ can be neglected. It has been argued that $|1-K(8)| \ll |1-K(27)|$ in order to justify this assumption as well as our more general conclusions. It is clear that the ratio L'/L also has considerable significance. (3) Hitherto, our remarks have been directed towards solutions of Eq. (3) in which the (1,1) dissymmetry predominates. We must also look at other solutions. In fact, Eq. (3) has so many solutions that a complete analysis would require extensive numerical computation. Since we are able to derive, by our present means, only qualitative information about the coefficients, such an investigation would be premature. Nevertheless, we should like to examine at least one example, and the discussion in the Introduction suggests a suitably simple one—that afforded by the dissymmetry $D(S)$; accordingly, we look at the coefficient L'' in the equation

$$[1-K(27)]a(S) = L''a(S)^2. \tag{35}$$

We commented in Sec. II that the second-order terms in the SU_2 model could not be easily estimated. There are similar effects here, which we must similarly ignore, but there are, in addition, some relatively simple second order terms, arising from the admixing of the (20) configuration, which can be obtained from Eq. (11). While we would not, perhaps, trust the magnitudes of

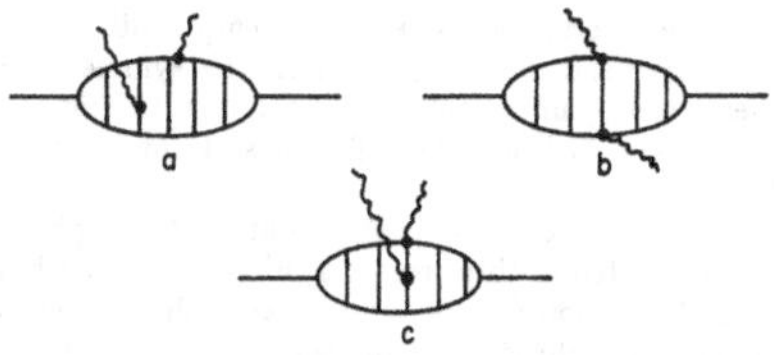

FIG. 4. Some graphs describing second-order perturbations.

the effects so derived, we might have more confidence in their ratios. For a picture of these effects, we refer to Fig. 4.

The graph 4(a), in which there are two perturbations that act in different sections of the ladder, describes an effect we estimate by standard second-order perturbation theory. The change in M^2 is proportional to the squares of the admixture coefficients; for the (1,1), $T=0$ perturbation

$$\Delta_\rho = -2th^2, \quad \Delta_M = -th^2, \quad \Delta_\varphi = 0. \tag{36}$$

There is a change in the average value of M^2 within the multiplet, which, however, is eliminated by a readjustment of the scale. If we express the deviations (36) in terms of the normalized operators $D(r,0)$, we find for the coefficients

$$a_{4a}(8,0) = (4/\sqrt{5})th^2, \quad a_{4a}(27,0) = -(3/\sqrt{30})th^2. \tag{37}$$

For the (S) perturbation, we have

$$\Delta_{\mathrm{i}} = 0, \quad \Delta_0 = -2th'^2, \quad a_{4a}(S) = -th'^2\sqrt{6}. \tag{38}$$

We now substitute into (37) and (38) the values of h and h' previously derived [Eqs. (21) and (29)], and obtain the results

$$\begin{aligned} L(4a) &= (9/5\sqrt{5})x(2\alpha'+\beta)^2, \\ L'(4a) &= -(27/20\sqrt{30})x(2\alpha'+\beta)^2, \\ L''(4a) &= -(36/19^2\sqrt{6})x(3\alpha'-\beta)^2, \end{aligned} \tag{39}$$

where $x = t/(t+t')^2$.

The graph 4(b) is calculated as the expectation values in the unperturbed states of the terms in v which are of second order in the f_i. Similarly, graph 4(c) represents the contribution to the expectation values of B which is of first order in the f_i. The results are given in Table II, along with (39). The parameter y represents $(t+t')^{-1}$. Both x and y are expected to be near to $\mathfrak{M}^{-2}$, or, likely, somewhat smaller.

TABLE II. Second-order perturbation coefficients associated with Fig. 4.

Graph	L	L'	L''
4(a)	$+0.805(2\alpha'+\beta)^2x$	$-0.246(2\alpha'+\beta)^2x$	$-0.041(3\alpha'-\beta)^2x$
4(b)	$+0.224(2\alpha'+\beta)^2y$	$+0.091(2\alpha'+\beta)^2y$	$+0.040(3\alpha'-\beta)^2y$
4(c)	$+0.179\beta(2\alpha'+\beta)y$	$+0.037\beta(2\alpha'+\beta)y$	$+0.115\beta(3\alpha'-\beta)y$

From Table II we see that graph 4(a) seems to be the most important. Note that the individual contributions to L are larger than those to L' or L''. Moreover, there seems to be some tendency toward cancellation among the contributions to L' and L'' which is not evident in L. The nature of these second order results, therefore, reinforces our contention that self-consistent deviations from SU_3 symmetry would necessarily be of the (1,1), or 8-fold, type.

Finally, we wish to remark on the influence of a true external perturbation, such as provided by electromagnetic interactions, on the self-consistent solutions. Let us denote by $\eta(0)$ and $\eta(1)$ the extra self-energy terms which are added to the right hand sides of Eq. (32). Solution (33a) is then perturbed to

$$a(0) = \eta(0)/(1-K), \quad a(1) = \eta(1)/(1-K), \tag{33a'}$$

to first order in the η's, while for (33b) we obtain

$$\begin{aligned} a(0) &= (1-K)/L - \eta(0)/(1-K), \\ a(1) &= \eta(1)/3(1-K). \end{aligned} \tag{33b'}$$

The extra contribution to the large dissymmetry $a(0)$ would, of course, be difficult to verify empirically. The feature of (33b') which we wish to point out is the greater stability of the unsymmetrical solution against an additional $T=1$ perturbation.

V. SUMMARY

The usual way to discuss the approximate symmetries of strong interactions is in terms of a zero-order symmetrical Hamiltonian and a perturbing addition having a specified structure. In this approach, one has the technical advantage that there is a well-known systematic procedure for deriving the consequences of the initial assumptions. We are now engaged in the construction of a new theory of the symmetries, in which it is assumed that these features do not reflect directly features of the Hamiltonian, but arise as special characteristic simplicities of the lowest lying states. Our method of investigation is to show first that there exists a self-consistent set of particles (in a certain approximation) which exhibits a full symmetry, and then, using this solution as the starting point, to examine the possibility of self-consistent sets of the same particles in which the mass ratios differ from unity. In studying the self-consistency, we trace the influence of a given dissymmetry among the interacting particles upon the coupling constants and masses calculated for the bound states. The technique of calculation in our self-consistency approach is exactly the same as in the standard one—all the familiar machinery of perturbation theory is evoked. Once it is realized that nothing useful of the conventional theory has been lost, there need be no difficulty about accepting the changed starting point.

In ascribing a dynamical origin to the symmetry, however, a great deal is gained; the possible dissymmetries become, so to speak, "quantized" by the self-consistency requirement, which determines not only the qualitative features of the allowable deviations from symmetry, but also their numerical magnitudes. In the model we have studied, in which eight vector mesons interact among themselves, self-consistency has led to a number of interesting results concerning the departure from SU_3 symmetry.

We found, first, that the model is very stable against a perturbation from symmetry which has the transformation properties of a 27-fold tensor, and much less stable against a perturbation of the 8-fold type.[18] This has the consequence that the model can be expected to have additional self-consistent solutions which have a small dissymmetry which is predominantly characterized by an 8-fold tensor, but does not have solutions with a small 27-fold dissymmetry. Since we consider a rather simplified model, and treat it only qualitatively, we do not attempt to calculate the numerical value of the dissymmetry. However, the fact that the magnitudes are determined by self-consistency leads at once, as we have shown, to retention of SU_2 symmetry. In other words, our model leads, in a naturalistic way, both to the Gell-Mann–Okubo mass formula and to the isotopic spin concept.

Finally, it should be pointed out that our present work is limited in three respects. First, we do not have a useful criterion for choosing between the completely symmetrical solution and the solution with perturbed symmetry; in fact, we have not even given an *a priori* reason for preferring SU_3 to any other group. Second, we have relied on qualitative arguments in estimating the parameters which describe the internal dynamical structure of the bound states. We should like to suggest, as a particularly useful program of numerical computation, the precise evaluation of bound state energies for a variety of input masses. This would determine these parameters more exactly, and also allow the exploration of the possibility of very unsymmetrical solutions to Eq. (1). Third, it is clear that the interrelations among the dissymmetries of different kinds of particles will be of particular interest. This last question we intend to discuss further in another paper.

[18] It would be quite wrong to speak of the symmetrical solution as being *unstable* against an 8-fold perturbation, since the magnitude of the deviation is, in fact, prescribed.

The Fifth Interaction: Origins of the Mass Breaking Asymmetry*

YUVAL NE'EMAN†
California Institute of Technology, Pasadena, California
(Received 17 January 1964; revised manuscript received 24 February 1964)

The relative success of the SU_3 mass formula, and the relevant choice of an asymmetric solution by spontaneous breakdown and bootstrap equations are explained by the suggestion of an additional interaction, similar to electromagnetism but some ten times stronger, mediated by a vector meson coupled to the strangeness $S=Y-B$ current. The possibility that this interaction may also be responsible for the muon mass is investigated.

THE CASE FOR THE FIFTH INTERACTION

THE success of the first-order mass formula[1,2] of[2,3] SU_3 has generally been received with mixed feelings. It is encouraging to find a symmetry producing such simple straightforward results—but it is also puzzling that what seems to be a first-order perturbation term, in the context of quantum field theory, should be experimentally exact even though it emerges via g_{strong}.

A bootstrap analysis, coupled with the symmetry formalism,[4] sheds some light upon the dynamical propagation of the mass breaking effect.[5,6] On the other hand, the bootstrapped system still has the choice between a fully symmetric solution and an asymmetric one. Once a perturbation is introduced, a self-consistent mass-breaking solution is allowed to exist—but one wonders what makes the equations vote for a nonsymmetric solution at all, and in the F_8 (or hypercharge) direction in particular. The only known "external" perturbation is provided by electromagnetism, and one would have expected the symmetry to break into U-spin multiplets[7]—that SU_2 subgroup of SU_3 which commutes with electric charge $Q=(\sqrt{3}F_3+F_8)/2$. Somehow, we have learned to compute the mass spectrum of the hadrons—without really understanding its origin. In a way, this is the beauty of the symmetry shortcut. Dynamically, it is unsatisfactory.

We would like to suggest a way out of this dilemma. Suppose there were just one more "type" of interaction, between the strong and the electromagnetic. The mass spectrum has always been assumed to derive from some graduation of very-strong and less-strong interactions, but what we now suggest is that this missing fifth is of a different nature than the strong interactions. It cannot be bootstrapped—just as one does not conceive at present a positron-electron pair to couple into a photon—they can only make positronium. This, of course, may be a temporary limitation—but it is just what is intended; if ever the "S-matrix approach" can cover all interactions, it will have evolved ways of breaking symmetries too. In the present context, we postulate a vector field, or particle, χ, with a coupling midway between the strong and electromagnetic interactions,

$$g_\chi^2/4\pi \sim 0.1-0.3$$

allowing us to use perturbation theory and believe in first-order terms (in g^2) as representing the main contribution to self-masses. [S^2 may contribute in part to the 27, but if the dynamical coefficient is small, the c and d of the general formula

$$\Delta M=a+bY+cY^2+dI(I+1)$$

will take on their first-order values $c=-d/4$ just as in the experimental situation for the decuplet masses.] This particle is coupled to the *strangeness* current,

$$S=F_8-B,$$

which will give the right rise of mass with increasing S^2 in a general way, conserve I and Y, and require all masses to obey the mass formula because it breaks the symmetry in the proper direction. Since its squared coupling is about 10–30 times larger than the photon's, it creates mass splits that are some 10–30 times larger than the electromagnetic mass splittings.

Does the χ have mass? If it is massless, it should be observed in radiative K-nucleon scattering, in ϕ, η decays, and perhaps in some χ-magnetic transitions in nuclei (as nucleons have no strangeness) and hyperfragments. It seems that most of these effects would be difficult to observe, and could have been easily confused with other neutral decay modes of emissions. Lee and Yang's[8] gravitational criterion wouldn't appear here as the earth is not strange.

On the other hand, χ may well be massive—as it could then also explain the muon's mass. This would entail assigning lepton-strangeness S_L to the muon, with $S=S_Y+S_L$. As its neutrino is not allowed to have a sizable mass, we would then think of π and μ decays

* Work supported in part by the U. S. Atomic Energy Commission.
† On leave of absence from Tel Aviv University, Tel Aviv, Israel and the Israel Atomic Energy Commission, Tel Aviv, Israel.
1 M. Gell-Mann, California Institute of Technology Synchrotron Laboratory Report No. CTSL-20, 1961 (unpublished).
2 S. Okubo, Progr. Theoret. Phys. (Kyoto) 27, 949 (1962).
3 Y. Ne'eman, Nucl. Phys. 26, 222 (1961).
4 E. Abers, F. Zachariasen, and C. Zemach, Phys. Rev. 132, 1831 (1963).
5 S. L. Glashow, Phys. Rev. 130, 2132 (1963).
6 R. E. Cutkosky and P. Tarjanne, Phys. Rev. 132, 1354 (1963).
7 C. A. Levinson, H. J. Lipkin, and S. Meshkov, Phys. Letters 1, 44 (1962).
8 T. D. Lee and C. N. Yang, Phys. Rev. 98, 1501 (1955).

as $|\Delta S_L|=1$ transitions, included in the weak interactions. On the other hand, μ would be coupled to the χ and participate directly in the fifth interaction. As to muon conservation, we would replace it by the Konopinsky-Mahmoud assignment[9] and a four-component neutrino where chirality conservation sets the count right with respect to the two neutrinos.[10] In the representation space

$$\begin{pmatrix}\mu^+\\ \nu\\ e^-\end{pmatrix}$$

the strangeness operator is

$$S_L=\begin{bmatrix}1 & \cdot & \cdot\\ \cdot & \cdot & \cdot\\ \cdot & \cdot & \cdot\end{bmatrix}.$$

The mass of χ would then have to be above the present lower bound as derived from the $(g-2)$ experiment.[11]

Computing the fifth interaction's matrix elements should be no trouble—it is fully renormalizable, like any singlet, uncharged vector meson, and obeys perturbation theory. It should also be worth checking on possible observable effects, e.g., χ creation through radiative scattering of highly energetic K mesons.

THE MUON AND THE FIFTH

Feinberg and Lederman[12] have summed up the situation with respect to an "anomalous" muon interaction. For an interaction with a vector field like our χ, they have (to first order in perturbation theory)

$$\delta m_\mu/m_\mu=\frac{3}{4\pi}\frac{g_\chi^2}{4\pi}\ln\left|\frac{\Lambda^2}{M_\chi^2}\right|,$$

which would now give us $\Lambda/M_\chi\geqslant 10^3$ for the muon and $\Lambda/M_\chi\sim 2$ for the baryons.

Considering that the χ is coupled to strangeness, we would have extremely small energy shifts in muonic atoms, there being no direct χ-mediated muon-nucleon interaction.

Serious limitations arise only from π-μ-ν decay and from the gyromagnetic ratio. The effective π-μ-ν coupling constant would be depressed by a factor

$$Z_2=1-\frac{\beta}{4\pi}\frac{g_\chi^2}{4\pi}\ln\left|\frac{\Lambda^2}{M_\chi^2}\right|$$

$$\sim 1-\beta/\pi,$$

where β is of the order 1. This could depress $g_{\pi\mu\nu}$ by anything up to some 30%, as against an experimental uncertainty of 2%.

As to the muon magnetic moment, with

$$10^{-5}>\delta_g=\frac{g_\chi^2}{4\pi}\frac{1}{3\pi}\left(\frac{m}{M}\right)^2,$$

we would get $m/M_\chi\leqslant 0.02$, i.e., $M_\chi\sim 5$ BeV for a coupling of 0.3 to about 3 BeV if $g_\chi^2/4\pi\sim 0.1$.

Nevertheless, we do not feel that these two limitations, mainly the Z_2 value, should be taken as definitive evidence. The χ may be massive enough to agree with δg, and wave function renormalization effects may involve cutoff-dependent quantities besides the first-order term given above. To account for the experimental value of the muon pair production by neutrinos passing through matter, we do not assume the muon neutrino to be coupled to χ, and have therefore chosen the Mahmoud-Konopinsky model.

BOOTSTRAPS AND THE FIFTH

Glashow[5] has analyzed the mechanism through which a "spontaneous" breakdown of the symmetry could occur. The result shows that three possibilities exist: (a) a symmetric solution; (b) a nonsymmetric one with the [27] contributing; (c) octet dominance of the symmetry breaking; even in case (c), it is not clear that the solution should conserve iso-spin and Y; it could also keep only I_3 and Y and break I. There is also no reason for the emergence of Y and I_3 at the existing angles to electric charge in the Cartan subalgebra diagonal plane of SU_3.

Cutkosky and Tarjanne[6] have obtained stability against b, thus restricting the "choice" to a symmetric solution versus an [8]. Again, why should a breakdown occur, and why in the Y, I direction? Any dynamical model that does it starts from some "contaminated" multiplet-ϕ-ω mixing as suggested by Gell-Mann,[1] Sakurai, Salam, Katz and Lipkin, and others; or one in which the mass breaking has already occurred, as in Cutkosky's[13] or Capps',[14] where the pseudoscalar mesons are taken to have their physical masses to start with. Cutkosky and Tarjanne have to assume Y conservation at the start, and have no way of choosing between the Y and Q directions.[15]

In our model, Y would be picked and the rest would follow, with the g_χ coupling doing it. It would give larger mass splittings in the Y direction as against Q it would determine the emergence of the nonsymmetric solution as the preferred one—an important point as most studies show that the symmetric one has "much

---

[9] E. J. Konopinsky and H. M. Mahmoud, Phys. Rev. 92, 1045 (1953).

[10] Y. Ne'eman, Nuovo Cimento 27, 922 (1963).

[11] G. Charpak, F. J. M. Farley, R. L. Garwin, T. Muller, J. C. Sens, and A. Zichichi, in *Proceedings of the 1962 International Conference on High Energy Physics at CERN*, edited by J. Prentki (CERN, Geneva, 1962), p. 476.

[12] G. Feinberg and L. M. Lederman, Ann. Rev. Nucl. Sci. 13, 431 (1963).

[13] R. E. Cutkosky, Ann. Phys. (N. Y.) 23, 415 (1963).

[14] R. H. Capps, Phys. Rev. (to be published).

[15] Nevertheless, the Y direction may yet prove to correspond to some limiting process where Q is iterated [R. E. Cutkosky (private communication)].

more probability" to occur. We would be left with the question of a reason for the existence of strangeness—having thus sizably reduced the extent of the mystery.

CAN THE ϕ BE THE XENODYNAMIC FIELD χ?

There is a slight chance that we may have already seen the χ field—as the ϕ (1020 MeV, 1^-) seems to be coupled to S indeed. The various ϕ-ω mixing theories, looking for a model in which $\phi \not\leftrightarrow 3\pi$ have arrived at an extremely small coupling to nucleons, of the order of 10^{-3} or 10^{-4}. If the ϕ is coupled to the strangeness current, it would have no ρ-π coupling, and the main virtual decay mode would thus disappear—as against the ω. The width of the ϕ is now thought to be about 3.1 ± 1.0 MeV, i.e., a minimum[16] of 2 MeV. This may be too large for a nonstrong interaction, though not by any appreciable factor; remembering the inconsistency between present theoretical computations and the π^0 width, we should not be surprised to have $\Gamma \sim 1$ to 2 MeV.

If the ϕ is our field, where is the eighth component of the vector meson octet? This question makes the ϕ-χ identification a highly speculative supposition. Nevertheless, it is worth investigating a new resonance reported by the Syracuse-Brookhaven group.[17] In a missing mass plot of

$$K^- + p \rightarrow \Lambda + \text{neutrals},$$

they see a peak at 930 MeV. This is exactly the value we would have expected the eighth component to have according to the mass formula. Can this be really it?

The 3π plots where the ω was found do show a very slight bump at[18] 930 MeV though this may be an ordinary statistical deviation with no significance. On the other hand, it could be the $\pi^+\pi^0\pi^-$ decay mode of the $\eta(930,1^-)$ of our speculation. Its width should be somewhat larger than the ω, as it can have all the ω decay modes and also go into 5π, plus additional electromagnetic decays. It is not clear, however, that these electromagnetic decays—the only neutral modes it has—would suffice to make it appear in the above-mentioned missing mass plot, where the ω itself is not clearly seen. A favorable example is the charged 2π decay mode of the ω, which is important enough to appear in ρ plots; and the neutral modes of the η, all of them electromagnetic also, and appearing in the missing mass plot at the sides of the 930-MeV bump. Again, if R symmetry is meaningful for mesons, its decay into 3π would be forbidden by it, which would increase the neutral modes.

These considerations, even though speculative, have the advantage of pointing at possible checks of the χ meson idea—making the χ in all the experiments where the ϕ is produced. It may be a matter of accumulating more statistics, to cope with the relative smallness of g_χ and the corresponding cross sections—all of them should be a fraction of the ϕ's.

The connection with the muons, if it exists, should appear in radiative scattering of high-energy muon beams, where $K\bar{K}$ pairs should be produced through χ (if this is the ϕ, they should be relatively accessible). Another check would consist in measuring $\phi \rightarrow \mu^+ + \mu^-$ and comparing it with the similar $e^+ + e^-$ result, where only electromagnetism operates.

Note added in proof. The possibility that the χ be massless would imply a relatively small xenomagnetic coupling, as pointed out to the author by Professor N. Ramsey. There may also be difficulties in reconciling its existence with the experimental limits on $\pi^0 \rightarrow \gamma + \chi$ and with the results of electromagnetic renormalization theory and its successful predictions. On the other hand, a mass of $2m_\pi < m_\chi$ seems consistent with most experimental results, i.e., the ϕ is a good candidate. Since the 930-MeV meson seems to have decay modes which do not fit a Γ octet's eighth component, we would then have to return the ω to this role. The muon problem would probably require a higher mass. The author is indebted to S. Frautschi, R. Dashen, and D. Beder who have studied the detailed experimental implications.

---

[16] R. H. Dalitz, Ann. Rev. Nucl. Sci. 13, 339 (1963).

[17] M. Goldberg, M. Gundzik, J. Leitner S. Lichtman, P. L. Connelley, E. L. Hart, K. W. Lai, G. London, G. C. Moneti, R. R. Rau, N. P. Samios, I. O. Skillicom, and S. S. Yamamoto, Bull. Am. Phys. Soc. 9, 23 (1964).

[18] G. Puppi, Ann. Rev. Nucl. Sci. 13, 287 (1963) (see Fig. 10a).

XI

THE PHYSICAL UNIQUENESS OF THE EIGHTFOLD WAY

The theory presented in this book is not in the nature of a postulate derived from abstract principles; neither does it correspond to some plausible assumption related to the geometrical substrate. The formalism of particle physics does have room for the inclusion of charge-like quantum numbers, either as symmetries of the Lagrangian or of the S-matrix, but the actual choice of a particular symmetry group is at present determined by the experimental situation only.

In the following review, Ne'eman points out the features of the group formulation; essentially, this is—like the rest of physics—a shorthand description of the phenomenology, perhaps somewhat more directly so than usual. The observation that all simple reactions allowed by the strangeness scheme actually seem to occur is abstracted into the mathematical condition that the Lie algebra should have two additive quantum numbers. This in turn leads to a list of some six main candidates; a discussion of assignments for the lowest states brings out numerous contradictions in every one but SU(3) in its baryon octet version. As explained in the article, this model corresponds to an invariant subgroup of SU(3); the case for and against hypothetical triplets fitting with SU(3) itself has been discussed in a previous section.

With the more recent discovery of the Omega-Minus, the exclusion of such semiplausible models as G(2) and the Sakata theory appears to be even more obvious, as can be seen from a look at the diagrams.

There remains the real possibility that SU(3) may be embedded in a larger symmetry group. In Part VIII, we have discussed how pseudoscalar generators may be adjoined to the algebra to give SU(3) $\times$ SU(3). Even the algebra of scalar generators may be enlarged, however, to yield the group SU(4) or a still bigger one. Any such enlargement brings with it the introduction of one or more new quantum numbers commuting with Y and I_z and corresponding conservation laws, exact or approximate.

 From: *Proc Intern Conf Nucleon Structure*, ed. R. Hoffstadter and L. I. Schiff (Stanford, 1963), pp. 172–187

The Symmetry Approach to Particle Physics

Y. NE'EMAN
Israel Atomic Energy Commission and Tel-Aviv University

1. Why Symmetries?

It is said that back in the fifties, one could still write down Lagrangians and fields without looking round to make certain that nobody was watching. It had been found that strong interactions obeyed exactly some constant-gauge symmetries—baryon number B, isospin T, and hypercharge Y. In the Gell-Mann–Nishijima scheme the hadrons (strongly interacting particles) [1] had all been assigned respective values of these additive quantum numbers, and reactions did exhibit evidence of the new conservation laws. Nevertheless, the interaction Lagrangian that combined all possible Yukawa terms allowed by these symmetries for the eight known baryons and seven bosons, contained eight independent couplings, g_1 to g_8 [2]. The quest for a higher symmetry was thus launched [3, 4, 5, 6, 7] when various schemes of "global" symmetry—e.g., equality of all pion couplings, or of all kaon couplings—were proposed. During the last two years, this search was systematized exhaustively by Behrends and Sirlin [8], and was continued by Shekhter [9] to cover the new η couplings as well. This is one motive for the search for a symmetry: to get a limitation, a ratio, for the forces between different sets of particles, just as isospin had generated charge independence. But from the beginning, Gell-Mann, for one, was also trying to get a symmetry-breaking effect from the kaon couplings, in order to generate mass differences among the four baryon multiplets. This is a second aspect that is present in the search for higher symmetries: perhaps they can explain the mass spectrum in terms of some kind of breakdown effect. Contributing much to the progress that has lately been made along these lines are the group-theoretical considerations developed by Gell-Mann and Okubo [10].

There was a third line of attack, which seemed to arise from a somewhat more detached and speculative interest. The existence of fifteen different hadrons raised the question—now less fashionable—of elementarity. Given some dynamical binding mechanism, what was the minimum number of different particles required to create all the rest? And what other particles would then be predicted by this model? This was the approach of Fermi and Yang [11], Goldhaber [12], Sakata [13], and Ikeda *et al.* [14]. Very early, the impact of the 33 resonance created a slight refinement. What was the status of this state? It could represent a typical non-elementary entity, even within the strong interactions themselves. The "model" approach has now been extensively surveyed by Speiser and Tarski [15], Gell-Mann and Glashow [16], Coleman [17], Ne'eman [18], and Behrends *et al.* [19].

The fourth motive was inspired by a desirable characteristic of electromagnetism—a local gauge. Could the strong interactions obey a gauge transformation that would exhibit the freedom of picking different phases at different points of space time? Yang and Mills [20] had investigated this question for isospin and had found that vector mesons were required as mediators* for the strong interactions, one for each parameter of the symmetry. When Nambu [21] showed that the Hofstadter nucleon structure experiments entailed the existence of a spin-1 boson, and when the weak interactions suddenly exhibited a conserved current nature and thus also seemed to lead to a local gauge and vector mesons, the search for a strong gauge began again. Salam and Ward [22], Sakurai [23], Gell-Mann and Glashow [16], and Ne'eman [24] contributed to this approach.

The fifth way to the symmetries comes from the weak interactions. Okubo and Ioffe [25] had shown that the near equality of the vector coupling constant in beta and mu decays could result from a symmetry effect: If the weak current were identical to the conserved current generated by the strong symmetry, the coupling would be unrenormalized by the strong interactions. Salam and Ward [26], Behrends and Sirlin [8], Gürsey [27], and Gell-Mann and Levy [28] followed this line.

Couplings and branching ratios, a mass spectrum, predictions of missing particles and quantum-number assignments, vector mesons, and a connection with the weak and electromagnetic interactions—every consideration of a symmetry has sought information on all five of these problems, even though it may have arisen out of work in only one or two of these lines. The emphasis has changed: an aspect that used to be an academic recreation—the model as a key to a list of particles—has, ever since the recent flood of resonances, become a high-priority concern, with careful experimental checking. The vector-meson approach has achieved considerable qualitative success; it gained field-theoretical respectability after Schwinger's demonstration [29] of the compatibility of a local gauge with massive vector bosons coupled to its currents. The connection with the weak interactions seems to imply more sophistication than was thought in the beginning. The branching ratios are of limited use, since they may be affected by the symmetry-breaking part of the strong interactions,† that part which is responsible for the mass spectrum. Still, every one of these approaches leads to some interesting predictions and can be checked by experiment. The reputation that the symmetries somehow acquired—of being able to predict only qualitative results—is libelous.

2. *Group-Theoretical Considerations*

The mathematical formulation of higher-symmetry schemes was provided by the theory of Lie groups, the main tools having been developed by Cartan and Weyl. In fact, the survey of available symmetries, including such oddities

* See also recent suggestions by K. E. Eriksson and S. A. Yngström, Univ. of Uppsala preprint (1963).

† The formalism has now been developed so as to include the symmetry-breaking effects (C. Dullemond, A. J. MacFarlane, and E. C. G. Sudarshan, Univ. of Rochester Report NYO-10260, 1963).

as the group $G2$, goes back to Cartan's thesis. It should be noted that dealing with Lie groups means that we are looking for additive symmetries (the rules of addition are more complicated here than for angular momentum), since all three basic quantum numbers B, Y, and T are additive [30]. Continuing our analysis of the mathematical formulation, it is easier to speak of the Lie algebras than of the groups [31]. In some cases, an algebra has led to different models, like the triplet and octet versions of unitary symmetry. This is roughly similar to the differences between a vectorial and a spinorial model in the case of three-dimensional rotations: the spinorial model will lead to all representations, the vectorial will yield only integer spin representations. Only one-third of the representations in the $SU3$ list will appear in $SU3/Z3$, which is the octet's group. But the infinitesimal generators are the same.

Algebras are characterized by their rank, which is the maximum number of linearly independent intercommuting elements they contain. In the case of the 16 Dirac gammas, for example, (looking only at their commutation relations, not at their anticommutators) this is four: $1, \gamma_0, \gamma_5\gamma_1, \gamma_2\gamma_3$ would be one such set. In the limit of equal masses, particles forming a basis in a single representation space are degenerate and have the same "external" quantum numbers—spin, parity, statistical behavior, and so forth. Thus the baryon number, effectively linked as it is to statistics, should be common to the whole representation; i.e., it should be an invariant of the group. In the algebra, it is realized for the basic baryon representation by the identity operator. Excluding this operator means that we are dealing only with traceless matrices when looking for an algebra containing the remaining diagonalizable elements of isospin and hypercharge. We are now dealing with either the "simple" algebras or their direct products. In the above example of the gammas, we would be left with a rank-three algebra, denoted $A3$.

Isospin contains one diagonal operator T_z; hypercharge adds a second. No other similar quantum numbers have been found to hold in the strong interactions. Every reaction that the above quantum numbers permit seems to take place; if a third such charge existed, it would thus have to be degenerate for the basic baryons, or, at most, it would have to distinguish only between the Λ and Σ^0, as pointed out by Behrends *et al.* [19]. Even then, models like global symmetry, which have included such a third commuting quantum number, have been shown to preclude known processes (e.g., $K^0 + p \rightarrow K^+ + n$). There is still some possibility that a charge that is degenerate for all eight baryons exists; such an algebra would have a sub-algebra without this charge, and, being less restrictive, the sub-algebra should also hold. Thus we first have to check the algebras of rank two, and if there is a perfect fit somewhere, continue and look for a larger and more restrictive rank-three symmetry.

3. *Candidate Symmetries*

In his thesis, Cartan classified all simple Lie algebras. These include four infinite sets and five exceptional algebras [32]: the algebras of traceless matrices A_r (r is the rank) in $(r + 1)$ dimensions; the orthogonal sets B_r in $(2r + 1)$, i.e., odd, dimensions, and D_r in $2r$ (even) dimensions; the symplectic series C_r representing rotations in even-dimensional ($2r$) spaces with an antisymmetric metric; and the isolated algebras $E6$, $E7$, $E8$, $F4$, $G2$.

There is only one rank-one algebra $A1 = B1 = C1$, i.e., isospin or three-dimensional rotations ($D1$ is Abelian). Of the rank-two algebras, in which we are interested, $D2$ is only semi-simple—we know that four-dimensional rotations split into two commuting $A1$ sets. The algebras $B2$ and $C2$ are isomorphic; we thus have only five possibilities:

1. $A1 \times D1$: T and Y alone, with no link. This is the simplest extension of the d'Espagnat-Prentki [33] scheme, and will add no new quantum numbers, links, or restrictions.

2. $D2$: the four-dimensional Salam-Polkinghorne [6] rotations symmetry. This is also a very loose scheme, both because the algebra contains a built-in free parameter and is not simple, and because the only supermultiplet is (N, Ξ). It reduces the eight veteran strong couplings to seven. Falling back on $D2$ is thus equivalent to reporting half-failure. There is also a new test that shows $D2$ to be insufficient in itself, and to require the addition of at least one discrete quantum number. This group has no way of differentiating between the two isoscalar vector mesons ϕ (1020 MeV) and ω (780 MeV), with $\eta_P = \eta_C = \eta_G = -1$. This, incidentally, could be a new criterion for selecting a supersymmetry, just as any lepton scheme should remove the μ-e degeneracy.

3. $B2 = C2$: There are three possible models that have been studied by Salam and Ward [34], by Glashow and Gell-Mann [16], and by Behrends *et al.* [19]. These correspond to the groups $Sp(4)$, with a double **4**-dimensional model (N, Ξ), $(K, \bar{K})$, the K mesons having to be introduced independently, since they cannot be generated otherwise; $O(5)$, with a **5**-dimensional model (N, Λ, Ξ); and $O(5)$, with a **10**-dimensional model $(N, \Lambda, \Sigma, \Xi, Z^{\pm})$, where $Z^{\pm}$ is a $T = 0$, $Y = \pm 2$ hyperon, here with $j = \frac{1}{2}^+$, which does not seem too plausible at this stage. The $Z^{\pm}$ particle is also required as a companion to the Σ in the two other models, so that it is a feature of all $B2$ or $C2$ models. Similarly, a $Z^{*\pm}$ with $j = \frac{3}{2}^{\pm}$ and another with $j = \frac{3}{2}^-$ always appear. For the mesons, we should have a $D^{\pm}$ with $T = 0$, $Y = \pm 2$ for both $j = 0^-$ and $j = 1^-$. Another typical feature is that the $33N^*$ resonance should be accompanied by a $T = \frac{1}{2}$, $j = \frac{3}{2}^+$ doublet, which seems excluded by experiment. The **5** model requires a $T = 2$, $Y = 0$, $j = \frac{3}{2}^+$ resonance, apparently nonexistent [35]. The $Sp(4)$ model has no room for the ω meson, since $\omega \to 3\pi$, and the product $\mathbf{5} \otimes \mathbf{5} \otimes \mathbf{5}$ does not contain isoscalars. A "redeeming" feature of $Sp(4)$ is that since $\mathbf{5} \not\subset \mathbf{5} \otimes \mathbf{5}$, it precludes $Y_1^* \to \Sigma + \pi$. On the other hand, it has for the same reason no room for Y_1^{**} (1660 MeV) [36], its $T = 1$, $Y = 0$ state having again no decay into $\Sigma + \pi$.

4. $G2$: this is the only exceptional algebra of rank two. It is contained in seven-dimensional rotations and thus lends itself to a combination with the weak interactions in which the latter follow the larger $O(7)$ symmetry. It has 14 parameters, with no $T = \frac{1}{2}$, $Y = \pm 1$ values in the corresponding adjoint **14** representation. For this reason, the existence of K^* with $j = 1^-$ precludes its use in local gauge theories of the Sakurai [23] type (the vector bosons should

always correspond to the adjoint representation in such a theory). There is also no room at all for $\phi \to K_1^0 + K_2^0$, since the antisymmetric scalar combination is not contained in $\mathbf{7} \otimes \mathbf{7}$. Otherwise it is nice. The model is $\mathbf{7}(N, \Xi, \Sigma)$, with the Λ appearing separately as a singlet. It was first suggested and investigated by Behrends and Sirlin [8].

There is an additional criterion which, if accepted, would eliminate both $B2$ (or $C2$) and $G2$. Behrends and Sirlin, and in a more general way Ruegg [37], have proved that these models would all rule out the existence of a magnetic moment for the neutron because the electromagnetic ($\Delta T_Z = \Delta Y = 0$) transitions contain only one amplitude, which is proportional to the electric charge of the particle. However, one could claim that this moment is created through a symmetry-breaking interaction, dominant at low momenta.

When the $j = \frac{3}{2}^+$ resonances are charted, N^* can go into the **14** representation, with no $T = \frac{1}{2}$ companion; it also requires a $Z^{*\pm}$, and a Ξ^* with $T = \frac{3}{2}$ instead of the experimental $T = \frac{1}{2}$ (if this is indeed a $j = \frac{3}{2}^+$ state). The alternative would be to put N^* in the **27** representation, which would allow Ξ^* to have $T = \frac{1}{2}$ but would require $Y = \pm 2$ triplets and the unseen $T = \frac{1}{2}N^{*\prime}$. Both assignments require another Y_1^* that does not decay into $\Lambda + \pi$, since **14** and $\mathbf{27} \not\subset \mathbf{1} \otimes \mathbf{7}$. The 1385-MeV resonance has to go into an additional **7** representation again, with $j = \frac{3}{2}^+$. On the other hand, we do need a Y_0^* with $j = \frac{3}{2}^+$ (perhaps the 1405-MeV resonance?).

5. $A2$: this algebra generates the three-dimensional unitary unimodular group $SU3$ and its adjoint group $SU3/Z3$. ($Z3$ is a set of three discrete elements forming the "center" of $SU3$, the three cubic roots of the identity [exp $(2\pi i/3)$, exp $(4\pi i/3)$, 1].) These two groups, when multiplied by an Abelian $U1$, lead to $U1 \otimes SU3 = U3$, the Sakata triplet (p, n, Λ) [14, 38, 39], and $U1 \otimes SU3/Z3$, the Gell-Mann–Ne'eman octet $(N, \Xi, \Sigma, \Lambda)$ [40, 24].

Both $A2$ models pass the neutron magnetic moment test [37, 41]. Both can also remove the ϕ-ω degeneracy.

The mesons and vector mesons lie in both models in the adjoint representation **8** and the scalar **1**; this fits the local gauge approach. The decay of the ϕ into K_1^0, K_2^0 pairs has been shown by Levinson, Lipkin, and Meshkov [42] to imply that this cannot be the singlet; it could be the $T = 0$, $Y = 0$ component of the octet, with ω as the singlet. Following independent suggestions by Gell-Mann, Salam, and Sakurai [43], we could alternatively visualize a mixing of ϕ and ω, which breaks the symmetry, with their linear combinations ($\alpha\omega \pm \beta\phi$) acting as the respective components of the octet and singlet representations. Incidentally, if there exists another $T = 0$, $j = 0^{++}$ (ABC or K_1^0, K_1^0) meson it could again be an $SU3$ singlet or octet (with the elusive ζ and 730-MeV $K^{*\prime}$ [44]).

Let us now look at the baryons and baryon resonances. These reveal many weaknesses in the triplet model. A quick glance at the Sakata diagrams shows that the Ξ takes $j = \frac{3}{2}^+$, with N^*; the same representation also requires for $j = \frac{3}{2}^+$ a $T = \frac{1}{2}$, $Y = 1 N^{*\prime}$, a $T = 1$, $Y = 2$ (K^+-p) resonance [45], and a Y_0^* singlet. Experimental evidence seems to favor $j = \frac{1}{2}^+$ for Ξ and to exclude all such $j = \frac{3}{2}^+$ manifestations. More difficulties arise when we inspect the companions of the Σ in the **6** $j = \frac{1}{2}^+$ representation: a further unseen nucleon-like entity N', and a Z^+ particle. The N' should have appeared in N-π scattering, and thus seems excluded. The existence of Ξ^* and Y_1^{**} (1660 MeV) would mean, for the Sakata model, that they are $j = \frac{3}{2}^-$ manifestations, with the second N-π resonance, and another Y_0^*. But we then require an unseen $j = \frac{3}{2}^-$ 33 resonance. There is no room for a $j = \frac{1}{2}^-$, $Y_0^* \to \Sigma + \pi$, since $\mathbf{1} \not\subset \mathbf{6} \otimes \mathbf{8}$. It is clear that this model lost its advantages when the Σ-Λ parity proved to be even [46]. The observed reaction $p + \bar{p} \to K^0 + \bar{K}^0$ is forbidden in this model (incidentally, it is also forbidden in $C2$).

Now for the octet. This is a very stringent model because it requires all eight baryons to have the same spin and parity; however this condition now seems to be satisfied [46]. On the other hand, the number of amplitudes in products is generally large, so that there is more freedom in the computation of quantities such as branching ratios. The baryon resonances seem to fill up a $j = \frac{1}{2}^-$ singlet, a $j = \frac{3}{2}^-$ octet, and a $j = \frac{3}{2}^+$ decuplet. This last assignment suffers from an allowed $Y_1^* \to \Sigma + \pi$ ($\sim$14 per cent);* on the other hand, if we assign the $\frac{3}{2}^+$ states to the **27** representation, the process is forbidden. Nevertheless, the **10** is favored because of its "modest" requirements: it lacks only a Z^-, which may still be found (it could be the 1954 Eisenberg event [47], as noted by Gell-Mann), whereas **27** would require unseen $T = 1$, $Y = \pm 2$ and $T = 2$, $Y = 0$ resonances.

In all the above identifications, we have checked only the assignments for $j = 0, \frac{1}{2}, 1, \frac{3}{2}$ because, according to the Chew-Frautschi conjecture [48] about the Regge-trajectory nature of the hadrons, they should all reappear at intervals of $\Delta j = 2$. This seems verified by the existence of the f_0 (1250 MeV), $j = 2^{++}$ meson [49] (a scalar in all the above symmetries, like the vacuum or the suspected $j = 0^{++}$ ABC or K_1^0, K_1^0 resonances [50]); the third N-π resonance at 1688 MeV ($T = \frac{1}{2}$, $j = \frac{5}{2}^+$) and the Y_0^* at 1815 MeV ($j \geq \frac{5}{2}$), which forms perhaps a repetition of the $j = \frac{1}{2}^+$ model itself (i.e., a new Sakaton or an octet); and the fourth N-π resonance ($T = \frac{3}{2}$, $j \geq \frac{7}{2}$), which is apparently a shadow thrown by the 33 resonance and should be completed similarly within the frame of the model.

Summing up, we see that as things stand now, the octet seems to have scored best as to identifications and the construction of a "periodic table" [51]. These results are summarized in Fig. 1. We shall now discuss some additional results based upon this model.

4. *Masses, Widths, and Other Experimental Tests*

Further confirmation seems to come from the Gell-Mann–Okubo [10] mass formula. This is an attempt to get the mass spectrum from assumptions about the $SU3$ tensor features of the mass operator. It should be given by some combination of the $T = 0$, $Y = 0$ components of the representations appearing in the self-product decomposition. The double-octet product contains three

* Cutkosky and Tarjanne have introduced corrections due to the symmetry-breaking mechanism—to first order—and have found this ratio to be 7 per cent.

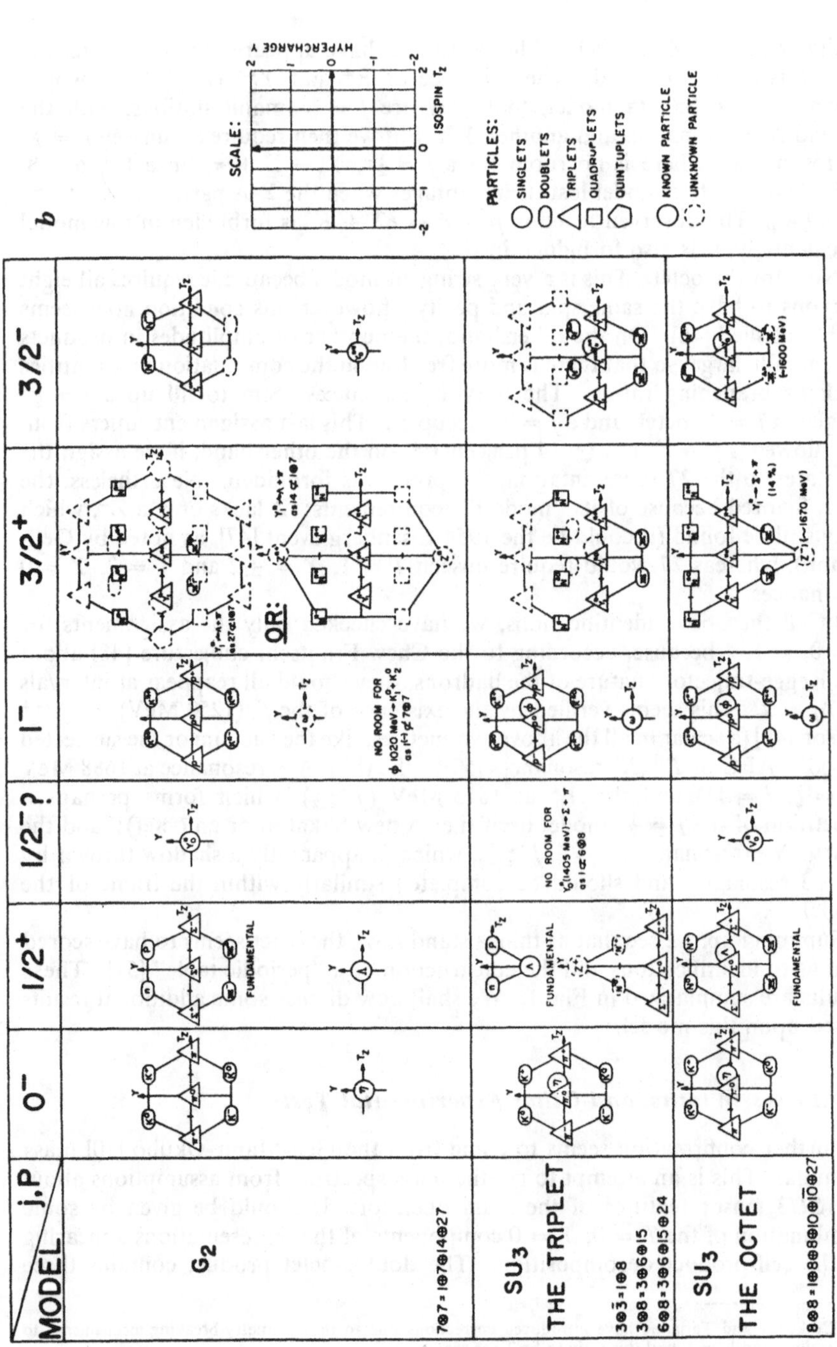

Fig. 1. *(concluded)*.

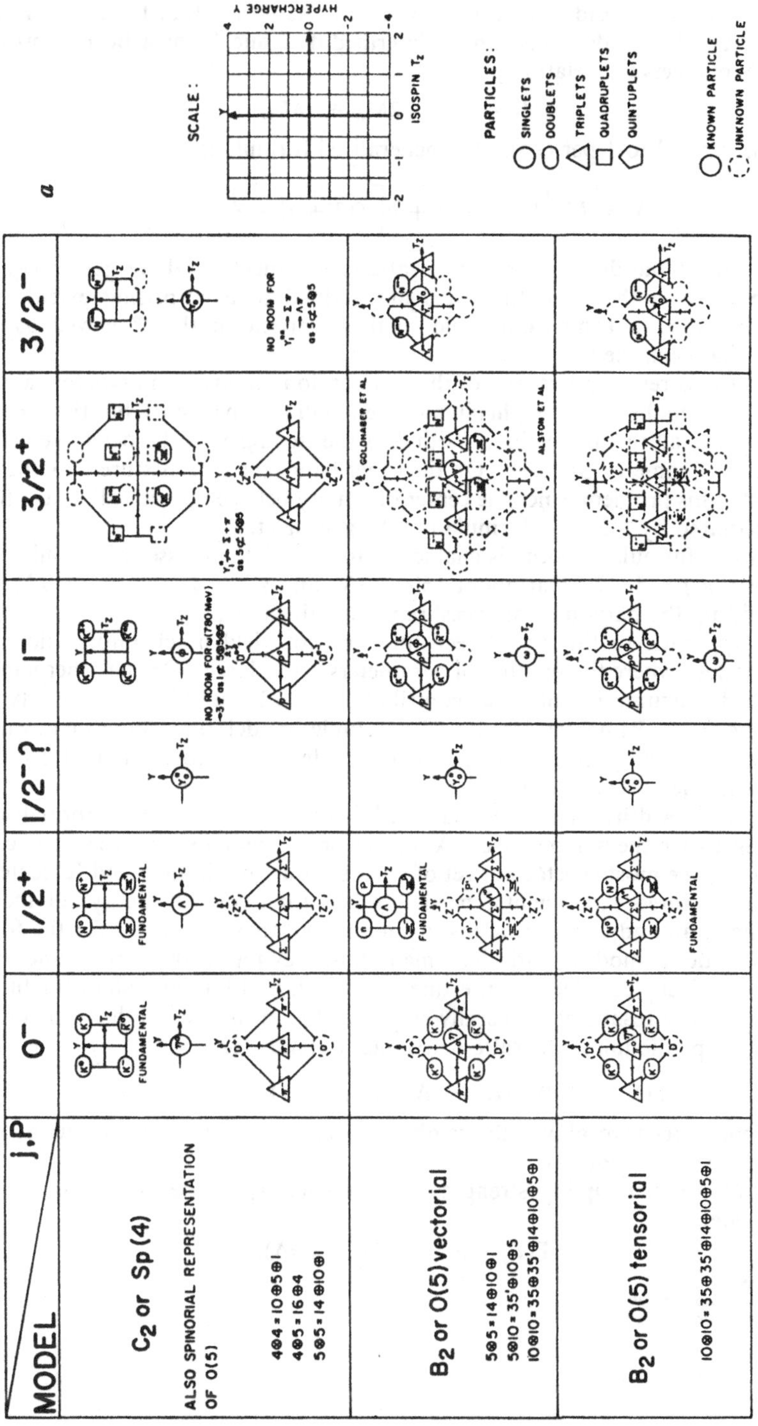

FIG. 1. The $C2$ and $B2$ models (a), and the $G2$ and $A2$ ($SU3$) models (b).

such components, besides the full $SU3$ scalar. Assuming that the **27** does not contribute to first order (this can be described dynamically as a Born approximation), one gets the relation

$$3M_\Lambda + M_\Sigma = 2M_N + 2M_\Xi,$$

which is verified to 1 per cent. The generalized formula is

$$M = M_0\left\{1 + aB\cdot Y + b\left[T(T+1) - \frac{Y^2}{4}\right]\right\}.$$

In the case of the decuplet, the quadratic term cancels, and we get an equal spacing between the isomultiplets, again verified to a surprising extent. It should be noted that this would hold for these particles in the **27** as well, as H. Harari* has remarked.

The mesons respond less well to this assumption about the masses (or rather masses squared; here). The η fits in nicely enough ($\sim$6 per cent), but the $j = 1^-$ case destroys all this perfection. With ϕ as the isosinglet in the octet, we get a difference of some 17 per cent. On the other hand, the ϕ-ω mixture hypothesis may settle this problem, since the center of gravity between ϕ and ω lies near the value required by the mass formula for the isosinglet.

The mass formula hypothesis enables us to predict the masses of the missing particles: a $j = \frac{3}{2}^-$, Ξ^{**} at about 1600 MeV, and a $j = \frac{3}{2}^+$, Z^- at 1680 MeV would fill up the baryon resonances' representations [51].

This approach to the mass spectrum does not yield much information in other models, where the number of parameters is too large. On the other hand, a first-order assumption of no **27** contribution in $G2$ would lead to linearity in Y for the (N, Σ, Ξ) heptet. In the $B2$ quintuplet model, one can write down a mass formula and check it for the mesons; Harari has found that it clashes badly with the observed values.†

Resonance widths have been examined in the light of the octet formalism, by using either the symmetry-breaking algebraic methods or dynamical calculations. The most spectacular set of results is due to Glashow and Rosenfeld [51]; using three experimental partial widths for two-body decays of the $\frac{3}{2}$ resonances, and adjusting a radius of interaction, they get values for the four remaining decay modes; with the same radius, they repeat the performance for the $j = \frac{3}{2}^+$ decuplet's decays, inserting one width and finding values for three processes. The results are generally very near the measured widths. The most interesting prediction is the ratio (the index γ stands for $j = \frac{3}{2}^-$)

$$\Gamma(\Sigma_\gamma \to \bar{K}N):\Gamma(\Sigma_\gamma \to \pi\Lambda):\Gamma(N_\gamma \to \pi N) = 1:4:22,$$

which compares favorably with the observed 1:4:27. (The moments are about the same in these modes.)

The $\frac{3}{2}^+$, or δ, coupling strengths are consistently greater by an order of magnitude; e.g.,

$$\Gamma(\Sigma_\delta \to \pi\Lambda) \approx 4\Gamma(\Sigma_\gamma \to \pi\Lambda),$$

* Private communication.
† Private communication.

even though the γ mode has three times as much available energy as the δ mode. This is consistent with the experimentally observed relatively copious production of δ resonances, as compared with the γ ones.

The same method has been applied to two-body decays of one baryon resonance into another baryon resonance plus a meson. The results can be checked for the "Regge recurrence" resonances, and, quoting the authors, "with satisfaction and relief we find that the calculated results are completely compatible with experiment." Let us note that work with this model has been simplified by the group-theoretical treatments of Edmonds, Levinson *et al.*, Matthews and Salam, Rashid, Dothan and Harari, and Goldberg [52].

Incidentally, it is worth noting that three years ago, an extensive series of predictions of that kind were prepared for the Sakata model by Yamaguchi, Sawada, and Yonezawa [53], and others. Many of these predictions—masses, parities, widths—clash badly with experiment.

Several recent papers have checked the dynamical self-consistency of the octet model. This was first done by assuming some simple static model interaction of the Chew-Low type and verifying that the baryons would really give attractive interactions in the right channels to produce the observed mesons, or the baryon resonances. Cutkosky, Kalcar, and Tarjanne [54] have initiated such verifications, the results agreeing with observations when the nucleon-meson coupling parameter (i.e., the ratio of "F" to "D" trilinear couplings) is $\theta = 35$ deg. Capps [55] has calculated the couplings $g_{\rho\pi\pi}$, $g_{\rho KK}$, $g_{\omega KK}$, $g_{K^*\pi K}$, $g_{K^*\eta K}$, assuming the vector mesons to be bound states of pairs of *p-s* mesons and using dispersion theory. The results are in exact agreement with the model. Other dispersion calculations of Capps' produce good values for the ratio

$$g^2_{K^*\pi K} : g^2_{K^*\eta K} : g_{\rho\pi\pi} g_{\rho KK}$$

and for the Ξ^* width. Sakurai [55] has shown that these results can be derived from perturbation theory when one assumes equality of the masses. Martin and Wali [56] have checked the decuplet assignments, using the N/D method; it did provide for the Ξ^*'s having $T = \frac{1}{2}$, whereas it should have had $T = \frac{3}{2}$ like the N^* had they used the Chew-Low static model without unitary symmetry assumptions. Similar checks have been made by Watson for processes such as hyperon-pair production [57]. In fact, Sakurai [55] has demonstrated that the appearance of mutually interacting octets and other $SU3$ multiplets, when fed into usual field theory, automatically imposes unitary symmetry. In a way this means that the real test of the symmetry lies in the charting of the interacting multiplets.

Using their geometrical simplification approach to the theory of representations of $SU3$, Levinson, Lipkin, and Meshkov [52] have lately derived a series of additional predictions for the octet.* For electromagnetic interactions they find

$$\frac{\langle \gamma P \mid N^{*0}\pi^+ \rangle}{\langle \gamma P \mid Y^{*0}K^+ \rangle} = -2,$$

* Similar results were derived by E. C. G. Sudarshan, A. J. MacFarlane, and C. Dullemond.

$Y_1^{*+} \to \Sigma^+ + \gamma$ and $N^{*+} \to P + \gamma$ are allowed and have equal matrix elements; and $Y_1^{*-} \to \Sigma^- + \gamma$ and $\Xi^{*-} \to \Xi^- + \gamma$ are forbidden. S. Okubo [10] has found

$$\langle \rho^+ | \pi^+\gamma \rangle = \langle \rho^0 | \pi^0\gamma \rangle = \langle K^{*+} | K^+\gamma \rangle = -\langle \omega | \eta + \gamma \rangle$$

$$= \frac{1}{\sqrt{3}} \langle \omega | \pi_0 + \gamma \rangle = \frac{1}{\sqrt{3}} \langle \rho_0 | \eta\gamma \rangle = -\frac{1}{2} \langle K^{0*} | K^0\gamma \rangle,$$

and

$$\langle \eta | 2\gamma \rangle = -\frac{1}{\sqrt{3}} \langle \pi^0 | 2\gamma \rangle ,$$

taking into account that mass-splitting interaction weakens some of these results, leaving only

$$\frac{1}{\sqrt{3}} [\langle \omega | \pi^0\gamma \rangle + \langle \rho^0 | \eta\gamma \rangle] = \langle \omega | \eta\gamma \rangle - \frac{4}{3} \langle K^{0*} | K^0\gamma \rangle + \frac{1}{3} \langle \rho^0 | \pi^0\gamma \rangle.$$

For the magnetic moments, Coleman and Glashow [41] have found for the octet

$$\mu(\Sigma^+) = \mu(p),$$
$$\mu(\Lambda) = \tfrac{1}{2}\mu(n),$$
$$\mu(\Xi^0) = \mu(n),$$
$$\mu(\Xi^-) = \mu(\Sigma^-) = -[\mu(p) + \mu(n)],$$
$$\mu(\Sigma^0) = -\tfrac{1}{2}\mu(n),$$
$$\mu(\Sigma^0, \Lambda) = -\tfrac{1}{2}\sqrt{3}\,\mu(n).$$

Okubo [10] has found that after the mass-splitting interaction is inserted, the symmetry predicts only

$$\mu(\Sigma^0, \Lambda) = \frac{\sqrt{3}}{6} [\mu(\Sigma^0) + 3\mu(\Lambda) - 2\mu(\Xi^0) - 2\mu(n)].$$

Other results have been found for baryon-antihyperon* production by Levinson, Lipkin, and Meshkov [52], by Goldberg, Nussinow, and Yekutieli [58], and by others. There is some freedom in these reactions, and the experiments may provide information on the dynamical mechanism involved. For proton-hyperon scattering, we find

$$\sigma(\Sigma^+, p \to Y^{*+}, p) = \sigma(\Sigma^+, p \to N^{*+}, \Sigma^+),$$
$$\sigma(\Sigma^+, p \to \Sigma^+, n, \pi^+) = \sigma(\Sigma^+, p \to p, \Xi^0, K^+),$$
$$\sigma(\Sigma^+, p \to 2p, \bar{K}^0) = \sigma(\Sigma^+, p \to 2\Sigma^+, K^0).$$

* The octet model has also been recently used to predict baryon–baryon resonances appearing in the same $\overline{10}$ as the deuteron; there is some experimental validation of these results. See R. J. Oakes, paper presented at the Topical Conference on Recently Discovered Resonant Particles, Ohio University, Athens, Ohio, April 1963.

It has also been shown by Levinson *et al.*, that $\Sigma^+ + p \rightarrow N^{*++} + Y^{0*}$ should show an angular distribution symmetric about 90 deg in the center-of-mass system, because its amplitude should be proportional to the system

$$p + p \rightarrow N^{*++} + N^{*0},$$

where the even and odd parity states have different spins and there is no interference between even and odd parity states.* Similarly, in the high-energy region, where the symmetry should be felt, we find

$$\langle \pi^- p \mid Y_1^{*-} K^{*+} \rangle = \langle K^- p \mid \Xi^{*-} K^{*+} \rangle = -\langle K^- p \mid Y_1^{*-} \rho^+ \rangle$$
$$= -\frac{1}{\sqrt{3}} \langle \pi^- p \mid N^{*-} \rho^+ \rangle,$$

and

$$\langle \pi^- p \mid Y_1^{*-} K^+ \rangle = \langle K^- p \mid \Xi^{*-} K^+ \rangle = -\langle K^- p \mid Y_1^{*-} \pi^+ \rangle$$
$$= -\frac{1}{\sqrt{3}} \langle \pi^- p \mid N^{*-} \pi^+ \rangle,$$

where we see that

$$\langle \pi^- p \mid Y_1^{*-} K^+ \rangle \quad \text{and} \quad \langle K^- p \mid Y_1^{*-} \pi^+ \rangle$$

should show similar angular distribution, which is indeed validated by the experimentally observed common peaking of the Y_1^* in the backward direction. The same angular distribution is predicted for Ξ^{*-} and N^{*-} as seen above. Another prediction is,

$$\sigma(K^-, p \rightarrow Y_1^{*0}, \rho^0) + \sigma(K^-, p \rightarrow Y_1^{*0} \omega) + \sigma(\pi^-, p \rightarrow Y_1^{*0}, K^{*0})$$
$$= \tfrac{1}{2}\sigma(K^-, p \rightarrow N^{*0}, \bar{K}^{*0}) + \sigma(K^-, p \rightarrow \Xi^{*0}, K^{*0})$$
$$+ \sigma(\pi^-, p \rightarrow N^{*0}, \rho^0) + \sigma(\pi^-, p \rightarrow N^{*0}, \omega).$$

It is worth noting that all these equalities are true only if the dynamics do not make a particular *s*-or-*t* channel resonate and enhance the effects of the symmetry-breaking interaction.

5. *Origins*

Several attempts have been made to produce a symmetry-breaking mechanism, mainly by means of "superconductivity" solutions. This method is still a bit vague, and does not yield quantitative results. We would like to get something like the Gell-Mann and Levy derivation of the Goldberger-Treiman value for the pion lifetime—some link between the symmetry-breaking mechanism and experiment.† An attempt in this direction has now been made by Sakurai [59],

* This approach has now been generalized by H. Ruegg and S. B. Treiman, (Princeton Univ. preprint, 1963).

† J. J. Sakurai and, independently, D. Horn have now suggested a connection between the strangeness-changing current's conservation and a 0^+ meson with $T = \frac{1}{2}$, $Y = 1$, perhaps the *K*-like resonance at 725 MeV.

who has derived a quantitative check of his ϕ-ω mixing theory. From the observed masses of ϕ and ω, as compared with the mass given by the mass formula for the ϕ' isosinglet component of a unitary octet of vector mesons, one gets the ratio of the mixtures

$$\phi = 61\,\%\,(\mathbf{8}, T = 0,\ Y = 0) + 39\,\%\,(\mathbf{1}, T = 0,\ Y = 0).$$

This now provides a prediction $\Gamma(\phi \to K' + \bar{K}) \approx 1.9$ MeV, which is compatible with experiment. Sakurai has also checked the self-consistency of his dynamical mass-breaking mechanism.

For the origin of the internal degrees of freedom, two approaches have been used lately. The de Broglie school [60] has tried to find it by assuming that the particle has an internal rotational movement of the orthodox spatial type; working with two sets of four axes, they obtain something similar to "global" symmetry. Yukawa has announced that one may even get unitary symmetry out of such a mechanical model.* On the other hand, Jauch and Piron [61] have pointed at weaknesses in this "hidden variables" approach. Alternatively, parastatistics have been proposed by Feshbach [62] and others to explain strangeness; this proposal can and should be tested experimentally. Another school has been working with quaternions, octonions, and so forth [63]. This again leads to either global symmetry or the $G2$ model.

At present it seems that the higher-symmetry approach is bearing dividends, and that the octet model seems to be nearest to reality. On the other hand, it is well to remember that many experimental tests are still due. It has become customary to compare our tabulation of particles and resonances with the inauguration of the Periodic Chart of the elements by D. I. Mendeleev in 1870; there is danger in this comparison: In 1862, Newlands discovered the periodicity of the elements but drew too crude a conclusion and proposed a theory that was afterwards replaced by Mendeleev's. Newland's theory was called "the theory of octaves."

Nevertheless, we feel that there is full justification in attempting to describe the strong interactions by means of symmetries. One really hopes that someday our dynamicist colleagues will provide us with a theory from which we shall trivially derive all our internal quantum numbers; but this still looks like wishful thinking. Imagine what would have happened if science had declined to use the concept of electric charge—another internal quantum number—until the last decade, when geometrodynamics at last found a way to explain its space-time origins; moreover, are we really satisfied with this derivation now that we know that electromagnetism and gravitation are only half the picture? My suggestion is that no harm will ensue if we do solve some problems "in the meantime" through our symmetries.

The author would like to thank Y. Dothan, Z. Fraenkel, H. Goldberg, E. Gotsman, D. Horn, H. Harari, S. Nussinow, Y. Rosen, and E. Weiss for their criticisms, suggestions, and assistance.

* The author thinks Yukawa's suggestion, based as it is on a (N, Λ, Z^+) quadruplet, would lead rather to a new $B2$ model.

References

[1] M. Gell-Mann, *Phys. Rev.*, **92**, 833 (1953); K. Nishijima, *Progr. Theoret. Phys.* (Kyoto), **12**, 107 (1954).

[2] B. d'Espagnat and J. Prentki, *Progress in Elementary Particles and Cosmic Ray Physics*, Vol. IV, J. C. Wilson and S. A. Wouthuysen, eds., Amsterdam: North Holland, 1958.

[3] M. Gell-Mann, *Phys. Rev.*, **106**, 1296 (1957).

[4] J. Schwinger, *Ann. Phys.*, **2**, 407 (1957).

[5] A. Pais, *Phys. Rev.*, **110**, 574 (1958).

[6] A. Salam and J. C. Polkinghorne, *Nuovo cimento*, **2**, 685 (1955).

[7] J. Tiomno, *Nuovo cimento*, **6**, 69 (1957).

[8] R. E. Behrends and A. Sirlin, *Phys. Rev.*, **121**, 324 (1961).

[9] V. M. Shekhter, *JETP*, **14**, 582 (1962).

[10] M. Gell-Mann, *Phys. Rev.* **125**, 1067 (1962); S. Okubo, *Progr. Theoret. Phys.* (Kyoto), **27**, 949 (1962), and *Phys. Lett.*, **4**, 14 (1963); H. Goldberg and Y. Ilamed, *J. Math. Phys.*, **4**, 501 (1963).

[11] E. Fermi and C. N. Yang, *Phys. Rev.*, **76**, 1739 (1949).

[12] M. Goldhaber, *Phys. Rev.*, **101**, 433 (1956).

[13] S. Sakata, *Progr. Theoret. Phys.* (Kyoto), **16**, 686 (1956); see also L. B. Okun, *ICHEP, CERN, 1958*, p. 223.

[14] M. Ikeda, S. Ogawa and Y. Ohnuki, *Progr. Theoret. Phys.* (Kyoto), **22**, 715 (1959); see also Y. Yamaguchi, *ibid.*, Supplement **11**, 1 (1959).

[15] D. R. Speiser and J. Tarski, "Possible Schemes for Global Symmetry," The Institute for Advanced Study, Princeton, New Jersey (1961).

[16] M. Gell-Mann and S. L. Glashow, *Ann. Phys.*, **15**, 437 (1961).

[17] S. Coleman, "Chiral Symmetries," Ph.D. Thesis, California Inst. of Technol. (1961) (unpublished).

[18] Y. Ne'eman, "Gauges, Groups and an Invariant Theory of the Strong Interactions," Israel A.E.C. Series Report IA-698, Rehovoth (1961).

[19] R. E. Behrends, J. Dreitlein, C. Fronsdal, and B. W. Lee, *Rev. Mod. Phys.*, **34**, 1 (1962).

[20] C. N. Yang and H. Mills, *Phys. Rev.*, **96**, 192 (1954); see also R. Shaw, Cambridge University dissertation (unpublished) 1954.

[21] Y. Nambu, *Phys. Rev.*, **106**, 1366 (1957).

[22] A. Salam and J. C. Ward, *Nuovo cimento*, **19**, 165 (1961).

[23] J. J. Sakurai, *Ann. Phys.*, **11**, 1 (1960).

[24] Y. Ne'eman. *Nuclear Phys.*, **26**, 222 (1961); see also H. Goldberg and Y. Ne'eman, *Nuovo cimento*, **27**, 1 (1963), and Y. Dothan, "Unitary Parity," Israel A.E.C. Series Report IA-838, Rehovoth (1963).

[25] S. Okubo, *Nuovo cimento*, **13**, 292 (1959); B. L. Ioffe, *Nuovo cimento*, **10**, 352 (1958).

[26] A. Salam and J. C. Ward, *Nuovo cimento*, **11**, 568 (1959).

[27] F. Gürsey, *Ann. Phys.*, **12**, 91 (1961).

[28] M. Gell-Mann and M. Levy, *Nuovo cimento*, **14**, 705 (1960).

[29] J. Schwinger, "Gauge Theories of Vector Particles," in *Theoretical Physics*, Trieste Seminar Lectures, International Atomic Energy Agency, Vienna, 1963.

[30] G. Feinberg, *Phys. Rev.*, **125**, 725 (1962).

[31] L. Michel, "Group Extensions," *Proceedings of the International Summer School in Theoretical Physics, Istanbul, 1962* (to be published by Gordon & Breach, London).

[32] E. B. Dynkin, The Structure of Semi-simple Algebras (*Uspekhi Mat. Nauk*, **2**, No. 4, 1947), American Math. Soc. Translations, No. 17 (1950).

[33] B. d'Espagnat and J. Prentki, *Nuclear Phys.*, **1**, 33 (1956).

[34] A. Salam and J. C. Ward, *Nuovo cimento*, **20**, 1228 (1961).

[35] M. H. Alston, A. Barbaro-Galtieri, A. H. Rosenfeld, and S. G. Wojcicki, *Bull. Phys.*

Soc., Series II, **8**, 348; G. R. Kalbfleisch, G. Alexander, O. I. Dahl, D. H. Miller, A. Rittenberg, and G. A. Smith, *Phys. Rev. Lett.*, **4**, 225 (1963).
[36] L. W. Alvarez, M. H. Alston, M. Ferro-Luzzi, D. O. Huwe, G. R. Kalbfleisch, D. H. Miller, J. J. Murray, A. H. Rosenfeld, J. B. Shafer, F. T. Solmitz, and S. G. Wojcicki, *Phys. Rev. Lett.*, **10**, 184 (1963).
[37] H. Ruegg, *Nuovo cimento*, **24**, 461 (1962).
[38] W. Thirring, *Nuclear Phys.*, **10**, 97 (1959); J. E. Wess, *Nuovo cimento*, **15**, 52 (1960); Y. Yamaguchi, *Progr. Theoret. Phys.* (Kyoto), **23**, 882 (1960); C. A. Levinson, H. J. Lipkin, and S. Meshkov, *Nuovo cimento*, **23**, 236 (1962).
[39] A. Salam and J. C. Ward, *Nuovo cimento*, **20**, 419 (1961).
[40] M. Gell-Mann, "The Eightfold Way—A Theory of Strong Interaction Symmetry," California Inst. of Technol. Synchrotron Lab. Report No. 20 (1961).
[41] S. Coleman and S. L. Glashow, *Phys. Rev. Lett.*, **6**, 1423 (1961); N. Cabibbo and R. Gatto, *Nuovo cimento*, **21**, 872 (1961).
[42] C. A. Levinson, H. J. Lipkin, and S. Meshkov, "Experimental Predictions from Unitary Symmetry," Weizmann Inst. preprint (1963).
[43] J. J. Sakurai, *Phys. Rev. Lett.*, **9**, 472 (1962).
[44] G. Alexander, G. R. Kalbfleisch, D. H. Miller, and G. A. Smith, *Phys. Rev. Lett.*, **8**, 447 (1962); S. G. Wojcicki, G. R. Kalbfleisch, and M. H. Alston, *Bull. Am. Phys. Soc.*, Series II, **8**, 341 (1963); see also A. Abashian, N. E. Booth, and K. M. Crowe, *ibid.*, **8**, 349 (1963).
[45] S. Goldhaber, W. Chinowsky, G. Goldhaber, W. Lee, T. O'Halloran, and T. F. Stubbs, G. M. Pjerrou, D. H. Stork, and H. K. Ticho, "K^+p Interactions from 140 to 642 MeV/c," *ICHEP, CERN, 1962*, p. 356.
[46] R. Burnstein, T. B. Day, F. Herz, B. Kehoe, M. Sakitt, N. Seeman, B. Sechi-Zorn, and G. A. Snow, R. Adair, H. Courant, H. Filthuth, P. Franzini, R. G. Glasser, A. Minguzzi, A. Segar, and W. Willis, *Bull. Am. Phys. Soc.*, Series II, **8**, 349 (1963).
[47] Y. Eisenberg, *Phys. Rev.*, **96**, 541 (1954).
[48] G. F. Chew and S. C. Frautschi, *Phys. Rev. Lett.*, **7**, 394 (1961); *ibid.*, **8**, 41 (1962); S. C. Frautschi, M. Gell-Mann, and F. Zachariasen, *Phys. Rev.*, **126**, 2204 (1962).
[49] W. Selove, V. Hagopian, H. Brody, A. Baker, and E. Leboy, *Phys. Rev. Lett.*, **9**, 272 (1962).
[50] A. R. Erwin, G. A. Hoyer, R. H. March, W. D. Walker, and T. P. Wangler, *Phys. Rev. Lett.*, **9**, 34 (1962); G. Alexander, O. I. Dahl, L. Jacobs, G. R. Kalbfleisch, D. H. Miller, A. Rittenberg, J. Schwartz, and G. A. Smith, *Phys. Rev. Lett.*, **9**, 460 (1962).
[51] S. L. Glashow and A. H. Rosenfeld, *Phys. Rev. Lett.*, **10**, 192 (1963).
[52] A. R. Edmonds, *Proc. Roy. Soc.* (*London*), **268**, 567 (1962); C. A. Levinson, H. J. Lipkin, and S. Meshkov, *Phys. Rev. Lett.*, **1**, 44 (1962); P. T. Matthews and A. Salam, *Proc. Phys. Soc.* (*London*), **80**, 28 (1962), and *Phys. Rev. Lett.*, **10**, 361 (1963); M. A. Rashid, *Nuovo cimento*, **26**, 118 (1962); Y. Dothan and H. Harari, "Reduction of a Triple Product of Octets," Israel A.E.C. Series Report IA-777, Rehovoth (1962); H. Goldberg, *Nuovo cimento*, **27**, 532 (1963).
[53] S. Sawada and M. Yonezawa, *Progr. Theoret. Phys.* (Kyoto), **23**, 662 (1960).
[54] R. E. Cutkosky, J. Kalckar, and P. Tarjanne, *Phys. Lett.*, **1**, 93 (1962); P. Tarjanne, *Ann. Acad. Scientiarum Fennicae*, Series A, VI, Phys., 105 (1962).
[55] R. H. Capps, *Nuovo cimento*, **27**, 1208 (1963); and *Phys. Rev. Lett.*, **10**, 312 (1963); J. J. Sakurai, *Phys. Rev. Lett.*, **10**, 446 (1963); Y. Hara and Y. Miyamoto, *Progr. Theoret. Phys.* (Kyoto), **29**, 466 (1963).
[56] A. W. Martin and K. C. Wali, "Coupled Channel Approach to $j = (3/2)^+$ Resonances in the Unitary Symmetry Model," Argonne Nat. Lab. Report TID-18295 (1962).
[57] H. D. D. Watson, "Hyperon Pair Production" (Imperial College, London).
[58] H. Goldberg, S. Nussinov, and G. Yekutieli, *Nuovo cimento*, **28**, 446 (1963).
[59] J. J. Sakurai, *Phys. Rev.*, **132**, 434 (1963).
[60] L. de Broglie, D. Bohm, P. Hillion, F. Halbwachs, T. Takabayasi, J. P. Vigier, *Phys. Rev.*, **129**, 438 (1963); see also *Phys. Rev.*, **129**, 451 (1963).

[61] J. M. Jauch and C. Piron, "Can Hidden Variables be Excluded in Quantum Mechanics?," CERN preprint.
[62] H. Feshbach, "Do Strange Particles Obey Parastatistics?," CERN Report Th. 321 (1962).
[63] A. Pais, *Phys. Rev. Lett.*, **7**, 291 (1961); see also D. Finkelstein, J. M. Jauch, S. Schimnovich, and D. Speiser, *J. Math. Phys.*, **3**, 207 (1962).

XII

NEW CONTRIBUTIONS

31

PARTICLE THEORY FROM S-MATRIX TO QUARKS

Murray Gell-Mann
Santa Fe Institute

It is a great pleasure to be here in Catalunya and to participate in this meeting on the important subject of the history of physical ideas. I am particularly pleased that this meeting is in a sense dedicated to my old friend Louis Michel. (It has been predicted by Professor Telegdi that some day in the distant future there will be a «station de métro» and even a «boucherie chevaline» named after our colleague, in the tradition of Louis Arago and others. For the moment it is wonderful to have this memorial to Professor Michel while he is still alive.)

This is called the first «trobada» on the history of scientific ideas. But there was another one on the history of elementary particle physics last summer, in Paris, presumably the zeroth. I was fortunate enough to be invited to give a talk there on some very early work of mine, on strangeness and related matters. For that talk as for this one, I do not pretend to have done any significant amount of historical research. I am far too lazy to have done more than look up a few papers here and there and glance at them. These talks should be considered as raw material for historical research, like reminiscences of some aged farmer discussing the rural landscape of his youth.

I had hoped to make the written report more systematic than the somewhat confused oral presentation, but that hope has not been fulfilled. However, the confusion has the merit that it reflects well the situation during the entire period that I am going to describe.

I find these reminiscences bittersweet, and the preparation of the talk has been somewhat painful for me. Sometimes, of couse, thinking about the evolution of my research in physics is amusing; and there is occassionally the delight of remembering past triumphs. But there is also the recollection of lost opportunities to see clearly what was going on.

There used to be a jingle posted on the wall of certain doughnut shops in the U.S. that ran:

As you ramble on through life, Brother,
Whatever be your goal,
Kepp your eye upon the doughnut,
And not upon the hole.

I am afraid that I am one of those that do not follow the injunction in the jingle; I tend to keep my eye on the hole in the toroidal doughnut. As Harald Fritzsch remarked to me, that means that you brood about having invested in stocks at the wrong time, you regret having taken the wrong job, and so forth. And so in physics I regret not having taken certain ideas seriously enough at certain times and being unnecessarily confused on many occasions when it would have been perfectly possible to think straight.

I have decided that in this talk I shall emphasize some of the confusions and hesitations in the progress of my own ideas and sometimes those of others.

At this point I should refer to a disagreement between two different schools of theoretical physics about how to report theoretical ideas. I belonged to one of these schools, not as a result of careful choice, but simply through personal predilection.

And Yang and Lee, in the middle fifties, exemplified for me the other school. I thought that it was not fair for a theorist to propose several contradictory theories at one time, that the theorist should save his money and then bet on one idea that he really thought was right. And I believed also that proposing a wrong theory counted as having written a wrong paper. The other school argued, at least at one time, many years ago, that this was not so, that a paper was wrong only if the author made a mathematical mistake, that the point of a theoretical paper was to demonstrate the consequences of a particular set of assumptions. The assumptions did not need to be correct in nature, and one could in different papers try different contradictory assumptions without disgracing one's self. My feeling was that a theorist should be judged by the correctness of his guesses about nature, that his reputation should be gauged by the number right minus the number wrong, or even the number right minus twice the number wrong.

I remember talking about somewhat the same methods of scoring with Henry Kissinger, when he started to work for Nixon. We discussed Dean Rusk, who is said to have distinguished himself during the war in Korea in the following way: Allen Whiting has discovered, through intelligence methods, that a Chinese army group was missing from Manchuria in September of 1950. He was told to go to our Assistant Secretary of State for Far Eastern Affairs and tell him that the Chinese army group was missing and was most likely in North Korea, in the path of General MacArthur. Whiting did as he was advised, and the Assistant Secretary for Far Eastern Affairs is said to have replied, speaking of the Chinese leaders: «They would not dare.» Was it on the strength of that brilliant prediction he was made Secretary of State a few years later? Kissinger agreed with me in 1968 that people in such positions should be scored by number right minus twice number wrong. I do not know whether he would still agree.

The reason I have brought up my reluctance to discuss several competing theories at once is that it explains why I rarely shared my agonies of choice with the readers of my articles. If I had felt differently I could in many cases have saved myself a good deal of hesitation and anguish. I could have explained in print the confusions that were bothering me and the choices that I believed we faced.

During this period, I enjoyed agreeable and productive collaborations with a number of distinguished colleagues. I have no wish to slight their contributions, which were of the greatest importance, but in these reminiscences I shall emphasize mainly the evolution of my own thinking, with all its fits and starts. In the same spirit, I shall not always distinguish sharply between my ideas that were published fairly promptly and those that were delayed a long time in submission for publication or never printed at all. Much of my communication with other physicists was in the form of lectures, seminars, and conversations.

1. The renormalization group and the possible failure of old-fashioned field theory

Let me begin my reminiscences with the very hot summer of 1953, thirty years ago. I talked about some events of that summer at the Paris meeting, but now I shall discuss others. I spent part of the summer sweltering in Urbana, where the University of Illinois is located, working with my good friend Francis Low on what is now called the renormalization group, what we called quantum electrodynamics at small distances. We thought of ourselves as working on a very far-out problem, because we were discussing phenomena that would occur at enormous energies like exp 137 times the electron mass. Nowadays, everybody talks about such energies, but at that time they seemed ridiculously high. Of course we hoped that in meson theories and so forth, the relevant energy would be very much lower, because the coupling constant would be larger, and so the results might have some immediate relevance, but we had justified doubts about the correctness of those theories.

Petermann and Stueckelberg, I understand, worked on something similar at around the same time. And I think they used the phrase «renormalization group», which subsequently became standard.

Also writing on this subject were Bogolyubov and Shirkov. Their work consisted, as far as I can tell, largely in copying our analysis. But they accused us of having used the wrong gauge, and said that therefore our work was wrong and needed to be replaced by theirs. Now, in a gauge invariant theory, I do not know why using a different gauge is a crime, but for many years people referred to Bogolyubov and Shirkov rather than to us. Even my student Ken Wilson learned about the renormalization group from their paper rather than ours! The situation was rectified only in the book by Bjorken and Drell, who by some historical research, discovered the actual situation. Now, I mention this not in order to complain but because it comes up importantly in a moment.

In simple glorified perturbation theory, where denominators are expanded in perturbation series, the vacuum polarization in quantum electrodynamics has this form:

$$d(k^2) = \frac{D(k^2)}{D_{free}(k^2)} \simeq \frac{1}{1 - \frac{\alpha}{3\pi}\ln\frac{k^2}{m^2}}, \tag{1}$$

where $D(k^2)$ is the photon propagator function. This formula is very well known; of course, if it were exact, it would contradict the positivity of the theory, because $d(k^2)$, which has to be positive, as we and others showed, would go to infinity and start again from minus infinity. The theory would therefore be self-inconsistent, and α would have to be zero.

Low and I found the results:

$$d(k^2) = \frac{e_0^{\,2}(k^2)}{e_1^{\,2}}, \tag{2}$$

and

$$\int_{e_1^2+O(e_1^4)}^{e_0^2(\Lambda)} \frac{dx}{\psi(x)} = \ln \frac{\Lambda^2}{m^2}, \tag{3}$$

where e_1^2 is the renormalized coupling constant, e_0^2 is the bare coupling constant as a function of the cut-off Λ, and ψ is the renormalization-group function. (For some mysterious reason, $\psi(x)$ has been re-christened $\sqrt{x}\,\beta(\sqrt{x})$ by some people.)

We noticed that there are two possibilities if the theory makes sense. One is for the integral to diverge at infinity, in which case $e_0^2(\infty)$ is infinity, and

$Z_3 \equiv \frac{e_1^2}{e_0^2(\infty)}$ is zero.

The other possibility is to have the integral diverging before x reaches infinity in which case the upper limit is finite for $\Lambda=\infty$, there is a finite unrenormalized charge in QED, and Z_3 is different from zero.

We did not list the remaining possibility, that the integral converges, in which case the theory is inconsistent. However, shortly afterward, Källén suggested that this kind of inconsistency might occur, that a formula with the properties of eq.(1) might actually hold in QED. Källén referred to this situation as a theory with a «ghost». Pauli, too, discussed such a possibility. And Landau and his collaborators in the Soviet Union claimed to prove it, calling it the «zeroness of the charge», because for consistency one would have to put $\alpha = \frac{e_1^2}{4\pi}$ equal to zero.

This question is still unresolved after thirty years, and is the subject of research and violent controversy right at this very moment. A lot of «pure people» (pure mathematical physics types, of whose work I cannot understand a single word) have been studying the matter, and indicate that they are «close to a proof» that this kind of thing actually happens, at least in $\lambda\varphi^4$ theory. Now how does a mathematically pure person get close to a proof? I do not know. «Computer people», theorists working with a salt crystal lattice for space-time, have also worked on the problem, and they claim too that they are close to showing the zeroness of the charge, as $\Lambda\rightarrow\infty$, at least for $\lambda\varphi^4$ theory. Nick Khuri has analyzed all these claims and says that he does not believe there are any good arguments yet. He finds that so far the pure people have proved a theorem that is inadequate, because it is based on a hypothesis that may not be true and that he believes may well be false. He also finds that the computer people have not yet computed just the right quantities, and that their evidence is therefore not useful either. So, if he is right, the question is still completely open and the arguments so far would appear to be no more conclusive than those given in the 1950's by Landau and collaborators. I am, of course, not saying that the conjecture is false, only that it is still unproved.

The question is still of importance today, because if $\lambda\varphi^4$ theory makes no sense by itself and QED makes no sense by itself, and if the standard model is embedded in a larger, presumably unified theory (say a Yang-Mills theory with formal scale invariance), then that unified theory would have to have very special properties in order to make sense. The restriction would be severe, probably a restriction to the case of asymptotic freedom, which itself may be related to the restrictions imposed by supersymmetry. Howard Georgi and I have done some thinking on this subject recently.

In 1956 I was with the first substantial group of Western physicists that visited Moscow after the war. Some of the people in the group had been there twenty years before, like Weisskopf. For the younger ones, like me, it was, of course, our first visit, and it was something like visiting another planet. We had inadequate information about what was going on there, and there were many surprises. Khrushchev had just delivered his speech about Stalin, millions of people had been released from captivity, and the atmosphere was extraordinary. Our plane from Copenhagen landed in the middle of the night at Vnukovo and we were met on the tarmac by Tamm, Landau, Pomeranchuk, and many other leaders of the physics section of the Academy of Sciences. Some of the oldtimers in our group knew them from twenty years before. The rest of us became acquainted with them instantly. Landau was a very interesting, exciting person but very full of his own ideas; it was difficult, in fact almost impossible, to argue with him. In the course of our discussions over the next few days, he maintained, along with his entire crew (Pomeranchuk, Abrikosov, Yoffe, and so on) that they had proved the zeroness of the charge, arguing from leading logs in perturbation theory.

I explained that $\psi(x)$ needed to be known for large values of x in order to answer the question, and that such information could not be obtained from leading logs in perturbation theory. But I made no impression, because they thought of our work as Bogolyubov's and Bogolyubov was the enemy! They would not talk about $\psi(x)$ and the discussion ended.

2. Dispersion relations and the «S-matrix» program

Now let us return to 1953. I was working not only on the renormalization group, but also on strange particles, as I described at the Paris meeting, and on dispersion relations, with Murph Goldberger. Murph and I were engaged in a program of extracting as many general results as possible from local field theory, by proof if possible but otherwise by finding rules that held to every order of perturbation theory. Thus we discovered (or perhaps re-discovered) the crossing relations among amplitudes on or off the mass-shell.

Around the same time, we found in the literature the dispersion relation for the forward scattering without spin-flip of a massless particle like the photon, and we set about generalizing that dispersion relation as much as possible. First, Murph and I found the spin-flip dispersion relation for photon scattering in the forward direction, and also the low energy limit of the real part of the same amplitude; together with the oddness of the amplitude under crossing, we had all the elements of a useful formula. Then, after I left for Caltech, Goldberger and some other collaborators found the forward dispersion relations for non-zero mass. Relations that had been thought to be good only for the photon now worked also for mesons, provided one extrapolated the mass-shell amplitudes to imaginary momenta over a small part of the range of integration. Then there were the non-forward dispersion relations, which were found by many different groups: by Goldberger and collaborators, by Capps and Takeda, by Polkinghorne and me at Caltech, and by many others in the course of 1955 and 1956.

The discovery of the non-forward relations made possible the observation that dispersion relations, together with other known properties of amplitudes, were almost enough to specify a field theory. I described that idea at the Rochester meeting in 1956. I showed that to each order of perturbation theory, if one uses crossing relations, unitarity (thus picking up lower orders of perturbation theory in the calculation of the absorptive part of each amplitude), and then dispersion relations (to calculate the dispersive part from the absorptive part), one can generate each order of perturbation theory from the lower orders, provided there are suitable boundary conditions in momentum space, especially at infinite momenta. Crossing, analyticity, and unitarity, together with such boundary conditions, would give all the scattering amplitudes. Furthermore, the whole procedure would be carried out on the mass shell, provided one generalized the mass shell to include imaginary momenta.

I then mentioned casually that the program I was outlining, if treated non-perturbatively, was reminiscent of Heisenberg's hope of writing down the S-matrix directly instead of calculating it from field theory. I indicated that I was talking about a mass-shell determination of the scattering amplitudes in a particular field theory, using certain conditions in order to specify the theory. (The part about the conditions did not get into the written report, which was in the third person, but I remember saying it.)

Goldberger and I tried to teach these notions to Geoffrey Chew. It was very difficult, because he resisted furiously. Among other things, he disliked the idea that there were mysterious boundary conditions (and perhaps other conditions) that would distinguish one theory from another and complete the information necessary to give the whole S-matrix.

In 1958 in Geneva, Geoffrey Chew quoted my Rochester remarks of 1956; apparently by then he had started to believe them. In the meantime, in the Soviet Union, Landau had caught on to these ideas. In his group they were now busily studying analyticity properties of the S-matrix, trying to obtain them from perturbation theory. Jon Mathews at Caltech was doing much the same sort of thing, but he did not reach as wide an audience as Landau. Landau, however, went on to enunciate a special dogma, starting from his impression that field theory was no damned good because all the coupling constants were zero. He proposed that while field theory was wrong, the program of using dispersion relations, crossing relations, and unitarity to calculate the S-matrix was right and could be used as a substitute for field theory. Whereas I had suggested the program as another way of dealing with field theory, Landau now proposed that it was correct and field theory incorrect. I could never understand that point of view and I still cannot.

Of course, the general program could allow for field theories (such as Yang-Mills theories, which were known at the time but not known to be renormalizable or asymptotically free) that one had overlooked and that might have better properties, but that is not the way the dogma was expressed. Presumably, in today's language, Landau wanted to say that non-asymptotically-free field theory was no good, but that was the only kind we knew about. In any case, we still do not know, as I mentioned earlier, whether non-asymptotically-free field theory has to have zero charges.

In Kiev in 1959, I had terrible arguments with Landau on the subject of condemning field theory and welcoming dispersion theory. As usual, it was impossible, really, to argue. He did not give an inch. Then Geoffrey Chew adopted the same sort of point of view in La Jolla in 1961.

I had suggested the La Jolla meeting and then turned it over to Keith Brueckner to organize as a way of celebrating the beginning of U.C.S.D. He had just moved to La Jolla, and it seemed to be a good way to put his department on the map. He did a beautiful job of organizing the meeting. Robert Oppenheimer attended, and I remember that Keith and I entertained him at the Hotel del Coronado, where there are (so to speak) heads of stuffed admirals on the walls. There was an aged, uniformed waiter who took our order. As a compliment to Robert because he fancied himself as a connoisseur of wine, I asked him to select the wine for lunch. He looked at the wine list, which did not give much information, and asked the waiter the year of a particular Bordeaux. The waiter replied: «Sir, all our wines are at least five years old.»

There were no Proceedings of that conference. That is a pity, because we heard Goldberger and Blankenbecler on Regge poles and Mandelstam on the Mandelstam representation, I talked about the current algebra in the Eightfold Way, and so forth and so on. There were many very interesting first reports of discoveries, all of which are lost to historians. But the most dramatic talk was that of Geoffrey Chew, who said, in effect: «I have always been a simple worker in physics: I do my calculations, I maintain a low profile, and I do not make grand pronouncements. But this time I have something important to say. Field theory is no damned good; instead we must use the S-matrix theory.» He did not say dispersion relations, crossing, unitarity and so on, but «S-matrix theory». He allowed that electromagnetism and gravity might be different. But for the strong interaction he insisted that we abandon field theory and go over to «S-matrix theory».

As I mentioned just now, I have never succeeded in understanding that point of view. While before 1961 one of the great pleasures of working in theoretical physics was the possibility of discussing theories with Geoffrey Chew, after 1961 it became very difficult.

His collaborator Stanley Mandelstam did not agree with the dogma. He did very important work on the «S-matrix» program, but he did not subscribe to the Chuvian religion. He thought, like me, that if the nonlinear system of unitarity and dispersion relations on the mass shell possessed a solution, then the set of linear equations that take us off the mass shell could well have a solution too, so there would be some kind of field theory for each «S-matrix» theory.

Associated with the dispersion theory or «S-matrix» approach to the hadrons was the «bootstrap» picture of the hadrons, as discussed by Chew and Frautschi; and about that I was quite enthusiastic. Using suitably smooth boundary conditions at infinite momenta in the system of unitarity, dispersion, and crossing relations, one was to calculate the properties of the hadrons in principle without introducing any fundamental objects. None of the observable hadrons would be any more fundamental than any other – we called that principle «nuclear democracy». Perhaps we should have called it «hadronic egalitarianism». Anyway, it was a good idea. In practice, one studied two colliding hadrons and demanded that the set of exchanged hadrons in the «t and u channels» be the same as the set of resonant and bound hadrons formed in the «s channel». I liked the approach, but I complained during the middle sixties that Geoffrey Chew and his group kept beating to death one state, the rho meson, and one kind of system, with two pions, rather than making approximate calculations with an infinite number of intermediate states in each channel and a correspondingly infinite number of systems. The approximation could involve assuming narrow resonances only. Some Caltech postdocs, Dolen, Horn, and Schmid, took this up in 1967 in a seminal paper on «duality», in which they suggested that one could in fact approximate the intermediate states by an infinite set of resonances in the s, t, and u channels and that these resonances could be the same as the external particles. This observation led to the Veneziano dual-resonance model, and later to the Neveu-Schwarz model. Remarkably, a recent variant of the Neveu-Schwarz model in ten dimensions is now being proposed as a possible universal theory of all the forces and all the elementary particles of nature, with ten-dimensional supergravity and ten-dimensional unified Yang-Mills theory as an approximation. Of course, the slope of the trajectory has changed a little bit, by a factor of 10^{38} or so. But apart from that trivial modification, the dual-resonance or string theories are again at the forefront of research.

In 1963, when I developed the quark proposal, with confined quarks, I realized that the bootstrap idea for hadrons and the quark idea with confined quarks can be compatible with each other, and that both proposals result in «nuclear democracy», with no observable hadron being any more fundamental than any other. This point is emphasized in my lecture to the Royal Institution in London in 1966.

What worried me about the bootstrap was that it distinguished hadrons sharply from leptons, while the weak and electromagnetic interactions treated them nearly alike. A feature that I liked about the quarks (the current quarks), when I found them, was that they presented an analogy with the leptons, in electromagnetic and weak interactions.

3. Hadron approximate symmetries and Yang-Mills theories for the electro-weak and strong interactions

In 1957 I tried an approximate symmetry for the hadrons, the «global symmetry», based on SO(4), which has the same algebra as SU(2)×SU(2). I shall not discuss it at length here, except to say that it was wrong, and that the mass formula for the N, Ξ, Λ, and Σ had the factors one and three in the wrong places. «Global Symmetry» gave the approximate mass formula:

$$\frac{1m_\Lambda + 3m_\Sigma}{4} \approx \frac{m_N + m_\Xi}{2}. \tag{4}$$

If we interchange the 1 and the 3, we get a much better formula, as I realized at that time. Nambu, I remember, noticed the same thing. But we did not know at that time what kind of theory would interchange them. A few years later, the *Eightfold Way* scheme gave the same approximate formula with the 1 and the 3 interchanged and was successful.

I shall not dwell on the complications of the discovery of the weak interaction theory in 1957. But it is well known that it required the rejection of a considerable number of experimental results. Here I shall begin to refer to Table 1, where I have listed a large number of confusions of the period, my confusions anyway, some of them shared by other people. These confusions are of several kinds. Some are mathematical questions about field theory (part A). Others are theoretical questions about how to explain crucial experimental results (parts C and D). Only a few involve confusions arising from wrong experimental results or wrong preliminary reports of experiments (part B). In the weak interaction domain there was a series of wrong experimental results, such as that of Rustad and Ruby on a scalar and tensor interaction in β-decay, the reported failure of $\pi^\pm$ to decay into electron and neutrino, and so forth and so on. (An example of a wrong result that held sway for a time in another part of particle physics is the report that the Σ-Λ relative parity was minus instead of plus.)

I was writing a review article on weak interaction with Arthur Rosenfeld during the spring and summer of 1957, when I came to the conclusion that there was a chance that the universal Fermi interaction could still be right, and have the form V−A, provided that we could ignore various experimental results. We included the idea in the review article, where we called it the «last stand» of the universal Fermi interaction.

Meanwhile, George Sudarshan, working with Bob Marshak, was making a similar suggestion, but in a more confident manner. I recall a summit meeting with them on the weak interaction at the RAND Corporation (not usually considered a place for that kind of activity). There was an exchange of views, and we mentioned our section on the «last stand», while they told us of their plans to write an article. Felix Boehm was there, too, describing his recent experimental results, which indicated V−A.

I then went on vacation, after further discussions of the «last stand» with Felix Boehm, who described it to Feynman when he returned from Brazil in my absence. Feynman got tremendously excited, expanded the idea somewhat, and wrote a long paper on it. When I returned we decided to modify it and sign it together, but it bore Feynman's stamp, including a notation that I would normally not have used. However, we made some useful proposals in that paper. We also did research that was not included and that Feynman and I presented in successive talks at the American Physical Society meeting at Stanford around Christmas of 1957.

That was the time when Okun' was visiting from Moscow. We taught him to play Monopoly, and sent him back to the Soviet Union with a Monopoly game. That may account for some of his difficulties in attending subsequent meetings. He was, by the way, a very good Monopoly player.

At that meeting I was chosen to talk about the intermediate boson, which we called X^+ and X^- (and which I still call X^+ and X^-!), and about the different possibilities for explaining the absence of $\mu \to e+\gamma$ decay. On behalf of Feynman and me, I described our calculation of $\mu \to e+\gamma$, using the intermediate boson, with electromagnetic properties derived from what is essentially the modern theory, of Yang-Mills type. We had found a finite result. The same result was obtained by Gerald Feinberg, who published it a little later. We showed that $\mu \to e+\gamma$ would go very quickly if there were a serious intermediate boson. Therefore there were only two possibilities, both of which I mentioned. One was that there was a cut-off below the mass of the intermediate boson or there was no intermediate boson at all. The other was that there were two kinds of neutrinos: «red» and «blue», for electrons and muons respectively. I wrote this speech up for publication, but we never submitted it, and Feynman gradually lost his enthusiasm for the red and blue neutrinos. When Bludman presented his work on two kinds of neutrinos the following year at Gatlinburg, Feynman did not like them any more. My worry about the red and blue neutrinos was not so much about whether they were a good idea as about how they could be verified. It did not occur to me, as it did later to Pontecorvo and to Lee and Yang, that it would soon be possible to do accelerator experiments with high energy neutrinos that could actually verify the existence of the two kinds of neutrinos.

During 1958, Feynman and I, together with the postdocs Richard Norton and Keith Watson, worked on the notion of a charged intermediate boson with dimensionless coupling constant approximately equal to e, and we saw that the mass of the intermediate boson had to be around 100 GeV. That was a very high energy, and we did not see any immediate possibility for verifying it. Also we were terribly concerned about the absence of the decay $\mu \to e+\gamma$ unless there were two different neutrinos. And of course, in connection with a neutral intermediate boson, we were worried about how to get a neutral current without a strangeness-changing term, the usual difficulty that was fixed years later by means of charm.

I played briefly with the idea of having two neutral intermediate bosons, as in the model that was published later by Lee and Yang, who called them schizons, because they behaved like $I=1$ for strangeness-preserving couplings and like $I=\frac{1}{2}$ for strangeness-changing couplings. Unfortunately, this model seemed to offer no explanation for the absence of a strangeness-changing neutral weak current, coupled to leptons, although it could explain the absence of a weak non-leptonic interaction with $|\Delta S|=2$. Later, Glashow and I were able to show that it did not fit in with Yang-Mills theory; there was no Lie algebra of the right kind to make it work.

In my list of confusions in Table 1, I should like to point out particularly the one (C 1.) referring to the electron with its neutrino and the muon with a different neutrino, making four leptons. (We did not, of course, know about the third family at that time.) In that case the intermediate boson would be O.K., and $\mu \to e+\gamma$ would not create any problems. And in that case the universality of the weak interactions required that the hadron analog of these doublets be something like proton and «neutron times cos θ plus Λ times sin θ» for a small angle θ. In 1959 Lévy and I worked out the size of this angle, and compared it successfully with a small discrepancy in the usual formulation of the universality of the Fermi constants. Our value of around 15 degrees has survived to this day. Later on, for quarks, the doublet became u and «d cos θ + s sin θ». But there was always lurking in the background the other possibility that there was only one neutrino, coupled to e + μ. That, of course, gave difficulties with the charged intermediate boson, because of the decay $\mu \to e+\gamma$, but we knew there were troubles with the neutral intermediate boson anyway, because of the strangeness-changing neutral weak current and $\Delta S=2$ problems. With a single neutrino, the hadron analog of ν and e+μ would be something like p and n+Λ, with the p Λ coupling conjectured to be much reduced by renormalization. I hesitated between these two possibilities for years, although I preferred the case of two neutrinos. In September of 1960 at Rochester, I actually explained the two alternatives carefully, with their consequences for the weak couplings of hadrons.

In our paper on the weak interaction, written in 1957, Feynman and I included the conserved vector current idea. That idea had actually been suggested some years before by Gershtein and Zeldovich in the Soviet Union, but we did not know that. The fact that the strangeness-preserving vector current for the charge-exchange weak interaction in the case of hadrons was the current for the plus and minus components of isotopic spin indicated to me that the entire charge-exchange weak current, including strangeness-changing and leptonic terms, must be the current of some weak charge operator and its Hermitian conjugate, with interesting algebraic properties. Not only that, but the hadronic part, when broken up into vector and axial vector charges and strangeness-changing and strangeness-preserving charges, would have to generate a larger algebra. I became very interested in the commutation relations and in how these commutation relations would close in each case, for the total weak charge and, in the case of hadrons, for the individual pieces.

I did not call the resulting algebra in either case a Lie algebra, because I had forgotten what a Lie algebra was. I had studied Lie algebras when I was an undergraduate, taking advanced mathematics courses at Yale. But you know what

Table 1. Some theoretical and experimental confusions of the late fifties and early sixties

(A) *Some mathematical questions:*

1. Is Yang-Mills theory generalizable beyond products of SU(2) and U(1) factors for the gauge group?
2. Is Yang-Mills theory with zero mass renormalizable?
3. Is Yang-Mills theory with hard masses, equal or unequal, renormalizable?
4. Is there a soft mass mechanism?
5. Are old-fashioned field theories consistent, or must their coupling constants be zero?
6. Is there another kind of field theory? (As it turned out, asymptotically free.)
7. Is the «S-matrix» approach different from field theory?

(B) *Some examples of confusion resulting from experimental error:*

1. Beta-decay weak interaction scalar and tensor?
2. $\pi \not\rightarrow e\nu$?
3. Σ-Λ parity odd?

(C) *Some physical confusions connected with missing charm:*

1. $\nu_e\ e;\quad \nu_\mu\ \mu$ $\qquad X^\pm$ OK $\qquad \dfrac{p,\ n+\Lambda\varepsilon}{\sqrt{1+\varepsilon^2}}$

versus

$\nu,\ e+\mu$ $\qquad X^\pm$ difficult $\qquad p,\ n+\Lambda$

with lack of fourth hadron flavor to correspond to four leptons in first alternative above.

2. Strangeness-changing neutral weak current in SU(2)×U(1) scheme in absence of fourth hadron flavor, and resulting problem of $\Delta S = 2$ transitions if there is only one such current.
3. Fourth hadron flavor versus TrQ=0 for quarks alone, and TrQ=0 for quarks alone versus TrQ=0 for quarks and leptons together.

(D) *Some physical confusions connected with missing quarks and color:*

1. Appeal of Yang-Mills theories for both strong and electro-weak interactions versus clashing of the two in chiral flavor space.
2. Flavor algebra for hadrons, besides being approximate symmetry, also generating gauge group for Yang-Mills theory of strong interaction?
3. Algebra generated by pieces of weak charge same as chiral flavor algebra for hadrons? (By the way, is the electro-weak gauge group just SU(2)×U(1) or is it some larger subgroup of the chiral flavor group?)
4. Appeal of hadron scheme with fundamental entities like p, n, Λ coupled to neutral vector boson(s) and with electro-weak couplings showing a parallel between these entities and leptons

versus

«nuclear democracy» (no observable hadrons distinguished as more fundamental than others and appeal of not having to put Σ and Ξ into separate 15 representation if they are in an octet with the nucleon.

(E) *Some points not clearly understood about quark statistics:*

1. Confinement of «mathematical quarks» connected with suppression of para-baryons and para mesons.
2. Parastatistics with suppression of observable para-hadrons equivalent to color with suppression of colored hadrons.

mathematics education is like, rarely emphasizing the intuitive and computational, concentrating on general results with theorems and proofs. While I followed the courses and did well in them, I never really understood what was going on. Mathematicians usually give only trivial examples; in the case of a Lie group, they might discuss U(1). A generous mathematician might include SU(2), but not SU(3) or E_6. Reduction of representations as one goes from a group to a subgroup is discussed, but again non-trivial examples are unlikely to be presented.

As a graduate student in physics, I struggled manfully to read the work of Professor Wigner published in 1936-1937: the Harvard tercentenary lecture on the nucleus and then his 1937 *Physical Review* paper on SU(4). I read his papers over and over again, I consulted books on group theory, but I still did not properly understand what it was all about.

In 1951 I attended the beautiful lectures by Giulio Racáh or Rácah. (He used the former pronunciation in Italian and the latter in Hebrew; if you were drinking the Italian wine with his name on the label, you would presumably put the stress on the second syllable.) I listened to his lectures, and received the notes, taken by my friends Park and Merzbacher, but I did not really understand the material. The reason was not that the lectures were not elegant, or that they were not explicit. The problem was his accent. Of course I understood the words; I have no trouble following English spoken with a foreign accent. But his accent was so remarkable that I could not hear the substance. Every English word was pronunced with a perfect Florentine accent. For example he would say: «Tay vah-loo-ay eess toh eeg». (The value is too high). So I never really learned about Lie algebras, and I had to rediscover them.

In any case, from 1957 through 1960 I worried about the commutation relations of the weak charge and their possible relation to a theory of the weak and electromagnetic interactions of the Yang-Mills type, and at the same time I worried about the presumably larger algebra generated by the separate pieces of the weak charge. I knew that universality of strength of the weak interaction had to depend on the commutation relations of the weak charges, which are nonlinear, and that universality of at least differences of electric charge would be similarly controlled since the weak charge operator always changes electric charge by one unit. But my thinking on the whole subject was hampered by my failure to realize that the possible algebraic systems were known and tabulated.

I spent the academic year 1959-60 in Paris, at the Collège de France and at other Parisian institutions. There I continued, with Lévy, work that I had started with Feynman at Caltech on the derivation of the Goldberger-Treiman relation. This led to the idea that the divergence of the axial vector current was proportional to a pion field. Then, with Lévy, Bernstein, Fubini, and Thirring, I progressed to a more sophisticated idea. After all, who knows what a pion field is? The main thing is that the divergence should be a «soft» operator, so that its matrix elements would be dominated at low frequencies by the contribution from the pion intermediate state. This was «PCAC», the principle of the partially conserved axial vector current.

According to PCAC, as the divergence of the current goes to zero, m_π tends to zero rather than m_N. In the limit, there is a realization of the «Nambu-Goldstone» mechanism, which those authors were developing independently around the same time.[1]

During 1959-60, Shelly Glashow came to visit me in Paris, and he explained his theory of SU(2)×U(1) Yang-Mills vector bosons, with mixing of the weak and electromagnetic interactions. We did not call it SU(2)×U(1), but that was what it was. Applied to leptons, it was like the present electro-weak theory except for the Higgs mechanism added by Weinberg in 1967 and discussed also by Salam. I gave the first easily intelligible presentation of this theory in September, 1960, attributing it of course to Glashow, at the Rochester meeting. The abstract was sent in longhand from Kisoro, Uganda, where I was looking at wild gorillas, with apologies for the lack of secretarial help.

All of this was very nice, provided the strangeness-changing contribution to the weak charge was omitted. But if it was included, then Shelly's theory seemed to be in trouble, because there would be a strangeness-changing neutral weak current and serious problems with $\Delta S=2$ transitions. Charm was, of course, the way out, but we did not use it, and consequently the puzzle remained unresolved.

I interested myself in the algebra that would be generated by the various pieces of the weak charge, those connected with the vector and axial vector, strangeness-preserving and strangeness-changing currents. If I had understood Lie algebras in 1959 and early 1960, I would have realized that (without charm) the system of commutation relations could close on the chiral algebra of SU(3)×SU(3). But I did not. The notion that all the relevant algebraic systems were tabulated and that I could look them up in a book was still strange to me.

In Table 1 (part A), we see some more confusions of that time. The Yang-Mills theory with zero vector boson mass, was it renormalizable or not? And what about the renormalizability with masses? For a while, Glashow claimed wrongly that a Yang-Mills theory was completely renormalizable, even with masses, even with different masses for different vector bosons.

Salam and Kumar were proving, in London during 1959-60, that this was false, but that the unbroken theory was renormalizable. Later on, in my *Eightfold Way* report, I advertised for a «soft mass» for vector bosons in Yang-Mills theory. Also, in the course of our PCAC work in Paris, Lévy and I had played with the sigma model, both linear and nonlinear (which had also been studied by Schwinger and by Gürsey, respectively, in different connections). By January 1961, I had all the materials necessary for inventing Higgs bosons as a mechanism for providing the «soft mass» needed for the renormalizability of a Yang-Mills theory with broken

1. Thus the algebra that we would now call the SU(2)×SU(2) chiral flavor algebra is realized as an approximate symmetry by having the charges of the vector currents (isotopic spin components) give approximate hadron degeneracies and having those of the axial vector currents give nearly massless pions. I was to extend this idea later to flavor SU(3)×SU(3). Even on this major point I wavered briefly in early 1964, when I tried using the charges of axial vector currents as very approximate degeneracy symmetries.

symmetry. However, I failed to see the relevance of the sigma model and I did not solve the problem.

I had continued to play with Yang-Mills theory for the weak interaction all during 1958 and 1959. Early in 1959 I decided that Yang-Mills theory must also be relevant to the strong interaction.[2] But it was very difficult to understand how the two could be related. Late in 1959, in Paris, I decided to find out what where the possible generalizations of the Yang-Mills theory —those two authors had given only the case of isotopic spin, what we would nowadays call SU(2). Quantum electrodynamics was, of course, an example of a gauge theory using U(1). We could readily conceive, therefore, of a gauge theory involving what we would call today a product of SU(2) factors and U(1) factors. When Shelly visited and presented his scheme, that fitted in easily because it corresponded to one factor of each type. The question was whether any other generalization existed. I worked and worked in my office in the Collège de France and finally I wrote down, as the necessary and sufficient conditions, the canonical relations:

$$[\mathbf{F}_i, \mathbf{F}_j] = ic_{ijk}\mathbf{F}_k \,, \tag{5}$$

where c_{ijk} is real and totally antisymmetric and the $\mathbf{F}_i$ are Hermitian charge operators. I had no idea what were the possible realizations of this formula.

Every day I would have lunch with my French friends and drink wine, and afterwards I would come back and struggle with drowsiness in my office. I worked through the cases of three operators, four operators, five operators, six operators, and seven operators, trying to find algebras that did not correspond to what we would now call products of SU(2) factors and U(1) factors. I got all the way up to seven dimensions and found none, of course. At that point I said: «That's enough!» I did not have the strength after drinking all that wine to try eight dimensions. Unfortunately I did not pay sufficient attention to the identity of one of my regular companions at lunch. It was Professor Serre, one of the world's greatest experts on Lie algebras. I knew, of course, that Serre was a famous pure mathematician, but I did not know what his specialty was. It never occurred to me to ask him about my equations, and I doubt whether he would have given me the answer, if I had asked him. Probably this canonical form for the commutation relations was far too explicit for Serre. He would have preferred something like:

$$T : s \rightarrow u \,. \tag{6}$$

At least that is my impression of the kind of formula that mathematicians like to write.

2. I had always thought that understanding classification and symmetry would lay the groundwork for discovering the dynamics of the weak (or electro-weak) interaction and of the strong interaction, but in Yang-Mills theory it was clear that the determination of the dynamics involved merely choosing the symmetries to be gauged. Thus the study of symmetry was more important than ever.

It was not until December 1960 at Caltech that a young mathematician called Block told me: «What you have there is a canonical form for the commutation relations of a Lie algebra generating a product of U(1) factors and simple compact Lie groups in a unitary representation.» I did not know that, of course; I had been reinventing those Lie algebras.

By this time Glashow was spending a year with me at Caltech, and we had been discussing how to generalize the Yang-Mills trick and how to apply it to weak and strong interactions. While he was away in Massachusetts, I received enlightenment about Lie algebras and immediately thought up the *Eightfold Way.* But on his return we resumed our collaboration.

Two pieces of work now emerged. Glashow and I wrote up our research on generalizing Yang-Mills theory from the group SU(2) to any product of simple compact Lie groups and U(1) factors; and we discussed our attempts to construct a Yang-Mills theory for the strong interaction and one for the weak and electromagnetic ones (such as Glashow's SU(2)×U(1) theory, which has since proved to be essentially correct). I wrote my Caltech Report on *The Eightfold Way,* in which I suggested flavor SU(3) as an approximate symmetry of the hadrons, with symmetry violation coming from the eighth component of an octet and leading to the baryon and meson mass formulae. The baryon octet mass formula was just the one, eq. (4) with the 1 and 3 interchanged, that I had wanted. In the Report I also proposed a Yang-Mills theory for the strong interaction based on flavor SU(3).

In both papers there was a fundamental difficulty, namely the absence of color space that permits us today to have the Yang-Mills gauge groups for strong and for electro-weak interactions operating on different sets of coordinates. Shelly and I had to cram our strong and electro-weak Yang-Mills theories into chiral flavor space, where they clashed in an intolerable way. While our work on generalizing the Yang-Mills idea was elegant, our attempts to construct both strong and electro-weak theories without color were necessarily clumsy.

I therefore had to face the question of whether to maintain or abandon my particular theory of the strong interaction based on flavor SU(3). I decided to abandon it as I converted my Caltech Report into an article for the *Physical Review;* the successful features of the theory could be replaced by the assumption of «vector dominance». Thus, the SU(3)×SU(3) chiral flavor algebra was to be used as an approximate symmetry, with the charges of the vector currents giving approximate degeneracy of hadron states and those of the axial vector currents giving approximately massless pseudoscalar mesons (the PCAC idea). At the same time, these currents would obey local current algebra relations, which are a kind of pre-condition for Yang-Mills theory, and thus they could include electromagnetic and weak currents that would be suitable sources for a Yang-Mills electro-weak theory. The pieces of the weak and electromagnetic charges would generate the whole chiral flavor algebra, while the total weak and electromagnetic charge operators would generate the electro-weak subalgebra.

I put off for a long time the construction of a Yang-Mills theory for the strong interaction. The abstract approach fitted in nicely with the assignment of the $J=\frac{1}{2}^+$ baryons to an octet. But then I wavered for a while on both the approach and the assignment.

Back in 1958 and 1959, I had played with the idea, which Sakata was to publish and elaborate, of treating the baryons p, n, and Λ as fundamental and then trying to explain others as composites of two «fundamental» baryons and one «fundamental» anti-baryon. Since at that time I did not understand SU(3), I used the permutation group S_3 to analyze the symmetry properties of the model. In this indirect manner I found the relation that in the language of SU(3) representations could be given as:

$$3 \times 3 \times \bar{3} \;=\; 3 + 3 + \bar{6} + 15\,. \tag{7}$$

The Σ and Ξ hyperons, along with some other hypothetical ones, would have to be put into the 15. At the time I did not find this possibility at all attractive, but Sakata adopted it.

After I wrote my Caltech Report on *The Eightfold Way* in January 1961, I became worried by the reports that the Σ and Λ hyperons had opposite parity. Also, my work on current algebra, involving currents for flavor SU(3)×SU(3), persuaded me that there should in some sense be three fundamental spin $\frac{1}{2}$ hadrons. Thus for a while I embraced what was by then the Sakata model, and submitted to the *Physical Review* a version of my work on flavor SU(3) in which the octet assignment of baryons was almost entirely suppressed.

During the summer of 1961, though, my faith in the octet assignment of baryons was restored; the odd parity rumor was receding and the beauty of the scheme continued to impress me. I added Section VIII to my *Physical Review* paper, in which I treated the octet assignment in detail. Since the early part of the paper used the Sakata model for explaining flavor SU(3) and SU(3)×SU(3), it was easy to emphasize the notion of abstracting these symmetries from a concrete model based on three fundamental objects. My agonizing over the correctness of the octet assignment for baryons thus led to a useful presentation of the subject. A very natural groundwork was being laid for the quark model of 1963.

In 1961, however, I had to cope with an additional confusion, the continuing puzzle about two neutrinos versus one and the small weak angle θ versus equal bare strength for strangeness-preserving and strangeness-changing weak currents, with the latter reduced by renormalization. In my early work on the *Eightfold Way,* I leaned away from the correct hypothesis of two neutrinos and a small weak angle and toward the wrong one, because of the parallel between the triplet substrate of SU(3) on the one hand and three leptons on the other. Of course, if we had suggested charm at that time, this confusion would have disappeared, along with the problem of the neutral strangeness-changing current; four flavors for leptons and four for hadrons would have been so neat. (I noted in my Paris talk of 1982 that Pais and I had presented a model at the Glasgow meeting in 1954 that contained what was essentially charm, but of course the «charmed» hadrons in the scheme were not found during the fifties and sixties, and I did not revive the idea.)

I should mention that there was a subtle reason why I stayed away from charm, even after it had been suggested by Bjorken and Glashow in 1964. Here is yet another source of confusion. In the *Eightfold Way* scheme, electric charge is a generator of the flavor SU(3) group, and correspondingly when we go over to quarks the sum of the quark charges $\left(\frac{2}{3} - \frac{1}{3} - \frac{1}{3}\right)$ is zero. If we bring in charm, then electric charge is no longer a generator of the resulting flavor SU(4), and the sum of the quark charges $\left(\frac{2}{3} + \frac{2}{3} - \frac{1}{3} - \frac{1}{3}\right)$ is no longer zero; the electric charge operator contains a piece that is outside the SU(4) algebra and commutes with it. I wanted to avoid that situation and to have the electric charge operator for quarks as a generator of the relevant algebra. I should have realized that such a narrow requirement was foolish since for the leptons the trace of electric charge is not zero either. In fact, in a modern unified theory of strong, electromagnetic, and weak interactions, we must consider each whole family of quarks and leptons together, including the three quark colors, for the trace to come out $(0 - 1) + 3 \times \left(\frac{2}{3} - \frac{1}{3}\right) = 0$, so that the electric charge can be a generator of a hypothetical unified Yang-Mills gauge group without a U(1) factor, such as SU(5) or SO(10). I did not appreciate this point fully until about 1973.

To return to the winter of 1960-61, Yuval Ne'eman invented the *Eightfold Way* independently and published it without much hesitation. David Speiser also thought up the SU(3) octet assignment for baryons, I understand, but was discouraged from publishing it. They will no doubt tell about their adventures.

In 1962, at the International Conference in Geneva, I heard a presentation of the experimental discovery of a baryon resonance that was an excited Ξ. It occurred to me immediately that along with the $J = \frac{3}{2}^+$, $I = \frac{3}{2}$ excited nucleon and a recently discovered excited Σ, it could form part of a decimet with $J = \frac{3}{2}^+$. I wrote down the mass formula and realized that in the case of the decimet the formula gave equal spacing. The three known masses were around 1240 MeV, 1385 MeV, and 1530 MeV, with the last mass being the one just reported. I became very excited and jumped up to address the meeting. I pointed out the likelihood that we were dealing with a $J = \frac{3}{2}^+$ decimet in the *Eightfold Way* scheme, and predicted the existence of Ω^- at around 1675 MeV, a metastable baryon of spin $\frac{3}{2}$ that could decay spectacularly into $\pi^- + \Xi^0$ or $K^- + \Lambda^0$. (I was so excited that at first I called it Ω^0 and had to correct myself.) At lunch, at the request of Leitner and Samios of Brookhaven, I wrote a note on a paper napkin to Maurice Goldhaber, the Director of Brookhaven, asking him to let them search for Ω^-. After two years and after scanning two million feet of film, they finally found, in 1964, the two events (one $\pi^- \Xi^0$ decay and one $K^- \Lambda^0$ decay) that confirmed the *Eightfold Way* scheme with flavor SU(3) as an excellent approximate symmetry.

4. Quarks

Now we get to quarks. In early 1963, lecturing at M.I.T., on leave from Caltech, I tried to work out for my lectures the minimal set of fundamental hadronic entities. I found various sets of four objects if I insisted on integral charges; no scheme of that kind looked particularly attractive. But then in March I went to Columbia on a visit, and at the Faculty Club there Bob Serber asked me why it was not possible to use my formula:

$$3 \times 3 \times 3 \ = \ 1 + 8 + 8 + 10\,, \qquad (8)$$

to obtain the baryons. I replied that I had tried to get the baryons that way, but that the fundamental entities would then have fractional charges –I showed him (again on a paper napkin) the fractional charges: $\frac{2}{3}$ and $-\frac{1}{3}$. He said in effect: «Oh well, I see why you do not do it then.» But thinking about the matter afterwards, that day and the next morning, it occurred to me that if the bootstrap approach were correct, then any fundamental hadrons would have to be unobservable, incapable of coming out of the baryons and mesons to be seen individually, and that if they were unobservable, they might as well have fractional charge.

Later on, after I had named the quarks,[3] I referred to these trapped fundamental entities as «mathematical quarks» –by which I always meant that they were permanently stuck inside the baryons and mesons. (Conceivably, they were almost trapped and very rarely, with great difficulty, they could come out and be seen in very sensitive experiments –then I would call them «real quarks».) But usually I thought of quarks as being permanently trapped, right from the first day.

Except for the bootstrap idea, the notion of «nuclear democracy», I do not know if I would have come so readily to the notion that fundamental hadrons ought to be unobservable and that therefore it was all right for them to have fractional charge. Of course, it was known that fractionally charged particles were absent or at least very rare in nature, and so it was particularly appropriate to think of the quarks as being unobservable; an alternative hypothetical set of integrally charged fundamental hadrons would not have this additional, experimental reason for being unobservable in isolation. That is what I meant, in one of my early papers on quarks, when I mentioned a less attractive model with a larger number of integrally charged fundamental objects and wrote that such objects «would be more likely to be real than quarks». I was not by any means rejecting the quark idea, merely remarking that fractionally charged quarks had an empirical reason for not being «real» (i.e., directly observable) that integrally charged counterparts would not have. At my lecture to the Royal Institution of London in 1966, I emphasized that «mathematical» (that is, confined) quarks would be compatible with the bootstrap idea. Later

3. By the way, in referring to the quotation about «three quarks» in *Finnegans Wake,* I was thinking of the three quarks in a baryon, corresponding, as we learned later, to the three colors. Thus the quotation is still apposite.

that same year, in my introductory review lecture at the International Conference in Berkeley, I said: «One may think of mathematical quarks as the limit of real light quarks confined by a barrier, as the barrier goes to an infinitely high one.» I was still using the terms «mathematical» and «real» to denote what today might be called «confined» and «incompletely confined» quarks. Now why did I use that language? Probably because I dreaded philosophical discussions about whether particles could be considered real if they were permanently confined. While a colleague of mine falsely claims to have a doctor's prescription forbidding him to engage in philosophical debates, I really do have one, given to me by a physician who was a student in one of my extension courses at U.C.L.A..

Even in my original letter on quarks, published early in 1964, I stressed their compatibility with the dispersion theory program and the likelihood that the quarks would not emerge. At that time I referred to the «mathematical» nature of the quarks as resulting from the limit of infinite mass rather than an infinite barrier.

The notion of confined quarks fitted in well with the abstract approach that I had taken in 1961 with respect to the octet assignment of baryons. In neither case did I insist on observable fundamental hadrons like p, n, and Λ. In 1961, except for a period of wavering, I took the position that SU(3) of flavor could be abstracted from something like the p, n, and Λ model and used as an approximate symmetry with whatever representations were needed; and in 1963 I maintained that quarks that were probably not «real» could be the substrate of the operation of SU(3).

On my first visit to Japan in the spring of 1964 I encountered with something of a shock the exactly contrary attitude of the group of Marxist theoretical physicists that included Sakata. Yukawa, who did not entirely share their views, had nevertheless helped to place many of these theorists in academic positions in the Kansai region, in the vicinity of his Institute in Kyoto. I was visiting Yukawa's Institute, and a meeting was arranged in my office with Taketani, Ohnuki, Maki, and some others. Sakata, who was in the building at the time, did not come. All of these men were strongly opposed to my abstract approach, which they condemned as smacking of «bourgeois or revisionist idealism». They had missed the octet assignment of baryons in flavor SU(3) on account of their *a priori* requirement of concreteness, and now they insisted that if there were basic hadrons they must be integrally charged and observable. Presumably, they wanted concrete basic objects that could be explained to the masses. It was interesting to see these very intelligent theoretical physicists, working on exactly the right problems with suitable mathematical methods, missing right answers because of their fixed philosophical positions. If I had designed a lesson in the virtues of pragmatism, I could not have found a better one, and I described that lesson in a public lecture in Tokyo sponsored by the newspaper *Yomiuri Shimbun*.

George Zweig arrived at the idea of quarks independently and somewhat later than I did, and his approach was significantly different. He had been a student of mine, and by 1963 he was on the faculty of Caltech. I was on leave from January to September of 1963, and so I did not have occasion to discuss my ideas about quarks with George after I first began to work seriously on them in March. When I returned to Pasadena in the fall, he had just left to spend the academic year 1963-64 at CERN in Geneva, and it was there that he started to think about quarks, under the name of «aces».

My only attempt to convey the notion of quarks to CERN during that fall was in a telephone call to my old teacher, Viki Weisskopf, who was the Director General at that time. After I had explained to him that I had been working for a few months on an exciting idea that each baryon was made of three fractionally charged particles, he said: «Please, Murray, let us be serious; this is an international call.» It is no wonder that the news of the idea never trickled down to George.

During that fall I wrote my first letter about the quarks. After several miserable experiences with *Physical Review Letters,* I had decided to use *Physics Letters* instead and they printed my communication with no difficulty. I shudder to think of the trouble I would have had if I had sent it to *Physical Review Letters.*

Meanwhile, George started to work on «aces». He never thought of them as confined, and to this day he is studying the theoretical chemistry of unconfined quarks and their possible role in catalyzing controlled thermonuclear reactions. Of course, by now we know that quarks must be pretty well confined, but there is still the possibility of a slight leak somewhere in the physics that could lead to the existence of a few isolated quarks in the environment; George is still betting on such a leak. It would be fascinating if he were right; a flourishing quarkonics industry would grow up, and money for particle physics would be easy to find.

Another difference between George Zweig's work and mine in the early days was the following: I particularly emphasized the «current quarks», the basic quark fields that enter into the weak and electromagnetic currents and present a close analogy with lepton fields. He concentrated more on «constituent quarks», the transformed objects out of which the hadrons are effectively constructed and on which the spectra of baryons and mesons depend. Another amusing difference is that I had sent my results voluntarily to a European journal, while George got into a furious argument with the head of the CERN Theory Division over a rule that required him to use a European journal, and ended up not publishing his work. Some of it appeared much later in the Proceedings of the 1964 Erice Summer School.

Now what about quark statistics? In September, 1963, I investigated whether para-Fermi statistics with index three would explain that fact that the three quarks in a baryon appeared to be in a symmetrical state of orbital, spin, and flavor variables. What I found was that the resulting baryons would include not only a normal one obeying Fermi-Dirac statistics but also some «para-baryons», which I was unwilling to buy. I therefore refused to commit myself on the matter of statistics. Greenberg later published the idea of parastatistics for the quarks, but as far as I know he did not resolve the problem of the para-hadrons.

In 1966, in my Royal Institution lecture, I alluded to the fact that «mathematical», i.e. confined quarks would make it easier to understand the puzzle of the statistics, but I did not clearly formulate the principle that confinement of paraquarks might involve the confinement of the parastatistics and result in hadrons that would be fermions and bosons. It was not until 1971, in collaboration with Harald Fritzsch, that I understood that color, with prohibition of the emergence of colored objects, is equivalent to parastatistics with prohibition of the emergence of para-objects. I wish that I had understood that much earlier, but it was only in 1971 that Fritzsch and I saw it clearly and proposed the confinement of color.

When, in the fall of 1963, I showed an early version of my first letter about quarks to Stanley Mandelstam, he exclaimed that I was trying to start a counter-revolution. I explained that if the quarks were not directly observable there would be no contradiction of the bootstrap approach to the scattering amplitudes («S-matrix») on the mass shell, and I believe Mandelstam was satisfied. Geoffrey Chew, however, was not, and he never approved of my work on the quarks and related subjects. There was a fundamental asymmetry in the situation, because I approved of his work on the «S-matrix» program and indeed I participated actively in that program (which I had proposed in 1956), doing research on Regge poles and other S-matrix singularities. In 1964, despite our disagreements, he and Arthur Rosenfeld and I wrote a joint article for the *Scientific American* on hadrons. We mentioned the bootstrap idea and «nuclear democracy», Regge poles, the classification of hadron states, and so forth. Flavor SU(3) was included as an approximate symmetry, but as I recall there was not a word about quarks.

As far as I know, something like the bootstrap program might still qualify as a valid description of hadrons, if the right boundary conditions at infinite momenta are imposed on the theory; but it has not proved as useful as the quark description, which led to the development of QCD.

Before discussing QCD, let me relate an anecdote that bears on the significance of the discovery of quarks. I gave a series of lectures on quarks in Cambridge, England in 1966. Dirac attended them all. As usual, he fell asleep shortly after each lecture began, awoke at the end, and then asked penetrating questions about the subject of each lecture. How he manages to do that I have never discovered. Anyway, his questions revealed that he was fond of the quark scheme, and one day I asked him why he liked quarks when most of my colleagues thought they were crazy. He replied by asking: «They do have spin one-half, don't they?» I am sure that he meant: «They do obey my equation, don't they?» He had grasped the essential point that the quarks, like the leptons, are Dirac particles that are coupled calculably, and in some sense weakly, to various fields. In other words, in discovering quarks we have found hadron building blocks that are just as elementary as the leptons. At very small distances, of course, there might turn out to be constituents for both, but surely the quarks and leptons are equally elementary or equally composite.

In 1966, in *Preludes to Theoretical Physics,* Nambu wrote an article in which some features of QCD were anticipated, especially the exchange of colored gluons. He did not, however, actually employ a Yang-Mills theory. Moreover, he used the Han-Nambu scheme, in which electric charge varied with color in such a way that the quarks were given integral charges, and color was not confined —the quarks could emerge singly. If I had known about his article during the next six years, especially after Fritzsch and I had started to collaborate, I would probably have proposed the relevant modifications: Yang-Mills color theory, no dependence of charge on color, fractional charges, and color confinement leading to quark and gluon confinement. Unfortunately, I never saw his article or heard of it. The volume in which it appeared was a *Festschrift* in honor of the sixtieth birthday of Viki Weisskopf, and I was so ashamed of not having managed to contribute to it that I never looked at it.

I did not learn about Nambu's article until after Fritzsch and I worked on QCD in 1971 and 1972. At that time, we had to overcome confusion about yet another matter, whether the fundamental hadronic theory, apart from quarks, would be a theory of colored strings or a local Yang-Mills field theory of gluons, with strings constituting only an approximation. We settled on the latter, in other words QCD, but only after passing through a period of uncertainty, principally caused by the properties of QCD in the limit of vanishing quark masses. We did not fully understand the subtle processes by which scale invariance and conservation of total quark helicity are violated in that limit. In both cases, anomalies explain the difficulties that bothered us.

Painful as it is to describe, this story of confusion, ignorance, mistakes, difficulties, and vacillations may be the most interesting aspect of our particular corner of the history of science. This mode of description is related to the approach through hindsight, which is condemned by many historians who work in the history of science, but is embraced by many scientists working in the field. To me it is important not only to situate scientific ideas in the context of their time but also to try to figure out what relation the scientists' thinking bore to what we now know to be correct, how close they came to the right answers, or why they missed them. I hope that my reflections, based as they are on looking less at the doughnut than at the hole, have been useful.

32

HADRON SYMMETRY, CLASSIFICATION AND COMPOSITENESS

Yuval Ne'eman
Tel Aviv University

1. The first thirty-five years

I was born in 1925 in Tel Aviv, Israel, then Palestine — or, officially and colloquially «Eretz-Israel» in the Hebrew version, even on the stamps issued by the Mandatory power. This was then a British Mandate charged with the mission of establishing that Jewish National Home which had been promised by Britain in the 1917 Balfour Declaration. Mine was an «Old Settlement» family, the equivalent of the American «Mayflower» group: three out of my four grandparents were born in the country, the fourth arriving in 1890 as an adolescent. Their own parents or grandparents had immigrated from Lithuania in 1807 with the Vilna Gaon's school (a group known as the «New Pharisees»). The Gaon's disciples represented the cutting edge of rationalistic and scholarly Judaism, the «Resisters» (as they were called) against the «sentimentalist» Hassidic view. Rabbi Eliyahu the Gaon (1720-97) had also been emphatic on the importance of science (he himself did some research in geometry and astronomy although, somewhat paradoxically, he had also been involved in Cabbalistic studies in his youth). The Gaon's school settled at Tsfat (Safed) in Galilee, thus counterbalancing the nearby Hassidic center established in Tiberias in 1777. After the 1837 earthquake, the family moved to Jerusalem later spreading all over the country. The critical approach characteristic of the Gaon's school lasted, and I could feel it in my environment as a child.

My paternal grandfather settled in Jaffa in 1900, establishing a small mechanical workshop producing pumps for drinking and irrigation.The firm was run by my father between 1939 and 1983, and my son is now the plant manager, though ownership has recently gone to a cousin, perhaps due to my defection... My grandfather was also to become one of the sixty founders of Tel Aviv in 1909. I graduated from Herzliyia High School (the oldest Hebrew High School) in 1940 aged 15. I had been something of a local prodigy, jumping classes and establishing a reputation in both the sciences and the humanities. In fact, my main interest between the ages of 10 and 13 had been in history and geography, with astronomy as a close third. It then switched to mathematics, especially after my first encounter with complex numbers. The idea of a generalizing algorithm captivated me and I developed on my own new arithmetical operations and new types of numbers. Physics came last. An older cousin — the same one who now owns the pumps factory — had mentioned Eddington's *Nature of the Physical World,* which I then read and re-read, returning to it time after time, gradually absorbing the ideas and falling in love with physics. Shmuel Sambursky of the Hebrew University gave a series of evening lectures in Modern Physics, which I attended in 1940. This also helped.

I had always been interested in taxonomy and classification, and one of my high school hobbies had consisted in spending afternoons in public libraries and charting tables in every field of knowledge. I can still quote some names from my lists of Chinese dynasties and emperors — or more to the point here in Catalonia, the intricate dynastic descendence of the Wisigoth Kings, with their Dukes of Tarragona, Septimania or Aquitania, or of the subsequent Kings of Navarra and of Aragón, the Counts of Barcelona or the Kings of León and Castille — or equivalently of the enlightened Ummayad Caliphs at Cordova, or of the intolerant and oppressive Almoravides or Almohades. I learned and tabulated everything, whether crystal structures, minerals or languages. Very early I became familiar with Linneus' classification of the plants, with the taxonomy of the animal kingdom and with Mendeleev's periodic table of the chemical elements. I had decided to study engineering, both in order to join the family pump factory and because I had no clear idea as to how a scientist managed to make a living. The Haifa Technion required a minimum entrance age of sixteen, and I had to wait a year. I spent that year working at the factory, attending those evening lectures in science, and joining the Hagana (Underground) paramilitary force.

It is at this point that I started my «bifocal» career and life, with those two ever intertwining threads: technology and science on the one hand, military and later political activity on the other. Both were (and still are) driven by powerful motivations. The pleasure of problem solving, with a particular fascination by those puzzles posed by nature and the excitement derived from an existentialistic feeling of partaking in a tremendous adventure, that of the evolutionary rise of conscious matter and its attempt to assimilate the universe — these power the scientist in me. A deep devotion to the Jewish People and a feeling of personal obligation to share in the shaping of its destinies, in this age when we have experienced both holocaust and statehood — these drive me as a struggler, whether soldier, politician or statesman. The engineer's enthusiasm may derive from both. Psychologically, the scientist's mental life is entirely in the contemplative mood; my other roles mostly require the man of action and fast decision. Still, some of my military tasks did bring out my best efforts in problem solving — in tactical thinking, in intelligence, in long-range national estimates and planning, in technological forecasting or in actual engineering design. All of that is outside the interest and scope of this report. Much of it is as yet an untold story, some parts have been amplified and distorted, but this is irrelevant here anyhow[1].

I studied mechanical and electrical engineering from 1941 to 1945 at the Haifa Technion, now the Israel Institute of Technology, graduating in the summer of 1945 and acquiring in addition, early in 1946, a Diploma of Engineering («Ingénieur Diplomé») modelled on the European tradition.

What captivated me most at the Technion were two courses: a brief introduction to Set Theory, by M. Grossman, the mathematics professor, and a course in Wave Mechanics by F. Ollendorf, professor of electrotechnics. Like most of the Technion professors, these two had arrived from Germany. Ollendorf was a charismatic teacher, had led a group of children out of Germany — perhaps the model for Leslie Howard's film «Pimpernel Smith». His Modern Physics course was considered to be «very tough» and towards the end I would be the only student in attendance. Still, he would go on, seemingly unperturbed. We reminisced, when in 1965 I received an Honorary Doctorate from the Technion, with him in attendance. He then remarked

1 See, for example DEACON 1977, p. 318.

that the fruits of that course, given to an audience of one, seemed to have been well worth the effort. This is often so in science (Chandrasekhar travelled weekly from Yerkes Observatory to Chicago to teach a course with two students: C.N. Yang and T. D. Lee...). When in 1946 (and again in 1951) I had thoughts of studying physics, I took letters of recommendation from Grossman and Ollendorf, but when I finally presented these letters to A. Salam at Imperial College in 1958, the stationary had yellowed and their names were, of course, unknown to him.

Except for a period of less than a year (in 1946/7) spent in hydrodynamical design at the family pumps factory, my years 1945-1957 were dedicated to Israel's defence. Indeed, I even felt some reluctance in mixing the two worlds: after graduation in 1945, I was interviewed, together with four other recent graduates who were also Hagana members, by General Avidar, Chief of Logistics of the Hagana (and after May 1948 of the Israel Defence Forces), and he suggested that we transfer to Hagana's industrial group and partake in establishing a military industry. All four others agreed and indeed became the founders of our weapons production. One of the four, M. Schur, has been the Director of Israel's Military Industries since 1974. I refused. «I am a Hagana member as a soldier, and intend to fight in the field — but as an engineer, I'd rather work at my family's plant», I answered. Somehow, I did not like to mix my emergency soldiering with a profession. It was only after my return to Israel from Imperial College in 1961 that I started using my scientific or technical background for national purposes directly, especially in 1961-63 as Scientific Director of the Atomic Energy Soreq Establishment; in 1974-76 as Defence Chief Scientist or in 1977-1983 as Chairman of the Steering Committee of the Mediterranean-Dead Sea Project; and to a smaller extent as Founding Dean of Tel Aviv University's School of Engineering in 1969-71, as University President in 1971-75 or as Minister of Science and Development since 1982.

Between 1945 and 1957 I made two attempts to reduce for a while my defence activities and couple them with studying physics, in 1946 and 1951. In both cases circumstances made me abandon my plans. Luckily, because (remembering that Wave Mechanics course) I had intented to go to Paris and work with de Broglie — a dead end by that time. Only towards the end of 1957, in the relative relaxation of defence tensions after the 1956 Sinai Campaign, did I try again. I felt time was running out, I was 32 and would perhaps soon be too old to start new studies. I asked General Dayan, then Defence Chief of Staff for a two-year-leave, thinking that I would return to the Technion where N. Rosen, Einstein's collaborator, had meanwhile established a Physics Department. This time I had decided to work in General Relativity, having been captivated by the aesthetic aspects of Einstein's theory, which I had been studying in my «spare time» (though there really wasn't any...). Dayan countered my request with a suggestion that I go to London as Defence Attaché and couple that post (seemingly a sinecure since the post-Suez British offishness) with studies at London University. I agreed, hoping to work with Bondi and collaborators, of whose Steady State Theory I had recently become aware.

Before leaving for London I attended the Rehovoth Conference in Nuclear

Physics, had glimpses of Pauli, T.D. Lee, Mrs. Wu and Ben Mottleson. It was all beyond my grasp, as I still had no notion of nuclear forces or nuclear structure (I have since made two modest contributions in that field: one with Dothan and Gell-Mann in 1965, on the excitations of deformed nuclei[2] and the second one in 1969, on hypercollapsed nuclei[3]).

Ivor Robinson (of the Robinson-Trautman metric etc.) was visiting the Weizmann Institute. He hitchhiked with me to Tel Aviv, heard of my future plans in London and warned me off General Relativity, which according to him, had become mostly a playground for mathematical sophistication, but with very little new physical content. When in addition I discovered in London that traffic made it quasi-impossible to combine studies at Kings' College, where Bondi taught, with my «job» at the Embassy in Kensington, I decided to look elsewhere.

Robinson was, of course, right at the time. The discovery of quasars in 1963 contributed to a first re-awakening (and I «rushed» to partake in it, suggesting[4] «lagging cores» or «white holes» in 1965). Semi-classical theory made great advances through the ideas of Beckenstein and Hawking after 1974, and research in Quantum Gravity came to life «with a bang» with the discovery of Supergravity in 1976. Again with Gell-Mann, we developed a group-theoretical preview of the theory, at the same time that it was being discovered in its dynamical form by the groups at Stony Brook and CERN[5].

Within half a mile from the Embassy in London, I found the Imperial College of Science and Technology. There I approached Salam with my two weathered Technion letters and a new one, signed by General Moshe Dayan. Salam was puzzled, but admitted me into his group. I took the graduate courses in 1958. Between July 1958 and May 1960 I barely managed to keep contact. Events in the Middle East had flared up again, and I was participating in the negotiations for the purchase of two S-class submarines and fifty Centurion Tanks... I spent the year 1959 organizing and supervising the training of the crews for both the submarines and the tanks. Finally, I complained, and General Laskow, Dayan's successor sent a replacement to the Embassy. As of May 1960, I went on a one-year full leave from the Defence Forces and received a modest grant enabling me to spend the year at Imperial. I was thirty-five, a Colonel on leave, the oldest student in Salam's group. At high school and at the Technion I had been the youngest... Would I still be able to make the transition? General Laskow regarded my escapade as an educational relaxation, and wanted me to found our National Defence College upon my return (I did indeed later send in a draft for that project). I gave myself a one-year chance, before deciding whether I had missed the physics boat or not.

2 DOTHAN, GELL-MANN and NE'EMAN 1965; WEAVER and BIEDENHARN 1972; BIEDENHARN, CUSSON, HAN and WEAVER 1972.
3 NE'EMAN 1968; NE'EMAN 1972; BODMER 1971; LEE and WICK 1974; LEE 1975.
4 NE'EMAN 1965; NOVIKOV 1964.
5 See for example, references and acknowledgements (of unpub. material); BRINK, GELL-MANN, RAMOND and SCHWARZ 1978; FREEDMAN 1977; FERRARA, SCHERK and ZUMINO 1977; NE'EMAN and REGGE 1978; THIERRY-MIEG and NE'EMAN 1979.

2. Symmetries and groups

The aim of physics is to describe observations and the results of experimentation and measurements as applied to natural phenomena by the smallest possible set of postulates and such rules as can be derived from the postulates through logical and mathematical inference. To achieve logical coherence between apparently distinct sets of phenomena, physics is forced into abstraction and generalization. The wider the field of application of a set of physical axioms, the further the abstraction and the simpler its content.

Symmetry laws are such postulates. Basically, they embody the idea of generalization through a negative statement, «postulates of impotence» in the words of Whittaker. Take, for example, Einstein's theory of gravitation. It rests on two postulates, one an axiom of symmetry, the principle of general covariance, and the other a dynamical one, the principle of equivalence. The principle of general covariance states that it is impossible to pick a preferred reference frame in space time. In other words — all frames are equally allowed, they are indistinguishable as far as the laws of physics are concerned. They may, of course, differ in practice because of the boundary conditions, and we can exploit these differences to simplify actual calculations. However, the laws of physics do not distinguish between them. As the French proverb says «la nuit, tous les chats sont gris» — all cats are grey at night. The impotence is represented by this indistinguishability.

This is symmetry, because we can think of all frames of reference—or all cats in the dark—as reflections of each other, symmetrically identical images, like in a hall with mirrors. Closing our eyes for a moment and reopening them we cannot be sure that this cat — or reference frame — we face is the same one as the one we were looking at before.

Suppose we turn to another cat on purpose, we replace one reference frame by another. This is an operation which is an element of a mathematical transformation group. It is a group because any two replacements make yet another replacement; the identity operation implies staring at the same cat again, and the inverse transition is also there in each case.

We thus have a defining vector space — finite or infinite — of reference frames V (the set of cat images), and a *transformation group G* acting on it. There is also a description of the physical system—the S-matrix, or the action, or just its 4-density, the Lagrangian (in Relativistic Quantum Field Theory), or perhaps the Hamiltonian, in a less explicit description. Our postulate of impotence then translates into an invariance of the action (or of the Lagrangian, Hamiltonian, S-matrix, path integral etc.).

The invariance can be observed directly in several ways. It appears in the *law of force,* as in charge independence: in the limit of isospin invariance, the *binding energies* of single protons or neutrons *in a nucleus are equal,* i.e. it takes the same amount of energy to extract either of them.

This is a mathematical consequence of the invariance of the interaction Lagrangian,

$$L = g^{I}\, \bar{p}\gamma_5\pi^0 p + g^{II}\, \bar{n}\gamma_5\pi^0 n + g^{III}\, \bar{p}\gamma_5\pi^+ n + g^{IV}\, \bar{n}\gamma_5\pi^- p, \qquad (1)$$

the group SU(2) fixing the *couplings*

$$g = g^{I} = -g^{II} = \frac{1}{\sqrt{2}}\, g^{III} = \frac{1}{\sqrt{2}}\, g^{IV}. \qquad (2)$$

We can indeed check that the *forces between any two nucleons* are the same:

$$\mathrm{p}\,\&\,\mathrm{p}: g^2; \qquad \mathrm{n}\,\&\,\mathrm{n}: g^2; \qquad \mathrm{p}\,\&\,\mathrm{n}: -g^2 + 2g^2 = g^2. \qquad (3)$$

The group G has irreducible representations R(G), some of which play a physical role, since the G-invariant equations of motion (derived from the Lagrangian, Hamiltonian, etc.) have solutions that are not G-invariant, only G-covariant. These solutions – the fields, the particles, the currents, etc. – thus have to span in each case a realization of G; in many cases – and these are the ones we exploit most – this realization is a *linear representation,* $\sum_i^{\oplus} R_i(G)$, sometimes even *irreducible* $R_i(G)$.

In the interaction Lagrangian (1), the nucleon field components N are assumed to span the $I = 1/2$ (fundamental) irreducible representation of SU(2), the pion components π span $I = 1$, and the couplings g^{I},...g^{IV} correspond to the Clebsch-Gordan coefficients of the first term of the r.h.s. in

$$(I = 1/2) \otimes (I = 1/2) \;\rightarrow\; (I = 1) \oplus (I = 0) \qquad (4)$$

The same G-invariance may manifest itself in other forms. To the extent that a mass term appears in the Lagrangian, we ought to get *equal masses* for the various fields or particles in one representation, i.e. in the SU(2) limit,

$$m(\mathrm{p}) = m(\mathrm{n}),$$
$$m(\pi^+) = m(\pi^0) = m(\pi^-). \qquad (5)$$

Anticipating on our story, we remind the reader that electromagnetic interactions are U-spin invariant within SU(3) multiplets. This is yet another SU(2), and we are applying it to the electromagnetic interaction Lagrangian instead of (1), so that the results constrain the electromagnetic mass differences or the magnetic moments. We can also derive *intensity rules* between amplitudes. The simplest way of understanding how they occur is to use Shmushkevich's method. Take the coupling

$N + \pi \longleftrightarrow \Delta$, where Δ is the $I = {}^3/_2$ resonance. We list all productions (or decays): (the Γ_i are partial widths)

$$
\begin{array}{llll}
\Delta^{++} & \longleftrightarrow & p + \pi^{+} & (\Gamma_1), \\
\Delta^{+} & \longleftrightarrow & p + \pi^{0} & (\Gamma_2), \\
\Delta^{+} & \longleftrightarrow & p + \pi^{+} & (\Gamma_3), \\
\Delta^{0} & \longleftrightarrow & p + \pi^{-} & (\Gamma_4), \\
\Delta^{0} & \longleftrightarrow & p + \pi^{0} & (\Gamma_5), \\
\Delta^{-} & \longleftrightarrow & p + \pi^{-} & (\Gamma_6).
\end{array} \tag{6}
$$

Symmetry in the Δ multiplet requires

$$\Gamma_1 = \Gamma_2 + \Gamma_3 = \Gamma_4 + \Gamma_5 = \Gamma_6,$$

whereas symmetry between the nucleon states imposes, once we rewrite those processes with the pions (or in fact their charge conjugates) on the left, i.e. $\Delta^{++} + \pi^{-} \leftrightarrow p \quad (\Gamma_1)$ etc.

$$\Gamma_1 + \Gamma_2 + \Gamma_4 = \Gamma_3 + \Gamma_5 + \Gamma_6,$$

and symmetry between the pion states, if we leave them on the left and transfer the charge-conjugates of Δ to the right, $\pi^{-} \leftrightarrow p + \overline{\Delta^{++}} \quad (\Gamma_1)$ etc., requires

$$\Gamma_1 + \Gamma_3 = \Gamma_2 + \Gamma_5 = \Gamma_4 + \Gamma_6.$$

The three equations are solved by

$$2\Gamma_1 = 3\Gamma_2 = 6\Gamma_3 = 6\Gamma_4 = 3\Gamma_5 = 2\Gamma_6 \tag{7}$$

and these are *intensity rules.*

Note that equations (6) could also be construed as manners of «making» the Δ from $N+\pi$. Equations (1) and (4) can be construed as ways of «making» the components of π from $\overline{N} + N$ as had indeed been suggested by Fermi and Yang in 1949. Thus the overall picture could be used to describe a theory in which only the two N states are «basic», with *composite* π or Δ.

The group has a *generator algebra.* If the group is not compact, there are two such algebras, one for action on the right, the other on the left, and they commute with each other. If the group realizes unitary transformations, the algebraic operators are

Hermitian, and Quantum Mechanics tells us they represent *observables*. Noethe theorem says they are conserved quantities

$$\frac{d}{dt}Q = 0 . \qquad (8$$

This can imply *selection rules*. The algebraic Lie brackets yield *sum rules*, derived saturating the bracketed operators between intermediate states.

In Relativistic Quantum Field Theory, the conserved algebraic generator can written as a volume-integral of a local density $\mathbf{j}(x)$, taken over the whole of 3-spa *Most dynamical postulates involve this density.* For instance, Einstein's theory gravitation couples the *gravitational field to the energy-momentum tensor density*. Si larly, all Yang-Mills local gauge theories have equations involving **j**, the curre density of the gauge group generators, (i.e. using forms and the notation of exterior calculus)

$$\mathbf{DF} = 0 \quad , \qquad \mathbf{D^*F} = {}^*\mathbf{j}.$$

Here **F** stands for the field-strength (or «curvature» in a geometric interpretation the field-potential **A** (or «connection» geometrically),

$$\mathbf{F} = d\mathbf{A} + {}^1/_2[\mathbf{A},\mathbf{A}] \qquad ($$

and * is the Hodge dual. Note that **Q**, **j**, **A** and **F** all transform as the adjo representation of G.

Summing up, we may encounter a symmetry realized through

a) the transformation group G → invariance of S-matrix etc.
→ law of force
→ ratios of binding energies
→ ratios between masses
→ intensity rules

b) representations → classification
→ composition relations («models»)

c) a generator algebra → observables
→ conservation laws
→ selection rules
→ sum rules

d) a local charge-current density → local current algebra sum rules
→ a gauge interaction

3. The clues

After the success of quantum electrodynamics in 1946-48, relativistic quantum field theory was tried for the strong interactions, as described by a Lagrangian based upon pseudoscalar meson exchange. Salam had shown that the theory was renormalizable. However, with a coupling $g^2/4\pi \sim 14$, perturbation methods were meaningless. As a result, M. Gell-Mann and M. Goldberger had introduced dispersion relations in 1954, which G. Chew had then proclaimed an alternative to relativistic quantum field theory, under the name of «S matrix theory». «Conservative» people still believed in symmetry and used the Lagrangian as a Born term hoping that its symmetries, even if approximate, would still hold for the full interaction. These hopes were strongly criticized by G. Chew and others.

Since the mid-fifties, one knew of 8 meta-stable baryons, grouped into 4 isospin-strangeness multiplets, plus the N-π(«3-3») Δ resonance discovered by Fermi and others in 1952, and 7 mesons: the three pions and four kaons. Omitting the Δ which appeared to be composed of N+π, the most general Lagrangian for the strong interaction, preserving isospin and strangeness, had eight independent *couplings,*

$$\begin{aligned} L_{\text{strong}} = {} & g_1 \bar{N}\gamma_5 \vec{\tau} N \cdot \vec{\pi} + g_2 \bar{\vec{\Sigma}} \gamma_5 \Lambda \cdot \vec{\pi} + \text{h.c.} + g_3 \bar{\vec{\Sigma}} \gamma_5 \wedge \vec{\Sigma} \cdot \vec{\pi} + g_4 \bar{\Xi} \gamma_5 \vec{\tau}\, \Xi \cdot \vec{\pi} \\ & + g_5 \bar{N}\gamma_5 K\, \Lambda + \text{h.c.} + g_6 \bar{N}\gamma_5 \vec{\tau} K \cdot \vec{\Sigma} + \text{h.c.} \\ & + g_7 \bar{\Xi}\gamma_5 \bar{K}\, \Lambda + \text{h.c.} + g_8 \bar{\Xi}\gamma_5 \vec{\tau} \bar{K} \cdot \vec{\Sigma} + \text{h.c.} \end{aligned} \tag{11}$$

In 1956, Gell-Mann and J. Schwinger had initiated attempts to constrain the g_1-g_8, suggesting «global symmetry» just as isospin had constrained the g^{I}-g^{IV} in eqs. (1)-(2).

A *higher symmetry* was expected to do the same in eq. (11). However, global symmetry itself had been declared a failure in 1959 at Kiev, when Salam had pointed out that the S-wave phase shifts deduced from experiments were considerably different in π-N from those in π-hyperon scattering, contrary to the theory.

J. Tiomno had shown earlier that global symmetry meant using rotations in 7 dimensions (SO(7)) as a higher internal symmetry. Salam and Ward developed a more advanced model, adapted to (chiral) weak interactions too, based upon SO(8) and SO(9). (Later on in 1960/61, when I had constructed my own solution, Lee and Yang published an amended SO(9), with a discrete extension.) I started looking at these schemes and tried to understand what was really needed. A higher symmetry would also provide a *classification,* and I noted the dimensionalities of the necessary representations (8 baryons perhaps, 7 mesons?) as clues to the group. I was thus hoping to replace the six multiplets by two only. Searching for groups with representations fitting the baryons and mesons was also the approach taken independently by Speiser and Tarski. They compiled a catalogue, finally recommending in 1962 the

7-dimensional representations of G(2), but mentioning the SU(3) octet (my own solution, as explained later) as well. At the 1962 CERN Conference, Speiser explained to me that he would rather set aside the Λ and have a basic 7, since the 8 in SU(3) was not fundamental and would be indicating compositeness of the baryons, an unwanted result as far as he was concerned. I had taken a different view. On the contrary, if these were *composite,* by chance, I might later even replace all of the multiplets in eq. (11) by just *one more fundamental one,* as Fermi and Yang had done for isospin.

Sakata had already added the Λ to the Fermi-Yang model, to provide for strangeness[6]. From p, n and Λ he could reproduce all other particles: Σ^+, for instance, would be made of ($\bar{p}n\Lambda$), and Ξ would be composed of ($\bar{p}\Lambda\Lambda$). The latter he predicted to haven spin $J = 3/2$, together with a state such as $\Delta^{++} = (\bar{n}pp)$. From a «Sakaton-anti-Sakaton» pair, one could also make various mesons, 7 of which looked very much like $\vec{\pi}$ and K,$\bar{K}$.

I was not happy with that approach. Why should the Λ be «more fundamental» than Σ? (Or less fundamental, in G(2)...) Setting apart the Ξ's was less objectionable, as their spin had not yet been measured, and they might really have been different. I had, in addition, some general dynamical misgivings. Where were the forces capable of such 3-baryon bindings? Of course, this was a somewhat circular argument. How could one decide about a structure and dynamics before understanding the pattern itself? It had taken 50 years to go from Mendeleev to Rutherford, to Bohr and to Pauli's exclusion principle. Still, in 1959/60, the Nagoya University group of M. Ikeda, S. Ogawa and Y. Ohnuki — at Yukawa's suggestion — had turned to group theory and applied SU(3) to the Sakata model[7]. The same idea occurred to Y. Yamaguchi, a Japanese physicist at CERN. I was not aware of all that work at the time. (Note indeed, that later on throughout 1961-64, the Sakata model, with an SU(3) symmetry, was the most serious competition encountered by *my own* model.)

A few years later I learned more about the motivation behind Sakata and his group. I have written elsewhere[8] about the influence of dialectical materialism and Marxist views on Japanese physics. Sakata and his colleague Taketani were strongly anti-American. They were also firm Marxists, suspicious of any more abstract approaches. They required a concrete model, little hard material particles as building blocks. This is why they fixed on (pnΛ), and why they did not notice – or perhaps chose to disregard – the octet pattern, even though they were using SU(3) and encountering octets all the time. Yamaguchi later confided that he had also thought of the octet for a while, but had finally rejected the idea and did not mention it in writing.

Thinking of the *currents* carrying the higher symmetry's *charges,* I looked at the *weak interactions.* Since 1958 and the work of Marshak and Sudarshan[9] one had the structure of the effective «V-A» Hamiltonian. It was a current × current structure,

6 SAKATA 1956; See also OKUN 1958.
7 IKEDA, OGAWA and OHNUKI 1959; See also YAMAGUCHI 1959.
8 NE'EMAN 1971.

and therefore carried a dimensional coupling (with the dimensions of an area, or of the inverse of a mass-quare term, as in a Feynman propagator for a boson). Again there was no way of using perturbation methods, as a dimensional coupling implies an infinite sequence of counter terms. Feynman and Gell-Mann[9] made the connection between this «V-A» current × current Hamiltonian and the Gershtein-Zeldovich[9] idea of the *conserved vector current* (CVC), which they had rediscovered independently. This explained the lack of renormalization of the vector current coupling, by identifying that current with a conserved current of the strong interaction symmetry. The idea resembled Einstein's equivalence principle. Einstein had explained the universal equality between the gravitational and inertial mass of a body, by assuming that the gravitational field interacted with the energy-momentum tensor, a «conserved current» provided by the Poincaré group (i.e. special relativity) when applying Noether's theorem. Similarly, there was no noticeable renormalization of the beta-decay coupling, when compared to the muon-decay coupling, even though the first involved hadrons and thus radiative corrections due to strong interactions. CVC explained the fact by assuming that the strong interactions could not produce their effects because the weak vector current happened to coincide with a Noether current of the free + strong Lagrangians. There was a similar dynamical principle discovered by Goldberger and Treiman[10] and reformulated by Gell-Mann and Lévy, Nambu and others[10], which became very useful in 1959-65, PCAC, with a similar postulate for the *axial-vector current's divergence*. Between 1966 and 1974 I discovered[11] and reformulated a similar postulate, for the *high-energy spin-2 Regge couplings*.

I was at once very attracted by these ideas. They impressed me as important dynamical clues. CVC could be used to test the patterns proposed for hadrons. One only had to study the *Noether current* for each case, and see whether or not it accounted for the transitions observed in *weak interactions*. For instance, the observed inappearance of transitions such as $\Sigma^+ \to n + e^+ + \nu$ was an indication that there should be no $\Delta S/\Delta Q = -1$ terms. Behrends and Sirlin[12] independently indeed, used that method in 1961. There was some (wrong) evidence for $|\Delta I| = 3/2$ transitions, and they were either led by it — or used it to justify their mathematical intuition — and like Speiser and Tarski[12] preferred the exceptional group G_2 as the higher symmetry, now induced by the weak current. G_2, indeed, has an $I = 3/2$ set of generators in the adjoint representation.

I used yet another set of related clues to identify the group through its *adjoint representation*. The work of Yang and Mills had become popular when experiments started bringing evidence for vector currents in the weak interaction and of vector

9 Marshak and Sudarshan 1958; Feynman and Gell-Mann 1958; See also Gershtein and Zeldovich 1956.
10 Goldberger and Treiman 1958; Taylor 1958; Gell-Mann and Lévy 1960; Nambu 1960; Goldstone 1961; Gell-Mann and Ne'eman 1964b.
11 Cabibbo, Horwitz and Ne'eman 1966; Ne'eman and Reddy 1975.
12 Behrends and Sirlin 1961; See also Shekhter 1962; Speiser and Tarski 1961.

bosons with strong interactions. Indeed, Nambu and after him Frazer and Fulco[13], showed that the (momentum-transfer)$^{-2}$ observed dependence of the electromagnetic form factors could be explained by *vector-boson dominance* of the dispersion relations. This was the first indirect evidence for the existence of the ω^0 and ρ, $J = 1^-$ mesons, although the ω^0 had already been postulated by Teller, to explain the hardcore relative incompressibility of nucleons.

One could thus try to replace eq. (11) as the *basic dynamics of strong interactions,* by a Yang-Mills gauge theory. Sakurai had just published a highly readable paper propagating these ideas[14] at the time I was starting my work at Imperial. I was very impressed by it. Later on in 1961, Glashow and Gell-Mann[15] worked out a general catalog of appropriate Lie groups. I was doing the same.

I had not heard about the Yang-Mills idea before my May 1960 reentrance into physics. T. Kibble was just giving a seminar on some work he had done on gravity as a Yang-Mills like gauge theory. I asked him at tea about the Yang-Mills paper he was mentioning. I read the paper, was very impressed and wrote a short essay for the class. I then came across Sakurai's paper, and learned about indirect experimental evidence for $J = 1^-$ bosons in strong interactions.

These then, were the clues I was using to look for the symmetry. In fact, my Ph. D. thesis at Imperial College utilized them all, searching for the «intersection» between their various sets of conclusions. Independently I had gone through the same considerations as those of the Sakata group[6], together with those of Speiser and Tarski[12], Behrends and Sirlin[12] and Gell-Mann and Glashow[15]. I was combining all these different approaches in my search, comparing how the various hypotheses were scoring on each of these separate criteria. I thus did compile my own catalog in my thesis, but I was after a definite selection rather than a catalog. In fact, once I had settled on one answer, after a certain period of comparative study, the rest of the data I gathered kept confirming my initial selection.

Let me add an anecdotal comment. Kibble's 1960 seminar had sounded incomprehensible at the time. Sixteen years later, working with Gell-Mann on our algebraic preview[5] of Supergravity (including its N=8 extension) I went back and read and enjoyed Kibble's published paper. In my 1977 work with Regge, I then managed to go beyond Kibble's original attempt and provided a deeper and more general extension of this approach[5].

4. Finding it

I told Salam that I was given a year at Imperial, and he suggested that I take on a calculation that I could be sure to complete within that time. Very soon, however, I was off in a different direction, inventing a classification, gauges, etc. I was rediscovering on my own some of the work that had taken place since 1956. I was encouraged,

13 Nambu 1957; Frazer and Fulco 1960.
14 Sakurai 1960.
15 Glashow and Gell-Mann 1961.

while Salam was getting worried about my possibly wasting that year. I explained that I felt a strong motivation to work on gauges and symmetries. He gave in, warning me «You are embarking on a highly speculative search». He had a good piece of advice «If you are going to do it, do it well. Don't be satisfied with the little group theory I taught you. Study the subject in depth». He mentioned having heard from Hamermesh that the Russian mathematician Dynkin[16] had compiled a very comprehensive list of groups and subgroups, identifying subgroups that «even Racah had missed in his applications to atomic spectroscopy». I had met Julio (Yoel) Racah back in 1953, when I was Director of Defense Planning, and represented the Ministry of Defense as an Observer on the Atomic Energy Board. Even the term «Group Theory» was not known to me at the time. Like Dorothy in the «Wizard of Oz», I was learning that I could have found the key to my search back there, at home... (I was even more surprised when at a funeral in Tel Aviv in 1967, I found I was standing next to the tomb of Issai Schur!)

Dynkin had been translated by the American Mathematical Society, but the volumes were not available in London. I ordered two copies from the U.S., one for Salam and one for me. When they arrived, I realized that this was already a classification of «Maximal semi-simple sub-algebras of all semi-simple Lie algebras». However, I did not even know the classification of the Lie algebras themselves!

That job had been performed by Killing and Cartan by the end of the XIXth century. I later learned that in 1951 Racah had given a concise but very comprehensive review of these topics in his Princeton lectures (they have been republished several times since). The lectures had been attended by Gell-Mann, Pais, Salam and many others. Apparently, they had not really absorbed the material, because in all their work in 1956-60 they had kept using the orthogonal groups only (rotations), going from 3 dimensions (isospin) to 4, then to 7, 8, 9, etc. Nobody had tried to generalize isospin as a unitary group, by going from SU(2) to SU(3).

Dynkin was referring to his own version of Cartan's thesis. He had improved on Cartan's proofs and developed a method of «Dynkin diagrams», an extremely helpful shorthand enabling you to construct any semi-simple Lie algebra and its representations. I found a heliograph of a U.S. Navy translation of Dynkin's thesis[17] in the British Museum. I read it thoroughly and finally understood what I should be looking for.

What was needed was a compact, simple Lie group of rank 2. I had noticed that all reactions allowed by the conservation of isospin and strangeness seemed to occur. In a rank 2 Lie algebra, I should be able to embed the SU(2) of isospin, and the U(1) of strangeness (or hypercharge) without fearing that another commuting U(1) would forbid some reaction. I should also check the resulting currents (for correct weak interactions transitions), vector bosons (for a gauge theory and Sakurai's phenomenology). I should list candidate representations for baryons and mesons, and the couplings imposed on them by the group. I should also study the way in which this

16 Dynkin 1957.
17 Dynkin 1947.

symmetry was being broken within the strong interactions themselves, since π and K had different masses, for instance.

Getting hold of Dynkin and learning about Lie groups and their classification took care of that 1960 summer. In September, I had drawn up the root-diagrams of A_2 (=SU(3)), B_2(=SO(5)), C_2(=Sp(4)), D_2(=SO(4)), G_2(one of 5 «exceptional» groups). I had a personal preference for the latter — its root-diagram had turned out to be a Star of David! However, I did not like some of its transitions. A_2 appeared to fit perfectly. The mesons would fit in an octet, with a prediction for an eighth 0^- particle (the η^0). I tried representations for the baryons — and the octet assignment seemed so obvious! The Ξ would then have $J=1/2$, rather than the 3/2 predicted in the Sakata model as I later learned. The Δ would fit a **10** or a **27**.

As to gauging, SU(3) reproduced just those vector mesons Sakurai had found essential, in order to explain phenomenology. Even his estimate of the relative strength (a factor 3) between some of the couplings was coming out automatically from the algebra.

There were problems. I could not guarantee $|\Delta I| = 1/2$ dominance in the weak Hamiltonian, without additional constraints. I could not reproduce single-π emission in a Yang-Mills picture. I worried less about the octet choice for the baryons implying that nucleons were already higher structures and not the basic objects. It went against the current belief at the time, but I felt sure of the assignment.

5. Publication

I wrote up the entire solution. Sometime in October, the new academic year had started. Salam returned in November from the 1960 Rochester Conference and some workshop where he had been working with J.C. Ward. I showed my finds to Salam. He was favourably impressed (I called my wife to tell her it had passed that test... She had encouraged me to move to science, and had willingly agreed to a loss in our living standards — from diplomat to research student. She ought to know the sacrifice had not been in vain).

Salam had just encountered SU(3) and had been impressed by it. He gave me preprints[7] by the Nagoya group, dealing with the Sakata model. Ohnuki had talked about it at Rochester. Salam had then considered gauging the group, which would reproduce the same vector meson multiplet that I had found.

Salam told me not to bother about those two points I could not get, and to publish the theory. I should prepare it for joint publication. He would add in his idea about gauging the Sakata model.

I was new and thought that if the Nagoya group had already introduced SU(3), I ought to stick to their basis and notation — which was awful (they had $1 \pm i$ as matrix elements...). I cut out my rather simple and didactic presentation of the algebra and rewrote my matrices directly, using the Nagoya basis. Luckily, I left in the 3×3 matrices with the particles as matrix-elements, a useful presentation of the model. I gave my draft to Salam (see Appendix I).

Several weeks went by, and I was becoming impatient. I inquired with Salam, and he returned the paper to me, telling me to go ahead and publish on my own. He had been asked to write a general review, and would include the SU(3)-gauged Sakata model as a note in it.

There was a long typing queue. After some days I went to the Embassy and asked my former secretary to type the article, which I then sent for publication[18] to *Nuclear Physics* («received 13th February 1961»).

I went on to work on the symmetry breaking[19] and on the way electromagnetic and weak interactions[20] would arise as additional gauges (my two next papers in *Nuclear Physics*). One morning in March, I found Salam waiting in the office for me, with a brochure in his hand. «Look at that! Gell-Mann is suggesting your model!» It was Caltech report CTSL-20, now reproduced in our book *The Eightfold Way,* carrying the same title as that report («First issuing of internal document, 20 January 1961»). Indeed, Gell-Mann was introducing precisely the same model, with similar arguments. Salam remarked that reading Gell-Mann's presentation, he had now become more impressed with the advantages and importance of the model. He was also sorry he had not yet published his Sakata gauge idea, with the same vector mesons.

Just then, my paper was returned by *Nuclear Physics,* with an angry letter to Salam from L. Rosenfeld, the editor. My Embassy secretary had never typed a manuscript before, and she had used single-spacing. The type-editor had worked very hard to put in his symbols and had finally given up.

I had the paper retyped and sent it back. At Salam's suggestion I added two remarks, noting Gell-Mann's recently arrived report and mentioning Salam's Sakata-gauge. Salam did write that up immediately, and sent it to *Nuovo Cimento.* I also sent a copy of my original to Gell-Mann. Since I had done the typing on my own, the preprint had not been sent out.

Some time later, Gell-Mann sent in a new version (issued 15 March 1961). This was also the version he submitted to the *Physical Review* for publication («received 27 March 1961»).[21,22]

As a result of some wrong preliminary experimental reports from Berkeley at the La Jolla Conference (June 1961), Gell-Mann soon withdrew his paper[22]. In these reports, the Σ-Λ relative parity appeared to be odd. Lorentz invariance would now require Σ and Λ to be assigned to two different representations. This fitted the Sakata model and was in contradiction to the octet. In our version, Σ and Λ were indistinguishable in the symmetry limit. When Gell-Mann resubmitted to the *Physical Review* a completely rewritten version of his paper in September 1961, the situation was still unclear, and he chose to present the octet and Sakata models as two equally plausible open possibilities. (By 1962, the relative parity was determined to be even, validating the octet assignments.)

18 NE'EMAN 1971a.
19 NE'EMAN 1961a.
20 NE'EMAN 1962.
21 GELL-MANN 1961.
22 GELL-MANN 1962a.

Salam had also attended that La Jolla meeting. In fact, he suggested in May that we work out a joint presentation of the representation structure for SU(3) in the two models. I worked on that, using a different notation which I found to be more transparent than that of the Nagoya group. I learned it in a paper by J. Wess[23] which was pointed out to me by Betty Hoffman, a newly arrived student who had been studying the Sakata model. From Wess' list I could see that if all particles should be reachable from octets, only representations whose Young tableaux were made of $3n$ boxes, n an integer, were relevant. «Eccentricity» was my appellation for that index Wess was using in his notation. Biedenharn replaced it by «triality» in 1963, a term that has since stuck.

I gave my draft of that new paper to Salam. With the Σ-Λ parity in doubt, he refrained from presenting it at La Jolla. It was never published as such, except for its inclusion in my Ph. D. thesis[24].

I left for Israel early in August 1961, returning to England in March 1962 to defend my thesis.

Before my return to Israel, favourable experimental evidence started to accumulate. The Alvarez group was discovering the entire vector-meson multiplet[25]. They were in contact with Gell-Mann and were aware of my work. Maglic passed through London. I was not at the college that day and found a note and preprints from him, with the ω^0 and ρ results. I heard the bad news about the Σ-Λ parity and decided it would go away.

At Johns Hopkins, Aihood Pevsner and his group found[26] the missing eighth 0^- meson, the η. Driving from London to Israel, we passed near Aix-en-Provence. A conference was being held there. I did not attend — no invitation— but the model was mentioned. By the time I arrived in Israel, the experimentalists (the only particle physicists in Israel at the time) had heard about the octet and its author.

6. SU(3): Gauge theory or «accident»?

One last point about «Derivation of Strong Interactions from a Gauge Invariance» — this is the title[18] I picked for my paper (see Appendix I). Notice the emphasis on a gauge theory. Gell-Mann's had a similar emphasis[21]. Within a year this had gone out of fashion. With the anti-field-theory drive from Berkeley, one had to apologize for using a Lagrangian. Even the fact that *the gauge predictions worked* in that *the 1^- multiplet really existed* and its *couplings were universal* did not matter. One even had to find an excuse for that fact. Gell-Mann's September paper contained just such an excuse[22]. He proved, using dispersion relations, that once a ρ-meson existed, with the quantum numbers of an isovector photon, it would have approximately universal couplings just because of the photon's universality. The Goldberger-Treiman rela-

23 WESS 1960.
24 NE'EMAN 1961b.
25 ALVAREZ 1972.
26 PEVSNER *et al.* 1961.

tion could be similarly derived from such arguments with respect to the weak axial current's divergence.

In 1963, at Gell-Mann's invitation, I went to Caltech where I stayed for two years. I certainly learned a lot of physics from Gell-Mann, Feynman and others there, but by the end of 1963 I was also refraining from writing Lagrangians. In the same defensive move, Gell-Mann had introduced in his 1961 paper the beautiful idea of current algebra. This was a way of having the results of a Lagrangian, without ever introducing one. It gave the hadron current an independent *de facto* definition, through its appearance in weak, electro-magnetic or gravitational interactions, all those interactions where one could claim that a Born term in a perturbation series was still meaningful. This answered the Berkeley critics and in fact induced a lot of excellent tests of SU(3) and of its chiral extension, using dispersion relations. The axial unitary symmetry current yielded all those beautiful soft-pion theorems (using the Nambu-Goldstone massless meson approach), just as the vector currents were yielding theorems about vector-meson couplings. Then in 1967, Weinberg showed how to redo it all in Lagrangian field theory language[27]. It had to be a «phenomenological» Lagrangian since pions and rho mesons were no more to be thought of as fundamental fields.

When the gauge theories proudly re-entered physics in 1971, these Lagrangians were demoted in «current thinking». One now had the «truly» fundamental Lagrangians, in terms of quarks, leptons, color, the electroweak group and GUTs. Weinberg himself consigned simple and chiral SU(3) to the obscurity of an «accidental» symmetry[28].

In recent years, since the discovery of the sequential nature of quarks and leptons[29], one is again looking for a new «fundamental» level («preons») underneath. Suddenly, color and the weak group are also «accidents» like SU(3). 't Hooft[30] has now suggested that «fundamental» is a relative term, and that some fields may be fundamental in one energy region and become composites when we describe the next higher region. The SU(3) vector mesons are being reinstated as true gauge mesons, in the lower energy world. This is still very speculative, of course.

One more remark on this phase. Between 1955 and 1971 the «consensus» in our field *erred* badly. In the USA, especially in the West, Relativistic Quantum Field Theory was considered as «plain wrong» and useless. It was only because Holland and the USSR were outside the direct influence of that consensus that work continued on field theory and the theory finally won. Beware of the consensus!

7. Triplets (quarks)

By the end of May 1961 in London, I was thinking along two lines radiating out of SU(3) and the octet choice.

27 Weinberg 1967.
28 Weinberg 1972.
29 Ne'eman 1979.
30 Hooft 1979 and 1980.

First — could SU(3) have anything to do with space-time and the Poincaré group? Could the «3» be due to the dimensionality of space? I returned to this question in Israel in 1963 and suggested, with Joe and Nathan Rosen, a Kaluza-Klein-like marriage, between the internal and external symmetries[31]. Several people attempted at the same time to connect the groups by identifying SU(3) with a subgroup of the complexified Poincaré group. The trouble was that spin and isospin would not commute, and rotating a proton might turn it into a neutron! Our model did not have such problems — I had already noticed that danger in London. In fact, ours contained the physics of spin and unitary spin independence, later known as SU(6) — but I did not notice it. Some years later, the same embedding was used for the spinning string of Ramond and Neveu-Schwarz, a ten-dimensional theory.

Secondly, I wondered what was more fundamental than the octet, and where were the triplets, etc. In reading Wess[23], I had realized I could make an octet out of 3 triplets. One was used to thinking of ($\nu e \mu$) as a Sakata-like triplet at that time. Okun, Thirring and Gamba-Marshak and Okubo had noticed that parallelism — which had to be abandoned when the octet replaced the Sakata model. Could a baryon be made of three leptons? I asked Salam. He laughed, but added that nothing is too stupid, provided it works...

The question kept bothering me. Leptons they probably were not, but why not a Sakata-like fundamental field? In Israel, the first person who came to work with me was Haim Goldberg. He knew group theory and had written a thesis in nuclear physics under Racah. I suggested to him that we develop that idea of a triplet model, existing underneath the observed octets. The results appeared in preprint form (Israel AEC report IA-725) in February 1962[32], and were sent for publication in *Nuovo Cimento* («received 22 February 1962») (See Appendix II).

We explained that rather than regarding the baryon octet as fundamental, it should be viewed as composite, *taking as a fundamental mathematical model a Sakata-like triplet with a baryonic charge $B=1/3$*. In Wess' notations $^{e}a(r, r')$, r is the order of the contravariant indices, r' that of the covariant ones. The e stands for a U(3) quantum number. In fact, for our model, baryon number would be given by

$$3B = r - r' + 3e$$

and the fundamental triplet (the future quark...) is $^{0}a(1,0)$, i.e. one contravariant index; the antitriplet is $^{0}a(0,1)$. We introduced an alternative notation (h_1, h_2, h_3) in which

$$B = \frac{1}{3}(h_1 + h_2 + h_3)$$

and $r = h_1 - h_2$, $r' = h_2 - h_3$. In fact, h_i is the number of boxes in the ith row of the

31 ROSEN, ROSEN and NE'EMAN 1964; KURSUNOGLU *et al.* 1964.
32 GOLDBERG and NE'EMAN 1962.

Young tableau of a given representation. The r and r' describe the «highest weight», with eigenvalues

$$I_3 = \frac{1}{2} r, \qquad U_3 = \frac{1}{2} r'$$

(U_3 is measured along the U-spin axis, e.g. from π^0 to K^0, with $U_3(K^0) = 1$, $U_3(K^+) = 1/2$.). Haim Goldberg and I thus explain that the baryon and meson octets are then given by different mathematical representations of U(3): for baryons, (2, 1, 0) — or ${}^1a(1,1)$ in Wess' notation[23], a «three pseudo-Sakātons» octet, corresponding to $3\times 3\times 3$; indeed,

$$\begin{array}{lcll} {}^0a(1,0) \times {}^0a(1,0) & = & {}^0a(2,0) + {}^1a(0,1) & (\text{«diquarks»}) \\ {}^0a(2,0) \times {}^0a(1,0) & = & {}^0a(3,0) + {}^1a(1,1) & (\text{3-quark baryons}) \\ {}^1a(0,1) \times {}^0a(1,0) & = & {}^1a(1,1) + {}^1a(0,0) & (\text{3-quark baryons}) \end{array}$$

in Wess' notation. For mesons, the octet is (1,0, -1) – or ${}^0a(1,1)$ in Wess' notation, a Sakaton-anti-Sakaton system or $3 \times \bar{3}$. Indeed,

$${}^0a(1,0) \times {}^0a(0,1) = {}^0a(1,1) + {}^0a(0,0)$$

However, the physical picture in that paper still appears somewhat ambivalent, because «production» from physical particles is sometimes confused with «composition» in terms of the mathematical triplets (the discussion of the bindings goes back to mesons as produced by combining baryons and antibaryons!). However, in seminars at the Weizmann Institute (May 1962), the physics were described more coherently and consistently. I explained that it is *«as if»* baryons are really *made* of three *fundamental triplets.* The octet was formally replaced by a «mathematical model» based upon a $B = 1/3$ field. H.J. Lipkin[33] (see Appendix II-1) summarizes that stage in his discussion in a comprehensive review of the quark model.

Report IA-725 was well disseminated and generated some interest, but the general view was very skeptical due to the fact that the octet assignment itself (which it was supposed to explain) was still in doubt in many eyes.

In the spring of 1963, M. Gell-Mann, and independently G. Zweig, arrived at the same model. Gell-Mann elaborated the idea during a visit to Columbia University in March 1963, in discussions with R. Serber, publishing it in a paper in *Physics Letters* («received 4 January 1964»). G. Zweig issued two CERN reports (unpublished) that also contained SU(6)! He has recently written up his story in an inspiring paper. Gell-Mann named the fundamental objects «quarks», a name that has stuck. The Gell-Mann paper appeared in the February issue of *Physics Letters,* at about the same time that the results of the Omega-Minus experiment were announced. These results[34] confirmed the decuplet assignment for the Ω-particle, and thereby also the

33 LIPKIN 1973. See discussion on p. 180.
34 BARNES 1964.

octet assignment for spin 1/2 baryons. The spectacular confirmation created enthusiasm for SU(3), and an interest in its structural foundations. Attention turned to the quark papers. The hunt for the quark was immediately launched with the highest priority.

The present view is that quarks are *confined,* so that the quark field perhaps fits somewhat in between our view[32, 33] of an abstract field and the more «straight» but daring view of a fractional-charge *free* particle, a possibility raised — cautiously but explicitly — in the Gell-Mann and Zweig papers[35, 36].

8. Experimental verification

In Israel, I had taken over the scientific management of the Atomic Energy Research Establishments in Rehovoth and Nahal Soreq. I learned a lot of new physics, mostly nuclear but also some of the physics of condensed matter. Amos de Shalit, whom I had come across in the mid-fifties, became a close friend. He was excited at the fact that I was now «importing» particle physics in Israel, and advised some of the best students of that year to work with me. I created a group at Soreq, with H. Harari, Y. Dothan, S. Nussinov. From the Technion, where I gave a course in 1961/62, came D. Horn and J. Rosen. I taught in Jerusalem in 1962/63, with Y. Frishman in that class. I collaborated with the experimentalists — Y. Eisenberg, D. Kessler, G. Yekuteili. H.J. Lipkin was applying Elliot's SU(3) in nuclear physics, together with C. Levinson, and S. Meshkov of the NBS. I gave a seminar and they transferred their skills, experience and drive to particle physics[37]. At first, they dealt only with the Sakata model. One of their papers even contained a remark about how much more complicated the octet computations were. Yet they got there too, in due time. In March 1962, Salam held a small conference at Imperial College. Lipkin attended (I was back in Israel) and as a result even published a joint paper with Salam and Munir Rashid. I never did...

The conference papers contained comparisons between the predictions of the Sakata model and those of the octet. The first answers came when the results of $p\bar{p}$ annihilation into $\pi\pi$ or $K\bar{K}$ became known. That paper of Lipkin's group together with Salam and Munir showed that an observed decay into $K^0\bar{K}^0$ was forbidden in the Sakata model. With my team, we showed that it did fit the octet[38] (see Appendix III). I wanted to present this paper at the CERN «Rochester» conference (which I attended in July 1962), but I encountered difficulties with the session organizers. They knew SU(3) only superficially and did not know about d_{ijk}-type couplings. They thus dismissed our work as erroneous. By the time I had explained it to them, the conference was over...

It was at that conference that I first met Shulamit and Gershon Goldhaber. When I learned that they could not find an N-K resonance, I realized that I could now assign

35 GELL-MANN 1964.
36 ZWEIG 1964; See also ZWEIG 1965.
37 LEVINSON, LIPKIN and MESHKOV 1962a, b and 1963.
38 DOTHAN, GOLDBERG, HARARI and NE'EMAN 1962.

the Δ and other $J = 3/2$ resonances to a **10**. The **27** contained positive strangeness states fitting N-K, the **10** did not. Moreover, one could then make a definite prediction for the last missing state in that **10** (the Σ^*, and Ξ^* had just been reported at the meeting), including its mass, spin, etc. There was no such state in the relevant Sakata assignments or in other models, so the test was a really good one. Gershon wanted the story in writing, and I gave it to him. On the next day, the *rapporteur* was summarizing the state of symmetries. I raised my hand to make a remark about that **10**, but the chairman said «Professor Gell-Mann». Murray had also lifted his hand! He went to the blackboard and explained it all, including (as usual) a name for that missing particle «Omega-Minus»[39]. Yehuda Eisenberg was sitting next to me. Returning to his seat, Gell-Mann passed near us, recognized Yehuda and exclaimed «this might also fit your cosmic-rays event!» Eisenberg had indeed probably observed an Ω^- back in 1954[40]. Alvarez, who was standing nearby, confirmed the data. Gell-Mann then noticed the label with my name on my jacket, and we went to have lunch together. It was the start of an exciting friendship.

Evidence favouring SU(3) continued to pour in. Gideon Alexander, belonging to the Israel AEC was spending 1961/64 at Berkeley with Alvarez (he had even been involved in the odd Σ-Λ parity story...). I had asked Alvarez for a report on Gideon's performance, for our appointments committee. In his reply in the Spring of 1963, Alvarez added a few sentences about new successes of the model — new multiplets and good fits. These were the Glashow-Rosenfeld assignments[41].

The spin of the Ξ was measured in 1964 at Berkeley[42]. The result was 1/2 with a very high probability, and almost sure not to be 3/2. Later experiments confirmed 1/2.

The model's successes attracted attention — and new criticism. Two groups were fighting back. On the one hand, partisans of the Sakata model and of G_2 (there was a G_2 revival in mid-1963!); on the other hand, real skeptics. Skepticism was based on a simple idea: SU(3) was an *approximate* symmetry, with successful predictions based upon the assumption that the symmetry was linearly broken in the 8th direction ($I=Y=0$). This gave the Gell-Mann Okubo mass formula, for instance, including the prediction of equal spacings of the masses in the **10**. Since the breaking itself was a strong-interaction effect, how could a Born-term produce a correct description? In fact, how come the entire multiplet structure was not messed up by the breaking? Sydney Drell visited Israel in 1962 and voiced these thoughts. They were only partly answered by current algebra and dispersion arguments.

A series of eight arguments in favour of odd Σ-Λ parity (again that red herring) had appeared in a humorously written paper by Nambu and Sakurai in 1961[43]. In a later paper[44], Sakurai was already stating that all eight arguments were dead. The last

39 Gell-Mann 1962b.
40 Eisenberg 1954.
41 Glashow and Rosenfeld 1963.
42 Carmony *et al.* 1964.
43 Nambu and Sakurai 1961.
44 Sakurai 1961.

counterattack (before capitulation...) against unitary symmetry was led by R.J. Oakes and C.N. Yang[45]. They went through a calculation purporting to show why even if there were an Ω^- at the right mass, it could still have nothing to do with our model. The paper received a remarkable reaction. Ross, Eden and Taylor in Cambridge (UK), Dalitz and Rajasekharan, etc. — I counted 11 answers within 2-3 months — all of them showing holes in our critics' argumentation and reaching opposite conclusions, sometimes even convincingly[46].

Then, in February of 1964, there came the Ω^- discovery[34] (see Appendix III). It settled the issue. I think that even Yang was convinced.

I had attended a conference at Coral Gables one week earlier, and listened to Maurice Goldhaber (the Brookhaven Director, and one of the last high energy physicists who qualifies as both a theoretician and an experimentalist) tell me that there was no Ω^- in 50,000 pictures. (They finally found it, after looking at 100,000 pictures...)

I was at Caltech at the time of the discovery. Nick Samios and the other experimentalists (with one Israeli, Yona Oren, later a member of my department in Tel Aviv) forgot to let me know about the result (although Maurice Goldhaber did tell us about it after we called to ask, having heard rumours). I sent in a request for a reprint, and received a nice set of bubble chamber pictures, with the note «Please excuse the over-sight, but you knew it existed before we did!».

During 1964, Feynman several times played a joke. He would come in, seemingly excited and announce: «Have you heard the news? The Omega Minus has a spin of one half!» Indeed, it was only in 1978 that its spin was finally measured and found to be 3/2 (see Appendix III).

9. Further applications

During 3-4 years it appeared as if the applications would be infinite. First, strong interaction branching ratios, couplings, mass formulae. Secondly, electromagnetic and weak matrix elements, especially after Cabibbo had identified the direction of the weak current in octet space (Gell-Mann had suggested a «Cabibbo angle» some years earlier, before SU(3)[47]).

Unitary symmetry turned out to have two regimes. At very low energy, it was carried by two quasi-conserved currents. One had an $SU(3)_L \times SU(3)_R$ group, chiral unitary symmetry. This is just as angular momentum is conserved separately in its spin and orbital parts, in non-relativistic situations (effectively, an $SU(2) \times SU(2)$). Once we go up in energy, S-L transitions occur and only $J=S+L$ is conserved, whereas $S - L$ conservation is lost. For chiral symmetry, again only the sum of the two algebras is conserved. This is unitary spin itself. The axial part's divergence coincides with the pion field.

45 Oakes and Yang 1963.
46 See, for example, Eden and Taylor 1963.
47 See Gell-Mann and Ne'eman 1964a, p. 210.

With Gell-Mann, we proved in 1964 that SU(3) × SU(3) was really the only possibility[48]. The applications, using current algebra methods, were very impressive[49,50].

Beyond all that, the «independent-particle» quark model gave astonishing results[51]. I was especially impressed by the predictions relating to the ratio between proton and neutron magnetic moments

$$\frac{\mu_p}{\mu_n} = -3/2$$

and the predicted ratio between the Nπ and NN total cross sections in the asymptotic limit

$$\frac{\sigma(\pi N)}{\sigma(NN)} \rightarrow 2/3$$

both predictions holding to within 1-2 %. With Cabibbo and Horwitz, I reproduced many of these results algebraically in 1966. Later in 1973/74, I could give a CVC-type derivation, using local current algebra[11]. I was doing my best to find derivations that would not depend on an understanding of the dynamics of quarks, since these appeared incomprehensible. Quarks had to be very weakly bound — to understand SU(6) type results[51]: but they also had to be very tightly bound, since they would not come out even at very high energies!

Until 1968, I tended to regard quarks as a basic field that on the other hand had no *direct* manifestation. The SLAC «deep-inelastic» electron scattering experiments on nucleons[52] revealed scaling, indicating the actual existence of point-charges within the extended nucleon. With more experiments, those «partons» looked more and more like quarks. Quarks are thus observable, provided you «look» inside the nucleon. However, it was only after the renormalization of Yang-Mills' theory, that G. 't Hooft himself, H.D. Politzer, and D.J. Gross and F. Wilczek discovered asymptotic freedom. This was just the kind of dynamics we now required: actual quarks, bound by a spring-like force roughly proportional to distance. Close quarks are thus weakly bound, whereas it becomes very difficult to pull them apart, beyond a certain distance. Can they even be completely confined? This is what current opinion nowadays believes, but it has yet to be finally proven. After 8 years of tremendous efforts, this has still not been done. Perhaps they are confined, but the QCD mechanism is not the complete story?

48 GELL-MANN and NE'EMAN 1964b.
49 See, for example, ADLER and DASHEN 1968.
50 See, for example, NE'EMAN 1967.
51 See, for example, KOKKEDEE 1969.
52 BLOOM *et al.* 1969; BREIDENBACH *et al.* 1969.

Between 1967 and 1970, our extended group in Israel was very successful in exploiting dispersion methods and learning more about the structure of hadrons[53]. First, Horn (with Dolan and Schmidt) invented FESR (Finite Energy Sum Rules). These were plausible bootstrap-type equations, Regge trajectories providing exchange forces resulting in making resonances, again sitting on Regge trajectories. Harari (and Freund) improved by accounting separately for elastic scattering and for the Pomeranchuk trajectory. Rosner and Harari fitted in quarks, and those «duality diagrams» looked convincingly correct. Veneziano found a «Standard solution» to the FESR equations, based on Gauss' beta function. Susskind (and Nambu) showed that this was equivalent to a string, with the quarks as endpoints.

This string structure is equivalent to a part of the Yang-Mills force. Is there perhaps a residual element, to be *added* to QCD? I am not sure we shall know the answers before we have to go on to «preons». After all, particle physics developed while nuclear structure was not yet understood. Maybe the answers here will even come from that still mysterious preon level? High energy experiments have revealed a repetitive sequential structure (an extension of the muon riddle), an indication — in my view — of such a further level[54]. One of these days, the pattern will emerge, and we shall embark on yet another pattern-structure-dynamics cycle...

53 See, for example, GOTSMAN 1971.
54 NE'EMAN 1979.

Appendix I

Fragmentary reproduction from *Nuclear Physics,* with the kind permission of North-Holland Physics Publishing.

8.B

Nuclear Physics **26** (1961) 222—229; © *North-Holland Publishing Co., Amsterdam*

DERIVATION OF STRONG INTERACTIONS FROM A GAUGE INVARIANCE

Y. NE'EMAN

Department of Physics, Imperial College, London

Received 13 February 1961

Abstract: A representation for the baryons and bosons is suggested, based on the Lie algebra of the 3-dimensional traceless matrices. This enables us to generate the strong interactions from a gauge invariance principle, involving 8 vector bosons. Some connections with the electromagnetic and weak interactions are further discussed.

1. Introduction

Following Yang and Mills [1]), two new theories deriving the strong interactions from a gauge invariance principle have been published lately, by Sakurai [2]) and by Salam and Ward [3]). Sakurai's treatment is based on three separate gauges — isospin, hypercharge and baryonic charge — unrelated from the point of view of group-theory; Salam and Ward postulate one unified gauge, an 8-dimensional rotation gauge, combining isospin and hypercharge through Tiomno's [4]) representation.

The following treatment is an attempt to formulate a unified gauge, while reducing the number of vector bosons. It does, indeed, generate a set of 8 mediating fields, seven of which are similar to the above seven, the eighth is rather like Sakurai's B_ν singlet.

To minimise the number of parameters of the gauge, and thus the number of vector bosons it will generate, we have adopted the following method: we abandoned the usual procedure of describing fields as vector components in a Euclidean isospace, and replace it by a matrix-algebra manifold. Fields still form vectorial sets only in the space of the group operators themselves, invariance of the Lagrangians being achieved by taking the traces of product matrices.

We have also abandoned rotations and use a group first investigated by Ikeda, Ogawa and Ohnuki [6]) in connection with the construction of bound states in the Sakata model. Our present use of this group is in an entirely different context, as our assumptions with regard to the representation of the fermions do not follow the prescriptions of the model.

using an octet with SU(3) rather than SU(8)

2. Matrix Formalism

We use an 8-dimensional linear vector space P spanned by the semisimple Lie algebra of the 3×3 matrices X_{ij} of ref. [6]). We have excluded the identity transformation and use as basis the 8 linearly independent $\mathbf{u}^i \in \mathbf{U}$ given by the following formulae:

$$\mathbf{U}\begin{cases} \mathbf{u}^1 = \frac{1}{2}\sqrt{2}(X_{(31)}-iX_{[31]}), & \mathbf{u}^4 = \frac{1}{2}\sqrt{2}(X_{(31)}+iX_{[31]}), \\ \mathbf{u}^2 = \frac{1}{2}\sqrt{2}(X_{(23)}-iX_{[23]}), & \mathbf{u}^3 = \frac{1}{2}\sqrt{2}(X_{(23)}+iX_{[23]}), \\ \mathbf{u}^5 = \frac{1}{2}\sqrt{2}(X_{(12)}+iX_{[12]}), & \mathbf{u}^6 = \frac{1}{2}\sqrt{2}(X_{(12)}-iX_{[12]}), \\ \mathbf{u}^7 = \frac{1}{2}(X_{11}-X_{22}), & \mathbf{u}^8 = \frac{1}{6}\sqrt{3}(X_{11}+X_{22}-2X_{33}), \end{cases} \tag{1}$$

using the Nagoya matrices

$$X_{ij}^{\alpha\beta} = \tfrac{1}{2}\delta_{i\alpha}\delta_{j\beta}(1-i)+\tfrac{1}{2}\delta_{i\beta}\delta_{j\alpha}(1+i),$$

$$X_{(ij)} = \tfrac{1}{2}(X_{ij}+X_{ji}), \qquad X_{[ij]} = \tfrac{1}{2}(X_{ij}-X_{ji}),$$

the indices α and β denoting the matrix elements. The X_{ij} are hermitian, whereas the basis matrices $\mathbf{u}_i$ are not, with the exception of $\mathbf{u}^7$ and $\mathbf{u}^8$, both diagonal.

When using our algebra for unitary transformations, we shall take the hermitian set V as a basis for the infinitesimal operators,

$$\mathbf{V}\begin{cases} \mathbf{v}^{14} = \frac{1}{2}\sqrt{2}(\mathbf{u}^1+\mathbf{u}^4), & \mathbf{v}^{41} = \frac{1}{2}i\sqrt{2}(\mathbf{u}^1-\mathbf{u}^4), \\ \mathbf{v}^{23} = \frac{1}{2}\sqrt{2}(\mathbf{u}^2+\mathbf{u}^3), & \mathbf{v}^{32} = -\frac{1}{2}i\sqrt{2}(\mathbf{u}^2-\mathbf{u}^3), \\ \mathbf{v}^{56} = \frac{1}{2}\sqrt{2}(\mathbf{u}^5+\mathbf{u}^6), & \mathbf{v}^{65} = -\frac{1}{2}i\sqrt{2}(\mathbf{u}^5-\mathbf{u}^6), \\ \mathbf{v}^7 = \mathbf{u}^7, & \mathbf{v}^8 = \mathbf{u}^8, \end{cases} \tag{9}$$

these are the same as the λ_i except for normalization (here to 1)

so that

$$\sum_{k=1}^{8} A_V{}^k B_V{}^k = \sum_{i=1}^{8} A_U{}^i B_{U,i}, \tag{10}$$

i.e. the scalar product (8) is Euclidean in the V system.

Under a unitary transformation $E^{(m_V)} = \exp(i\varepsilon^{m_V}\mathbf{v}^{m_V})$ (m_V is the single or double index in V), the component $A^k\mathbf{u}^k$ transforms like

$$\sum_{l=1}^{8} \delta_{(m_V)} A^l\mathbf{u}^l = i\varepsilon^{m_V} A^k[\mathbf{v}^{m_V}, \mathbf{u}^k] = i\varepsilon^{m_V} A^k \sum_{l=1}^{8} f^l_{m_V,k}\mathbf{u}^l$$

and for

$$E = \exp(i\sum_{m_V} \varepsilon^{m_V}\mathbf{v}^{m_V}) \tag{11}$$

we get variations

$$\delta A^l = i\sum_{m_V} \varepsilon^{m_V} \sum_{k=1}^{8} f^l_{m_V,k} A^k. \tag{12}$$

The $f^l_{m_V,k}$ define an 8×8 representation of our algebra in P space,

$$C^{l,k}_{m_V} = f^l_{m_V,k}, \tag{13}$$

so that (12) becomes in P

$$\delta A^l = i\sum_{m_V=1}^{8} \varepsilon^{m_V} \sum_{k=1}^{8} C^{l,k}_{m_V} A^k,$$

3. Fields and Interactions

We define the quantum operators

$$\mathbf{I}(C_5, C_6, C_7), \qquad I_z = C^7, \qquad Q = \tfrac{2}{3}\sqrt{3}\, C^{8'}, \qquad Y = \tfrac{2}{3}\sqrt{3}\, C^8, \tag{15}$$

and write the fields as vectors in P space

$$\begin{array}{ll} \psi(\mathrm{p}, \mathrm{n}, \Xi^0, \Xi^-, \Sigma^+, \Sigma^-, \Sigma^0, \Lambda), & \bar{\psi}(\overline{\Xi^-}, \overline{\Xi^0}, \bar{\mathrm{n}}, \bar{\mathrm{p}}, \overline{\Sigma^-}, \overline{\Sigma^+}, \overline{\Sigma^0}, \bar{\Lambda}) \\ \varphi(\mathrm{K}^+, \mathrm{K}^0, \overline{\mathrm{K}^0}, \overline{\mathrm{K}}^-, \pi^+, \pi^-, \pi^0, \pi^{0'}), & \bar{\varphi} = \varphi, \end{array} \tag{16}$$

or in matrix form

$$\psi = \tfrac{1}{2}\sqrt{2}\left|\begin{array}{c|c|c} \tfrac{1}{2}\sqrt{2}\,\Sigma^0 + \tfrac{1}{6}\sqrt{6}\,\Lambda & \Sigma^+ & \mathrm{p} \\ \hline \Sigma^- & -\tfrac{1}{2}\sqrt{2}\,\Sigma^0 + \tfrac{1}{6}\sqrt{6}\,\Lambda & \mathrm{n} \\ \hline \Xi^- & \Xi^0 & -\sqrt{\tfrac{2}{3}}\,\Lambda \end{array}\right|,$$

$$\bar{\psi} = \tfrac{1}{2}\sqrt{2}\left|\begin{array}{c|c|c} \tfrac{1}{2}\sqrt{2}\,\overline{\Sigma^0} + \tfrac{1}{6}\sqrt{6}\,\bar{\Lambda} & \overline{\Sigma^-} & \overline{\Xi^-} \\ \hline \overline{\Sigma^+} & -\tfrac{1}{2}\sqrt{2}\,\overline{\Sigma^0} + \tfrac{1}{6}\sqrt{6}\,\bar{\Lambda} & \overline{\Xi^0} \\ \hline \bar{\mathrm{p}} & \bar{\mathrm{n}} & -\sqrt{\tfrac{2}{3}}\,\bar{\Lambda} \end{array}\right|, \tag{17}$$

$$\varphi = \tfrac{1}{2}\sqrt{2}\left|\begin{array}{c|c|c} \tfrac{1}{2}\sqrt{2}\,\pi^0 + \tfrac{1}{6}\sqrt{6}\,\pi^{0'} & \pi^+ & \mathrm{K}^+ \\ \hline \pi^- & -\tfrac{1}{2}\sqrt{2}\,\pi^0 + \tfrac{1}{6}\sqrt{6}\,\pi^{0'} & \mathrm{K}^0 \\ \hline \overline{\mathrm{K}^-} & \overline{\mathrm{K}^0} & -\sqrt{\tfrac{2}{3}}\,\pi^{0'} \end{array}\right|,$$

The free field Lagrangians are

$$\mathscr{L}_\psi{}^0 = -\bar{\psi}\cdot(\gamma^\mu\partial_\mu + m_\psi)\psi, \qquad \mathscr{L}_\varphi{}^0 = -\tfrac{1}{2}(\partial^\mu\varphi\cdot\partial_\mu\varphi + m_\varphi{}^2\varphi\cdot\varphi). \tag{18}$$

We postulate the invariance of these Lagrangians under the unitary gauge transformation

$$E = \exp(i \sum_{m\nu} \varepsilon^{m\nu}(x) C^{m\nu}) \tag{19}$$

The total Lagrangian becomes

$$\mathscr{L}_{\text{total}} = \mathscr{L}_\psi{}^0 + \mathscr{L}_\varphi{}^0 + \mathscr{L}_\psi{}^{\mathscr{I}} + \mathscr{L}_\varphi{}^{\mathscr{I}} + \mathscr{L}_B{}^0, \tag{20}$$

$$\mathscr{L}_\psi{}^{\mathscr{I}} = -\sum_{i=1}^{8}\sum_{n=1}^{8} \bar{\psi}^n \gamma^\mu C_i \psi_n B_\mu{}^i, \tag{21}$$

$$\mathscr{L}_\varphi{}^{\mathscr{I}} = -\sum_{\substack{i=1\\ j=1}}^{8}\sum_{n=1}^{8} (\partial_\mu\bar{\varphi}_n + C_i\bar{\varphi}_n B_\mu{}^i)C_j\varphi^n B^{j\mu}. \tag{22}$$

The $B_\mu{}^i$ is a set of 8 vector bosons, with the following isobaric and strangeness qualities:

$$\begin{array}{llll} B_\mu{}^1 \to \mathrm{K}^+, & B_\mu{}^2 \to \mathrm{K}^0, & B_\mu{}^3 \to \overline{\mathrm{K}^0}, & B_\mu{}^4 \to \overline{\mathrm{K}}^-, \\ B_\mu{}^5 \to \pi^+, & B_\mu{}^6 \to \pi^-, & B_\mu{}^7 \to \pi^0, & B_\mu{}^8 \to \pi^{0'}. \end{array} \tag{23}$$

$$\mathscr{L}_B{}^0 = -\tfrac{1}{4}(\mathbf{F}_{\mu\nu}\cdot\mathbf{F}^{\mu\nu}), \tag{24}$$

$$\mathbf{F}_{\mu\nu} = \mathbf{H}_{\mu\nu}+\mathbf{G}_{\mu\nu}, \tag{24'}$$

$$\mathbf{H}_{\mu\nu} = \partial_\mu \mathbf{B}_\nu - \partial_\nu \mathbf{B}_\mu, \tag{24''}$$

$$\left.\begin{aligned} G^Z_{\mu\nu} &= \tfrac{1}{2}\{Z_\mu(\mathbf{V}_\nu\cdot\boldsymbol{\tau}+\sqrt{3}\,X_\nu{}^0)-(\mathbf{V}_\mu\cdot\boldsymbol{\tau}+\sqrt{3}\,X_\mu{}^0)^\sim Z_\nu\},\\ G^{\bar Z}_{\mu\nu} &= -\tfrac{1}{2}\{\bar Z_\mu(\mathbf{V}_\nu\cdot\boldsymbol{\tau}+\sqrt{3}\,X_\nu{}^0)^\sim-(\mathbf{V}_\mu\cdot\boldsymbol{\tau}+\sqrt{3}\,X_\mu{}^0)\bar Z_\nu\},\\ G^V_{\mu\nu} &= i\mathbf{V}_\mu\wedge\mathbf{V}_\nu+\tfrac{1}{2}\{\bar Z_\mu\boldsymbol{\tau}Z_\nu-Z_\mu\boldsymbol{\tau}\bar Z_\nu\},\\ G^X_{\mu\nu} &= \tfrac{1}{2}\sqrt{3}\,\{\bar Z_\mu Z_\nu-Z_\mu\bar Z_\nu\}. \end{aligned}\right\} \tag{24'''}$$

$$\mathbf{B}_\mu = \tfrac{1}{2}\sqrt{2}\begin{vmatrix} \tfrac{1}{2}\sqrt{2}\,V_\mu{}^0+\tfrac{1}{6}\sqrt{6}\,X^0 & V_\mu{}^+ & Z_\mu{}^+ \\ V_\mu{}^- & -\tfrac{1}{2}\sqrt{2}\,V_\mu{}^0+\tfrac{1}{6}\sqrt{6}\,X_\mu{}^0 & Z_\mu{}^0 \\ \bar Z_\mu{}^- & \bar Z_\mu{}^0 & -\sqrt{\tfrac{2}{3}}\,X_\mu{}^0 \end{vmatrix}. \tag{25}$$

4. Discussion

In its general features, our Lagrangian reflects a certain similarity with Sakurai's theory [2]). The V_μ is similar to the $B_T{}^\mu$ (isospin-current boson) of the latter, and the $X_\mu{}^0$ is similarly related to its $B_Y{}^\mu$ (the hypercharge-current boson singlet). On the other hand we have no $B_B{}^\mu$ (baryon-current singlet) and do have a Z_μ set which has no place in ref. [2]).

We note that we do get directly from our group structure a ratio between the couplings; for the V and X fields, this is $f_X = f_V\sqrt{3}$, a value that fits Sakurai's phenomenological conclusion (from KN and $\overline{\text{K}}$N at low energies) that $(1/4\pi m_X{}^2)f_X{}^2 \approx (3/4\pi m_V{}^2)f_V{}^2$ if we assume the masses of be similar. Our X^0 field does not interact with the (Σ, Λ) set, and V does not interact with Λ, so that we get a split (N Ξ), Σ, Λ but though the interactions of X and Z with N and Ξ have opposite signs, lacking $B_B{}^\mu$ we cannot repeat here Sakurai's simple interpretation of the origin of the N-Ξ mass split. The arguments explaining the π-N S-wave scattering exist in our gauge. We also note that $\mathscr{L}_B{}^0$ in (24) with its $G^i_{\mu\nu}G_i^{\mu\nu}$ provides us with effective mass terms (in the sense of the mass of $A_\mu{}^\pm$ in a former work [8]) of Salam and Ward) for V_μ, Z_μ and $X_\mu{}^0$ (from $G^Z_{\mu\nu}$ and $G^{\bar Z}_{\mu\nu}$), whereas ref. [2]) lacks such terms for the singlets B_Y and B_B. From $\mathscr{L}_\varphi{}^g$ we see that provided the masses are sufficient, there exist fast decays

$Z \to \text{K}+\pi,$ — *now K**

$V \to 2\pi$ or $V \to \text{K}+\overline{\text{K}}$ (the even G combination), — *now ρ*

$X \to \text{K}+\overline{\text{K}}$ (odd G combination). — *now φ*

Note the possibility that $m_X \approx m_V$.

From the point of view of the Lagrangian formalism, it seems preferable to us to have what is in fact one conservation law for a "charge" that behaves like a 2nd rank tensor in three dimensions, than three separate unrelated conservation laws; this is even more important in view of the necessity to bring in at some further stage the electromagnetic and the weak interactions. In Sakurai's theory, these seem to imply two new independently conserved quantities Q and l, though the relation $Q = I_z + \frac{1}{2}Y$ seems to indicate that the interactions are not wholly independent (and so does the $|\Delta \mathbf{I}| = \frac{1}{2}$ rule of the weak interactions). We think that the "aesthetic' value of Sakurai's theory and the "Urschmiere" approach would be enhanced if there proved to be only one kind of "Urschmiere" instead of five.

I am indebted to Prof. A. Salam for discussions on this problem. In fact, when I presented this paper to him, he showed me a study he had done on the unitary theory of the Sakata model, treated as a gauge, and thus producing a similar set of vector bosons [9]).

Shortly after the present paper was written, a further version, utilizing the 8-representation for baryons, as in this paper, reached us in a preprint by Prof. M. Gell Mann.

added in March

References

1) C. N. Yang and H. Mills, Phys. Rev. **96** (1954) 192
2) J. J. Sakurai, Ann. of Phys. **11** (1960) 1
3) A. Salam and J. C. Ward, Nuovo Cim. **19** (1961) 167
4) J. Tiomno, Nuovo Cim. **6** (1957) 1
5) J. Schwinger, Ann. of Phys. **2** (1957) 407
6) M. Ikeda, S. Ogawa, Y. Ohnuki, Progr. Theor. Phys. **22** (1959) 5, 719
7) R. Utiyama, Phys. Rev. **101** (1956) 1597
8) A. Salam and J. Ward, Nuovo Cim. **11** (1959) 4, 569
9) A. Salam and J. Ward, Nuovo Cim., to be published

Appendix II

Fragmentary reproductions from *Physics Reports* and *Il Nuovo Cimento,* with the kind permissions of North-Holland Physics Publishing and La Società Italiana di Fisica.

PHYSICS REPORTS (Section C of Physics Letters) 8, no. 3 (1973) 173–268. NORTH-HOLLAND PUBLISHING COMPANY

QUARKS FOR PEDESTRIANS

Harry J. LIPKIN
The Weizmann Institute of Science, Rehovot, Israel

Received 29 March 1973

The Sakata model has difficulties with baryon states. To make more baryons we need two sakatons plus an antisakaton. This gives negative parity states of spin $\frac{1}{2}$ or $\frac{3}{2}$. However, the next low-lying baryon states have positive parity and spin $\frac{1}{2}$ like the nucleon. They seem to belong on the same diagram as the Sakata triplet to form an octet which looks like the meson diagram as shown in fig. 1.2c. This led Gell-Mann [1.8] and Ne'eman [1.9] to the octet model of unitary symmetry or the Eightfold Way [1.10]. The particle spectrum is now expressed in terms of the allowed values of isospin and strangeness obtained from a composite model with three basic objects, but the three basic objects are no longer there. There are eight basic baryons with a "mysterious symmetry" governing their interactions, but no good reason for the number three.

Goldberg and Ne'eman [1.11] then pointed out that the octet model was consistent with a composite model constructed from a basic triplet with the same isospin and strangeness quantum numbers as the sakaton, but with baryon number $B = \frac{1}{3}$. The baryon octet was constructed from three triplets. However, equations (1.8) and (1.9) show that particles having third-integral baryon number must also have third-integral electric charge and hypercharge. At that time the eightfold way was generally considered to be rather far fetched and probably wrong. Any suggestion that unitary symmetry was based on the existence of particles with third-integral quantum numbers would not have been considered seriously. Thus the Goldberg–Ne'eman paper presented this triplet as a mathematical device for construction of the representations in which the particles were classified.

Several years later, new experimental data forced everyone to take SU(3) more seriously. The second baryon multiplet was found, with structure shown in fig. 1.2d including the Ω^-, and with spin and parity $\frac{3}{2}^+$. Gell-Mann [1.12] and Zweig [1.13] then proposed the possible existence of the fundamental triplet as a serious possibility and gave it the name of quark.

IL
NUOVO CIMENTO
ORGANO DELLA SOCIETÀ ITALIANA DI FISICA
SOTTO GLI AUSPICI DEL CONSIGLIO NAZIONALE DELLE RICERCHE
E DEL COMITATO NAZIONALE PER L'ENERGIA NUCLEARE

VOL. XXVII, N. 1 — *Serie decima* — 1° Gennaio 1963

Baryon Charge and R-Inversion in the Octet Model.

H. GOLDBERG and Y. NE'EMAN (*)
Israel Atomic Energy Commission - Rehovoth

(ricevuto il 22 Febbraio 1962)

Summary. — A method for the application of the group $U(3)$ to the octet version of unitary symmetry is suggested, circumventing the difficulty implicit in the tracelessness of the $SU(3)$-representations linked with this model. In this way, baryon number is naturally introduced. The meaning of R-inversion (introduced by M. GELL-MANN) is generalized.

1. – Baryon charge.

The unitary symmetry approach (1) to the strongly interacting particles makes use of irreducible representations of $SU(3)$. In the octet model (2), as it now stands, one has to attach the *same* representation, namely (1 1), whether it be to the baryons, to the antibaryons, or to the mesons. Thus, the model does not reflect baryonic charge, which has to be introduced as an external feature.

To overcome this defect, it is suggested that the group $U(3)$ be used instead of $SU(3)$ (as in the Sakata model). This enables us to attach to the 3 octets three different irreducible representations, distuinguishable exactly by the one extra infinitesimal operator. In H. WEYL's notation (3), it is plausible to associate the representation $[2\ 1\ 0]$ with the baryon octet, $[0\ -1\ -2]$ with the antibaryons, and $[1\ 0\ -1]$ with the mesons.

In this correspondence, the baryon-number operator becomes

(1) $$B = \frac{1}{3}\begin{pmatrix} 1 & 0 & 0 \\ 0 & 1 & 0 \\ 0 & 0 & 1 \end{pmatrix}.$$

Define

$$(2)\qquad H_1 = Y + B\,,\quad H_2 = T_z - \tfrac{1}{2}Y + B,\quad H_3 = -T_z - \tfrac{1}{2}Y + B\,,$$

(where Y is the hypercharge operator). These operators, which commute with one another, belong to $U(3)$. If Ψ is an eigenvector of H_1, H_2 and H_3 with the eigenvalues λ_1, λ_2 and λ_3, respectively, we say that Ψ corresponds to the *weight* $(\lambda_1\,\lambda_2\,\lambda_3)$ (4). (Weights are considered as 3-dimensional vectors.) The correspondence between particles and weights is as follows:

	[2 1 0]		[1 0 −1]		[0 −1 −2]
p	(2 1 0)	K^+	(1 0 −1)	$\overline{\Xi^-}$	(0 −1 −2)
n	(2 0 1)	K^0	(1 −1 0)	$\overline{\Xi^0}$	(0 −2 −1)
Σ^+	(1 2 0)	Π^+	(0 1 −1)	$\overline{\Sigma^-}$	(−1 0 −2)
Λ	(1 1 1)	Π^{00}	(0 0 0)	$\overline{\Lambda}$	(−1 −1 −1)
Σ^0	(1 1 1)	Π^0	(0 0 0)	$\overline{\Sigma^0}$	(−1 −1 −1)
Σ^-	(1 0 2)	Π^-	(0 −1 1)	$\overline{\Sigma^+}$	(−1 −2 0)
Ξ^0	(0 2 1)	$\overline{K^0}$	(−1 1 0)	$\overline{\text{n}}$	(−2 0 −1)
Ξ^-	(0 1 2)	K^-	(−1 0 1)	$\overline{\text{p}}$	(−2 −1 0)

The baryon-, meson- and antibaryon-octets are thus identical, respectively, with the representations ${}^{1}a(1\;1)$, ${}^{0}a(1\;1)$ and ${}^{-1}a(1\;1)$ of J. E. WESS (5), whereas $B = 1/3N$.

after that physical "composition" is confused with "production metho

(*) On leave of absence from the Department of Physics, Imperial College, London.

(1) See for example, M. IKEDA, S. OGAWA and Y. OHNUKI: *Progr. Theor. Phys.*, **22**, 715 (1959) and **23**, 1073 (1960). In these papers the Sakata model is investigated.

(2) M. GELL-MANN: Report CTSL-20, California Institute of Technology, Pasadena; Y. NE'EMAN: *Nucl. Phys.*, **26**, 222 (1961).

(3) H. WEYL: *The Classical Groups*, 2nd ed. (Princeton, 1946), pp. 132-135.

(4) See, for example, G. RACAH: *Group Theory and Spectroscopy*, lectures 3 and 4 (Mimeographed notes by E. MERZBACHER and D. PARK, 1951).

(5) J. E. WESS: *Nuovo Cimento*, **15**, 52 (1960).

Appendix III

Fragmentary reproductions from *Physics Letters* and *Physical Review Letters*, with the kind permission of North-Holland Physics Publishing and The American Physical Society.

Volume 1, number 4 PHYSICS LETTERS 15 May 1962

A REACTION FORBIDDEN BY THE SAKATA MODEL OF UNITARY SYMMETRY *

C. A. LEVINSON, H. J. LIPKIN and S. MESHKOV **
Weizmann Institute of Science, Rehovoth, Israel

and

A. SALAM and R. MUNIR
Physics Department, Imperial College, London

Received 11 April 1962

The reaction

$$p + \bar{p} \rightarrow K_1^0 + K_2^0 \tag{1}$$

has recently been observed in anti-proton annihilation at rest [1]. The purpose of this letter is to point out that this reaction is strictly forbidden according to the Sakata [2,3] model of unitary symmetry. The experimental observation can therefore be interpreted as evidence against the Sakata model.

The selection rule forbidding reaction (1) can be obtained by calculating amplitudes for proton-antiproton annihilation into the various two-meson channels. It can also be obtained by examining the "parity" of the states of (1) under a particular transformation of SU3; namely the one which interchanges the neutron and the lambda. This transformation is called "axis 2 reflection" in ref. [3]. Under this transformation,

$$\begin{aligned} p &\rightarrow p \\ \bar{p} &\rightarrow \bar{p} \\ K_0 &\rightarrow \bar{K}_0 \\ \bar{K}_0 &\rightarrow K_0 \,. \end{aligned} \tag{2}$$

The proton-antiproton system remains invariant under the transformation (2); all proton-antiproton states therefore have even parity under this transformation. For neutral K-mesons, however, the transformation (2) is equivalent to charge conjugation; the parity of neutral K-meson states under this transformation is therefore the same as the parity under charge conjugation. If strong interactions are assumed to be invariant under unitary transformations, then parity under the transformation (2) must be conserved. Since the proton-antiproton system always has even parity, transitions to neutral K-meson states which are odd under (2) are forbidden.

The state on the right hand side of the reaction (1) is odd under charge conjugation, since K_1^0 and K_2^0 are eigenstates of the charge conjugation operator with opposite parity. This state is therefore odd under the transformation (2) and the reaction (1) is forbidden.

No corresponding selection rule exists for the octet model [4], in which the proton does not remain invariant under axis 2 reflection, but goes into Σ^+. There is no selection rule forbidding charged kaon or pion production in either model. The reactions

$$p + \bar{p} \rightarrow 2K_1^0 \quad \text{or} \quad p + \bar{p} \rightarrow 2K_2^0 \tag{3}$$

are not forbidden since the K states are even under charge conjugation and the transformation (2).

The observation of the reaction (1) with rates comparable to those of charged kaon and pion production must be considered as being in serious disagreement with the predictions of unitary symmetry on the Sakata model.

* The research reported in this document has been sponsored in part by the Air Force Office of Scientific Research, through the European Office, Aerospace Research, United States Air Force.

** Supported in part by the United States National Science Foundation; on leave from the Department of Physics, University of Pittsburgh, Pittsburgh, Pa.

References

1) Preliminary results of CERN and Collège de France groups, communicated at the symposium held at Imperial College, London, March 1962 by D.R.O. Morrison.
2) S. Sakata, Progr. Theor. Phys. 16 (1956) 686.
3) C. A. Levinson, H. J. Lipkin and S. Meshkov, Nuovo Cimento 23 (1962) 236.
4) Y. Neeman, Nuclear Phys. 26 (1961) 222.
M. Gell-Mann, Phys. Rev. 125 (1962) 1067.

* * * * *

Volume 1, number 8 PHYSICS LETTERS 15 July 1962

* * * * *

PROTONIUM TWO-MESONS ANNIHILATION

Y. DOTHAN, H. GOLDBERG, H. HARARI and Y. NE'EMAN
Israel Atomic Energy Commission Laboratories, Rehovoth

Received 18 June 1962

Morrison has recently reported * 369 observed events in which the products of $\overline{p}p$ annihilation were 2 mesons. Of these, 56 events were K^+K^-, 63 events were $K^0\overline{K}^0$ and 250 events were $\pi^+\pi^-$; no $\pi^0\pi^0$ was observed. As the energy available for the two mesons is about 2 GeV, considerations of phase space alone cannot explain such a difference between σ_{KK} and $\sigma_{\pi\pi}$.

We would like to show that unitary symmetry [2)] in its "octet" version [3,4)] can explain the above results. The analysis is based on the assumption that the main contribution to the process includes a final interaction mediated by a vector-

* Preliminary results of CERN and Collège de France groups, communicated at the symposium held at Imperial College, London, March 1962 by D. R. O. Morrison. These results were discussed from the point of view of the Sakata model by Levinson et al. [1)].

In order to get a fit with the experiment, the annihilation is assumed to be due to the Φ ([10-1]; $T = 1$) part. Momentum space considerations, using the Born approximation, will favor $\overline{p}p \to \pi^-\pi^+$ over $\overline{p}p \to \overline{K}^0K^0$ or $\overline{p}p \to K^-K^+$ by the ratio 1.56 : 1. Taking into account the coefficients of the F-coupling in $L_I(\Phi, \partial_\mu\Phi, B^\mu)$ one gets

$$\sigma_{\pi^+\pi^-} : \sigma_{K^0\overline{K}^0} : \sigma_{K^+K^-} : \sigma_{\pi^0\pi^0} = 6.25 : 1 : 1 : 0 .$$

This is not far from the experimental results which are approximately 4.46 : 1.125 : 1 : 0 *.

is because these reactions cannot occur through a virtual vector meson belonging to an octet.

2) The branching ratios in table 1 are associated with the corresponding annihilations. Momentum space considerations in the Born approximation have been always taken into account. f is the momentum space factor, which is a function of the mass of π^{00}. If we assume that $\pi^{00} = \eta$ (550 MeV, $J = 0^-$), then $f = 0.74$. (Cross sections which do not appear in this table vanish, e.g. $\sigma(\overline{n}n \to \pi^{00}\pi^{00}) = 0$.)

Table 1

		$\overline{p}n$	$\sigma(\pi^0\pi^-) : \sigma(K^0K^-) = 3.12 : 1$
$\overline{n}p$	$\sigma(\pi^+\pi^0) : \sigma(K^+\overline{K}^0) = 3.12 : 1$	$\overline{n}n$	$\sigma(\pi^+\pi^-) : \sigma(K^+K^-) : \sigma(K^0\overline{K}^0) : \sigma(\pi^0\pi^0) = 6.25 : 1 : 1 : 0$
$\overline{\Lambda}p$	$\sigma(K^0\pi^+) : \sigma(K^+\pi^0) : \sigma(K^+\pi^{00}) = 2 : 1 : 3f$	$\overline{\Lambda}n$	$\sigma(K^+\pi^-) : \sigma(K^0\pi^0) : \sigma(K^0\pi^{00}) = 2 : 1 : 3f$
$\overline{\Sigma^0}p$	$\sigma(K^0\pi^+) : \sigma(K^+\pi^0) : \sigma(K^+\pi^{00}) = 2 : 1 : 3f$	$\overline{\Sigma^0}n$	$\sigma(K^+\pi^-) : \sigma(K^0\pi^0) : \sigma(K^0\pi^{00}) = 2 : 1 : 3f$
$\overline{\Sigma^+}p$	$\sigma(K^+\pi^-) : \sigma(K^0\pi^0) : \sigma(K^0\pi^{00}) = 2 : 1 : 3f$	$\overline{\Sigma^-}n$	$\sigma(K^0\pi^+) : \sigma(K^+\pi^0) : \sigma(K^+\pi^{00}) = 2 : 1 : 3f$

* * * * *

VOLUME 12, NUMBER 8   PHYSICAL REVIEW LETTERS   24 FEBRUARY 1964

OBSERVATION OF A HYPERON WITH STRANGENESS MINUS THREE*

V. E. Barnes, P. L. Connolly, D. J. Crennell, B. B. Culwick, W. C. Delaney,
W. B. Fowler, P. E. Hagerty,† E. L. Hart, N. Horwitz,† P. V. C. Hough, J. E. Jensen,
J. K. Kopp, K. W. Lai, J. Leitner,† J. L. Lloyd, G. W. London,‡ T. W. Morris, Y. Oren,
R. B. Palmer, A. G. Prodell, D. Radojičić, D. C. Rahm, C. R. Richardson, N. P. Samios,
J. R. Sanford, R. P. Shutt, J. R. Smith, D. L. Stonehill, R. C. Strand, A. M. Thorndike,
M. S. Webster, W. J. Willis, and S. S. Yamamoto
Brookhaven National Laboratory, Upton, New York
(Received 11 February 1964)

$$K^- + p \to \Omega^- + K^+ + K^0$$
$$\llcorner \Xi^0 + \pi^-$$
$$\llcorner \Lambda^0 + \pi^0$$
$$\llcorner \gamma_1 + \gamma_2$$
$$\llcorner e^+ + e^-$$
$$\llcorner e^+ + e^-$$
$$\llcorner \pi^- + p.$$

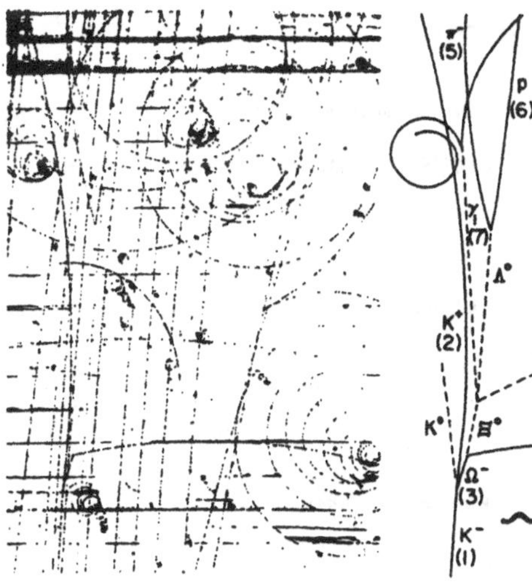

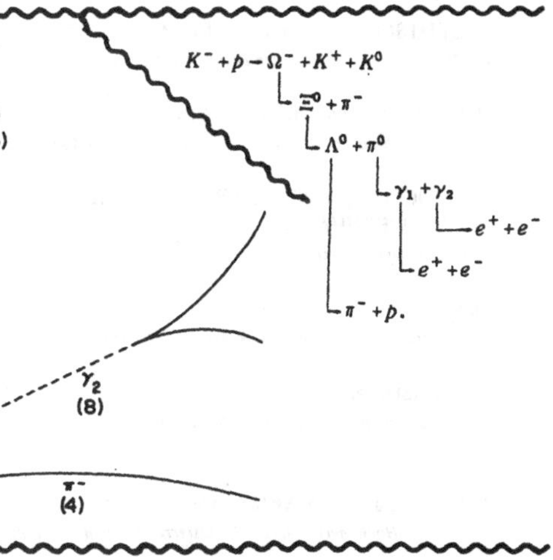

In view of the properties of charge ($Q = -1$),
strangeness ($S = -3$), and mass ($M = 1686 \pm 12$
MeV/c^2) established for particle 3, we feel justified in identifying it with the sought-for Ω^-.
Of course, it is expected that the Ω^- will have
other observable decay modes, and we are continuing to search for them. We defer a detailed
discussion of the mass of the Ω^- until we have
analyzed further examples and have a better understanding of the systematic errors.

The observation of a particle with this mass
and strangeness eliminates the possibility which
has been put forward[6] that interactions with ΔS
= 4 proceed with the rates typical of the strong
interactions, since in that case the Ω^- would decay very rapidly into $n + K^0 + \pi^-$.

*Work performed under the auspices of the U. S.
Atomic Energy Commission and partially supported
by the U. S. Office of National Research and the Na-
tional Science Foundation.
†Syracuse University, Syracuse, New York.
‡University of Rochester, Rochester, New York.

[1]M. Gell-Mann, Proceedings of the International Conference on High-Energy Nuclear Physics, Geneva, 1962
(CERN Scientific Information Service, Geneva, Switzerland, 1962), p. 805; R. Behrends, J. Dreitlein,
C. Fronsdal, and W. Lee, Rev. Mod. Phys. 34, 1
(1962); S. L. Glashow and J. J. Sakurai, Nuovo Cimento 25, 337 (1962).
[2]A possible example of the decay of this particle was
observed by Y. Eisenberg, Phys. Rev. 96, 541 (1954).
[3]M. Gell-Mann, Phys. Rev. 125, 1067 (1962); Y. Ne'e-
man, Nucl. Phys. 26, 222 (1961).
[4]See, however, R. J. Oakes and C. N. Yang, Phys.
Rev. Letters 11, 174 (1963).
[5]M. Gell-Mann, Synchrotron Laboratory, California
Institute of Technology, Internal Report No. CTSL-20,
1961 (unpublished); S. Okubo, Progr. Theoret. Phys.
(Kyoto) 27, 949 (1962).
[6]G. Racah, Nucl. Phys. 1, 302 (1956); H. J. Lipkin,
Phys. Letters 1, 68 (1962).

Volume 73B, number 1 PHYSICS LETTERS 30 January 1978

SPIN AND LIFETIME OF THE Ω^- HYPERON

Aachen–Berlin–CERN–Innsbruck–London–Vienna Collaboration

M. DEUTSCHMANN and G. OTTER
III. Physikalisches Institut der Technischen Hochschule, Aachen, Germany

H. BÖTTCHER and H.J. SCHREIBER
Institut für Hochenergiephysik der Akademie der Wissenschaften der DDR, Berlin-Zeuthen, DDR

V.T. COCCONI, G. KELLNER, D.J. KOCHER [1], D.R.O. MORRISON, H. SAARIKKO [2] and K.L. WERNHARD [1]
CERN, Geneva, Switzerland

D. KUHN
Institut für Experimental Physik der Universität Innsbruck, Innsbruck [3], Austria

R.M. EASON and B. POLLOCK
Physics Department, Imperial College, London, UK

and

F. MANDL and M. MARKYTAN
Institut für Hochenergiephysik der Österreichischen Akademie der Wissenschaften, Vienna, Austria

Received 21 September 1977
Revised manuscript received 13 December 1977

A total of 101 Ω^- decays have been found in K^-p interactions at 10 and 16 GeV/c. The decay angular distribution has been fitted under the assumptions that the Ω^- has spin either 1/2 or 3/2. It has been found that the probability of isotropy (spin 1/2) is less than one in a thousand, whereas the probability for spin 3/2 is about 70%. Thus, we exclude the spin 1/2 assignment. The lifetime of the Ω^- is found to be $(1.41^{+0.15}_{-0.24}) \times 10^{-10}$ s, in agreement with our earlier result based on about 1/5 of the present sample.

A search for Ω^- hyperons has been performed in two K^-p experiments at incident momenta of 10 GeV/c and 16 GeV/c, consisting of 240 000 and 350 000, respectively, inelastic interactions of all types observed in the 150 cm British and 200 cm CERN H_2 bubble chambers.

The analysis yielded 101 events in which an Ω^- hyperon was produced. This represents the world's largest sample presently available and is summarized in table 1. The 10 GeV/c experiment has been split into two parts, corresponding to different selection criteria. The numbers given for the 16 GeV/c experiment come from the analysis of about 25% of the expected final statistics.

[1] CERN Fellow from the University of Innsbruck, Austria.
[2] Now at Institute of Nuclear Physics, University of Helsinki, Finland.
[3] Work supported by Fond zur Förderung der Wissenschaftlichen Forschung, Proj. 2867.

Volume 78B, number 2,3 PHYSICS LETTERS 25 September 1978

A STUDY OF THE LIFETIME AND SPIN OF Ω^- PRODUCED IN K^-p INTERACTIONS AT 8.25 GeV/c

Birmingham[a]–CERN[b]–Glasgow[c]–Michigan State[d,1]–Paris LPNHE[e,2] Collaboration

M. BAUBILLIER[e], I.J. BLOODWORTH[a], G.J. BOSSEN[b], A. BURNS[c], J.N. CARNEY[a], M.J. CORDEN[a], C.A. COWAN[a], G.F. COX[a], C.J. De LIMA[a], D. DIXON[c], Ph. GAVILLET[b], J.B. KINSON[a], K. KNUDSON[b], F. LEVY[c], H. McCANN[c], M. MacDERMOTT[a], P.J. NEGUS[c], B.H. OH[d], M. PRATAP[d], E. QUERCIGH[b], M. RIVOAL[e], J.M. SCARR[c], J.C. SHIERS[a], G.A. SMITH[d], D. TEODORO[b], O. VILLALOBOS BAILLIE[a], M.F. VOTRUBA[a], J. WHITMORE[d] and R. ZITOUN[e]

Received 20 June 1978

Using the decay mode $\Omega^- \to \Lambda K^-$, we have obtained a measurement of the Ω^- lifetime $\tau_\Omega = (0.80 \pm ^{0.16}_{0.12}) \times 10^{-10}$ s. The Ω^- decay angular distribution is consistent with $W(\cos\theta^*) \propto 1 + 3\cos^2\theta^*$. The probability of consistency with a flat distribution is ~1/300 indicating $J \neq \frac{1}{2}$.

1. Introduction. A study is presented on the decay of the Ω^- hyperon. The results are based on 41 bubble chamber events of the type

$$K^-p \to \Omega^- + \text{anything}, \quad \Omega^- \to \Lambda^0 K^-, \quad \Lambda^0 \to p\pi^- . \tag{1}$$

where both the Ω^- and the subsequent Λ decay were observed. For these events, a kinematic fit of the whole decay chain and a visual estimate of the ionization allowed a unique interpretation of the decay as $\Omega^- \to \Lambda K^-$; i.e. the competing hypothesis $\Xi^- \to \Lambda\pi^-$ had a probability $< 10^{-4}$ or it was inconsistent with the ionization estimate.

These events have been found in the analysis of the first 1/3 (corresponding to about 50 events/μb) of a 5.3×10^6 pictures exposure of the CERN 2m hydrogen bubble chamber to an RF separated beam of 8.25 GeV/c K^- mesons at the CERN PS. All events with at least one neutral decay are being measured and processed through standard bubble chamber analysis chains and all the candidates for reaction (1) have been carefully examined on a scanning table by physicists.

In order to calibrate the analysis methods and to study the effect that a kinematic ambiguity between Ω^- and Ξ^- can have on the results, we have also analysed all 2032 events of the type

$$K^-p \to \Xi^- + \text{anything}, \quad \Xi^- \to \Lambda^0 \pi^-, \quad \Lambda^0 \to p\pi^- \tag{2}$$

that have been found on the same film sample together with 58 events which are ambiguous between (1) and (2) (i.e. $\Omega^- \to \Lambda K^-$ and $\Xi^- \to \Lambda\pi^-$ decay hypotheses had both a kinematic fit probability $> 10^{-4}$.

In the following, we determine the Ω^- mass, lifetime, decay distribution and weak decay parameter α_Ω from the unique Ω^- events [‡1]. It will be shown below that the additional 58 events ambiguous between Ω^-

[1] Work supported by the National Science Foundation, USA.
[2] Laboratory associated to IN2P3.

[‡1] Ω^- production characteristics will be presented in a later paper.

References

ADLER, S.L. and R.F. DASHEN
1968 *Current Algebra and Applications to Particle Physics* (W.A. Benjamin, Pub., New York 1968).

ALVAREZ, L.W.
1972 *Nobel Lectures, Physics 1963-1970* (Elsevier Pub., New York 1972) 241-290.

BARNES, V.E. *et al.*
1964 «Observation of a Hyperon with Strangeness Minus Three»; *Phys. Rev. Lett. 12,* 204 (1964).

BEHRENDS, R.E. and A. SIRLIN
1961 «Weak-Coupling Currents and Symmetries of Strong Interactions»; *Phys. Rev. 121,* 324 (1961).

BIEDENHARN, L.C., R.Y. CUSSON, M.Y. HAN and O.L. WEAVER
1972 «Hadronic Regge sequences as primitive realizations of SL(3,R) Symmetry»; *Phys. Lett. B42,* 257 (1972).

BLOOM, E.D. *et al.*
1969 «High-energy inelastic *e-p* scattering at 6° and 10°»; *Phys. Rev. Lett. 23,* 930 (1969).

BODMER, A.R.
1971 «Collapsed Nuclei»; *Phys. Rev.* D4, 1601 (1971).

BREIDENBACH, M. *et al.*
1969 «Observed behavior of highly inelastic electron-proton scattering»; *Phys. Rev. Lett. 23,* 935 (1969).

BRINK, L., M. GELL-MANN, P. RAMOND and J. SCHWARZ
1978 «Supergravity as geometry of superspace»; *Phys. Lett. 74B,* 336 (1978).

CABIBBO, N., L. HORWITZ and Y. NE'EMAN
1966 «The Algebra of Scalar and Vector Vertex Strengths in Regge Residues»; *Phys. Lett. 22,* 336 (1966).

CARMONY, D.D. *et al.*
1964 «Properties of Ξ hyperons»; *Phys. Rev. Lett. 12,* 482 (1964).

DEACON, R.
1977 *The Israeli Secret Service* (Taplinger Pub. Co., New York 1977).

DOTHAN, Y., H. GOLDBERG, H. HARARI and Y. NE'EMAN
1962 «Protonium two-mesons annihilation»; *Phys. Lett. 1,* 308 (1962).

DOTHAN, Y., M. GELL-MANN and Y. NE'EMAN
1965 «Series of hadron energy levels as representations of non-compact groups»; *Phys. Lett. 17,* 148 (1965).

DYNKIN, E.B.
1947 «The structure of semi-simple algebras»; *Uspekhi Mat. Nauk 2,* N.4 (1947); *Am. Math. Soc. Transl.* N.17 (1950).
1957 «Semi-simple subalgebras of semi-simple Lie algebras», «Maximal subgroups of the classical groups» *Am. Math. Soc. Transl.,* Series 2. *6,* pp. 111-224 and 245-368 (1957).

EDEN, R.J. and J.R. TAYLOR
1963 «Resonance multiplets and broken symmetry»; *Phys. Rev. Lett. 11,* 516 (1963).

EISENBERG, Y.
1954 «Possible Existence of a New Hyperon»; *Phys. Rev. 96,* 541 (1954).

FERRARA, S., J. SCHERK and B. ZUMINO
1977 «Supergravity and local extended supersymmetry»; *Phys. Lett. 66B,* 35 (1977).

FEYNMAN, P.P. and M. GELL-MANN
1958 «Theory of the Fermi Interaction»; *Phys. Rev. 109,* 193 (1958).

FRAZER, W.R. and J.R. FULCO
1960 «Effect of a Pion-Pion Scattering Resonance on Nucleon Structure»; *Phys. Rev. 117,* 1609 (1960).

FREEDMAN, D.Z.
1977 «SO(3)-Invariant Extended Supergravity»; *Phys. Rev. Lett. 38,* 105 (1977).

GELL-MANN, M.
1961 «The Eightfold Way: A Theory of Strong Interaction Symmetry»; *Caltec Report* CTSL-20, unpub. (1961).
1962a «Symmetries of Baryons and Mesons»; *Phys. Rev. 125,* 1067 (1962).

1962b «Strange Particle Physics. Strong Interactions»; *Proc. Internat. Conf. High Energy Physics,* CERN, p. 805 (1962).

1964 «A Schematic Model of Baryons and Mesons»; *Phys. Lett 8,* 214 (1964).

GELL-MANN, M. and M. LÉVY

1960 «The Axial Vector Current in Beta Decay»; *Nuovo Cim. 16,* 705 (1960).

GELL-MANN, M. and Y. NE'EMAN

1964a *The Eightfold Way* (W.A. Benjamin, Pub. New York 1964).

1964b «Current-Generated Algebras»; *Ann. Phys.* (N.Y.) *30,* 360 (1964).

GERSHTEIN, S.S. and J.B. ZELDOVICH

1956 «Meson Corrections in the Theory of Beta Decay»; *Sov. Phys.* JETP *2,* 576 (1956).

GLASHOW, S.L. and M. GELL-MANN

1961 «Gauge theories of Vector Particles»; *Ann. Phys.* (N.Y.) *15,* 437 (1961).

GLASHOW, S.L. and A.H. ROSENFELD

1963 «Eightfold-Way Assignments for $Y^o{}_1$ (1660) and Other Baryons»; *Phys. Rev. Lett. 10,* 192 (1963).

GOLDBERG, H. and Y. NE'EMAN

1962 «Baryon Charge and R-Inversion in the Octet Model»; *Israel AEC Report* IA-725 (February 1962) *Nuovo Cim., 27,* 1 (1963).

GOLDBERGER, M.L. and S.B. TREIMAN

1958 «Decay of the Pi Meson»; *Phys. Rev. 110,* 1178 (1958).

GOLDSTONE, J.

1961 «Field Theories with 'Superconductor' Solutions»; *Nuovo Cim. 19,* 155 (1961).

GOTSMAN, E. (ed.)

1971 *Duality and Symmetry in Hadron Physics* (Weizmann Science Press, Jerusalem 1971).

HOOFT, G.T.

1980 *Recent Developments in Gauge Theories* (Cargese 1979), (Plenum Press, New York 1980).

IKEDA, M., S. OGAWA and Y. OHNUKI

1959 «A possible Symmetry in Sakata's Model for Bosons-Baryons System»; *Prog. Theoret. Phys.* (Kyoto) *22,* 715 (1959).

KOKKEDEE, J.J.J.

1969 *The Quark Model* (W.A. Benjamin, Pub., New York 1969).

LEE, T.D.

1975 «Abnormal nuclear states and vacuum excitation»; *Rev. Mod. Phys. 47,* 267 (1975).

LEE, T.D. and G.C. WICK

1974 «Vacuum stability and vacuum excitation in a spin-0 field theory»; *Phys. Rev. D9,* 2291 (1974).

LEVINSON, C.A., H.J. LIPKIN and S. MESHKOV

1962a «Experimental Tests of Unitary Symmetry»; *Nuovo Cim. 23,* 236 (1962).

1962b «Unitary Symmetry of Strong Interactions»; *Phys. Lett. 1,* 44 (1962).

1963 «Verification of the tenfold assignment of the baryon resonances»; *Phys. Rev. Lett. 10,* 361 (1963).

LIPKIN, H.J.

1973 «Quarks for Pedestrians»; *Physics Reports 8* (1973) N.3.

MARSHAK, R.E. and E.C.G. SUDARSHAN

1958 *Proc. Padua-Venice Conf. (1957),* Soc. Ital. di Fisica, Padua-Venice 1958; *Phys. Rev. 109,* 1860 (1958).

NAMBU, Y.

1960 «Quasi-Particles and Gauge Invariance in the Theory of Superconductivity»; *Phys. Rev. 117,* 648 (1960).

1957 «Possible Existence of a Heavy Neutral Meson»; *Phys. Rev. 106,* 1366 (1957).

NAMBU, Y. and J.J. SAKURAI

1961 «Odd $\Lambda\Sigma$ Parity and the Nature of the $\pi\Lambda\Sigma$ Coupling»; *Phys. Rev. Lett. 6,* 377 (1961).

NE'EMAN, Y.

1961a «Derivation of Strong Interactions from a Gauge Invariance»; *Nucl. Phys. 26,* 222 (1961).

1961b «Yukawa Terms in the Unitary Gauge Theory»; *Nucl. Phys. 26,* 260 (1961).

1961c *The Applications of Lie Groups to Invariance Principles in Physics,* Univ. of London Thesis. Reprinted as *Gauges, Groups and an Invariant Theory of the Strong Interactions,* Israel AEC Report IA-698 (August 1961).

1962 «Unified Interactions in the Unitary Gauge Theory»; *Nucl. Phys. 30,* 347 (1962).
1965 «Expansion as an Energy Source in Quasi-Stellar Radio Sources»; *Astrophys. J. 141,* 1303 (1965).
1967 *Algebraic Theory of Particle Physics* (W.A. Benjamin, Pub., New York 1967).
1968 *Proc. Fifth Coral Gables Conf. on Symmetry Principles at High Energy,* A. Perlmutter *et al.* eds. (W.A. Benjamin Pub., New York 1968) 149.
1971 *The Interaction between Science and Philosophy,* Y. Elkana ed. (Humanities Press, Atlantic Heights 1974) 1-26; Japanese version in *Shizen,* 71-12, pp. 94-105 (1971).
1972 *Proc.* (Boulder 1972) *IAU Symposium on Physics of Dense Matter* (Reidel Pub.) pp. 111-115.
1979 «Primitive particle model»; *Phys. Lett. 82B,* 69 (1979).

NE'EMAN, Y. and V.T.N. REDDY
1975 «Universality in the algebra of vertex strengths as generated by bilocal currents»; *Nucl. Phys. B84,* 221 (1975).

NE'EMAN, Y. and T. REGGE
1978 «Gauge theory of gravity and supergravity on a group manifold»; *Rivista d. Nuovo Cim. 1,* N.5 (1978).

NOVIKOV, I.C.
1964 «Delayed Explosion of a part of the Fridman Universe, and Quasars»; *Astron. Zh 41,* 1075 (1964).

OAKES, R.J. and C.N. YANG
1963 «Meson-baryon resonances and the mass formula»; *Phys. Rev. Lett., 11,* 174 (1963).

OKUN, L.B.
1958 *International Conference on High Energy Physics;* CERN, p. 223 (1958).

PEVSNER, A. *et al.*
1961 «Evidence for a three-pion resonance near 550 Mev»; *Phys. Rev. Lett.* 7, 421 (1961).

ROSEN, J., N. ROSEN and Y. NE'EMAN
1964 *Proc.* (1964) *Coral Gables Conf.,* B. Kursunoglu *et al.,* eds. (W.H. Freeman Pub., San Francisco 1964) 93.

SAKATA, S.
1956 «On a Composite Model for the New Particles»; *Prog. Theoret. Phys.* (Kyoto) *16,* 686 (1956).

SAKURAI, J.J.
1960 «Theory of Strong Interactions»; *Ann. Phys.* (N.Y.) *11,* 1 (1960).
1961 «New resonances and strong interaction symmetry»; *Phys. Rev. Lett.* 7, 426 (1961).

SHEKHTER, V.M.
1962 «Symmetry Properties of Strong Interactions»; *Sov. Phys. JETP 14,* 582 (1962).

SPEISER, D.R. and J. TARSKI
1961 *IAS* (Princeton) *report* (1961).

TAYLOR, J.C.
1958 «Beta Decay of the Pion»; *Phys. Rev. 110,* 1216 (1958).

THIERRY-MIEG, J. and Y. NE'EMAN
1979 «Extended Geometric Supergravity on Group Manifolds with Spontaneous Fibration»; *Ann. of Phys.* (N.Y.) 123, pp. 247-273 (1979).

WEAVER, L. and L.C. BIEDENHARN
1972 «Nuclear rotational bands and SL (3,R) symmetry»; *Nucl. Phys. A185,* 1 (1972).

WEINBERG, S.
1967 «Dynamical approach to current algebra»; *Phys. Rev. Lett. 18,* 188 (1967).
1972 «Approximate Symmetries and Pseudo-Goldstone Bosons»; *Phys. Rev. Lett. 29,* 1698 (1972).

WESS, J.E.
1960 «Investigation of the Invariance Group in the Three Fundamental Fields Model»; *Nuovo Cim. 15,* 52 (1960).

YAMAGUCHI, Y.
1959 «A Composite Theory of Elementary Particles»; *Prog. Theoret. Phys.* (Kyoto), Supplement *11,* 1 (1959).

ZWEIG, G.
1964 CERN *Reports* TH401, 402 (unpub.) (1964).
1965 «Fractionally charged Particles and SU_6»; *Symmetries in Elementary Particle Physics,* Acad. Press, New York, A. Zichichi, ed., p. 192 (1965).

Printed in the United States
by Baker & Taylor Publisher Services